Water-Resources
Engineering

Water-Resources Engineering

RAY K. LINSLEY

Professor of Hydraulic Engineering
Stanford University

JOSEPH B. FRANZINI

Professor of Civil Engineering
Associate Chairman, Department of Civil Engineering
Stanford University

SECOND EDITION

McGRAW-HILL BOOK COMPANY

New York   San Francisco   St. Louis   Düsseldorf   Johannesburg
Kuala Lumpur   London   Mexico   Montreal   New Delhi
Panama   Rio de Janeiro   Singapore   Sydney   Toronto

To **Eugene Lodewick Grant**
Pioneer in engineering economy
Teacher, colleague, and friend

Preface

With world population growing rapidly, the water resources of the world are becoming one of its most important assets. Water is essential for human consumption and sanitation, for the production of many industrial goods, and for the production of food and fiber. Water is an important means of transport in many parts of the world and a significant factor in recreation. Even a valuable resource can be a hazard and excessive water—floods—cause substantial damage and loss of life throughout the world. Water is unequally distributed about the earth, and its availability at any place varies greatly with time. Finally, in his use of this resource, man pollutes much of the available fresh water and degrades it so that it is unfit for many or all uses.

Skilled planning and careful management are essential to achieve the level of efficiency in water use which will be required in the future. These efforts, however, must be substantially broader in scope than the common concept of engineering would visualize. Investments in water-resources development are influenced by economic, social, and political considerations as well as the basic engineering facts. In revising and updating their book "Elements of Hydraulic Engineering," the authors elected to increase the emphasis on planning and to adopt the title "Water-Resources Engineering" as more descriptive of its intended scope. Planning in the true sense of the word is a complex operation, which is perhaps nowhere more complex than in the planning of water resources for many competing uses while subject to both economic and physical constraints. The engineer cannot, however, shirk the planning task if he wishes to maintain his position in the profession which "directs the forces of nature for the benefit of man."

Water-resources engineering draws on the background of the student in engineering science, design, the humanities, and social sciences. A course in water-resources engineering by its nature must be a terminal course. It seems important, therefore, to present the material in a unified course treating the subject in its broadest scope and emphasizing the "why" instead of the "how." The authors hope that students will be stimulated by the broad perspective of water-resources engineering as a unified field and that the instructor will find the text both interesting and effective.

The first five chapters present the subject of hydrology, basic to all water-management efforts. Chapter 6 discusses the legal aspects of water use which often provide important constraints on the planning effort. Chapters 7 through 12 discuss the physical works—dams, canals, pipelines, etc.—which are utilized in almost all types of water-resources projects. Chapter 13 reviews the important principles of engineering economy basic to water management. Chapters 14 through 20 then discuss the principal water uses in more detail, with particular attention to the aspects of each use which differ from the other uses. The final chapter summarizes the planning procedure for single- and multipurpose projects.

The authors wish to express special appreciation to Professor Eugene L. Grant for his review of Chapter 13, which he originally prepared, and for his many other helpful suggestions. The sections on water purification in Chapter 16 and wastewater treatment in Chapter 19 have been extensively rewritten with the assistance of Professor Perry McCarty. We are grateful to Professor Irwin Remson for his review of Chapter 4, to Professor Charles Meyers for constructive comments on Chapter 6, to Professor Robert L. Street for review of Sec. 7-10, and to Professor Kaare Hoëg for help with the discussion of earth dams in Chapter 8. Our thanks must also be extended to other colleagues at Stanford and elsewhere who have contributed many helpful ideas to this and earlier editions of the book. The fine work of Mrs. Dawn Cone, who painstakingly assisted in preparing the manuscript for this edition, must also be acknowledged.

Ray K. Linsley
Joseph B. Franzini

Contents

Chapter 1

Introduction

The development of water resources requires the conception, planning, design, construction, and operation of facilities to control and utilize water. It is basically a function of civil engineers, but the services of specialists from other fields are required. Water-resources problems are also the concern of economists, political scientists, geologists, electrical and mechanical engineers, chemists, biologists, and other specialists in the natural and social sciences. Each water-development project encounters a unique set of physical conditions to which it must conform, hence standard designs which lead to simple, handbook solutions can rarely be used. The special conditions of each project must be met through an integrated application of the fundamental knowledge of many disciplines.

1-1 Fields of water-resources engineering Water is controlled and regulated to serve a wide variety of purposes. Flood mitigation, land drainage, sewerage, and highway culvert design are applications of water-resources engineering to the *control of water* so that it will not cause excessive damage to property, inconvenience to the public, or loss of life. Water supply, irrigation, hydroelectric-power development, and navigation improvements are examples of the *utilization of water* for beneficial purposes. Pollution threatens the utility of water for municipal and irrigation uses and seriously despoils the aesthetic value of rivers—hence pollution control or *water-quality management* has become an important phase of water-resources engineering. Finally the potential of nonstructural measures such as zoning for flood mitigation and the preservation of natural beauty are factors which the water-resources engineer must consider. There has been a tendency toward specialization within these applications in the water-resources field, but actually the problems encountered and the solutions to these problems have much in common. Table 1-1 summarizes the problems which may be

TABLE 1-1   Problems of Water-resources Engineering

Studies and facilities required	Control of excess water				Conservation (quantity)				Conservation (quality)
	Flood mitigation	Storm drainage	Bridges, culverts	Sewerage	Water supply	Irrigation	Hydro power	Navigation	Pollution control
How much water is needed?	...	...	...	...	X	X	X	X	X
How much water can be expected?									
Minimum flow	...	...	...	X	X	X	X	X	X
Annual yield	...	...	...	X	X	X	X	X	X
Flood peaks	X	X	X	...	X	X	X	X	
Flood volume	X	X							
Groundwater	...	X	...	X	X	X	...	...	X
Who may use the water?	...	...	...	...	X	X	X	X	X
What kind of water is it?									
Chemical	...	X	...	X	X	X	...	...	X
Bacteriological	...	X	....	X	X	X	...	...	X
Sediment	X	X	X	X	X	X	X	X	X
What structural problems exist?									
Geology	X	X	X	X	X	X	X	X	X
Dams	X	...	...	...	X	X	X	X	X
Spillways	X	...	...	...	X	X	X	X	X
Gates	X	X	...	X	X	X	X	X	X
Sluiceways	X	...	...	...	X	X	X	X	
Intakes	...	...	...	...	X	X	X		
Channel works	X	X	X	X	...	...	...	X	
Levees	X	X	X						
Pipelines	...	X	...	X	X	X	X	...	X
Canals	X	X	...	...	X	X	X	X	
Locks	...	...	...	...	...	...	...	X	
Pumps	X	X	...	X	X	X	X	X	X
Turbines	...	...	...	...	...	...	X		
Purification	...	...	...	X	X	X	...	...	X
Does project affect wildlife or natural beauty?	X	X	X	X	X	X	X	X	X
Is the project economic?	X	X	X	X	X	X	X	X	X

encountered within the nine main functional fields of water-resources engineering.

1-2 Quantity of water   At some risk of oversimplification, the job of the water-resources engineer may be reduced to a number of basic questions. Since the water-resources project is for the control or use of water, the first questions naturally deal with the quantities of water. Where utilization is proposed, the first question is usually *How much water is needed?* This is probably the most difficult of all the design problems to answer accurately because it involves social and economic aspects as well as engineering. On the basis of an economic analysis, a decision must also be made concerning the span of years for which the proposed works should be adequate.

Table 1-2 summarizes 1968 total water use in the United States in relation to gross water supply–precipitation. In discussing water use it is important to distinguish between *diversion* (withdrawal), or water taken into a system, and *consumption*, water that is evaporated or combined in a product and is no longer available for use.

Almost all project designs depend on the answer to the question *How much water can be expected?* Peak rates of flow are the basis of design of projects to control excess water, while volume of flow during longer periods

TABLE 1-2   *Water Balance of the Conterminous United States**

Component	In./yr	Millions of acre-ft/yr	Percentage
Precipitation	30	4,750	100
Evapotranspiration†	21.5	3,410	72
Streamflow	8.5	1,340	28
Irrigation diversion	0.80	127	2.7
Public use diversion	0.18	29	0.6
Industrial diversion	0.79	124	2.6
Diversions (total)‡	1.77	280	5.9
Irrigation consumption	0.47	75	1.6
Public use consumption	0.05	8	0.2
Industrial consumption	0.03	5	0.1
Consumption (total)	0.55	88	1.9
Return flow (total)	1.22	192	4.0
Outflow to oceans	7.95	1,262	26.1

*Adapted from data in "The Nation's Water Resources," U.S. Water Resources Council, Washington, D.C., 1968.
†11.6 in. (1,840 maf) used beneficially where it falls for crops, forage and forests.
‡0.39 in. (62 maf) supplied from groundwater.

of time is of interest in designing projects for use of water. The answers to this question are found through the application of *hydrology*, the study of the occurrence and distribution of the natural waters of the earth. The principles of hydrology are outlined in Chaps. 2 to 5.

The water flowing in a natural stream is not necessarily available for use by any person or group desiring it. The right to use water has considerable value, especially in regions where water is scarce. Like other things of value, water rights are protected by law, and a legal answer to the question *Who may use this water?* may be required before the quantities of available water can be evaluated. Diversion of natural streamflow may cause property damage, and alterations in natural flow conditions are governed by legal restrictions which should be investigated before completion of the project design.

1-3 Water quality In addition to being adequate in quantity, water must often withstand certain tests of quality. Problems of water quality are encountered in planning water-supply and irrigation projects and in the disposal of wastewater. Polluted streams create problems for fish and wild-life, are unsuited for recreation, and are often unsightly and sometimes odorous. Chemical and bacteriological tests are employed to determine the amount and character of impurities in water. Plant and human physiologists must evaluate the effect of these impurities on crops or human consumers and set standards of acceptable quality. The engineer must then provide the necessary facilities for removing impurities from the water by mechanical, chemical, or bacteriological methods. Governmental agencies having the authority to regulate the disposal of wastes are required to safeguard our waters against pollution.

1-4 Hydraulic structures Structural design of facilities for water-resources projects utilizes the techniques of civil engineering. The shape and dimensions of the structure are often dictated by the hydraulic characteristics which it must possess and hence are determined by application of the principles of fluid mechanics. Many hydraulic structures are relatively massive as compared with buildings and bridges, and the structural design involves much less fine detail. However, hydraulic structures frequently involve complex curved and warped surfaces and sometimes intricate detail for gates, valves, control systems, etc. Almost all the conventional engineering materials are employed in hydraulic structures. Earth, mass and reinforced concrete, timber, clay tile, asphaltic compounds, and most of the common metals are found in such structures.

Largely because of topographic controls, it is not always possible to select the most satisfactory location for a hydraulic structure from the structural viewpoint. Hence, geologic investigations are an important part of the preliminary planning. These investigations should be aimed at selecting the best of the otherwise suitable sites, predicting the structural problems which will result from the particular conditions at the site, and locating sources of native material suitable for use in the proposed structure.

1-5 Economics in water-resources engineering Little skill is required to design a structure for some purpose if unlimited funds are available. The special ability of the engineer is reflected in the planning of projects which serve their intended purpose at a cost commensurate with the benefits. An economic analysis to determine the best of several alternatives is required in planning most projects. It must usually be demonstrated that the project cost is sufficiently less than the expected benefits to warrant the required investment. In many cases the estimated benefits serve also as a basis for determining a schedule of payments by the beneficiaries who will repay the project cost to the construction agency.

Precipitation and streamflow vary widely from year to year. It is usually uneconomic to design a project to provide protection against the worst possible flood or to assure an adequate water supply during the most severe drought which could conceivably occur. Instead the project design is gaged against a scale of probability so that the probability of the project failing to serve its purpose is small but still positive. Economic analysis (Chap. 13) is dependent on statistical analysis of the probability of occurrence of extreme floods or droughts (Chap. 5).

1-6 Social aspects of water-resources engineering Most water projects are planned for and financed by some governmental unit—a municipal water-supply or sewerage system, a state highway department, or a federal irrigation or flood-mitigation project—or by a public utility. Many such projects become controversial political issues and are debated at length by people whose understanding of the basic engineering aspects of the problem is very limited. It is a clear responsibility of an engineer who has the necessary facts concerning such a project to take a firm position in the public interest, if the final decision is not to be made on political and emotional grounds. It is particularly important that the engineer carefully analyze the facts and present a sound case in simple terms. He should avoid championing a "pet" project which is of limited benefit to the public. Throughout any negotiations concerning a publicly financed project, the engineer should adhere carefully to the code of ethics of the professional society that represents the civil engineering profession in his country. Failure to do so prejudices his case and the entire profession in the eyes of the public.

1-7 Planning of water-resources projects Planning is an important step in the development of a water-resources project. The planning of a project (Fig. 1-1) generally involves a *political incentive* or recognition of the need for a project. This is followed by the conception of alternative

FIG. 1-1   Steps in the planning of a water-resources project.

technically feasible solutions which would satisfy the need. The alternative proposals are subjected to an economy study that analyzes their benefits and costs and thus determines their economic feasibility. Finally, *financial feasibility* (can the project be paid for?) and *political practicality* (is the project acceptable to the public?) enter the picture and play an important role in the choice of alternatives. A detailed discussion of planning for water-resources development is presented in Chap. 21.

   1-8   History of water-resources engineering   The first water project has been lost in the mist of prehistoric time. The importance of water to human life justifies the supposition that some ancient man conceived the idea of diverting streamflow from a natural channel to an artificial one in order to convey water to some point where it was needed for crops or humans. The Old World contains numerous evidences of water projects of considerable magnitude. The earliest large-scale drainage and irrigation works are attributed to Menes, founder of the first Egyptian dynasty, about 3200 B.C. These works were followed by many varied projects in the Mediterranean and Near East area, including dams, canals, aqueducts, and sewer systems. Some 381 mi of aqueducts were constructed to bring water to the city of Rome. An irrigation project in Szechuan Province of China dating from about 250 B.C. is still in use. Even in the New World, projects of considerable scope antedate the coming of the white man. Ruins of an elaborate and extensive irrigation project constructed about A.D. 1100 by Hohokam Indians in what is now Arizona and similar Aztec works in Mexico indicate flourishing irrigation economies.

   These early works were not designed and built by engineers in the modern sense of the word. The ancient builders were master craftsmen and technicians (the Greek *architekton*, or archtechnician) who employed amazing intuitive judgment in planning and executing their works. Rules of thumb developed through experience guided the leading builders, but these trade secrets were not necessarily conveyed to other men. The great thinkers of the Greek era contributed much to science, but since manual labor was considered demeaning, the application of their knowledge in practical pursuits was retarded. Many erroneous concepts and gaps in understanding delayed the development of engineering as it is known today. It was not until the time of Leonardo da Vinci (about A.D. 1500) that the idea that precipitation was the source of streamflow received any real support and many years later before it was definitely proved. The limitations of available construction materials also influenced early engineering works. Since no materials suitable for large pressure pipes were available to the Romans, their aqueducts were designed as massive structures to carry water under atmospheric pressure at all times.

The first effort at organized engineering knowledge was the founding in 1760 of the École des Ponts et Chaussées in Paris. As late as 1850, however, engineering designs were based mainly on rules of thumb developed through experience and tempered with liberal factors of safety. Since that date, utilization of theory has increased rapidly until today a vast amount of careful computation is an integral part of most project designs. It must not be assumed that the rule of thumb has been completely eliminated from engineering practice, for there are still many facets of engineering so imperfectly understood that theoretical solutions are not feasible. A considerable lag seems to exist between research and application. The answers to many professional problems are available in laboratory records and even published papers, but they have not yet been extensively employed by practicing engineers.

1-9   The future of water-resources engineering   Laymen, unfamiliar with engineering problems, often view the enormous activity in flood mitigation, irrigation, and other phases of water-resources engineering with the thought that opportunities for further work must be negligible. Actually modern civilization is far more dependent on water than were the civilizations of the past. Modern medical science together with modern sanitary engineering has reduced death rates and increased life expectancy, and the world's population is increasing rapidly. Modern standards of personal cleanliness require vastly more water than was used a century ago. The increasing population requires expanded acreage for agriculture, much of which must come through land drainage and irrigation. Increasing urban populations require more attention to storm drainage, water supply, and sewerage. Industrial progress finds increasing uses for water in process industries and for electric-power production. The emphasis of water-resources engineering shifts more or less continuously. The major work in this field during the early years of the United States was the construction of canals for transport. Other modes of transportation have made the canal-boat obsolete, but these new means of transport have introduced new problems of drainage for highways, railroads, and airports.

The development of civilization has increased the importance of water-resources engineering, and there is no prospect of a decline of activity in this field in the foreseeable future. In fact, the increasing pressure for water is forcing the development of marginal projects which would not have been considered only a few years ago. If a project of marginal value is to be successful, it must be planned with more care and thought than was required for the more obvious projects of the past. More accurate hydrologic methods must be employed in estimating available water. More efficient methods and better construction material must be utilized to reduce costs so that difficult projects may become economically feasible.

The water-resources engineer of the future will find himself deeply involved with new technology and new concepts. Desalting of salt water, reclamation of wastewater, weather modification, land management to improve water yield, and new water-saving techniques in all areas of water use are topics of increasing interest and research. An expanding world population is changing ecologic patterns in many ways, and water planning must include evaluation of ways to minimize undesirable ecologic consequences. Concern for the preservation of the natural environment will be increasingly important in water planning of the future.

The conflict between preserving our ecosystems and meeting the "needs" of people for water management must certainly lead to new approaches in water management and quite possibly to new definitions of "need." It will not be sufficient to attack water problems of the future by simply copying methods of the past.

BIBLIOGRAPHY

Ackerman, Edward A., and George O. G. Löf: "Technology in American Water Development," Johns Hopkins Press, Baltimore, 1959.
Biswas, Asit K.: "A History of Hydrology," North Holland Publishing Company, Amsterdam, 1970.
Chow, Ven Te (ed.): "Handbook of Applied Hydrology," McGraw-Hill, New York, 1964.
Finch, James K.: "Engineering and Western Civilization," McGraw-Hill, New York, 1951.
Huberty, Martin R., and Warren L. Flock: "Natural Resources," McGraw-Hill, New York, 1959.
Langbein, W. B., and W. G. Hoyt: "Water Facts for the Nation's Future," Ronald, New York, 1959.
Merdinger, Charles J.: Civil Engineering through the Ages, *Trans. ASCE*, Vol. CT, pp. 1–27, 1953.
"The Nation's Water Resources," U.S. Water Resources Council, Washington, D.C., 1968.
Stewart, G. R., and M. Donnelly: Soil and Water Economy of the Pueblo Southwest, *Sci. Monthly*, Vol. 56, pp. 31–44, January, 1943, pp. 134–144, February, 1943.
"Water Resources Activities in the United States," Senate Select Committee on National Water Resources, 1961.
Wilbur, J. B.: Future Possibilities in Civil Engineering, *Trans. ASCE*, Vol. CT, pp. 125–132, 1953.

Chapter 2

Descriptive hydrology[1]

The world's supply of fresh water is obtained almost entirely as precipitation resulting from evaporation of seawater. The processes involved in the transfer of moisture from the sea to the land and back to the sea again form what is known as the *hydrologic cycle*. An understanding of these processes is important to the water-resources engineer.

2-1 The hydrologic cycle The first stage in the hydrologic cyle is the evaporation of water from the oceans. This vapor is carried over the continents by moving air masses. If the vapor is cooled to its dew point, it condenses into visible water droplets which form cloud or fog. Under favorable meteorological conditions the tiny droplets grow large enough to fall to earth as precipitation.

Cooling of large masses of air is brought about by lifting. The resulting decrease in pressure is accompanied by a temperature decrease in accordance with the gas laws. *Orographic lifting* occurs when air is forced to rise over a mountain barrier. For this reason the windward slopes of mountains are usually regions of high precipitation. Air may also rise over a cooler air mass. The boundary between these air masses is called a *frontal surface*, and the lifting process is called *frontal lifting*. Finally, air heated from below may rise by convection through cooler air (*convective lifting*) to cause the isolated convective thunderstorm characteristic of summer climate in much of the United States. Frequently two or more of these lifting processes take place simultaneously.

---

[1] "Hydrology is the science that treats of the waters of the Earth, their occurrence, circulation, and distribution, their chemical and physical properties, and their reaction with their environment, including their relation to living things." (From "Scientific Hydrology," Federal Council for Science and Technology, June, 1962.)

About two-thirds of the precipitation which reaches the land surface is returned to the atmosphere by evaporation from water surfaces, soil, and vegetation and through transpiration by plants. The remainder of the precipitation returns ultimately to the ocean through surface or underground channels (Table 1-2). The large percentage of precipitation which is evaporated has often led to the belief that increasing this evaporation by construction of reservoirs or planting of trees will increase the moisture available in the atmosphere for precipitation. Actually only a small portion of the moisture (usually much less than 10 percent) which passes over any given point on the earth's surface is precipitated.[1] Hence, moisture evaporated from the land surfaces is a minor part of the total atmospheric moisture.[2]

The hydrologic cycle is depicted diagrammatically in Fig. 2-1. No simple figure can do justice to the complexities of the cycle as it occurs in nature. The science of hydrology is devoted to a study of the rate of exchange of water between phases of the cycle and in particular to the variations in this rate with time and place. This information provides the data necessary for the hydraulic design of physical works to control and utilize natural water.

2-2   The river basin   A river basin is the area tributary to a given point on a stream, and is separated from adjacent basins by a *divide*, or

---

[1] G. S. Benton, R. T. Blackburn, and V. O. Snead, The Role of the Atmosphere in the Hydrologic Cycle, *Trans. Am. Geophys. Union*, Vol. 31, pp. 61–73, February, 1950.

[2] F. A. Huff and G. E. Stout, A Preliminary Study of Atmospheric-moisture-precipitation Relationships over Illinois, *Bull. Am. Meteorol. Soc.*, Vol. 32, pp. 295–297, 1951.

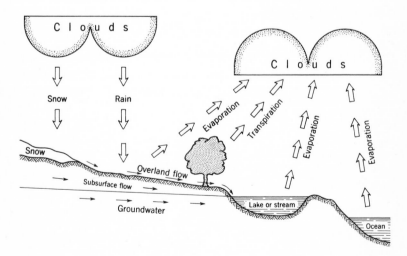

FIG. 2-1   *Schematic diagram of the hydrologic cycle.*

ridge, which can be traced on topographic maps. All surface water originating in the area enclosed by the divide is discharged through the lowest point in the divide through which the main stream of the basin passes. It is commonly assumed that the movement of groundwater conforms to the surface divides, but this assumption is not always correct, and large quantities of water may be transported from one basin to another as groundwater.

PRECIPITATION

2-3   Types of precipitation   Precipitation includes all water which falls from the atmosphere to the earth's surface. Precipitation occurs in a variety of forms which are of interest to the meteorologist, but the hydrologist is interested in distinguishing only between liquid precipitation (rainfall) and frozen precipitation (snow, hail, sleet, and freezing rain). Rainfall runs off to the streams soon after it reaches the ground and is the cause of most floods. Frozen precipitation may remain where it falls for a long time before it melts. Melting snow is rarely the cause of major floods although, in combination with rainfall, it may contribute to major floods such as that on the upper Mississippi River in 1969. Mountain snowpacks are often important sources of water for irrigation and other purposes. The snowfields serve as vast reservoirs which store the winter precipitation until spring thaws release it near the time it is required for irrigation.

2-4   Fog drip and dew   Fog consists of water droplets so small that their fall velocities are negligible. Fog particles which contact vegetation may adhere, coalesce with other droplets, and eventually form a drop large enough to fall to the ground. This process, *fog drip*, is an important source of water for native vegetation during the rainless summers of the Pacific Coast. Some 200 in. of annual precipitation[1] was attributed to fog drip from Norfolk pines on the slopes of the island of Lanai, Hawaii.

On clear nights the loss of heat by radiation from the soil causes cooling of the ground surface and of the air immediately above it. Condensation of the water vapor present in the air results in a deposit of *dew*. The small quantities of dew and fog drip deposited in any day do not contribute to streamflow or groundwater. They do, however, offer a source of water which may be exploited locally. Research in Israel[2] has shown that broadleaved crops such as cabbage may be efficient dew collectors which can be grown in an arid region with little or no irrigation.

---

[1] N. K. Carlson, Fog and Lava Rock, Pines and Pineapples, *Amer. Forests*, pp. 8–10, 58–59, February, 1961.

[2] D. Ashbel, Frequency and Distribution of Dew in Palestine, *Geog. Rev.*, Vol. 39, pp. 291–297, April, 1949.

2-5 Precipitation measurement Amount of precipitation is expressed as the depth in inches or millimeters which falls on a level surface. This may be measured as the depth of water deposited in an open, straight-sided container. The standard gage[1] used in the United States (Fig. 2-2) consists of a funnel 8 in. in diameter discharging into a tube 2.53 in. in diameter. The area of the inner tube is 0.1 that of the funnel, and a stick graduated in inches and tenths can be used to measure precipitation to the nearest 0.01 in. Precipitation in excess of 2 in. overtops the inner tube and collects in the overflow can. By removing the funnel and inner tube from the gage, the 8-in.-diameter overflow can may be used to collect snowfall, which is melted and measured in the inner tube. Large *storage gages* are used in remote areas to catch and store precipitation for periods of 30 days or more. If snowfall is expected, an initial charge of calcium chloride brine is placed in the gage to melt the snow and to prevent the freezing of the liquid in the gage. A thin film of oil is used to prevent evaporation from the gage between observations.

Wind sets up air currents around precipitation gages which usually cause

---

[1] World-wide, a variety of different types of gages are used. Practically, there is little difference in accuracy in measuring rain, but smaller gages are not suitable for snowfall.

FIG. 2-2   Standard 8-in. precipitation gage. (National Weather Service)

the gages to catch less precipitation than they should.[1] The low fall velocity of snowflakes makes this effect even more marked for snowfall than for rain. The deficiency in catch may vary from 0 to 50 percent or more depending on the type of gage, wind velocity, and local terrane. The National Weather Service[2] uses an *Alter shield* consisting of a series of metal slats pivoted about a circular ring near the top of the gage and joined by a chain at the bottom. The tops of the slats are about 2 in. above the top of the gage. The flexible construction is intended to permit wind to move the slats and minimize the accumulation of snow on the shield.

In order to determine rates of rainfall over short periods of time, recording rain gages are used. The most common type is the *weighing rain gage*, in which a bucket is supported by a spring or lever balance. Movement of the bucket is transmitted to a pen which traces a record of the increasing weight of the bucket and its contents on a clock-driven chart. Recorders utilizing punched paper tape which can be processed automatically are also used. The *tipping-bucket gage* consists of a pair of buckets pivoted under a funnel in such a way that when one bucket receives 0.01 in. of precipitation it tips, discharging its contents into a reservoir and bringing the other bucket under the funnel. A recording mechanism indicates the time of occurrence of each tip. The tipping-bucket gage is well adapted to the measurement of rainfall intensity for short periods, but the more rugged construction of the weighing-type gage and its ability to record snowfall as well as rain make it preferable for most purposes.

Subsequent to the development of radar in World War II it was found that microwave radar (1 to 20 cm wavelength) would indicate the presence of rain[3] within its scanning area. The amount of reflected energy is dependent on the raindrop size and the distance from the transmitter. Drop size is roughly correlated with rain intensity, and the image on the radar screen (isoecho map) can be interpreted as an approximate indication of rainfall intensity. A calibration may also be determined from actual rain-gage measurements in the area scanned by the radar. Radar offers a means of obtaining information on rainfall distribution which would be only roughly defined by the usual network of rain gages.

2-6 Computation of average precipitation Large differences in precipitation are observed within short distances in mountainous terrane or during showery precipitation in level country. The average density of rain

---

[1] C. C. Warnick, Experiments with Windshields for Precipitation Gages, *Trans. Am. Geophys. Union*, Vol. 34, pp. 379–388, June, 1953.

[2] The U.S. Weather Bureau was changed to the National Weather Service in 1970.

[3] M. G. H. Ligda, Radar Storm Observation, "Compendium of Meteorology," pp. 1265–1282, American Meteorological Society, Boston, Mass., 1951.

gages in the United States is about one per 250 sq mi, and the data so obtained represent only a scattered sample of precipitation over large areas. It is therefore necessary to consider methods of computing the average precipitation over a given area. The simplest method of estimating average precipitation is to compute the arithmetic average of the recorded precipitation values at stations in or near the area. If the precipitation is nonuniform and the stations unevenly distributed within the area, the arithmetic average may be quite incorrect. To overcome this error, the precipitation at each station may be weighted in proportion to the area the station is assumed to represent.

A common method of determining weighting factors is by use of the *Thiessen network* (Fig. 2-3). A Thiessen network is constructed by connecting adjacent stations on a map by straight lines and erecting perpendicular bisectors to each connecting line. The polygon formed by the perpendicular bisectors around a station encloses an area which is everywhere closer to that station than to any other station. This area is assumed to be best represented by the precipitation at the enclosed station. This is often a reasonable assumption but may not always be correct. Since the perpendiculars bisect the sides of triangles with stations at each apex, three bisectors must meet at a point. To compute the average rainfall, the area represented by each station is expressed as a percentage of the total area. The average rainfall is the sum of the individual station amounts, each multiplied by its percentage of area. An alternative method is shown on Fig. 2-3. If the stations are uniformly distributed in the area, the Thiessen areas will be equal and the computed average rainfall will equal the arithmetic average.

The sole basis for the Thiessen method is the assumption that a station best represents the area which is closest to it. If precipitation is controlled by topography or results from intense convection, this assumption may not be valid. An *isohyetal map* (Fig. 2-4) showing contours of equal precipitation may be drawn to conform to other pertinent information in addition to the

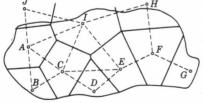

Station	Thiessen area, sq mi	Precipitation, in.	Product, sq mi in.
A	72	3.50	252
B	34	4.46	152
C	76	4.28	325
D	40	5.90	236
E	76	6.34	482
F	92	5.62	517
G	46	5.20	239
H	40	5.26	211
I	86	3.86	332
J	6	3.30	20
Total	568	47.72	2766

Average precipitation = $\frac{\Sigma \text{ Product}}{\Sigma \text{ Area}} = \frac{2766}{568} = 4.87$ in.

FIG. 2-3   Thiessen network.

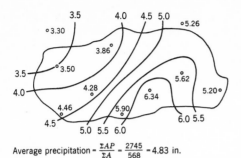

Isohyets	Area between isohyets, sq mi	Average precipitation, in.	Product $A \times P$ sq mi in.
3.0			
	19	3.45	66
3.5			
	106	3.75	398
4.0			
	102	4.25	434
4.5			
	60	4.75	285
5.0			
	150	5.25	788
5.5			
	84	5.75	483
6.0			
	47	6.20	291
6.5			
Total	568	—	2745

Average precipitation $= \dfrac{\Sigma AP}{\Sigma A} = \dfrac{2745}{568} = 4.83$ in.

FIG. 2-4   Isohyetal map.

precipitation data and thus present a more accurate picture of the rainfall distribution. Since precipitation usually increases with elevation, the isohyets may be made to conform approximately with the contours of elevation.

To compute average precipitation from an isohyetal map, the areas enclosed between successive isohyets are measured and multiplied by the average precipitation between the isohyets. The sum of these products divided by the total area is the average precipitation. If the isohyets are interpolated linearly between stations, the computed average precipitation will not differ appreciably from that computed with a Thiessen network.

2-7   Snow   The measurement of snowfall has been discussed in Sec. 2-5. Snow on the ground is measured in terms of its depth in feet and inches. Shallow depths are measured with any convenient scale, while large depths are measured on a *snow stake*, a post graduated in feet and inches and permanently installed at the desired site. Because of variations in snow density, a depth measurement is not sufficient to tell how much water is contained in the snow pack. The *water equivalent*, or depth of water which would result from melting a column of snow, is measured by forcing a small tube into the snow, withdrawing it, and weighing the tube to determine the weight of the snow core removed. There are a number of types of snow samplers but the most common type is the Mt. Rose pattern with an internal diameter of 1.485 in. so that each ounce of snow in the core represents 1 in. of water equivalent. The specific gravity of freshly fallen snow is usually about 0.1. Thus, its water equivalent is 0.1 in. for each inch of snow depth. The specific gravity increases with time as the snow remains on the ground and may reach a maximum of about 0.5 in heavy mountain snowpacks. The term *density* of snow is often used synonymously with specific gravity although this usage is not in accord with the common meanings of the two terms.

2-8   Variations in precipitation   The complex pattern of precipitation in the United States (Fig. 2-5) reflects several interacting influences. In

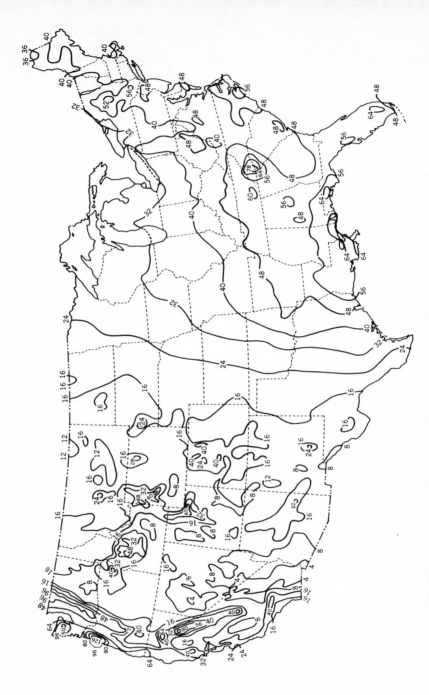

FIG. 2-5 Average annual precipitation in the United States. (National Weather Service)

general, precipitation decreases with increasing latitude because decreasing temperatures reduce atmospheric moisture. A more important control in the United States is distance from a moisture source, as evidenced by the concentration of precipitation along the coasts and to some extent to the leeward of the Great Lakes. The importance of mountains as a factor in the production of precipitation by orographic lifting is evident in the isohyetal pattern in the Western states and along the Appalachian Mountains. Heavier precipitation normally occurs along the windward slope of a mountain range with a *rain shadow* on the leeward slope.

From the engineering viewpoint, time variations in precipitation may be more important than regional variations. The most marked of these variations is the annual precipitation cycle shown for selected stations in Fig. 2-6. In the Far West precipitation is at a minimum during the summer because a large high-pressure area in the Pacific blocks the path of storms. In contrast, a summer maximum of precipitation is observed in the Great Plains, where the cold continental high-pressure center recedes northward during the summer. An essentially uniform distribution of precipitation prevails in the Eastern states. The dry summers of the West make irrigation a necessity for many crops and emphasize the importance of storage reservoirs.

Variations in precipitation from year to year make it important to design reservoirs that are adequate during years of low rainfall. In some cases reservoirs must carry water in storage for a period of several years. Over 100 different cycles in precipitation with periods up to 700 yr in length have

FIG. 2-6   *Typical monthly distribution of precipitation in various climatic regions.*

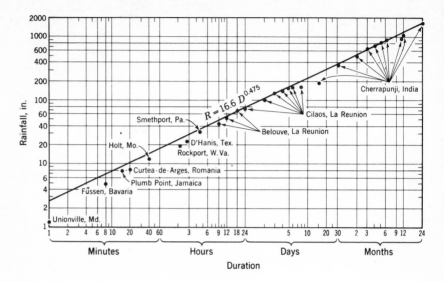

FIG. 2-7   Maximum observed world rainfalls for various durations.

been reported by various investigators. Sunspots and planetary configurations have been among the factors suggested as controlling these cycles. No one has been successful, however, in employing cycles for forecasting precipitation several years in advance. The best evidence now available suggests that the occurrence of a series of wet or dry years is purely random, such as might be expected by successive tosses of a coin. Accurate precipitation records are too short for a really satisfactory analysis of cyclic variation, which, if it does occur, must be a complex variation consisting of several superimposed cycles of differing period.

Figure 2-7 shows the world's record rainfalls for various durations.[1] Most of the observations are derived from cooperative and unofficial stations. A similar plot for record rainfalls at U.S. First Order stations would fall at about one-third the magnitudes shown on Fig. 2-7. This difference reflects the more effective sampling of the large number of cooperative and unofficial stations and emphasizes the importance of a careful search for data when a study requires information on rainfall intensities. Table 2-1 presents maximum observed intensities for various durations at a number of geographically distributed stations in the United States.

---

[1] J. L. H. Paulhus, Indian Ocean and Taiwan Rainfalls Set New Records, *Monthly Weather Review*, Vol. 93, No. 5, pp. 331–335, May, 1965.

TABLE 2-1 *Maximum Recorded Point Rainfall (In inches)*

Station		Duration							
		Min				Hr			
		5	15	30	60	2	6	12	24
Portland, Ore.	Amt.	0.40	0.83	1.10	1.31	1.74	...	...	7.66
	Date	8/8/00	8/8/00	8/8/00	6/7/27	6/7/27	...	...	12/12/82
Los Angeles, Calif.	Amt.	0.44	1.05	1.51	1.87	2.11	3.37	5.02	7.36
	Date	1/14/08	11/19/67	11/19/67	11/19/67	11/19/67	3/2/38	1/26/56	12/31/33
Boise, Idaho	Amt.	0.34	0.59	0.67	0.98	1.23	...	...	2.46
	Date	5/22/42	8/9/63	5/18/21	7/30/12	6/12/58	...	...	3/28/04
Phoenix, Ariz.	Amt.	0.68	1.14	1.27	1.72	2.20	2.41	3.51	4.98
	Date	9/16/69	9/16/69	9/16/69	8/18/66	9/4/39	9/4/39	7/1/11	7/1/11
Amarillo, Tex.	Amt.	0.68	1.58	2.57	3.36	3.56	5.27	6.06	6.75
	Date	8/26/49	7/8/43	6/24/48	6/24/48	7/8/43	6/9/60	6/9/60	5/15/51
Galveston, Tex.	Amt.	0.85	1.94	3.06	5.31	7.58	11.79	12.75	14.35
	Date	4/13/29	4/13/29	10/6/10	10/22/13	4/22/04	10/8/01	10/8/01	7/13/00
Rapid City, S. Dak.	Amt.	0.97	1.39	1.93	2.43	2.76	3.49	4.44	5.57
	Date	7/19/58	7/19/58	7/29/44	7/22/10	7/22/10	10/8/01	5/27/26	5/27/26
New York, N.Y.	Amt.	0.75	1.63	2.34	2.97	3.52	4.44	6.30	9.55
	Date	8/12/26	7/10/05	8/12/26	8/26/47	8/26/47	10/1/13	10/8/03	10/8/03
Pittsburgh, Pa.	Amt.	0.72	1.23	1.46	2.00	2.29	2.53	2.93	4.08
	Date	6/26/31	6/26/31	7/4/03	7/27/43	7/27/43	7/27/43	9/29/36	9/17/76
Miami, Fla.	Amt.	0.71	1.89	2.92	4.53	6.11	10.64	13.86	15.10
	Date	6/14/33	10/11/47	10/11/47	6/14/33	11/30/25	11/30/25	11/30/25	11/29/25

STREAMFLOW

In the streamflow phase of the hydrologic cycle, the water from a given drainage basin is usually concentrated in a single channel, and it is possible to measure the entire quantity of water in this phase of the cycle as it leaves the basin.

2-9 Measurement of streamflow A continuous record of stream-flow requires the establishment of a relation between rate of flow and water level in a channel. In small channels this may sometimes be accomplished by use of a weir or measuring flume (Chap. 10) for which a head-discharge relation can be determined in the laboratory. Measurements of head may then be converted to rates of flow.

In large streams, the use of laboratory-rated flow-measuring devices becomes impracticable, and measurements of discharge are made with a *current meter*. A *stage-discharge relation*, or *rating curve*, is constructed by plotting the measured discharge against the *stage* (water-surface elevation) at the time of measurement (Fig. 2-8). If the station is located just upstream from a rapids or other natural control which fixes a definite relation between stage and discharge, an accurate and permanent rating is obtained. An artificial control consisting of a low weir of concrete or masonry is sometimes constructed on small streams. It may be necessary to accept the rating curve fixed by the channel geometry of large streams, and erosion or deposition of sediment may change the rating from time to time. Under these conditions a satisfactory streamflow record can be obtained only by frequent current-meter measurements to fix the position of the rating curve at any time. Even

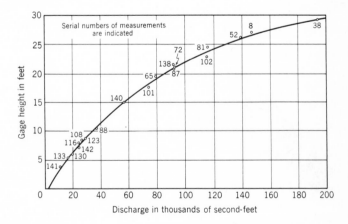

FIG. 2-8   Stage-discharge relation for the Willamette River at Albany, Oregon. (Data from the U.S. Geological Survey)

where a good control exists, routine check measurements are considered desirable.

The simple stage-discharge relation of Fig. 2-8 is typical of most streamflow-measuring stations. At some stations, however, backwater from an intersecting stream or from a reservoir may affect the rating curve. If the channel slope is flat, variations in water-surface slope resulting from rising or falling stages may also affect the rating curve. Under either of these conditions a *slope-stage-discharge relation* may be employed. For this purpose an auxiliary stage record is required near the main station. The distance between the two stations should be such that the difference in water-surface elevation is at least 1 ft, so that errors in reading the gages will be small compared with the difference. The difference in elevation or *fall* is used as an index of the slope between the stations. A rating curve is then drawn relating discharge and stage for some fixed value of fall, Δz_0. If the discharge at any stage as read from this curve is Q_0, then the discharge Q_a for the same stage and a fall of Δz_a is

$$Q_a = \left(\frac{\Delta z_a}{\Delta z_0}\right)^n Q_0 \qquad (2\text{-}1)$$

With reasonably uniform flow between the two stations, the exponent n will be about 0.5, but in practice it often varies slightly from this value.

The Price current meter (Fig. 2-9) is most widely used in the United States. The meter assembly consists of a cup wheel rotating about a vertical axis, tail vanes to keep the meter headed into the current, and a weight to keep the meter cable as nearly vertical as possible. In deep water the meter is suspended from a bridge, cable car, or boat by a cable which serves also as a conductor to transmit electrical contacts made by the cup wheel to a counter or earphones worn by the hydrographer. In shallow water the meter may be attached to a rod for wading measurements.

The velocity variation with depth in most streams is logarithmic, and the average of the velocities at 0.2 and 0.8 of the depth is very nearly equal to the mean velocity in a vertical section. A single measurement at 0.6 depth below the surface is only slightly less accurate as an estimate of the mean velocity. A streamflow measurement is usually made by determining the mean velocity in a number of vertical sections across the stream. The velocity in any vertical section is assumed to represent the velocity in a portion of the total cross section extending halfway to the adjacent vertical sections. The discharge in this portion of the cross section is computed by multiplying its area by the mean velocity. The total discharge of the stream is the sum of the discharges in the several partial sections.

Discharge may also be estimated by application of open-channel formulas (Chap. 10), use of weir formulas for dams or spillways (Chap. 9), calculation

FIG. 2-9   *Price current meter and 30-lb C-type sounding weight. (U.S. Geological Survey.*

of flow through a contracted opening at a bridge (Chap. 18), or timing the travel of floats in the stream. If surface floats are used, mean velocity is ordinarily assumed to be 0.85 times the float velocity. These methods are dependent on the selection of proper coefficients and are often inaccurate. They are normally used for reconnaissance purposes or for computing flood flow in the absence of meter measurements.

If a tracer solution is injected into a stream at a constant rate and samples are taken downstream at a point where turbulence has achieved complete mixing, the steady flow rate Q in the stream is given by[1]

$$Q = Q_t \frac{C_1 - C_2}{C_2 - C_0} \tag{2-2}$$

[1] H. Addison, "Applied Hydraulics," 4th ed., pp. 583–584, Wiley, New York, 1954.

where Q_t is the steady dosing rate, C_0 the concentration of the tracer in the undosed flow, C_1 the concentration of the tracer in the dose, and C_2 the concentration of the tracer in the dosed flow. The tracer may be a salt evaluated by titration, a dye evaluated colorimetrically, or a radioactive element evaluated with a suitable counting device. The procedure is well adapted to boulder-strewn streams where use of a conventional meter is difficult.

The accuracy of streamflow records depends upon the physical features of the cross section, the frequency of measurement, and the quality of the stage-measuring equipment. The adjective classification used by the U.S. Geological Survey is given in Table 2-2.

TABLE 2-2   Accuracy of Streamflow Data*

Adjective classification	Individual measurements	Published records
Excellent	<2	<5
Good	<5	<10
Fair	<8	<15
Poor	>8	>15

*Figures are probable errors in percent.

2-10   Measurement of river stage   The simplest device for measuring river stage is a *staff gage*, a scale graduated in feet or meters. Staff gages are usually read by an observer once or twice a day, but on streams subject to rapid changes in stage it is not possible to get a reliable record without use of recording equipment. The most common type of recording gage uses a float connected to the recording mechanism in such a way that motion of the float is recorded on a paper chart. A gage house and stilling well (Fig. 2-10) of corrugated steel, concrete, or timber is required to protect the recording equipment and float. A simple diaphragm-type pressure cell and recorder constitute a less costly installation which is useful in many circumstances.[1] Bubbler gages which measure the pressure required to force gas from the end of a submerged pipe have also proved accurate and relatively inexpensive.[2] The stage-recording equipment must be upstream from the control; but current-meter measurements may be made at any con-

---

[1] R. A. Duncan and G. Whitehouse, A Low Cost Streamflow Recorder, *Bull.* 3, Water Research Foundation of Australia, Sydney, 1959.

[2] H. R. Mortley, Gas-Electric Transducer for Transmission of River Heights, *J. Inst. Engrs., Australia*, pp. 191–195, June, 1961.

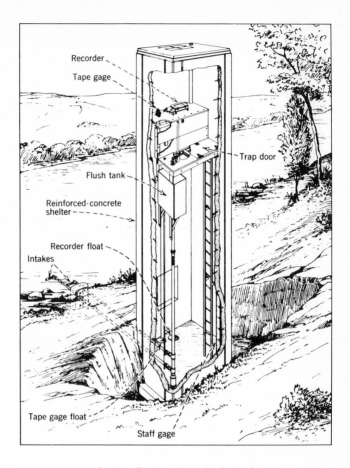

Recorder

Tape gage

Trap door

Flush tank

Reinforced-concrete
shelter

Recorder float

Intakes

Tape gage float

Staff gage

FIG. 2-10   Typical water-stage recorder installation. (U.S. Geological Survey)

venient section along the stream, provided there is no large difference in
discharge between the measuring section and the gaging station.

Numerous inexpensive *crest-stage gages* have been designed to provide a
record of the highest stage observed at a station. One simple gage[1] (Fig.
2-11) consists of a piece of pipe with holes for entry of water. A wooden
staff gage is placed in the pipe together with a small quantity of ground cork.
The cork floats on the water, and some adheres to the staff as the water level
falls. After a period of high water, an observer removes the staff from the

---

[1]G. E. Ferguson, Gage to Measure Crest Stages of Streams, *Civil Eng.* (*N.Y.*), Vol. 12,
pp. 570–571, October, 1942.

pipe, notes the highest mark with adhering cork grains, brushes the cork off the stick, and replaces it in the pipe with additional cork until the next rise.

2-11   Streamflow units   In the United States, rate of flow is usually expressed in *cubic feet per second*, sometimes called *second-feet* or *cusecs* and abbreviated *cfs*.

Volume of flow may be expressed in *second-foot-days* or *acre-feet*. The second-foot-day (abbreviated *sfd*) is the volume represented by a discharge of 1 cfs for 24 hr, or 86,400 cu ft. The acre-foot is the quantity required to cover an acre to a depth of 1 ft, or 43,560 cu ft. The second-foot-day is 1.98 acre-ft, but a conversion factor of 2.00 is widely used. In waterworks studies water volumes are often expressed in cubic feet or millions of gallons. In countries using the metric system flow rates are usually expressed in cubic meters per second and volumes in cubic meters. For comparison with rainfall it is convenient to express flow volumes in inches or millimeters of depth over the contributing area. A 1-in. depth over 1 sq mi equals 26.9 sfd, or 53.3 acre-ft. Conversion tables are given in the Appendix.

It is generally desirable that annual values of runoff represent a period beginning and ending during a time of low flow. In this way the total runoff for a single rainy season is included in the runoff year. The *water year* commonly used in the United States is the period from October 1 to September 30 of the following calendar year.

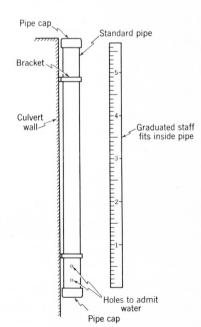

FIG. 2-11   Crest-stage gage used by the U.S. Geological Survey.

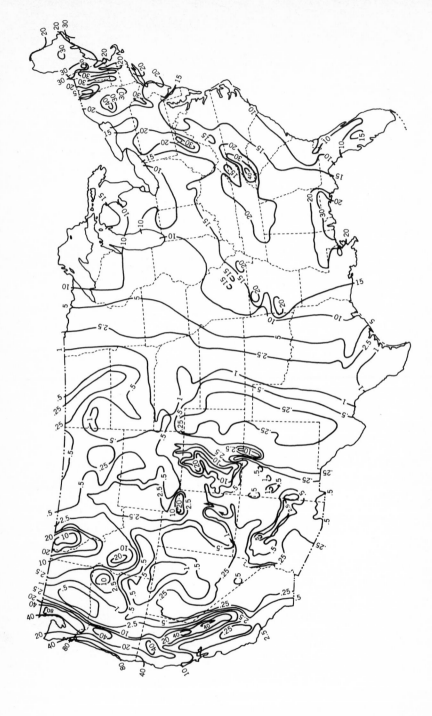

FIG. 2-12　Normal annual runoff in the United States in inches. (U.S. Geological Survey)

2-12   Variations in streamflow   The general pattern of normal annual runoff (Fig. 2-12) is quite similar to that of precipitation, but modified by soil and geologic characteristics and other factors. Relatively more precipitation appears as runoff in the cool, moist regions of the country than in the dry, warm regions where evaporation is high. Variations in streamflow throughout the year (Fig. 2-13) are controlled by the precipitation distribution. In most of the country the ratio of runoff to precipitation is lowest in summer, but in the West heavy runoff from melting snow occurs during the spring and early summer even though precipitation during this period is light.

A plot of streamflow against time is called a *hydrograph*. Hydrographs for a year at three selected stations are shown in Fig. 2-14. The upper graph is for a small stream subject mainly to rainfall. Note the irregular pattern of flow with isolated peaks corresponding to days of heavy precipitation. The middle graph is for a station far downstream in the same river system as the preceding station. Here the irregularities have been smoothed out, rates of rise and fall are slower, and the highest peaks result from sustained periods of rainfall lasting from several days to a week or two. The lower graph depicts the flow at a station in the Sierra Nevada of California. The period of winter floods resulting from rainfall is similar to the upper graph, but the late-season period of sustained, moderately high flows is characteristic of a region of heavy snows.

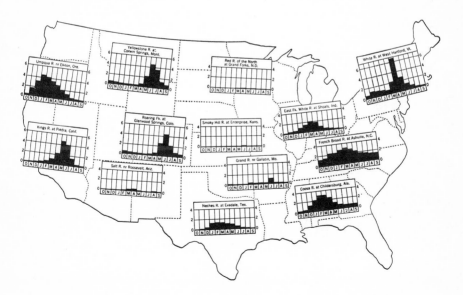

FIG. 2-13   *Median monthly runoff in inches at selected stations in the United States.*

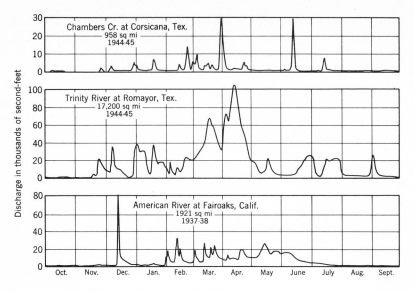

FIG. 2-14   Annual hydrographs for selected stations.

EVAPORATION AND TRANSPIRATION

Evaporation is the transfer of water from the liquid to the vapor state. *Transpiration* is the process by which plants remove moisture from the soil and release it to the air as vapor. More than half of the precipitation which reaches the land surfaces of the earth is returned to the atmosphere by the combined processes, *evapotranspiration*. In arid regions evaporation may consume a large portion of the water stored in reservoirs.

2-13   Factors affecting evaporation   The rate of evaporation from a water surface is proportional to the difference between the vapor pressure at the surface and the vapor pressure in the overlying air (Dalton's law). In still air, the vapor-pressure difference soon becomes small, and evaporation is limited by the rate of diffusion of vapor away from the water surface. Turbulence caused by wind and thermal convection transports the vapor from the surface layer and permits evaporation to continue.

Evaporation of a pound of water at 68°F requires[1] about 1050 Btu (585 cal/g at 20°C), and unless a heat supply is available, there can be no evaporation. Hence, total evaporation over a period of time is controlled by the available energy. This is a feedback process—a high rate of evaporation

---

[1] Heat of vaporization is a function of temperature. See Appendix.

at any time utilizing energy which might have caused evaporation at some other time.

Evaporation may also be controlled by conditions at the surface from which evaporation is occurring. Loss from soil cannot exceed the water available in the soil. Evaporation from snow or ice can occur only if the dew point of the overlying air is less than the temperature of the snow surface (which cannot exceed 32°F). If the dew point of the air is over 32°F, condensation of vapor on the snow surface will occur. Dissolved salts reduce the vapor pressure of a water surface. Hence saline water will evaporate less readily than fresh water, the reduction being about 1 percent for each percent of dissolved salts.

2-14   Determination of evaporation   A direct determination of evaporation can be made by applying the equation of continuity to a reservoir. This requires accurate knowledge of all items of inflow, outflow, and storage. Unless the evaporation is of the same order of magnitude as the other items and there are no unmeasured gains or losses from subsurface flows, the method is generally too inaccurate to calculate evaporation which is the residual including all errors in other terms.

In principle, if one measured humidity, temperature, and wind at two levels above a water surface, it should be possible to compute the upward vapor transport by use of turbulence theory, and many complex equations have been derived to express this relation.[1] Tests at Lake Hefner,[2] Oklahoma, showed that simple empirical equations such as

$$E = 0.00241(e_s - e_8)V_8 \qquad (2\text{-}3)$$

were more satisfactory than the derived equations. In Eq. (2-3), E is evaporation in inches per day, e_s is the vapor pressure (inches of mercury) at the water surface, and e_8 and V_8 are the vapor pressure and the wind velocity (miles per day) 8 m above the surface. The terms V and e must be measured carefully, otherwise large errors will result.

Another approach to the problem of estimating lake evaporation is the *energy-balance method*. This method is expressed by the equation

$$E = \frac{H_i - H_o - \Delta H}{\rho[\lambda(1 + R)]} \qquad (2\text{-}4)$$

where H_i is the total heat input to the lake including solar radiation and heat

[1] E. R. Anderson, L. J. Anderson, and J. J. Marciano, A Review of Evaporation Theory and Development of Instrumentation, *U.S. Navy Electron. Lab. Rept.* 159, San Diego, Calif., February, 1950.

[2] Water-loss Investigations: Lake Hefner Studies, Technical Report, *U.S. Geol. Surv., Profess. Paper* 269, 1954.

entering with inflowing water, H_o is the heat leaving the lake as reflected and back radiation and the heat content of the outflowing water, ΔH is the change in heat content of the reservoir water, ρ is the density of the evaporated water, λ the latent heat of vaporization, and R a ratio of the heat used for evaporation to that transferred to the air as sensible heat. Known as Bowen's ratio,[1] R is given by

$$R = \frac{0.61 p_{atm}(T_s - T_a)}{1000(e_s - e_a)} \tag{2-5}$$

where T_s and T_a are the surface and air temperatures in degrees centigrade and pressures are in millibars.[2] Equation (2-4) states that that portion of the total available heat not stored in or taken from the lake by other means divided by the latent heat of vaporization indicates the volume of evaporation. This approach is theoretically sound but difficult to utilize because of the problems in gathering the necessary data.

The oldest method of estimating lake evaporation is by use of *evaporimeters* or *evaporation pans* from which the water loss can be accurately measured. The most common pan is the Weather Service Class A pan (Fig. 2-15), which is 4 ft in diameter and 10 in. deep. Evaporation from the pan is measured daily with a hook gage to the nearest 0.001 in. Theory and experiment have shown that evaporation from a pan is considerably different than that from a reservoir surface, largely because of the difference in the water temperature of the two surfaces. The small mass of water in the pan and the exposed metal of the pan favor wide fluctuations in water temperature as the air temperature and solar radiation vary. The large mass of water in a lake and the stabilizing effects of convection currents and of the earth around the reservoir result in much smaller temperature fluctuations. Numerous attempts have been made to devise a pan which would be a thermal model of a lake. These efforts have included increasing the size of the pan and burying it in the soil up to its rim (Bureau of Plant Industry pan—6 ft in diameter, 2 ft deep); floating the pan in the reservoir; and covering the pan with screen wire to reduce the effect of solar radiation. Although these devices do decrease the differences between lake and pan evaporation, it is impossible to design a pan which is thermodynamically and aerodynamically similar to all lakes under all climatic conditions.

A recent analysis of the relationship between annual lake evaporation E_r

---

[1] I. S. Bowen, The Ratio of Heat Losses by Conduction and Evaporation from Any Water Surface, *Phys. Rev.*, Series 2, Vol. 27, pp. 779–787, June, 1926. See also E. R. Anderson, Energy-budget Studies, *U.S. Geol. Surv.*, Circ. 229, pp. 71–119, 1952.

[2] A millibar is a unit of pressure equal to a force of 1000 dynes/sq cm. The standard sea-level atmosphere (14.7 psi) is equivalent to 1013.2 millibars.

FIG. 2-15   *Class A land pan showing hook gage and anemometer.* (*National Weather Service*)

and annual pan evaporation E_p found that the ratio E_r/E_p, known as the *pan coefficient*, averaged very nearly 0.7 for all reliable determinations (about 10 cases) of annual evaporation based on Class A pans. The range of the coefficient was from about 0.67 to 0.81 for the Class A pan. It appears that the use of an average coefficient of 0.7 should provide estimates of annual reservoir evaporation within about 15 percent if the lake and pan are subjected to similar climatic conditions.[1] Monthly ratios of E_r/E_p at Lake Hefner, Oklahoma, varied from about 0.13 to 1.31. The higher ratios are observed in late fall, when the heat stored in the lake during the summer is contributing to evaporation and the pans are relatively cool, while the lower values occur in early spring, when the pans warm up more rapidly than the lake.

It has been shown that it is possible to correct for the heat losses through the walls of a Class A pan and for differences in advected energy between the pan and a reservoir so that reliable estimates of the evaporation for short periods of time can be made from the pan evaporation record. Space does

[1] M. A. Kohler, Lake and Pan Evaporation, "Water-loss Investigation," Vol. 1, Lake Hefner Studies, *U.S. Geol. Surv. Circ.* 229, pp. 127–150, 1952.

not permit the reproduction of the necessary charts, but they can be obtained from the original reference.[1]

There is no simple solution for estimates of evaporation from a proposed reservoir. Field measurements at the site will not yield data which can be used in Eq. (2-3) or Eq. (2-4) since the completion of the reservoir will considerably alter the microclimate of the site. Until further study is complete, there seems no better solution than to use pan data reduced by the coefficient appropriate to the pan.

2-15 Transpiration Plants remove water from soil through their roots, transport the water through the plant, and eventually discharge it through pores (stomata) in their leaves. The *transpiration ratio* (Table 2-3) is determined by dividing the weight of water transpired by a plant during its growth by the weight of dry matter produced by the plant exclusive of roots.

TABLE 2-3   Transpiration Ratios

Plant	Ratio
Corn*	349
Wheat*	557
Rice*	682
Flax*	783
Fir trees†	145
Oak trees†	220
Birch trees†	375

*From H. L. Shantz and L. N. Piemeisel, The Water Requirements of Plants at Akron, Colo., *J. Agr. Research*, Vol. 34, pp. 1093–1190, 1927.
†From O. Raber, Water Utilization by Trees with Special Reference to the Economic Forest Species of the North Temperate Zone, *U.S. Dept. Agr. Misc. Pub.* 257, pp. 1–97, 1937.

Transpiration[2] is essentially the evaporation of water from the leaves of plants. Rates of transpiration will, therefore, be about the same as rates of evaporation from a free water surface if the supply of water to the plant is not limited. Estimated free-water evaporation may, therefore, be assumed to indicate the *potential evapotranspiration* from a vegetated soil surface.

The total quantity of transpiration by plants over a long period of time is limited primarily by the availability of water. In areas of abundant rainfall

[1] M. A. Kohler, T. J. Nordensen, and W. E. Fox, Evaporation from Pans and Lakes, *U.S. Weather Bur. Res. Paper* 38, May, 1955.

[2] D. W. Hendricks and V. E. Hansen, Mechanics of Evapotranspiration, *J. Irrigation and Drainage Div., ASCE*, Vol. 88, No. IR2, pp. 67–82, June, 1962.

well distributed through the year, all plants will transpire at about the same rates and the differences in total will result from the differences in the length of the growing seasons for the various species. Where water supply is limited and seasonal, depth of roots becomes very important. Here, shallow-rooted grasses wilt and die when the surface soil becomes dry while deep-rooted trees and plants will continue to withdraw water from lower soil layers. The deeper-rooted vegetation will transpire a greater amount of water in the course of a year. The evidence[1] indicates that the rate of transpiration is not materially reduced by decreases in soil moisture until the wilting point of the soil is reached (see Chap. 14).

2-16 Evapotranspiration *Evapotranspiration*, sometimes called *consumptive use* or *total evaporation*, describes the total water removed from an area by transpiration and by evaporation from soil, snow, and water surfaces. Estimates of the actual evapotranspiration from an area are usually made by subtracting measured outflow from the area (surface and subsurface) from the total water supply (precipitation, surface and subsurface inflow, and imported water). Change in surface and underground storage must be included when significant.

Several attempts[2] have been made to relate evapotranspiration to climatological data through simple equations such as[3]

$$U_c = 0.9 + 0.00015\Sigma(T_{max} - 32) \tag{2-6}$$

where U_c is the consumptive use and $\Sigma(T_{max} - 32)$ is the sum of the growing season maximum temperatures less 32°F. Such formulas agree fairly well with average values of annual evapotranspiration over a long period of years, but it is clear that the evaporative process is too complex to be well defined by a simple temperature function.

As indicated in Sec. 2-15, the potential evapotranspiration from any area can be estimated from the free-water evaporation. Actual evapotranspiration equals the potential value E_{pot} as limited by the available moisture.[4] On a natural watershed with many vegetal species, it is reasonable to assume that evapotranspiration rates do vary with soil moisture since shallow-rooted

[1] F. J. Veihmeyer and A. H. Hendrickson, Does Transpiration Decrease as Soil Moisture Decreases? *Trans. Am. Geophys. Union*, Vol. 36, pp. 425–448, June, 1955.

[2] Jerald E. Christiansen, Pan Evaporation and Evapotranspiration from Climatic Data, *J. Irrigation and Drainage Div.*, ASCE, pp. 243–265, June, 1968; George H. Hargreaves, Consumptive Use Derived from Evaporation Data, *J. Irrigation and Drainage Div.*, ASCE, pp. 97–105, March, 1968.

[3] R. L. Lowry and A. F. Johnson, Consumptive Use of Water for Agriculture, *Trans. ASCE*, Vol. 107, pp. 1243–1302, 1942.

[4] C. W. Thornthwaite, The Moisture Factor in Climate, *Trans. Am. Geophys. Union*, Vol. 27, pp. 41–48, February, 1946.

species will cease to transpire before deeper-rooted species. A moisture-accounting[1] procedure can be established using the continuity equation

$$P - R - G_o - E_{act} = \Delta M \tag{2-7}$$

where P is precipitation, R is surface runoff, G_o is subsurface outflow, E_{act} is actual evapotranspiration, and ΔM is the change in moisture storage. E_{act} is estimated as

$$E_{act} = E_{pot} \frac{M_{act}}{M_{max}} \tag{2-8}$$

where M_{act} is the computed soil moisture storage on any date and M_{max} is an assumed maximum soil moisture content. A moisture-accounting procedure of this type may be used to calculate runoff[2] as well as to estimate evapotranspiration.

2-17   Variations in evaporation and transpiration   Figure 2-16 shows average annual lake evaporation values[3] for the United States. The figure might also be described as showing the *evaporation potential* (Sec. 2-15). In many parts of the country evaporation potential substantially exceeds annual precipitation. In these areas construction of a reservoir means a large loss of water through evaporation. Where evaporation potential is high, runoff (Fig. 2-12) tends to be low, since runoff is essentially the residual after evapotranspiration requirements are subtracted from precipitation. The pattern of transpiration in the United States would be quite similar to that of evaporation, modified by the vegetal characteristics of the different regions of the country. In a desert region with little or no vegetation, transpiration is necessarily low, although evaporation potential is high. In no case can evapotranspiration exceed precipitation except when water from another basin is available to augment the local supply.

Variation in evaporation from year to year is much less than the variations in streamflow or precipitation. The data of Lowry and Johnson show extreme variations in annual consumptive use to be about ±25 percent of the mean annual value. Variations in measured pan evaporation generally fall within this range. The limited variation might be expected since mean annual humidity, temperature, and wind vary only moderately from year

[1] M. A. Kohler, Meteorological Aspects of Evaporation, *Int. Assn. Sci. Hydr. Trans.*, Vol. III, pp. 423–436, General Assembly, Toronto, 1958.

[2] R. K. Linsley and N. H. Crawford, Computation of Synthetic Streamflow Records on a Digital Computer, *Int. Assn. Sci. Hydrology*, *Publ.* 51, 1960; N. H. Crawford and R. K. Linsley, Digital Simulation in Hydrology: Stanford Watershed Model IV, *Tech. Rept.* 39, Department of Civil Engineering, Stanford University, 1966.

[3] "Climatic Atlas of the United States," U.S. Weather Bureau, 1968.

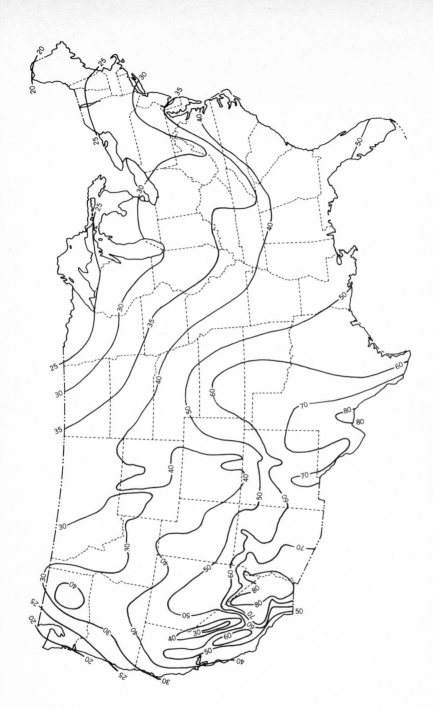

FIG. 2-16 *Average annual lake evaporation in inches. (National Weather Service)*

to year. Typical variations of evaporation on a monthly basis are presented in Table 2-4.

2-18 Sampling hydrologic data   Chapter 2 emphasizes the considerable amount of data required in hydrology, and subsequent chapters will demonstrate the ways in which these data are used. Hydrology is highly data dependent and requires good samples of the hydrology of a watershed in both time and space. For meaningful answers, these samples must be *representative*. In addition, the time samples must be *homogeneous*. The precise meaning of the words "representative" and "homogeneous" are somewhat obscured by the fact that the requirements are dependent on the purpose for which the data are to be used and the method of analysis to be employed.

With respect to time, "representative" means that the sample must be long enough to include an adequate range of the information to be used. For a flood frequency analysis (Chap. 5) using only the highest flood each year, 15 or 20 yr of record may be minimal. On the other hand for a storm-rainfall runoff relation (Chap. 3) 2 or 3 yr of record may be adequate in a humid climate with frequent storms. No absolute limit can therefore be set for length of record. However, it must be apparent that some data are essen-

TABLE 2-4   *Mean Monthly and Annual Class A Pan Evaporation at Selected Stations**
(*In inches*)

Month	Newark, Calif.	W. Palm Beach, Fla.	Vicks-burg, Miss.	Seattle, Wash.	Norris, Tenn.	Ithaca, N.Y.	Lincoln, Nebr.	Hoaeae, Hawaii	Bartlett Dam, Ariz.
January	1.36	3.42	1.67	†	1.01	†	†	3.56	4.09
February	1.90	3.73	2.10	0.89	1.32	†	†	3.85	4.50
March	3.42	4.99	3.79	1.76	2.65	†	†	4.73	7.10
April	5.05	6.11	4.96	2.91	4.08	†	5.73	5.44	10.43
May	7.19	6.54	5.95	4.40	5.52	4.29	7.00	5.99	14.55
June	8.27	6.20	6.60	4.77	6.15	5.16	8.58	6.37	17.03
July	8.75	6.88	7.13	6.28	5.88	5.87	10.54	7.00	17.23
August	7.73	6.37	6.68	4.97	5.24	4.94	8.78	7.00	14.50
September	6.60	5.18	5.06	3.25	4.33	3.35	6.94	5.88	12.70
October	4.32	4.87	3.91	1.55	2.96	2.14	4.63	5.28	9.25
November	2.35	3.60	2.34	0.65	1.61	†	†	3.88	6.06
December	1.26	2.98	1.42	0.53	0.94	†	†	3.57	4.19
Year	58.20	60.87	51.61	...	41.69	...	...	62.55	121.63

*Mean Monthly and Annual Evaporation from Free Water Surface, *U.S. Weather Bur. Tech. Paper* 13, 1950.
†Pan inoperative because of ice.

tial, and even short records initiated when the project is first conceived may prove valuable.

The concept of homogeneity implies that the records should have a common meaning throughout the period of record. Inhomogeneities are most commonly introduced into a precipitation record by moving the station. In mountainous terrane or in cities, even a small move may cause a marked change in catch. A change in observer or type of equipment may also cause a shift of several percent. The history of a station should be checked before a record is used; and, if any doubt remains, double mass analysis[1] should be used to test for inhomogeneities and determine the magnitude of the change.

Inhomogeneities in streamflow data are most commonly caused by construction of a dam, levees, or diversions upstream of the station. In small watersheds, urbanization, other major changes in land use, or forest fires may be significant. A change in station location may also cause an inhomogeneity in the record. Corrections for changes should be made before the record is used; or, alternatively, the analysis can be limited to the portion of the record before or after the change.

The problem of representivity in areal sampling is primarily encountered in dealing with precipitation data. The requirement is that the data adequately represent the true precipitation over the watershed. If a water balance (i.e., continuity analysis) is to be made, the data should be true values. On the other hand, if a regression analysis is contemplated, the data need only be in fixed ratio to the true precipitation. Since the "true" precipitation is rarely known, precise confirmation of representivity is difficult. A successful relation between precipitation and streamflow is a pragmatic test. If a good relation capable of accurately reconstructing historic flows is derived, the precipitation must clearly be representative. The converse is not necessarily true. A poor relation may result from omission of important parameters even though the data are quite satisfactory.

Much can be learned before analysis. Short and fragmented records can be adjusted to longer time periods and used to construct a mean annual (or seasonal) precipitation map which will show the effect of topography. Many short records can be found in most watersheds by a careful search. Analysis of the climatology of the watershed is also helpful. If convective storms are important, the required density of stations will be higher than if frontal storms are the primary source of rain. Because each watershed is different, no general rules about the required density of stations have ever proved successful.

[1] R. K. Linsley, M. A. Kohler, and J. L. H. Paulhus, "Hydrology for Engineers," p. 33, McGraw-Hill, New York, 1958.

BIBLIOGRAPHY

Chow, Ven Te (ed.): "Handbook of Applied Hydrology," McGraw-Hill, New York, 1964.
"Climatic Atlas of the United States," U.S. Weather Bureau, 1968.
Corbett, D. M., and others: Stream Gaging Procedure, *U.S. Geol. Surv. Water Supply Paper* 888, 1945.
Kolupaila, S.: "Bibliography of Hydrometry," University of Notre Dame Press, Notre Dame, Ind., 1961.
Kurtyka, J. C.: Precipitation Measurement Study, *Illinois State Water Surv. Rept. Invest.* 20, 1953.
Langbein, W. B., and others: Annual Runoff in the United States, *U.S. Geol. Surv. Circ.* 52, June, 1949.
Linsley, R. K., M. A. Kohler, and J. L. H. Paulhus: "Applied Hydrology," McGraw-Hill, New York, 1949.
Penman, H. L.: "Vegetation and Hydrology," Commonwealth Agricultural Bureaux, Farnham Royal, Bucks, England, 1963.
Thiessen, A. H.: "Weather Glossary," U.S. Weather Bureau, 1946.
Trewartha, G. T.: "An Introduction to Climate," 3d ed., McGraw-Hill, New York, 1954.

PROBLEMS

2-1 Determine from National Weather Service publications the mean annual precipitation at all available stations within a river basin assigned by your instructor. Using these data, find the average annual precipitation over the basin by the arithmetic average, Thiessen network, and isohyetal map. Compare your results by the three methods with the class average.

2-2 Repeat Prob. 2-1, using precipitation during a specified storm.

2-3 Plot the annual precipitation at some selected station as a time series. Are any regular cycles evident? It may help to plot the mass curve of departure from normal precipitation or curves of 5- or 10-yr running averages. Five-year running averages are computed by averaging the annual precipitation values in overlapping 5-yr periods. The average value is usually plotted at the middle year of the period.

2-4 Compute the streamflow for the measurement data below:

Distance, ft	0	2	4	6	8	10	12	14	16	18	20	22
Depth, ft	0	1.0	4.3	7.2	8.5	7.4	5.6	4.7	3.5	2.1	1.4	0
Velocity, ft/sec:												
$0.2d$	0	1.4	1.9	2.6	2.9	2.7	2.5	2.3	2.1	1.8	1.5	0
$0.8d$	0	0.7	1.2	1.8	2.0	1.9	1.7	1.5	1.3	1.1	1.0	0

Express answer in cubic feet per second and in cubic meters per second.

2-5 Annual precipitations at four weather stations in the vicinity of Salt Lake City, Utah, are as indicated. Analyze the data for homogeneity. That is, determine whether or not any of the stations had been moved and in what year the move was made. Homogeneity may be checked by plotting a double mass curve, i.e., plotting Σ precipitation at Station A versus Σ precipitation at Station B, etc.

Adjust the data to make them homogeneous with respect to the present location of the stations. Also determine the most likely value of the annual precipitations at Station A in 1961 and 1962. Finally, determine the mean annual precipitation of one of the moved stations in terms of its present location for the years 1950 to 1958, 1955 to 1964, and 1950 to 1967. Note that one item of data is incorrect. Which one is it? What do you think its true value may have been?

Year	Station A	Station B	Station C	Station D
1950	21.0	32.5	23.9	29.1
1951	37.0	37.7	27.7	33.3
1952	19.8	29.8	22.6	27.6
1953	24.6	37.1	28.1	34.6
1954	22.8	34.8	26.2	31.5
1955	16.9	25.5	19.2	23.4
1956	23.8	32.1	24.5	29.9
1957	22.4	31.4	23.1	27.8
1958	23.1	31.0	23.8	29.4
1959	25.0	33.4	25.5	30.5
1960	27.7	37.5	28.2	32.4
1961		34.2	25.7	27.8
1962		42.3	31.8	34.4
1963	24.3	33.8	24.8	27.4
1964	27.4	36.1	27.8	30.1
1965	26.2	35.5	26.9	29.2
1966	21.0	28.4	21.4	23.5
1967	25.1	34.0	25.6	27.8

2-6 Tabulated below are measured discharges, stages, and stages at an auxiliary station 2000 ft downstream from the main gage. Develop a slope-stage-discharge relation from these data. Determine the exponent n [Eq. (2-1)] by plotting Q_a/Q_0 vs. $\Delta z_a/\Delta z_0$ on logarithmic paper. The slope of the line is the exponent n. Find the discharge for a stage of 24.95 ft and a fall of 1.43 ft. Note that it is convenient to take $\Delta z_0 = 1$ ft.

Measured Q, cfs	Stage, ft	
	Main gage	Auxiliary gage
93,900	11.71	10.81
80,000	9.73	8.73
242,000	16.11	15.10
27,000	6.00	4.99
553,000	24.95	23.97
417,200	21.02	19.99
560,000	21.75	20.19
421,000	23.51	22.77
241,000	14.48	13.02
112,700	10.75	9.53
365,000	18.30	17.00
702,000	27.22	26.07
570,000	26.45	25.58
297,600	21.60	20.98
129,800	13.72	13.02

2-7 With a stage of 20 ft and a water-surface slope of 1 ft in 4000, the flow rate in a river is 2500 cfs. Approximately what would the flow rate be if the stage were 20 ft and the water-surface slope 1 ft in 6000?

2-8 On a river the following data were obtained by stream gaging:

Main staff, ft	Auxiliary staff, ft	Flow rate, cfs
30.0	29.0	250
30.0	27.0	470

Estimate as accurately as possible the flow rate when the main staff reads 30.0 ft, and the auxiliary staff reads 28.0 ft.

2-9 Lake Mead behind Hoover Dam has a capacity of approximately 30 million acre-ft. For how many years would this water supply a city with a population of 800,000 if the daily consumption is 150 gal per person? Neglect the effect of evaporation.

2-10 A certain Asiatic city with a population of 400,000 uses 24 million cu m of water per year. What is the mean consumption (a) in cubic meters per capita per day, (b) in gallons per capita per day?

2-11 The average daily streamflows resulting from a heavy storm on a basin of 1347 sq mi are tabulated below. Compute the total flow volume in second-foot-days, acre-feet, inches, and millions of gallons.

Day	1	2	3	4	5	6
Mean daily flow, cfs	3200	10,450	6580	3230	1560	670

2-12 What is the volume of rainfall in second-foot-days if 2.47 in. occurs over an area of 943 sq mi? How many acre-feet? How many tons?

2-13 For a stream selected by your instructor, find the mean monthly flows for a 10-yr period from the *U.S. Geological Survey Water Supply Papers*. On the average, what percent of the annual flow occurs in each month? Compare these percentages with the percentage of annual precipitation in the corresponding month. What explanations can you see for the apparent differences?

2-14 For a stream basin selected by your instructor, determine the normal annual runoff and normal precipitation in inches. Express the extreme values of runoff as percents of normal. What is the average variation of streamflow and precipitation? Average variation is computed as the sum of the departures from normal without regard to sign divided by the length of record. Compute the annual values of water loss (evapotranspiration) by subtracting streamflow from precipitation. Note that this assumes no significant change in surface or groundwater storage during each year. What are the extreme and average variations of evapotranspiration? How do these values compare with the corresponding values for runoff and precipitation?

2-15 A reservoir is located in a region where the normal annual precipitation is 24.0 in., and the normal annual *pan* evaporation is 60 in. If the average area of the reservoir water surface is 3000 acres and if, under natural conditions, 20 percent of the rainfall on the land flooded by the reservoir runs off into the stream, what is the net increase or decrease of streamflow as a result of the reservoir?

2-16 Repeat Prob. 2-15 using precipitation and evaporation values appropriate to your locality.

⌄2-17   What evaporation rate would be indicated by Eq. (2-3) when the reservoir water surface is 60°F, the air temperature at 8 m is 75°F, the relative humidity is 80 percent, and the wind velocity at 8 m is 12 mph? If the relative humidity at 8 m were only 20 percent, what would be the evaporation rate, all other factors being the same?

   2-18   What daily evaporation is indicated by Eq. (2-4) on a day when the total insolation is 600 cal/sq cm and 15 percent of the insolation is reflected? Compute back radiation from the Stefan-Boltzmann equation $H_b = 0.82 \times 10^{-10} \ T^4/\pi$ where T is in degrees Kelvin and H_b is in calories per square centimeter per minute. Assume water temperature constant at 60°F for the day. Compute Bowen's ratio from the data of the second part of Prob. 2-17. Assume no change in heat storage in the reservoir and standard sea-level atmosphere.

   2-19   During the month of June the evaporation from a Class A land pan at a certain lake was 10 in. The surface area of the lake decreased from 4400 to 3000 acres during the month. Approximately how many acre-feet of water were evaporated from the lake during this month? State assumptions.

   2-20   Using temperature data for your locality, compute the annual consumptive use for some year by use of Eq. (2-6).

   2-21   A stream valley contains 30,000 acres of irrigable land. It is estimated that this will be used as follows:

Crop	Area, acres	Consumptive use, acre-ft/acre
Orchards	6000	0.8
Small grains	15,000	1.8
Truck crops	9000	1.3

If the average annual precipitation on the valley is 18 in., of which approximately 8 in. is available for crops, what quantity of irrigation water must be applied annually?

   2-22   The mean annual precipitation and mean annual runoff at several locations are approximately as follows:

Location	Precipitation, in.	Runoff, in.
Western Washington	80	40
Southwestern Arizona	6	0.3
Nebraska	22	1
Central Georgia	46	15
Central Ohio	38	11
Maine	40	25

Convert these values to centimeters per year. Approximately what is the mean annual precipitation at your home town? Express answer in inches per year and in centimeters per year.

   2-23   Estimate the number of e's on page 37 of this book by the following sampling procedures:

(a) Count off the letters consecutively and sample every hundredth letter. Include numerals and punctuation marks.

(b) Repeat (a), but sample every fortieth letter.

(c) Repeat (a), but sample every tenth letter.

These samples represent approximately 1 percent, 2.5 percent, and 10 percent of the population.

Now determine the actual number of e's on page 37 by sampling each letter. Draw some conclusions concerning the effect of sample size.

2-24 A statistician with offices in Boston was asked to estimate the number of male citizens in the United States over the age of twenty-one having the last name O'Brien. He did this by counting the number of O'Briens in the Boston telephone directory and multiplying by the ratio (population of the United States/population of Boston). Mention at least five fallacies in this approach.

Chapter 3

Quantitative hydrology

Occasionally the hydrologist may find a streamflow record at the site of a proposed project. More often, however, the nearest available record is elsewhere on the stream or on an adjacent stream. The hydrologist must, therefore, be prepared to transfer such data as are available to his problem area with appropriate adjustments for differences in the hydrologic characteristics of the two basins. In addition to transposition in space, the hydrologist may be asked to estimate the magnitude of an event greater than anything observed (extrapolation in time). Many techniques, some empirical, some rational, have been devised to meet these problems of space and time adjustment. Type examples of commonly used procedures are presented in this chapter. References indicate sources of further information on other solutions.

    3-1 Basin recharge and runoff As rain falls toward the earth, a portion of it is intercepted by the leaves and stems of vegetation. The water so retained, *interception*, together with *depression storage* and *soil moisture*, constitutes *basin recharge*,[1] the portion of precipitation which does not contribute to streamflow or groundwater. Depression storage includes the water which is retained as puddles in surface depressions. Soil moisture is held as capillary water in the smaller pore spaces of the soil or as hygroscopic water adsorbed on the surface of soil particles (Sec. 14-6).

    Rainwater or melting snow, exclusive of the water withheld as basin recharge, may follow three paths to a stream. A portion travels as *overland flow (surface runoff)* across the ground surface to the nearest channel. Still other water may infiltrate into the soil and flow laterally in the surface soil

---

[1] For problems concerned with surface streamflow only, basin recharge may be assumed to include groundwater accretion as well.

to a stream channel as *interflow*. A relatively impermeable stratum in the subsoil favors the occurrence of interflow. A third portion of the water may percolate downward through the soil until it reaches the groundwater. Vertical percolation of rainwater results in groundwater accretion only if the soil is highly permeable or if the groundwater is near the surface. Low soil permeability encourages overland flow, while a thick soil mantle, even though permeable, may retain so much water as soil moisture that none can reach the groundwater.

It is convenient but inaccurate to discuss recharge and runoff as if runoff began only after recharge of the basin was complete. While the potential rate of recharge is at a maximum at the beginning of a storm, recharge normally continues at decreasing rates as long as the storm lasts. A condition of complete saturation, i.e., all moisture-storage capacity of the basin fully used, occurs very rarely. The distinction between the three types of runoff is also somewhat artificial. Water moving as surface runoff may infiltrate and become interflow or groundwater, while infiltrated water may come to the surface and finally reach a channel as surface flow. These concepts do, however, permit a rational approach to hydrology.

Overland flow and interflow are frequently grouped together as *direct runoff*. This water reaches the stream shortly after it falls as rain and is discharged from the drainage basin within a few days. Much of the low water flow of streams is derived from groundwater. Stream channels which have perennial flow are below the groundwater table and are called *effluent streams*. *Intermittent streams*, which go dry if much time elapses between rains, are usually *influent streams*, i.e., their channels are above the level of the groundwater, and percolation from the stream channel to the groundwater occurs. Most river basins contain streams which fall into both categories, and some streams may be either influent or effluent depending upon the rate of flow and the existing groundwater levels.

3-2 Hydrograph analysis The characteristics of direct and groundwater runoff differ so greatly that they must be treated separately in problems involving short-period, or storm, runoff. There is no practical means of differentiating between groundwater flow and direct runoff after they have been intermixed in the stream, and the techniques of hydrograph analysis are rather arbitrary. The typical hydrograph resulting from a single storm (Fig. 3-1) consists of a rising limb, peak, and recession. The recession represents the withdrawal of water stored in the stream channel during the period of rise. Double peaks are sometimes caused by the geography of the basin but more often result from two or more periods of rainfall separated by periods of little or no rain.

Numerous methods of hydrograph separation have been used. The method illustrated by *ABC* in Fig. 3-1 is simple and as easily justified as any

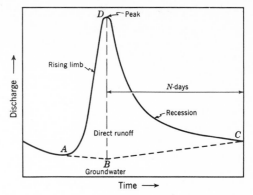

*FIG. 3-1   Typical flood hydrograph
showing method of
separating direct and
groundwater runoff.*

other. The recession of flow existing prior to the storm is extended to point
B under the crest of the hydrograph. The straight line BC is then drawn to
intersect the recession limb of the hydrograph N days after the peak. The
value of N is not critical and may be selected arbitrarily by inspection of
several hydrographs from the basin. The selected value should, however, be
used for all storm events analyzed to conform to the unit hydrograph concept
(Sec. 3-12). The time N will increase with size of drainage basin since a
longer time is required for water to drain from a large basin than from a
small one. A rough guide to the selection of N (in days) is given by

$$N = A_d^{0.2} \qquad\qquad (3\text{-}1)$$

where A_d is the drainage area in square miles. Large departures from Eq.
(3-1) may be appropriate in some basins.

ESTIMATING VOLUME OF RUNOFF

The discussion of Sec. 3-1 suggests the simple equation

$$R = P - L - G \qquad\qquad (3\text{-}2)$$

where R is direct runoff, P is precipitation, L is basin recharge, and G is
groundwater accretion, all in units of depth over the drainage area. Computation of average precipitation P is discussed in Sec. 2-6. The runoff volume
R is the area $ABCD$ of Fig. 3-1. Thus, two of the factors in Eq. (3-2) may be
evaluated from observed data. Accurate estimates of R therefore depend on
estimates of basin recharge L and groundwater accretion G.

3-3   Runoff coefficients   In the design of storm drains and small
water-control projects, runoff volume is commonly assumed to be a percentage of rainfall. If Eq. (3-2) is correct, then an equation of the form

$$R = kP \qquad\qquad (3\text{-}3)$$

TABLE 3-1   *Values of the Runoff Coefficient k [Eq. (3-3)] for Various Surfaces*

Surface	Value of k
Urban residential	
Single houses	0.30
Garden apartments	0.50
Commercial and industrial	0.90
Forested areas depending on soil	0.05–0.20
Parks, farm land, pasture	0.05–0.30
Asphalt or concrete pavement	0.85–1.0

cannot be rational since the runoff coefficient k must vary with both recharge and precipitation. The reliability of Eq. (3-3) improves as the percentage of impervious area increases and k approaches unity. The percentage or coefficient approach is most suitable for urban drainage problems where the amount of impervious area is large. For moderate rainfalls, all runoff may come from the impervious area making k the percent of impervious area.[1] Customary values of k are given in Table 3-1. The coefficient approach should be avoided in rural areas and for analysis of major storms.

3-4   Infiltration   Infiltration is the movement of water through the soil surface and into the soil. The *infiltration capacity* of a soil at any time is the maximum rate at which water will enter the soil. Infiltration capacity depends on many factors. A loose, permeable soil will have a higher capacity than a tight clay soil. If much of the pore space is filled with water, infiltration capacity is generally less than when the soil is relatively dry. If the pore space of the surface soil is completely filled with water, further downward movement of moisture is controlled by the subsoil permeability. A hard, driving rain may pack surface dirt into soil pores and reduce infiltration. A good vegetal cover provides protection against raindrop impact, and, in addition, plant roots and organic plant litter help to increase soil permeability. Theoretically, if the infiltration capacity of a soil were known, the volume of runoff resulting from a given rainfall could be computed by subtracting infiltration and surface retention (interception plus depression storage) from total rainfall.

The *infiltration rate* is the rate at which water actually enters the soil during a storm; and it must equal the infiltration capacity or the rainfall rate, whichever is lesser. Infiltration rates or capacities are estimated experimentally by measuring the surface runoff from a small test plot subjected

---

[1] This should be impervious area connected to the drainage system. House roofs and patios which drain onto soil should be excluded.

to either natural or artificial rain.[1] If the plot is subjected to rainfall rates in excess of the infiltration capacity, the capacity will vary with time in the manner shown in Fig. 3-2. Different capacity curves will be obtained for different values of initial soil moisture.

Many thousands of infiltration tests have been conducted. Infiltrometers may consist of small plots of ground sprayed with water to simulate rainfall or tubes partially embedded in the soil and filled with water (Sec. 14-8). These tests[2] have indicated that the infiltration capacity of bare soil under average summer conditions and after 1 hr of rain will vary from 0.01 in./hr for heavy clay soils to 1.0 in./hr for loose sandy soil. A permanent forest or grass cover in good condition will increase these rates three to seven times.

Natural rain of varying intensity, sometimes below and sometimes above the prevailing infiltration capacity, results in a distortion of the capacity-time curve. The decrease in infiltration capacity during periods with rainfall rates less than capacity is not so great as it is when infiltration takes place at capacity rates. It is often assumed that the infiltration capacity at any time is determined by the mass infiltration which has occurred up to that time. Thus, if a rain begins at low rates and rainfall during the first hour is one-half the infiltration capacity, the capacity at the end of the hour would be taken as that at about 0.5 hr on the applicable time-capacity curve.

---

[1] This statement includes surface retention with infiltration. Unless the plot surface contains large depressions or is covered with heavy vegetation, surface retention will be quite small and computed infiltration rates not greatly in error.

[2] See "Hydrology Handbook," *ASCE Manual* 28, pp. 47–51, American Society of Civil Engineers, New York, 1949.

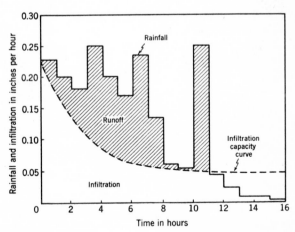

FIG. 3-2 *Typical infiltration curve superimposed on a rainfall diagram to illustrate method of calculating runoff.*

3-5   Infiltration indices   The direct application of infiltration curves as described in the previous section to large heterogeneous areas is difficult. At any instant both infiltration capacity and rainfall rate may differ greatly from point to point. Moreover, interflow is often a substantial portion of the total runoff; and, since interflow is a part of infiltration, it will not normally be included in the runoff computed with infiltration-capacity curves determined on test plots. Estimates of runoff volume from large areas are sometimes made by use of *infiltration indices.*

One common index is the *average infiltration rate* (*loss rate*, or *W*-index) which may be computed by

$$W = \frac{P - R}{t_R} \tag{3-4}$$

where t_R is the duration of rainfall in hours. A second index is the Φ index, which is defined as that rate of rainfall above which the rainfall volume equals the runoff volume (Fig. 3-3). If rainfall intensity is reasonably uniform or if rainfall is heavy, the two indices will be nearly equal. In the usual case of moderate rain at nonuniform intensities, the Φ index will be somewhat higher than the W index. These indices vary with initial soil moisture, with changes in the depression storage and interception capacity of the area, and with amount of precipitation. Infiltration indices are not infiltration rates but merely measures of potential basin recharge.

In some design problems, it is assumed that maximum runoff conditions will prevail, and the lowest observed values of W or Φ are used to compute runoff. The determination of the frequency of the resulting runoff is difficult (see Sec. 5-9). For flood forecasting or reconstruction of a historic storm the use of minimum indices is not satisfactory and appropriate index values must be derived by correlation with those factors which determine the

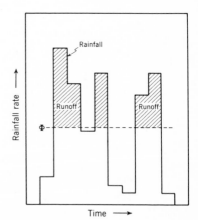

FIG. 3-3   Schematic diagram showing the meaning of the Φ index.

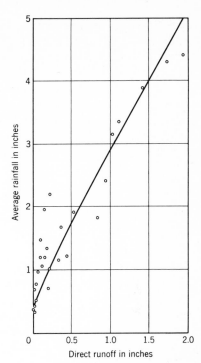

FIG. 3-4   *Simple rainfall-runoff relation for*
the Valley River at Tomotla,
North Carolina.

index at any time. There seems to be no advantage in such an approach over the use of direct correlations between rainfall and runoff (Sec. 3-6). Information on infiltration can be used to estimate the runoff coefficient k from

$$k = \frac{i - W}{i} \tag{3-5}$$

where i is rainfall intensity. This may be useful where the computational procedure is simplified by the use of a runoff coefficient.

3-6   Rainfall-runoff correlations   The simplest rainfall-runoff correlation is a two-variable plotting of average rainfall versus resulting runoff (Fig. 3-4). Typically the relation is slightly curved, indicating an increasing percentage of runoff at the higher rainfalls. Such simple relations do not account for variations in initial conditions which may affect runoff, and there is usually considerable scatter of the points about the mean line.

A third variable may be introduced to explain departures from the simple relationship. This is done by plotting rainfall against runoff and noting the value of the third variable for each point.[1] Lines of best fit are

---

[1] The graphical method is discussed here because it is more easily visualized. Multiple regression methods can also be used.

then drawn for various values of the third variable. In humid regions the initial flow in the stream reflects antecedent conditions rather well and often serves as an effective parameter (Fig. 3-5). Another parameter is antecedent precipitation, which serves as an index to the moisture condition of the soil. Since the most recent rainfall has the greatest effect on soil moisture, precipitation values used in an antecedent-precipitation index should be weighted according to time of occurrence. This is conveniently accomplished by assuming that the index value P_{a_N} at the end of the Nth day is given by

$$P_{a_N} = bP_{a_{N-1}} + P_N \tag{3-6}$$

where $P_{a_{N-1}}$ is the precipitation index on the previous day, P_N is the precipitation recorded on the Nth day, and b is a coefficient. When there is no rain for t days, Eq. (3-6) becomes

$$P_{a_{N+t}} = P_{a_N} b^t \tag{3-7}$$

The weight assigned for rainfall t days prior to a given time is thus b^t. Values of b are usually between 0.85 and 0.95. Actual measurements of soil moisture would probably be superior to either of the parameters discussed above, but systematic records of soil moisture are difficult to obtain for large areas.

Soil moisture is not the only factor influencing basin-recharge conditions and an antecedent-precipitation index or an initial-flow index does not always completely explain the scatter of points in a rainfall-runoff plot. Week of the year has proved to be a useful parameter as it indicates approxi-

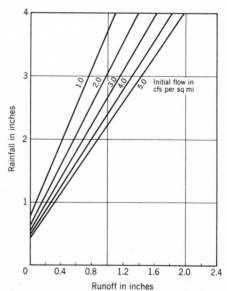

FIG. 3-5   *Three-variable runoff relation for the French Broad River at Newport, Tennessee.*

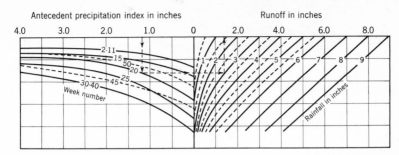

FIG. 3-6   *Coaxial rainfall-runoff relation for several tributary areas of the Ohio River basin. (After Lundquist and Fulk)*

mately the stage of vegetal development, which influences interception and the condition of the ground surface as affected by agricultural operations. Week of the year also reflects typical evapotranspiration conditions, which determine soil moisture jointly with antecedent precipitation. Such a relation is shown in Fig. 3-6.[1] Duration of rainfall has also proved useful in some correlations,[2] as might be expected from the fact that infiltration is a time phenomenon. Simple rainfall-runoff relations, infiltration indices, and runoff coefficients are normally applicable only to a single small river basin. The more complex rainfall-runoff relations have, however, been applied to large areas, including a number of basins. Figure 3-6 is applicable to several of the larger tributaries of the Ohio River.

3-7   Moisture-accounting procedures   Equation (3-2) suggests that runoff might be computed by a moisture-accounting procedure using a modified form of Eq. (2-7):

$$R + G_o = P - E_{act} - \Delta M \tag{3-8}$$

Actually the procedure must compute a running sequence of values of soil moisture and then, with appropriate rules, divide each increment of rainfall into runoff and basin recharge. This latter requires that infiltration be expressed as a function of soil moisture. Such a process would be exceedingly tedious if performed manually, but highly successful results have been obtained[3] using digital computers to perform the computation and to determine the significant constants. Figure 3-7 illustrates the flow diagram employed in such a model.

---

[1] For an explanation of the method of coaxial graphical correlation used to derive relations such as Fig. 3-6 see R. K. Linsley, M. A. Kohler, and J. L. H. Paulhus, "Hydrology for Engineers," Appendix A, McGraw-Hill, New York, 1958.

[2] M. A. Kohler and R. K. Linsley, Predicting the Runoff from Storm Rainfall, *U.S. Weather Bur. Res. Paper* 34, September, 1951.

[3] N. H. Crawford and R. K. Linsley, Digital Simulation in Hydrology: Stanford Watershed Model IV, *Tech. Rept.* 39, Department of Civil Engineering, Stanford University, 1966.

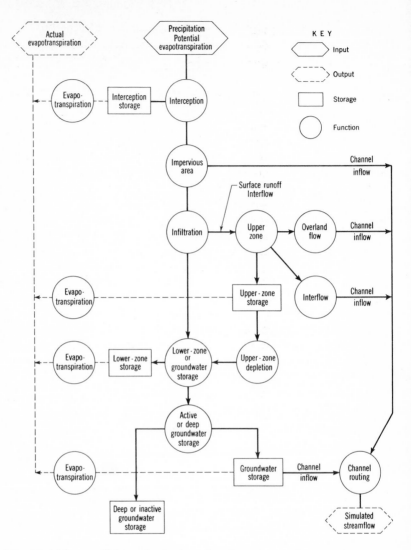

FIG. 3-7   *Model for runoff determination by soil-moisture accounting.*

3-8   Long-period runoff relations   The discussion thus far has been concerned with estimating the volume of direct runoff from single storms. It is occasionally necessary to estimate monthly or annual streamflow from precipitation data. Such estimates may be needed to extend a short record of streamflow or to forecast future runoff for planning reservoir operation. Forecasts for several months in advance are feasible only when precipitation is largely in the form of snow which remains on the ground during the winter months. In estimating runoff volumes for long periods, the distinction between direct and groundwater runoff is usually of no concern. The most accurate method of estimating long-term runoff is probably as a summation of storm runoff amounts, but such a procedure is tedious in the extreme unless digital computers are employed.

Over the period of a year, variations in antecedent conditions tend to average out, and the refinements necessary in storm rainfall-runoff relations become less important. Often a simple plotting of water-year precipitation against water-year runoff such as Fig. 3-8 is sufficient. In regions of heavy snowfall, summer runoff is commonly correlated with average water equivalent of snow on the ground at the end of the snowfall season. In some areas, there is a substantial lag between precipitation and the subsequent discharge of that portion of the precipitation which recharges the groundwater. In this case, a parameter such as precipitation or streamflow during the previous year may be used as an index of groundwater carry-over. The seasonal distribution of precipitation may be important in determining the runoff. This is particularly true where snowfall occurs during the winter. Scattered summer showers usually produce less runoff than general rains. Consequently, it may be necessary to use monthly or seasonal precipitation data as separate parameters in an annual rainfall-runoff relation.

In order to estimate monthly runoff from precipitation, it is usually necessary to develop separate relations for each month of the year. The relationships for two or more months may be the same, but distinct seasonal

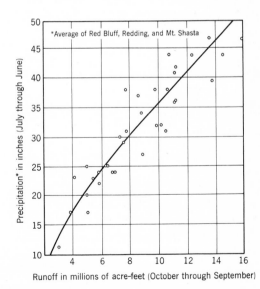

FIG. 3-8   Relation between annual runoff and precipitation for the Sacramento River above Red Bluff, California.

differences will be evident. In some months a storm near the end of the month may produce a large volume of runoff in the subsequent month. Such data must be either adjusted or neglected in developing the rainfall-runoff relation. Snowfall remaining on the ground at the end of the month will create a similar problem.

RUNOFF FROM SNOW

Liquid precipitation is not the sole source of water for streamflow. Snow may remain on the ground for some time, but eventually it melts and contributes to runoff. In parts of the West, melting snow produces the major portion of the annual runoff. Rates of runoff from snow are dependent on the heat available for melting, and computation of snowmelt runoff is a much different problem from the computation of runoff from rainfall.

3-9 Physics of snowmelt The heat required to melt snow comes from several sources. The most obvious source is direct solar radiation. The amount of radiation effective in melting snow is dependent on the reflectivity, or *albedo*, of the snow. Almost 90 percent of the radiation incident on clean, fresh snow is reflected without causing melt, while lesser amounts are reflected from old, dirty snow. The heat from warm air is a second important heat source. Because of the low heat conductivity of air very little melt results from conduction in still air. However, turbulence resulting from wind brings large quantities of warm air into contact with the snow, where heat exchange can take place. If the vapor pressure of the air is higher than that of ice at 32°F, turbulence also brings moisture which can condense on the snow surface. Since the heat of condensation of water at 32°F is 1073 Btu/lb (596 cal/g) and the heat of fusion of ice is only 144 Btu/lb (80 cal/g), the condensation of 1 in. of moisture on the snow surface results in the melting of approximately 7.5 in. of water from the snow. Since melt by convection of warm air and by condensation is dependent on turbulence, wind speed is an important factor in determining melting rates.

Rainfall also brings heat to the snow since the rain must be at a temperature above freezing. The amount of melt M_s in inches of water caused by a rainfall of P in. can be calculated by a simple calorimetric equation

$$M_s = \frac{P(T_w - 32)}{144} \tag{3-9}$$

where T_w is the wet-bulb temperature in degrees Fahrenheit (assumed to equal the temperature of the rain). If $T_w = 50°F$, 1 in. of rain will melt only about 0.12 in. of water from the snow. Rainfall is less important as a melting agent than is commonly believed. Actually it is the warm air, strong winds, and high humidity which accompany rainfall that are responsible for cases of rapid melt during rainstorms.

3-10 Snowmelt computation Equations and charts expressing snowmelt as a function of radiation, air temperature, vapor pressure, and wind have been prepared on the basis of theoretical concepts.[1] Utilization of these relations in a practical computation of snow melting rates is difficult because of the wide variation in the important factors over a typical river basin and because the necessary data are usually lacking. Forest cover and land slope materially affect the amount of radiation which reaches the snow. Forest cover and topography influence wind and to some extent the temperature and humidity of the air. Methods of snowmelt computation are therefore approximations to the ideal conditions represented by the theoretical approach.

Snowmelt in basins with little range in elevation The common procedure for estimating rates of snowmelt in areas where the basin is covered with a fairly uniform depth of snow is the use of degree-day factors. A degree day is defined for this purpose as a departure of 1° in mean daily temperature above 32°F. Thus a day with a mean temperature of 40°F is said to have 8 melting degree days. The degree-day factor is the depth of water melted from the snow in inches per degree day and may be determined by dividing volume of streamflow (in.) produced by melting snow within a given time period by the total degree days for the period. Degree-day factors usually range between 0.05 and 0.15 in./degree day with an average value of about 0.08 in./degree day. Since snow melting depends upon humidity, wind, and solar radiation as well as air temperature, some variation in the degree-day factor from day to day must be expected. Frequently the factor seems to increase as a melting period progresses.

Snowmelt in basins with a wide range of altitude No wholly satisfactory basis for estimating rates of snowmelt from basins of high relief has been developed. Figure 3-9 shows the typical situation in such a basin. The snowpack is not uniform in depth but is shallower at the lower elevations. The line of zero snow depth is known as the *snow line*. As temperature fluctuates, varying areas of snow are subjected to melting influences. In addition, because of the variation of temperature with altitude, the records of a single station do not indicate the actual degree days over the melting zone except when the station is about at midelevation of this zone. Since the snow line moves downslope when new snow falls and retreats upslope as melting occurs, no single station provides a constant index of temperature in the melting zone. The ultimate solution to the problem of computing melting rates in mountainous regions appears to require the establishment

---

[1] Walter T. Wilson, An Outline of the Thermodynamics of Snow Melt, *Trans. Am. Geophys. Union*, Vol. 22, pp. 182–195, 1941; P. Light, Analysis of High Rates of Snow Melting, *Trans. Am. Geophys. Union*, Vol. 22, pp. 195–205, 1941; "Snow Hydrology," Northern Pacific Div., Corps of Engineers, Portland, Oreg., June 30, 1956.

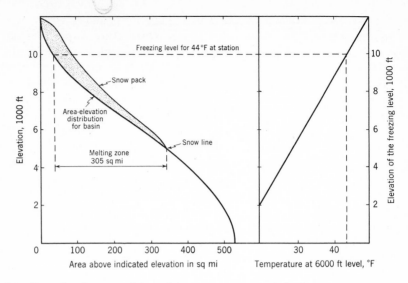

FIG. 3-9   Snow distribution and the melting zone in a mountain basin.

of systematic observations of the location of the snow line or, more precisely, the portion of the basin covered by snow. The snow line is rarely a contour of elevation but is usually lower on northern and forested slopes and higher on southern and bare slopes. Computer simulation[1] has proven reasonably successful for snowmelt computation. The computer permits detailed hourly computations as a substitute for observed data.

If the variation of temperature with altitude is assumed and the average elevation of the snow line is estimated, it is possible to compute the area subject to melting on any day.

ILLUSTRATIVE EXAMPLE 3-1   The area-elevation distribution in a basin is shown in Fig. 3-9. The average snow line is at 5000-ft elevation, and the temperature index station is at 6000 ft. Assume a temperature decrease of 3°F per 1000-ft increase in elevation and a degree-day factor of 0.10. Compute the snowmelt in second-foot-days for a day when the mean daily temperature at the index station is 44°F.

With a temperature of 44°F at 6000 ft the freezing level is at

$$6000 + \frac{44 - 32}{3} \times 1000 = 10,000 \text{ ft}$$

The area between the snow line (5000 ft) and the freezing level is 305 sq mi, from Fig. 3-9. The average temperature over this area is

[1] Eric A. Anderson, The Synthesis of Continuous Snowmelt Runoff Hydrographs on a Digital Computer, *Tech. Rept.* 36, Department of Civil Engineering, Stanford University, 1964.

$$\frac{47 + 32}{2} = 39.5°F$$

and the average degree days above 32°F is

$$39.5 - 32 = 7.5 \text{ degree days}$$

The total melt is therefore

$$7.5 \times 0.10 \times 305 = 229 \text{ in. sq mi}$$

or

$$26.9 \times 229 = 6150 \text{ sfd}$$

It has already been indicated that degree-day factors will vary with time. If an estimate of runoff is required for a short period and actual flows immediately prior to this period are available, the process of Illustrative Example 3-1 may be reversed to compute a degree-day factor for the period of observed flow. This computed factor may then be applied to the succeeding period. This approach assumes a persistence of existing conditions and should not be carried too far beyond the last observed data. In some areas, degree-day factors show a systematic variation with date[1] throughout the melting season (Fig. 3-10). Successful correlations between degree-day factors and accumulated runoff since the beginning of active snowmelt have also been derived.

Snowmelt concurrent with rain   If a light snow cover is completely melted during a rainstorm, the combined runoff from rain and snow may be estimated by entering a rainfall-runoff correlation with the sum of the rainfall and the water equivalent of the snow at the beginning of rain. If the depth of snow is such that it will not completely melt during the storm, the melt caused by rain can be estimated from Eq. (3-9) and the melt caused by other

---

[1] R. K. Linsley, A Simple Procedure for Day-to-day Forecasts of Runoff from Snow Melt, *Trans. Am. Geophys. Union*, Vol. 24, pp. 62–67, 1943.

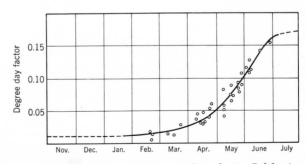

FIG. 3-10   *Degree-day factor variation in the lower San Joaquin River basin, California.* (*After Linsley*)

factors may be approximated by the methods outlined in the preceding paragraphs. Snow can retain up to about 5 percent by weight of liquid water. Thus if the snowpack is deep, much of the meltwater may be retained in the snow with little runoff.

HYDROGRAPHS OF BASIN OUTFLOW

For some purposes an estimate of total runoff volume from a basin within a given time period is adequate. More often, however, an estimate of the instantaneous peak-flow rate is required for design, and in many cases the complete hydrograph is needed. Hydrologic methods must therefore include techniques for converting estimates of runoff volume to estimates of rate of flow.

3-11   The rational method   If rainfall were applied at a constant rate to an impervious surface, the runoff from the surface would eventually reach a rate equal to the rate of rainfall. The time required to reach this equilibrium is the *time of concentration* t_c and for small, impervious areas one may assume that if rain persists at a uniform rate for a period at least as long as t_c, the peak of runoff will equal the rate of rainfall. This is the basis of the *rational formula*

$$Q_p = kiA_d \qquad \text{10 Ac or less}$$ (3-10)

where Q_p is the peak rate of runoff in acre-inches per hour, k is a runoff coefficient (Table 3-1), i is the intensity of rainfall in inches per hour for a duration equal to t_c, and A_d is the watershed area in acres. One acre-in./hr is equal to 1.008 cfs, and therefore Eq. (3-10) is commonly assumed to give peak flow in cubic feet per second. The rational formula is used for design of storm drains, culverts, and other structures conveying runoff from small areas although it has serious deficiencies[1] (see Sec. 5-9).

For small plots without defined channels and from which runoff occurs as laminar overland flow, Izzard[2] found the time to equilibrium t_e in minutes to be

$$t_e = \frac{41bL_o^{1/3}}{(ki)^{2/3}}$$ (3-11)

where L_o is the length of overland flow in feet. The coefficient b is given by

$$b = \frac{0.0007i + c_r}{S_0^{1/3}}$$ (3-12)

where S_0 is the slope of the surface and c_r is a retardance coefficient (Table

---

[1] L. A. V. Hiemstra and B. M. Reich, Engineering Judgement and Small Area Flood Peaks, *Hydrology Paper* 19, Colorado State University, April, 1967.

[2] C. F. Izzard, Hydraulics of Runoff from Developed Surfaces, *Proc. Highway Res. Board*, Vol. 26, pp. 129–150, 1946.

TABLE 3-2   *Values of Retardance Coefficient c_r in Eq. (3-12)*

Surface	Value of c_r
Smooth asphalt surface	0.007
Concrete pavement	0.012
Tar and gravel pavement	0.017
Closely clipped sod	0.046
Dense bluegrass turf	0.060

3-2). Equations (3-11) and (3-12) are applicable only when the product iL_o is less than 500.

Time of concentration for a small basin would be equal to the longest combination of overland flow time and channel flow time which exists anywhere in the basin. Channel flow time is commonly taken as the length of the longest channel divided by the average velocity of flow in the channel at about bankfull stage.

When rates of overland or channel flow increase, depth of flow must also increase. As the depth increases, a quantity of water is stored temporarily until the runoff rate decreases. Equilibrium conditions cannot occur until a sufficient amount of runoff has occurred to supply this *channel* or *valley storage*. This is analogous to the case of a barrel filled by a uniform inflow while outflow occurs through an orifice. Equilibrium is attained when the depth in the barrel is sufficient to cause discharge through the orifice at a rate equal to the inflow. If the diameter of the barrel is increased, the time to equilibrium is increased because of the additional volume which must be filled before the necessary depth is reached. Thus as drainage area increases, the true value of t_c becomes much larger than the transit time computed by dividing channel length by flow velocity. Moreover, as time increases, the likelihood of continuous rain at uniform intensity decreases. Both factors cause the accuracy of Eq. (3-10) to diminish very much when it is applied to large areas. The limiting area beyond which the assumptions of Eq. (3-10) are inadequate depends on slope, type of surface, basin shape, and the accuracy required. This formula should be used with caution for areas over 10 acres.

3-12   Unit hydrographs   If two identical rainstorms could occur over a drainage basin with identical conditions prior to the rain, the hydrographs of runoff from the two storms would be expected to be the same. This is the basis of the *unit-hydrograph* concept.[1] Actually the occurrence of

---

[1] L. K. Sherman, Streamflow from Rainfall by the Unit-graph Method, *Eng. News-Record*, Vol. 108, pp. 501–505, 1932.

identical storms is very rare. Storms may vary in duration, amount, and areal distribution of rainfall. A *unit hydrograph* is a hydrograph with a volume of 1 in. of runoff resulting from a rainstorm of specified duration and areal pattern. Hydrographs from other storms of like duration and pattern are assumed to have the same time base, but with ordinates of flow in proportion to the runoff volumes.

A unit hydrograph may be constructed from the rainfall and streamflow data of a storm with reasonably uniform rainfall intensity and without complications from preceding or subsequent rainfall. The first step in the derivation is the separation of groundwater flow from direct runoff. The volume of runoff (area *ABCD*, Fig. 3-11) is determined and the ordinates of the unit hydrograph are found by dividing the ordinates of the direct runoff by the volume of direct runoff in inches. The resulting unit hydrograph should represent a volume of 1 in. of runoff. The final step is the assignment of an effective storm duration from a study of the rainfall records. Periods of low rainfall at the beginning and end of the storm are omitted if they do not contribute substantially to the runoff. Special methods[1] are available for the analysis of complex storms in which two or more streamflow peaks occur. The results obtained from study of these complex storms are rarely as

[1] R. K. Linsley, M. A. Kohler, and J. L. H. Paulhus, "Applied Hydrology," pp. 446-449, McGraw-Hill, New York, 1949.

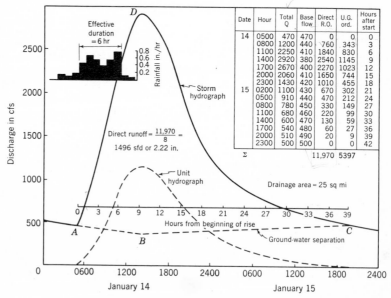

Date	Hour	Total Q	Base flow	Direct R.O.	U.G. ord.	Hours after start
14	0500	470	470	0	0	0
	0800	1200	440	760	343	3
	1100	2250	410	1840	830	6
	1400	2920	380	2540	1145	9
	1700	2670	400	2270	1023	12
	2000	2060	410	1650	744	15
	2300	1430	420	1010	455	18
15	0200	1100	430	670	302	21
	0500	910	440	470	212	24
	0800	780	450	330	149	27
	1100	680	460	220	99	30
	1400	600	470	130	59	33
	1700	540	480	60	27	36
	2000	510	490	20	9	39
	2300	500	500	0	0	42
Σ				11,970	5397	

Direct runoff $= \dfrac{11,970}{8} =$ 1496 sfd or 2.22 in.

Drainage area = 25 sq mi

FIG. 3-11   *Derivation of a unit hydrograph.*

$$\frac{11,970 \times 3^{hr} \times 60 \times 60 \times 12}{25(5280)(5280)^2} = 2.2$$

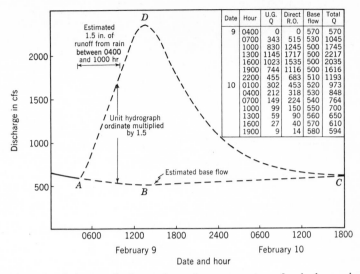

Date	Hour	U.G. Q	Direct R.O.	Base flow	Total Q
9	0400	0	0	570	570
	0700	343	515	530	1045
	1000	830	1245	500	1745
	1300	1145	1717	500	2217
	1600	1023	1535	500	2035
	1900	744	1116	500	1616
	2200	455	683	510	1193
10	0100	302	453	520	973
	0400	212	318	530	848
	0700	149	224	540	764
	1000	99	150	550	700
	1300	59	90	560	650
	1600	27	40	570	610
	1900	9	14	580	594

FIG. 3-12   *Application of the unit hydrograph to the construction of a hydrograph resulting from rainfall of unit duration.*

satisfactory as those from the simple isolated storm event, and the latter data are much to be preferred.

The use of a unit hydrograph to estimate the streamflow for a storm of like duration is illustrated in Fig. 3-12. The flow existing prior to the storm serves as a starting point for the construction of line *ABC* representing the estimated groundwater flow during the storm. The ordinates of the unit hydrograph multiplied by the estimated volume of direct runoff are added to this groundwater flow to obtain the total hydrograph *ADC*. The estimate of direct runoff is made by any of the methods described earlier in this chapter.

The number of unit hydrographs for a given basin is theoretically infinite since there may be one for every possible duration of rainfall and every possible distribution pattern in the basin. Practically, only a limited number of unit hydrographs can be used and it is common practice to ignore variations in rainfall distribution within a basin. This is reasonable for small basins, but the variations over large areas are usually too great to be ignored. Some basins are so large that a storm covers only a portion of the basin. It is not advisable to utilize unit hydrographs for basins much over 2000 sq mi in area, unless an approximate answer is sufficiently accurate. It is generally preferable, however, to divide a large basin into subareas, utilize unit hydrographs for each subarea independently, and combine the resulting hydrographs by routing techniques (Sec. 3-17).

The application of the 6-hr unit hydrograph of Fig. 3-11 to a storm of

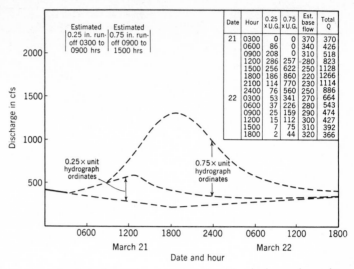

Date	Hour	0.25 x U.G.	0.75 x U.G.	Est. base flow	Total Q
21	0300	0	0	370	370
	0600	86	0	340	426
	0900	208	0	310	518
	1200	286	257	280	823
	1500	256	622	250	1128
	1800	186	860	220	1266
	2100	114	770	230	1114
	2400	76	560	250	886
22	0300	53	341	270	664
	0600	37	226	280	543
	0900	25	159	290	474
	1200	15	112	300	427
	1500	7	75	310	392
	1800	2	44	320	366

FIG. 3-13   *Application of unit hydrograph to construction of a hydrograph resulting from two periods of rainfall.*

longer duration is illustrated in Fig. 3-13. The storm is divided into two parts, and the hydrographs of each part are computed separately and added. The hydrograph for the second part of the storm starts 6 hr later than that for the first part. This example shows how a few unit hydrographs for various durations might be used to synthesize the hydrograph of a long-duration storm. A tolerance of as much as 25 percent of the adopted unit-hydrograph duration can ordinarily be accepted without serious error. Thus a 6-hr unit hydrograph might be applied for storms of 4.5 to 7.5 hr duration. Conversely, unit hydrographs having a similar range in duration may be averaged. The resulting average unit hydrograph is generally preferred over those obtained from a single storm.

Large variations in rainfall intensity during the unit period may materially affect the accuracy of the unit-hydrograph approach. Errors from this source can be minimized by using unit hydrographs for relatively short time periods. These short-period unit hydrographs can be used to develop the hydrograph resulting from a long rain of varying intensity in the manner illustrated in Fig. 3-13. Experience has shown that the best unit period is about one-fourth of the *basin lag*, the time from the center of mass of rainfall to the peak of the hydrograph.

The basis of the unit-hydrograph approach is the assumption of linearity. There is considerable evidence that this assumption is not completely valid. In general, unit-hydrograph peaks tend to be higher and earlier as the volume of runoff increases. This probably results from the differing proportions of interflow and surface runoff and the improved hydraulic efficiency of channels at high flows. This trend may reverse at very high overbank flows. In the ordinary range of data usually available, the nonlinear effects may not be detectable; and by the same token the unit-hydrograph procedure can be used with success as long as it is not extrapolated beyond the range of data from which the unit hydrographs are derived. Application of unit hydrographs for estimating spillway design floods or other extreme events may result in an underestimate.

3-13 Estimating annual hydrographs   In some design problems it is necessary to estimate the complete hydrograph for a year or period of years. Such a problem may be encountered when an outstanding flood or drought is known to have occurred prior to the beginning of streamflow record. In most cases the monthly flow volumes as estimated from precipitation-runoff relations will be sufficient. If greater detail is necessary, runoff may be calculated storm by storm and unit hydrographs used to construct a detailed hydrograph. Still another procedure is to estimate the annual runoff volume with a precipitation-runoff relation and distribute this volume throughout the year by comparison with records of neighboring streams. A rough relation between annual or seasonal runoff and annual peak flow may be possible on streams where the annual peak results from melting snow (Fig. 3-14) or seasonal floods on large rivers. It is also possible in some cases to

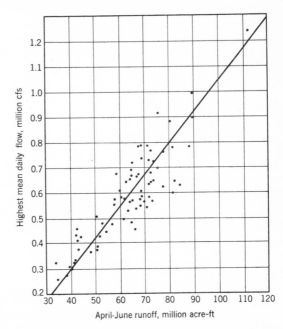

FIG. 3-14   *Relation between highest mean daily discharge and April-June flow for the Columbia River at The Dalles, Oregon.*

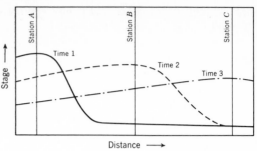

FIG. 3-15   *Successive profiles of a flood wave, showing changes in shape.*

develop useful relations between annual or seasonal flow volume and the minimum flow during the year. Neither type of relation can be expected to have a high order of accuracy.

STORAGE ROUTING

The earlier sections of this chapter outline methods for estimating stream-flow hydrographs from small areas. A hydrograph is really a record of the movement of a wave past a gaging station. As the wave moves downstream, its shape is changed by the addition of flow from tributaries and also because velocities at various points along the wave are not the same. Without additional inflow the modification in shape consists of an attenuation or lengthening of the time base of the wave (Fig. 3-15) and a lowering of the peak flow. With additional inflow the attenuation effect is still present, but the increase in total volume makes it less obvious.

3-14   The routing process   Theoretical computation of the change in shape of a flood wave on the basis of wave mechanics (unsteady flow) is extremely difficult when applied to the complex conditions encountered in natural channels. Computer solutions are feasible using numerical solutions of the differential equations,[1] although large computers are normally required. For manual computation, a solution based on the principle of continuity applied to a short section (reach) of the stream is commonly used. This principle is expressed in the storage equation

$$\bar{I}\Delta t - \Delta s = \bar{O}\Delta t \tag{3-13}$$

where $\bar{I}$ and $\bar{O}$ are the average rates of inflow and outflow for the time interval Δt and Δs is the change in volume of water in the channel between the inflow and outflow sections during time Δt. Since I is the measured

---

[1] E. J. Isaacson, J. J. Stoker, and B. A. Troesch, Numerical Solution of Flood Protection and River Regulation Problems, *New York Univ. Inst. Math. Sci. Rept. I*, 1953, *Rept. II*, 1954, and *Rept. III*, 1956.

inflow to the reach, the solution of the equation for O depends on a determination of Δs.

If the average rate of flow during a given time period is equal to the average of the flows at the beginning and end of the period, Eq. (3-13) can be written

$$\frac{I_1 + I_2}{2}\Delta t - \frac{O_1 + O_2}{2}\Delta t = s_2 - s_1 \qquad (3\text{-}14)$$

where the subscripts 1 and 2 refer to the beginning and end of the period Δt, respectively. The assumption of a linear variation in flow during the period is satisfactory if Δt is sufficiently short. In a practical problem the inflows, I_1 and I_2, and the initial outflow and storage, O_1 and s_1, are known or can be estimated with little error. Since there remain two unknowns, O_2 and s_2, a second equation is necessary. This equation must relate storage to some measurable parameter, preferably outflow rate.

3-15 Routing through uncontrolled reservoirs A reservoir is an enlargement of a river channel, and storage in reservoirs may modify the shape of a flood wave more markedly than an equivalent length of natural channel. If the reservoir has no gates, discharge takes place over a weir or through an uncontrolled orifice in such a way that O is a function of the reservoir level. In short, deep reservoirs where water velocity is low, the water surface will be nearly horizontal and the volume of water in the reservoir is directly related to the reservoir elevation. Hence storage and outflow can be directly related (Fig. 3-16). Storage volumes are determined by

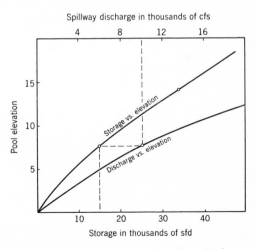

FIG. 3-16 Relation between reservoir surface elevation, storage, and spillway discharge for a reservoir with an ungated spillway.

planimetering a contour map of the reservoir area (Sec. 7-1). Equation (3-14) may be rewritten as

$$I_1 + I_2 + \frac{2s_1}{\Delta t} - O_1 = \frac{2s_2}{\Delta t} + O_2 \qquad (3\text{-}15)$$

The second relation required for a solution is a graph of values of $(2s/\Delta t) \pm O$ as functions of O (Fig. 3-17). At the beginning of a routing period (time 1) all terms on the left-hand side of Eq. (3-15) are known, and a value for the term on the right-hand side may be computed (Table 3-3).

Entering Fig. 3-17 with this value, a value of O_2 and the corresponding value of $(2s/\Delta t) - O$ may be determined.

If the reservoir surface has a considerable slope, storage becomes a function of inflow as well as outflow and the outflow-storage curve of Fig. 3-16 must be replaced by a family of curves with inflow as a parameter. Consequently, the routing curves of Fig. 3-17 must also be replaced by curve families with inflow as a parameter. The routing operation is unchanged.

3-16   Routing in controlled reservoirs   The relation between storage and outflow for a reservoir with spillway gates or outlet valves is dependent on the number of gates or valves which are open. The solution is much the same as that for the case of the reservoir with the sloping water surface. In the case of a gated spillway with all gates the same size, the elevation-discharge curve may be represented by a curve family with the number of gates open as a parameter (Fig. 3-18). Hence, curves relating $(2s/\Delta t) \pm O$ and O must be replaced by curve families with number of gates open as parameter. The routing operation is similar to that shown in Table 3-3 except that the number of gates open must be tabulated and $(2s/\Delta t) \pm O$ inter-

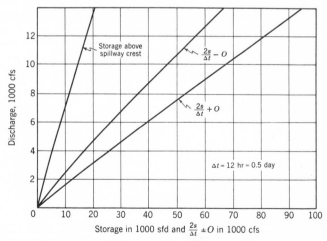

FIG. 3-17   Routing curves for an uncontrolled reservoir.

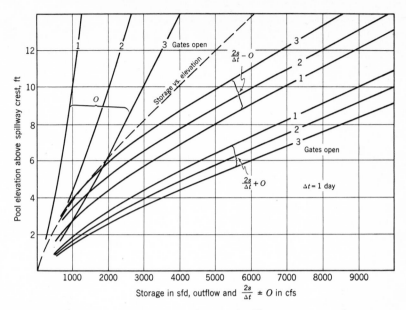

FIG. 3-18   Routing curves for a reservoir with a gated spillway.

polated from the curves in accordance with these values. If there is no change in gate opening during the period of the study, the procedure is identical with that of Table 3-3 since all values are read from the pair of curves representing the constant gate opening.

3-17   Storage routing in natural channels   The volume of water in a channel at any instant is called *channel*, or *valley*, *storage s*. The most direct determination of *s* is by measurement of channel volume from topographic maps. However, lack of adequately detailed maps plus the need to assume or compute a water-surface profile for each possible condition of flow in the

TABLE 3-3   Routing with the $2s/\Delta t \pm O$ Curves of Fig. 3-17

Date	Hour	I, cfs	$\dfrac{2s}{\Delta t} - O$, cfs	$\dfrac{2s}{\Delta t} + O$, cfs	O, cfs
1/8	Noon	2000	*8,500*	*12,500*	2000
	Midnight	2800	*8,900*	*13,300*	*2200*
1/9	Noon	4000	*10,900*	*15,700*	*2400*
	Midnight	5200	*13,700*	*20,100*	*3200*
1/10	Noon	6000	*17,300*	*24,900*	*3800*
	Midnight	5700	*20,000*	*29,000*	*4500*

Note: Computed values shown in italics.

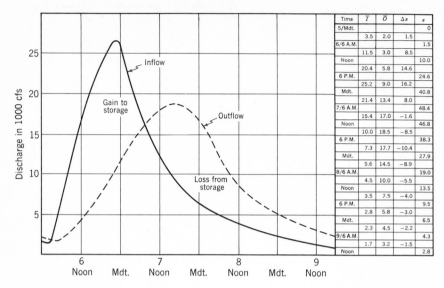

Time	$\bar{I}$	$\bar{O}$	Δs	s
5/Mdt.				0
	3.5	2.0	1.5	
6/6 A.M.				1.5
	11.5	3.0	8.5	
Noon				10.0
	20.4	5.8	14.6	
6 P.M.				24.6
	25.2	9.0	16.2	
Mdt.				40.8
	21.4	13.4	8.0	
7/6 A.M.				48.4
	15.4	17.0	−1.6	
Noon				46.8
	10.0	18.5	−8.5	
6 P.M.				38.3
	7.3	17.7	−10.4	
Mdt.				27.9
	5.6	14.5	−8.9	
8/6 A.M.				19.0
	4.5	10.0	−5.5	
Noon				13.5
	3.5	7.5	−4.0	
6 P.M.				9.5
	2.8	5.8	−3.0	
Mdt.				6.5
	2.3	4.5	−2.2	
9/6 A.M.				4.3
	1.7	3.2	−1.5	
Noon				2.8

FIG. 3-19   *Inflow and outflow hydrographs for a reach of river, showing method of calculating channel storage.*

channel makes this approach generally unsatisfactory. Since Eq. (3-13) involves only Δs, absolute values of storage need not be known. Values of Δs can be found by solving Eq. (3-13), using actual values of inflow and outflow (Fig. 3-19). The hydrographs of inflow and outflow for the reach are divided into short time intervals, average values of I and O are determined for each period, and values of Δs computed by subtracting $\bar{O}$ from $\bar{I}$. Storage volumes are computed by summing the increments of storage from any arbitrary zero point.

When values of s computed as just described are plotted against simultaneous outflow (Fig. 3-20), it usually appears that storage is relatively higher during rising stages than during falling stages. As a wave front passes through a reach, some storage increase occurs before any increase in outflow. After the crest of the wave has entered the reach, storage may begin to decrease although the outflow is still increasing. Nearly all methods of routing streamflow relate storage to both inflow and outflow in order to allow for these variations. In some cases, a family of curves relating storage, outflow, and inflow is developed; and the routing equation is solved in the same manner as for a reservoir with a sloping water surface.

Another very widely utilized assumption is that storage is a function of weighted inflow and outflow as given by

$$s = K[xI + (1 - x)O] \tag{3-16}$$

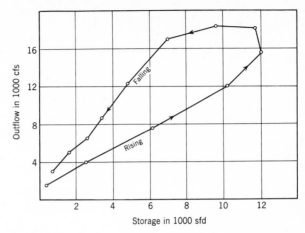

FIG. 3-20   *Relation between outflow and storage for the data of Fig.* 3-19.

where s, I, and O are simultaneous values of storage, inflow, and outflow, respectively, x is a dimensionless constant which indicates the relative importance of I and O in determining storage, and K is a storage constant with the dimension of time. The value of K approximates the time of travel of the wave through the reach. The constant x varies from 0 to 0.5. Since $ds/dt = I - O$, differentiating Eq. (3-16) yields

$$I - O = \frac{ds}{dt} = K\left[x\frac{dI}{dt} + (1 - x)\frac{dO}{dt}\right] \tag{3-17}$$

If $I = O$, then

$$x = \frac{dO/dt}{dO/dt - dI/dt} \tag{3-18}$$

which permits estimating x from concurrent inflow and outflow records. For a reservoir where $O = f(s)$, ds/dt and dO/dt must be zero when $I = O$. Therefore x for this case is zero. A value of zero indicates that the outflow alone determines storage (as in a reservoir). When $x = 0.5$, inflow and outflow have equal influence on storage. In natural channels x usually varies between 0.1 and 0.3. Equation (3-16) is the basis of the Muskingum method[1] of routing. Values of K and x for a reach are usually determined by trial. Values of x are assumed, and storage is plotted against $xI + (1 - x)O$. The value of x which results in the data conforming most closely to a straight line is selected (Fig. 3-21). The factor K is the slope of the line relating s to $xI + (1 - x)O$.

---

[1]"Engineering Construction—Flood Control," pp. 147–156, Engineer School, Ft. Belvoir, Va., 1940.

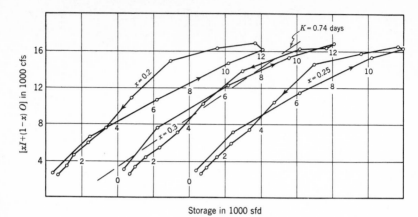

FIG. 3-21   *Method of determining K and x for the Muskingum method of routing.*

The Muskingum routing equation is found by substituting Eq. (3-16) for s_1 and s_2 in Eq. (3-15) and solving for O_2,

$$O_2 = c_0 I_2 + c_1 I_1 + c_2 O_1 \tag{3-19}$$

$$c_0 = -\frac{Kx - 0.5\Delta t}{K - Kx + 0.5\Delta t} \tag{3-20a}$$

$$c_1 = \frac{Kx + 0.5\Delta t}{K - Kx + 0.5\Delta t} \tag{3-20b}$$

$$c_2 = \frac{K - Kx - 0.5\Delta t}{K - Kx + 0.5\Delta t} \tag{3-20c}$$

$$c_0 + c_1 + c_2 = 1 \tag{3-20d}$$

The significance of Eq. (3-20d) may be seen if it is noted that, for steady flow $(I_1 = I_2 = O_1 = O_2)$, Eq. (3-19) can be correct only when the sum of the constants is unity. It is important that K and Δt be in the same units when used in Eqs. (3-20). When storage is computed in 6-hr increments, the units are quarter second-foot-days. If these units are not changed in plotting Fig. 3-21, the units of K are quarter days or a slope of unity represents $K = 6$ hr.

The application of the Muskingum method is illustrated in Table 3-4. Values of c_0, c_1, and c_2 were computed by substituting $K = 0.74$ day, $x = 0.3$ (Fig. 3-21), and $\Delta t = 6$ hr in Eqs. (3-20a) to (3-20c). Values of I are tabulated, and the products $c_0 I_2$ and $c_1 I_1$ are computed. With an initial value of O given or estimated, the product $c_2 O_1$ is calculated and the three products added to obtain O_2. The computed value of O_2 becomes O_1 for the next routing period, and another value of O can be determined. The process continues as long as values of I are known. This routing is easily performed on a digital computer.

TABLE 3-4   Application of the Muskingum Method

Date	Hour	I, cfs	$c_0 I_2$	$c_1 I_1$	$c_2 O_1$	O, cfs
4/9	6 A.M.	1000	...	...	...	1000
	Noon	2400	−360	+540	+610	790*
	6 P.M.	3900	−585	+1296	+482	1193
	Mdt.	5000	−750	+2106	+726	2082
4/10	6 A.M.	4900	−735	+2700	+1270	3235
	Noon	4000	−600	+2646	+1974	4020

Note: Computed values are italicized. Coefficients used are $c_0 = -0.15$, $c_1 = 0.54$, and $c_2 = 0.61$.

*The first computed outflow often drops when inflow increases sharply. This is simply disregarded.

Storage in a river reach actually depends on water depths. The assumption that storage is correlated with rates of flow is a valid approximation only when stage and discharge are closely correlated. On streams where slope-stage-discharge relations are necessary, more complex routing methods[1] are required to obtain satisfactory accuracy. The methods described in the preceding sections also assume that the longitudinal profile of the water surface in a reach is the same every time a given combination of inflow and outflow occurs. This is also an approximation but is usually sufficiently precise if the reach is not excessively long. In general the methods which have been described are satisfactory on the great majority of streams.

3-18   Local inflow   The previous discussion has considered the routing of inflow entering at the head of a reach. In almost all streams there is additional inflow from tributaries which enter the main stream between the inflow and outflow points of the reach. Occasionally this *local inflow* is small enough to be neglected, but in most cases it must be considered. The conventional procedures are (1) add the local inflow to the main-stream inflow, and consider the total as *I* in the routing operation or (2) route the main-stream inflow through the reach, and add the estimated local inflow to the computed outflow. The first method is used when the local inflow enters the reach near its upstream end, while the second method is preferred if the greater portion of the tributary flow joins the main stream near the lower end of the reach. The local inflow might also be divided into two portions, one part combined with the main-stream inflow and the remainder added to the computed outflow.

[1] B. R. Gilcrest, Flood Routing, chap. 10 in H. Rouse (ed.), "Engineering Hydraulics," Wiley, New York, 1950; H. A. Thomas, "The Hydraulics of Flood Movements in Rivers," Carnegie Institute of Technology, 1937.

The hydrograph of local inflow may be estimated by comparison with streamflow records on tributary streams or by use of rainfall-runoff relations and unit hydrographs. In working with past data, the total volume of local inflow should be adjusted to equal the difference between the reach inflow and outflow, with proper allowance for any change in channel storage during the computation period. Since local inflow may be a small difference between two large figures, slight errors in the streamflow record may result in large errors in local inflow, even to the extreme of indicating negative local inflows.

3-19   Electrical analogy to streamflow   A condenser stores electricity in proportion to the voltage across the plates. In a circuit with fixed resistance, the voltage varies with the current. Hence a circuit may be constructed so that it has storage characteristics in accord with Eq. (3-16). Such a circuit is analogous to a stream channel within the limitations of the equation and may be used for the solution of routing problems.[1] The analogy has the advantage of producing a solution rapidly; but, more important, it solves the differential form of the storage equation. In this form the assumptions involved in transforming Eq. (3-13) to Eq. (3-14) are avoided.

3-20   Computer simulation   The digital computer has made possible a new approach to hydrology called computer simulation. Because of the computing speed of modern computers it is possible to program the runoff cycle (Fig. 3-7) in its entirety and to solve it continuously using short time increments. It is necessary to write functions describing each step in the cycle and to define the parameters of these functions. The Stanford Watershed Model,[2] the earliest of the simulation programs, uses hourly rainfall and potential evapotranspiration as input data. Interception, surface retention, infiltration, overland flow, interflow, groundwater flow, and soil-moisture storage are simulated to calculate inflow to the channels; and routing is used to simulate the channel system. The basin may be divided into segments having different rainfall or other characteristics. Hourly ordinates of the hydrograph, mean daily flows and monthly totals of the water balance are output.

The model is calibrated to a watershed by trial until the observed flows are reproduced accurately. Most of the input parameters are determined from maps and other data on the watershed. Only three key parameters must be fixed by trial. In general, a good calibration can be obtained with 3 to 5 yr of data. The value of computer simulation lies in (1) the ability to calculate in detail for short time intervals thus permitting a complete

---

[1] R. K. Linsley, L. W. Foskett, and M. A. Kohler, Electronic Device Speeds Flood Forecasting, *Eng. News-Record*, Vol. 141, pp. 64–66, Dec. 23, 1948.

[2] N. H. Crawford and R. K. Linsley, Digital Simulation in Hydrology: Stanford Watershed Model IV, *Tech. Rept.* 39, Department of Civil Engineering, Stanford University, 1966.

evaluation of the complex process of runoff and (2) the use of all the data available. The second point is particularly important. Conventional hydrologic analysis is based on selected storms, and substantial quantities of data are ignored. By utilizing the entire range of data, simulation interprets the nonlinearities in both the land and channel phases and is therefore a safer base for extrapolation.

BIBLIOGRAPHY

Chow, Ven Te (ed.): "Handbook of Applied Hydrology," McGraw-Hill, New York, 1964.

Linsley, R. K., M. A. Kohler, and J. L. H. Paulhus: "Applied Hydrology," McGraw-Hill, New York, 1949.

Linsley, R. K., M. A. Kohler, and J. L. H. Paulhus: "Hydrology for Engineers," McGraw-Hill, New York, 1958.

Meinzer, O. E. (ed.): "Hydrology," Vol. IX, Physics of the Earth Series, McGraw-Hill, New York, 1942 (reprinted Dover, New York, 1949).

Williams, G. R.: Hydrology, chap. 4 in Hunter Rouse (ed.), "Engineering Hydraulics," Wiley, New York, 1950.

Wisler, C. O., and E. F. Brater: "Hydrology," 2d ed., Wiley, New York, 1959.

PROBLEMS

3-1 Using data from a *U.S. Geological Survey Water Supply Paper* reporting on a flood in your area, plot a flood hydrograph, separate groundwater flow by the method outlined in Sec. 3-2, and compute the volume of direct runoff in second-foot-days and in inches over the drainage area.

3-2 Tabulated below are data for a flood on the South Branch of the Potomac River near Springfield, West Virginia. The drainage area is 1471 sq mi. Separate the groundwater flow, and compute the direct runoff volume in second-foot-days and inches over the drainage area.

Date	Hour	Flow, 1000 cfs	Date	Hour	Flow, 1000 cfs
4/18/43	2400	1.5	4/21/43	2400	8.5
19	0600	1.5	22	1200	7.0
	1200	1.6		2400	5.7
	1800	2.0	23	1200	4.7
	2400	3.1		2400	3.9
20	0600	5.2	24	1200	3.2
	1000	7.4		2400	2.8
	1200	9.6	25	1200	2.4
	1400	12.0		2400	2.2
	1800	12.45	26	1200	2.0
	2400	12.1		2400	1.9
21	0600	11.1	27	2400	1.7
	1200	10.2	28	2400	1.5
	1800	9.3	29	2400	1.4

3-3   Compute the runoff volume for the data of Fig. 3-2. Compute the Φ index.

3-4   Using the infiltration-capacity curve of Fig. 3-2, compute the runoff volume for the precipitation sequence tabulated below. Compute the Φ index.

Hour	1	2	3	4	5	6
Rain, in.	0.10	0.15	0.25	0.20	0.10	0.07

3-5   The following data were gathered during a single 6-month period from individual rainstorms on a certain basin where there was no snowfall.

Date of storm	Average basin rainfall, in.	Basin runoff, in.	Duration of rain, hr
Feb	2.2	0.6	12
Feb	0.8	0.0	12
Mar	2.6	1.2	12
Apr	1.4	0.4	12
May	0.4	0.0	12
May	3.0	1.8	12
Jun	1.2	0.6	12
Jun	2.8	2.0	12
July	2.0	1.4	12

From these data answer the following questions:

(a) About how much runoff would there have been from a 12-hr 2.0-in. rain in May?

(b) If the storm mentioned in (a) above had occurred in 6 hr instead of 12 hr would the runoff have been greater or less than that indicated by your answer to (a) above? Why?

(c) What would have been the approximate value of the coefficient of runoff from a 12-hr 3.0-in. rainfall in February?

(d) What would have been the coefficient of runoff from a 12-hr 3.0-in. rainfall in August?

(e) From the given information what can one say about the January rainfall?

(f) Make a quantitative statement about the infiltration rate during the May storm which had no runoff.

(g) Make a rough plot of infiltration capacity vs. month of the year, i.e., show trend. You are not expected to show actual values of infiltration capacity.

(h) Make an accurate plot of coefficient of runoff versus month of the year with storm rainfall as the parameter. From this plot estimate the coefficient of runoff for a 12-hr 1.0-in. rainfall in March.

(i) Discuss briefly whether or not your answer to (a) above is generally applicable to any 12-hr 2.0-in. rainfall that is likely to occur in May of any year.

3-6   Listed below are data from storms on a given basin.

Storm	A.P.I., in.	Precipitation, in.	Duration, hr	Runoff, in.
1	3.8	1.0	6	0.8
2	2.4	0.8	4	0.5
3	2.6	1.2	8	0.5
4	2.0	1.9	4	1.0
5	1.8	2.2	8	0.8
6	2.8	2.0	6	1.2
7	1.6	3.2	4	1.9
8	2.8	2.8	4	2.0
9	2.4	3.1	8	1.5
10	1.4	2.5	6	1.0
11	1.4	3.0	6	1.4
12	3.2	1.6	6	1.0
13	3.3	3.0	6	2.0
14	1.7	1.0	6	0.1

Construct a coaxial relation between rainfall, runoff, A.P.I., and storm duration. Do this by plotting in the upper right-hand quadrant A.P.I. (ordinate) vs. runoff (abcissa) with storm precipitation as parameter. Then by eye sketch curves for precipitations of 1, 2, and 3 in. Next, for each storm enter the upper right-hand quadrant with the given A.P.I., interpolate the given precipitation between the sketched curves, and plot the given runoff positively downward in the lower right-hand quadrant with storm duration as parameter. Finally, sketch curves in the lower right-hand quadrant for storm durations of 4, 6, and 8 hr.

3-7 Using the coaxial relation developed in the preceding problem determine the runoff, ratio of runoff to precipitation, and average loss rate for each of the following storms:

Storm	A.P.I., in.	Precipitation, in.	Duration, hr
A	3.0	2.4	8
B	3.0	2.4	6
C	3.0	2.4	4
D	2.0	2.4	8
E	2.0	2.4	6
F	2.0	2.4	4

3-8 The average annual precipitations and the average annual runoffs from a certain basin are as given. Plot the annual runoffs vs. the annual precipitations using the previous year's precipitation as a parameter. Using this plot estimate the average annual runoffs from the basin in 1971, 1972, and 1973 if the average annual precipitations on the basin in those years were 18.5, 24.2, and 30.1 in. respectively. Assume no change in land use. What would be the runoffs if the 1971 to 1973 precipitations occurred in the reverse order?

Year	Annual precipitation, in.	Annual runoff, in.
1960	16.2	4.2
1961	24.7	6.4
1962	18.1	5.1
1963	30.2	10.0
1964	24.6	9.8
1965	18.0	5.4
1966	26.2	7.6
1967	28.7	10.5
1968	32.1	14.2
1969	18.3	6.1
1970	20.4	4.9

3-9   Using Fig. 3-9, determine the melt on a day with an index temperature of 38°F, a snow line at 4000 ft, and a degree-day factor of 0.08.

3-10   If the degree-day factor in a certain region is 0.08 in./degree day, how much would the depth of a snowpack decrease on a warm spring day where the maximum temperature is 77°F and the minimum temperature 45°F? Assume the specific gravity of the snow is 0.20.

3-11   The time of concentration for a rectangular area is 30 min. The direction of overland flow is parallel to the longer sides of the rectangle. Should one expect a greater peak rate of runoff from this area from a storm of intensity 4 in./hr of 10 min duration or from a storm of intensity 1 in./hr 40 min of duration? Why?

3-12   A rectangular parking lot 400 ft wide and 800 ft long has an estimated time of concentration of 20 min. Of the 20-min concentration time, 15 min is required for overland flow across the pavement to the longitudinal gutter along the center of the lot. A rain of 2 in./hr intensity falls on the lot for 5 min and then stops abruptly. If the runoff coefficient is 0.85 determine the peak rate of outflow past point A. Express answer in cfs.

3-13   For the rectangular parking lot of Prob. 3-12 construct the outflow hydrographs at A for storms of durations 5, 10, 15, 20, and 30 min, all storms having an intensity of 1.5 in./hr. In this case assume a runoff coefficient of 1.0.

3-14   Information on the 10-yr storm in a given area is as follows:

Time, min.	Depth, in.
5	0.30
10	0.49
20	0.70
30	0.90
60	1.02

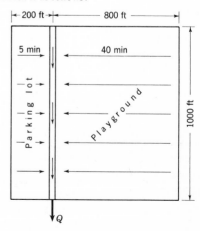

Find the 10-yr design flow at the outlet from this composite area: parking lot ($k = 0.9$) and playground ($k = 0.3$). The lateral flow times may be assumed to be 5 min for the parking lot and 40 min for the playground. Take gutter flow velocity as 3 ft/sec.

3-15   An area of 12 acres in a single-family residence district has an average length of overland flow of 125 ft, average slope of the lots of 0.003, and the design rainfall is given by $i = 25/t^{0.5}$. Ignoring the intensity term in Eq. (3-12) and using $c_r = 0.05$, find the time of concentration of the overland flow from this area. Assuming gutter flow time to add 10 min, find the peak rate of runoff to be expected.

3-16   Construct the unit hydrograph for the data of Prob. 3-2.

3-17   Assuming the effective rainfall duration for the unit hydrograph of Prob. 3-16 is 24 hr, find the peak flow from a storm producing 1.0 in. of runoff in 48 hr. Consider this as two successive storms, each producing 0.5 in. of runoff. Assume a constant base flow of 1500 cfs.

3-18   Using the unit hydrograph of Prob. 3-16, compute the peak rate of flow resulting from four successive 24-hr periods with runoff values of 0.3, 0.7, 0.1, and 1.4 in., respectively. Assume zero base flow.

3-19   A storm lasting 8 hr has 1.40 in. of rain and produces the following flows:

Time, hr	Flow, cfs
0	20
4	100
8	200
12	250
16	200
20	160
24	130
28	100
32	80
36	60
40	40
44	30
48	20

The basin has an area of 14 sq mi. Find the peak of the unit hydrograph. What fraction of the precipitation appeared as runoff? Assume zero base flow.

3-20   If a 4-hr rainstorm producing 2 in. of storm runoff will create a peak rate of flow of 60 cfs, what peak flow might one expect from an 8-hr storm on the same basin also producing 2 in. of storm runoff? Assume that base flow is negligible and that both storms are uniform in intensity and distribution. In the case of the 4-hr storm the peak flow of 60 cfs occurred 10 hr after the first noticeable rise in flow. An approximate answer (to the nearest 5 cfs) will be satisfactory.

3-21   In a typical 4-hr storm producing 2.0 in. of runoff from the Big Bear Creek basin, the flows in the stream are as follows:

Time, hr	Flow, cfs
0	0
2	90
4	300
6	500
8	420
12	250
16	100
20	0

Estimate as accurately as possible the peak flow and the time of its occurrence in a flood created by an 8-hr storm which produces 1.0 in. of runoff during the first 4 hr and 1.5 in. of runoff during the second 4 hr. Assume that the base flow is negligible.

3-22 The rainfall-runoff relation for the 300-sq mi Beaver Creek basin under dry-soil conditions is as follows:

Rainfall, in.	Runoff, in.
0.6	0.0
2.0	0.8
3.0	1.6
4.0	2.6
5.0	3.5

A storm of 2-hr effective duration occurred over this basin at a time when the soil was dry. The resulting hydrograph was as follows:

Time, hr	Flow, cfs
0	0
3	12,000
6	20,000
9	16,000
12	10,000
15	3,000

Determine the number of inches of runoff from the basin resulting from this storm. Assume zero base flow. Approximately how much rain was there in this storm? What was the average coefficient of runoff during this storm? Predict the peak flow that would result from a 4-hr storm during which there were 4.5 in. of rainfall. Once again assume dry-soil conditions, i.e., zero base flow. Plot the resulting hydrograph showing the approximate time of peak flow.

3-23 For a certain reservoir the elevation vs. surface-area data are as follows:

Elevation, ft	60	58	56
Area, acres	520	450	400
Outflow, cfs	156	153	150

Using these data, determine as accurately as possible the time required to lower the water surface from elevation 60 to elevation 56. During the entire period there is a steady inflow of 100 cfs. Neglect evaporation and seepage.

3-24 The elevation-discharge and elevation-area data for a small reservoir without spillway gates are tabulated below. If the reservoir inflow during a flood corresponds to the flows given in Prob. 3-2, determine by routing the maximum water level in the reservoir and the peak rate of outflow. Use a 6-hr routing period, and assume that the reservoir level just reaches the spillway crest (elev. 0) at 2400 hr, Apr. 18, 1943. Be sure your units are correct.

Elevation, ft	0	1	2	3	4	5	6	7	8	9
Area, acres	1000	1020	1040	1050	1060	1080	1100	1120	1,140	1,160
Outflow, cfs	0	525	1490	2730	4200	5880	7660	9620	11,800	14,300

3-25 Given below are the relations between discharge, storage, and pool elevation for a reservoir with a fixed outlet. Compute the outflow hydrograph corresponding to the inflows given assuming the reservoir to be initially at elevation 100. Outflow from the reservoir commences at reservoir elevation 103.0.

Elevation, ft	Storage, sfd	Outflow, cfs	Time, days	Mean flow, cfs
100	50	0	1	10
101	60	0	2	20
102	70	0	3	30
103	80	0	4	35
104	92	8.0	5	30
105	105	16.6	6	22
106	120	26.6	7	15
107	140	40.0	8	10
			9	10
			10	10

3-26 The inflow hydrograph to a reach of river is as follows:

Time	Flow, cfs
0	1,000
6	2,000
12	7,500
18	10,000
24	7,800
30	6,000
36	4,700
42	3,600
48	2,700
54	2,000

Route this flood down the reach of river assuming $Q = 1000$ cfs at both upper and lower ends of the reach at time zero. Consider these four cases:

(a) $K = 18$ hr, $x = 0.3$.
(b) $K = 18$ hr, $x = 0.1$.
(c) $K = 9$ hr, $x = 0.3$.
(d) $K = 9$ hr, $x = 0.1$.

Use a routing period of 6 hr.

3-27 Tabulated below are the inflow and outflow hydrographs for a reach of a river. Calculate the channel storage at 6-hr intervals, and determine the best values of K and x for the Muskingum routing method.

Date	Time	Inflow, cfs	Outflow, cfs
1	0600	1,000	1,000
	1200	4,000	1,300
	1800	9,550	1,500
	2400	13,720	3,100
2	0600	12,450	6,020
	1200	10,200	7,900
	1800	8,200	8,800
	2400	6,500	8,700
3	0600	5,500	8,200
	1200	4,700	7,500
	1800	4,100	6,750
	2400	3,600	6,150
4	0600	3,100	5,800
	1200	2,700	5,100
	1800	2,400	4,500
	2400	2,100	3,900

3-28 Taking the hydrograph of Prob. 3-2 as the inflow, find the peak and time of peak of the outflow hydrograph for a reach for which $K = 1.5$ days and $x = 0.15$. Plot inflow and outflow hydrographs.

3-29 In a given watershed the loss rate may be assumed to have a constant value of 0.2 in./hr. What percentage increase in runoff is to be expected from a 6-hr storm of uniform intensity in which cloud seeding increased the precipitation by 10 percent? In this storm there normally would have been 2.40 in. of precipitation.

3-30 (a) What is the effect of urbanization on the hydrologic cycle? Relate your discussion to such things as interception, infiltration, direct runoff, evaporation, transpiration, and groundwater recharge.

(b) What is the effect of urbanization on flood peaks?

(c) What is the effect of urbanization on total volume of runoff?

(d) What is the effect of urbanization on sediment loads in the major water courses?

3-31 Over a period of 30 yr a particular agricultural watershed was urbanized. Before urbanization with particular antecedent conditions the average loss rate may be assumed to have been 0.30 in./hr. After urbanization this value dropped to 0.10 in./hr.

The 1-hour unit graphs for this watershed were determined to be approximately as follows:

Before urbanization

After urbanization

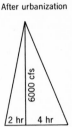

(a) Determine the area of the watershed in acres.
(b) Show the hydrographs of flow for the "before" and "after" urbanization cases from a 2-hr storm in which 0.50 in. of rain fell during the first hour and 0.60 in. during the second hour. Assume a uniform loss rate and neglect base flow. In plotting the hydrographs use the following scales:
Horizontal: 1 in. = 2.5 hr
Vertical: 1 in. = 1000 cfs
(c) Compute the coefficient of runoff before and after urbanization for this particular storm with the given antecedent conditions.

Chapter 4

Groundwater

An estimate[1] in 1965 placed groundwater use in the United States at about 58 billion gallons daily—roughly one-fifth of the total water use in the country. Groundwater is a vital source of water supply, especially in areas where dry summers or extended droughts cause streamflow to stop. The discussion of groundwater in a chapter apart from surface water should not be construed as indicating that the two sources of water are independent of each other. On the contrary, many surface streams receive a major portion of their flow from groundwater. Elsewhere, water from surface streams is the main source of recharge for the groundwater. The two sources of supply are definitely interrelated, and use of one may affect the water available from the other source. Both surface-water and groundwater problems should be considered together in plans for water-resources development.

OCCURRENCE OF GROUNDWATER

4-1 Zones of underground water Figure 4-1 shows in more detail than Fig. 2-1 the water conditions below the ground surface. Immediately below the surface, the soil pores contain both water and air in varying amounts. After a rain, water may move downward through this *zone of aeration*. Some water is dispersed through the soil to be held by capillary forces in the smaller pores or by molecular attraction around the soil particles. Water in the upper layers of the zone of aeration is known as *soil moisture*. If the retention capacity of the soil in the zone of aeration is satisfied, water moves downward into a region where the pores of the soil or

---

[1] "The Nation's Water Resources," U.S. Water Resources Council, p. 3–2–7, Washington, D.C., 1968.

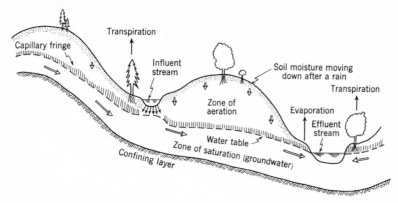

FIG. 4-1 *Schematic diagram illustrating the occurrence of groundwater.*

rock are filled with water. The water in this *zone of saturation* is called the *groundwater*. Above the zone of saturation is a *capillary fringe* in which the smaller soil pores contain water lifted by capillary action from the zone of saturation.

4-2 Sources of groundwater   The main source of groundwater is precipitation, which may penetrate the soil directly to the groundwater or may enter surface streams and percolate from these channels to the groundwater. The disposition of precipitation which reaches the earth is discussed in Sec. 3-1. It should be emphasized that the groundwater has the lowest priority on the water from precipitation. This low priority is an important factor in limiting the rates at which groundwater may be utilized. Interception, depression storage, and soil moisture must be satisfied before any large amount of water can percolate to the groundwater. Only prolonged periods of heavy precipitation can supply large quantities of water for groundwater recharge (Fig. 4-2). Groundwater recharge is an intermittent and irregular process.

Geologic conditions determine the path by which water from precipitation reaches the zone of saturation. If the water table is near the surface, there may be considerable percolation through the soil. Relatively impermeable layers above the water table may prevent such direct percolation. Stream channels which cut through permeable alluvial deposits offer a path for water to reach the groundwater, provided the stream is above the level of the groundwater (Fig. 4-1). The rate of percolation from an influent stream is limited by the extent and character of the underlying material, and flood flows in excess of the limiting percolation rate may discharge into the ocean or to lakes.

Other sources of groundwater include water from deep in the earth which is carried upward in intrusive rocks and water which is trapped in sedimen-

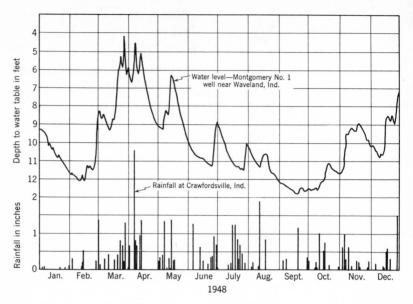

FIG. 4-2   Comparison of groundwater levels and precipitation.

tary rocks during their formation. The quantities of such waters are small, and they are often so highly mineralized as to be unsuited for use. These deep groundwaters may, however, contaminate other useful waters. For example, considerable boron is added to groundwater in the San Joaquin Valley, California, by water rising through faults in the Franciscan rocks of the Coast Ranges.

4-3   Aquifers   Formations which contain and transmit groundwater are known as *aquifers.* The amount of groundwater which can be obtained in any area depends on the character of the underlying aquifer and the extent and frequency of recharge. The capacity of a formation to contain water is measured by the *porosity*, or ratio of the pore volume to the total volume of the formation. Pores vary in size from submicroscopic openings in clay and shales to large caverns and tunnels in limestone and lava. The porosity of a material can be determined by oven-drying an undisturbed sample and weighing it. The sample is then saturated with some liquid such as kerosene and weighed again. Finally the saturated sample is immersed in the same liquid, and the weight of liquid displaced is noted. The weight of liquid required to saturate the sample divided by the weight of the displaced liquid gives the porosity as a decimal. It is sometimes necessary to inject the liquid under pressure to completely displace all air in the voids. Table 4-1 indicates the variation in porosity for the more common formation materials.

A high porosity does not indicate that an aquifer will yield large volumes of water to a well. The only water which can be obtained from the aquifer is that which will flow by gravity. The *specific yield* (Table 4-1) is the volume of water (expressed as a percent of the total volume of the aquifer) which will drain freely from the aquifer. Specific yield is always less than porosity since some water will be retained in the aquifer by molecular or capillary forces. The specific yield of fine-grained materials is much less than that of coarse materials. Clay, although having a high porosity, is so fine-grained that it ordinarily yields little water to wells. In contrast, a cavernous limestone or fractured sandstone with low porosity may yield almost all the water it contains. The most important aquifers economically are deposits of sand and gravel, which have a fairly high specific yield.

4-4 The water table The static level of water in wells penetrating the zone of saturation (Fig. 4-1) is called the *water table*. The water table is often described as a subdued replica of the surface topography. It is commonly higher under the hills than under valleys, and a contour map of the water table in an area may look much like the surface topography. The water table is the surface of a water body which is constantly adjusting itself toward an equilibrium condition. If there were no recharge to or outflow from the groundwater in a basin, the water table would eventually become horizontal. Few basins have uniform recharge conditions at the surface. Some areas receive more rain than others. Some portions of the basin have more permeable soil. Thus, when intermittent recharge does occur, mounds and ridges form in the water table under the areas of greatest recharge. Subsequent recharge creates additional mounds, perhaps at other points in the basin, and the flow pattern is further changed. Superimpose upon this fairly simple picture variations in permeability of the aquifers, impermeable strata, and the influence of lakes, streams, and wells, and one obtains a picture of a water table constantly adjusting toward equilibrium. Because of the low

TABLE 4-1   Approximate Average Porosity, Specific Yield, and Permeability of Various Materials

Material	Porosity, percent	Specific yield, percent	Permeability, K_s, gpd/sq ft	Intrinsic permeability, darcys
Clay	45	3	0.01	0.0005
Sand	35	25	1000	50
Gravel	25	22	100000	5000
Gravel and sand	20	16	10000	500
Sandstone	15	8	100	5
Limestone, shale	5	2	1	0.05
Quartzite, granite	1	0.5	0.01	0.0005

flow rates in most aquifers this equilibrium is rarely attained before additional disturbances occur.

When water occurs in cracks, fissures, and caverns, the situation is somewhat different. Flow in large openings is usually turbulent, and adjustments take place fairly rapidly. Water is usually found at about the same level anywhere within a system of interconnected openings. Water levels may vary considerably, however, between entirely separate openings in the same formation (Fig. 4-3). Wells driven into such formations will yield little water unless they intersect one of the fissures or caverns.

4-5 Artesian aquifers The discussion thus far has dealt with aquifers in which the upper surface of the water is unconfined. Sometimes an aquifer is confined by strata of low permeability (Fig. 4-4). Such *artesian aquifers* are analogous to pipelines. The static pressure at a point within the aquifer is equivalent to the elevation of the water table in the recharge area less the loss in head through the aquifer to the point in question. A well piercing the confining stratum acts much like a piezometer in a pipe, and water will rise in the well to the level of the local static pressure (artesian head). If the pressure is sufficient to raise water above the ground, the well is called a *flowing well.* The surface defined by the water level in a group of artesian wells is called the *piezometric surface* and is the artesian equivalent of the water table.

The shape of the piezometric surface may be visualized in much the same manner as the hydraulic grade line of a pipe. If no flow takes place through the aquifer, the piezometric surface will be level. As discharge increases, the surface slopes more steeply toward the discharge point. The slope of the gradient is steep through areas of low permeability and relatively flat where permeability is high. Because of the low velocities of flow in groundwater, velocity head is negligible and minor variations in the cross section of the aquifer are not reflected in the artesian levels.

When water is withdrawn from an artesian well, a local depression of the piezometric surface results (Sec. 4-10). This decrease in pressure permits a

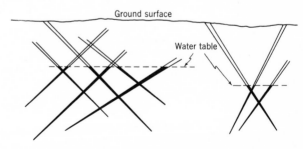

FIG. 4-3   Water table in fractured rock.

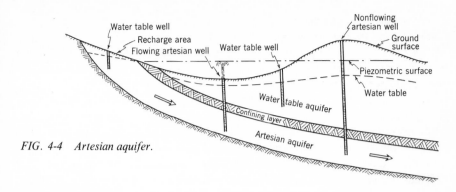

FIG. 4-4   Artesian aquifer.

slight expansion of the water and a compaction of the aquifer. The lowered pressure around the well increases the flow toward the well, and after sufficient time has elapsed, this increased flow is reflected in a lowering of the water table in the recharge area. In extensive aquifers the elapsed time may be measured in years.

Artesian aquifers usually have relatively small recharge areas as compared with water-table aquifers and generally yield less water. The economic importance of artesian aquifers lies in the fact that they transmit water substantial distances and deliver it above the level of the aquifer, thus minimizing pumping costs. The Dakota sandstones provide water from the Black Hills to much of South Dakota. Initially this aquifer transmitted large flows with ground-surface pressures as high as 130 psi. Heavy withdrawal of water has resulted in lower pressures and yields. Little effect of the withdrawal has been noted in the recharge area, and it is felt that much of the water used during the last 50 yr has come from storage as a result of compaction of the aquifer and expansion of the water. This suggests that hydraulic head is not the sole source of pressure in artesian aquifers but that the weight of the overlying formations is also a factor. Pumping from artesian aquifers has resulted in subsidence of the ground in some areas. In the western San Joaquin Valley[1] of California ground-surface elevations dropped 10 ft between 1932 and 1954. During this period the piezometric surface had dropped 190 ft.

4-6   Discharge of groundwater   Groundwater in excess of the local capacity of an aquifer is discharged by evapotranspiration and surface discharge. Whenever the capillary fringe reaches the root systems of vegetation, a route for direct transpiration to the atmosphere is provided. Some desert-type plants (phreatophytes) have root systems which extend downward

[1] J. F. Poland and G. H. Davis, Subsidence of the Land Surface in the Tulare-Wasco (Delano) and Los Banos–Kettleman City Area, San Joaquin Valley, California, *Trans. Am. Geophys. Union*, Vol. 37, pp. 287–296, 1956.

more than 30 ft to reach underground water. In some instances a diurnal fluctuation of the water-table elevation is noted as a result of daytime transpiration. As the capillary fringe nears the ground surface, increasing quantities of water may be evaporated directly from the soil.

If the water table or an artesian aquifer intersects the ground surface, water is discharged as surface flow. If the discharge rate is low or the flow is spread over a large area, *diffuse seepage* may occur and the water does little more than wet the ground from which it evaporates. Diffuse seepage along the banks of streams or lakes may, however, aggregate into a considerable volume and is often the main source of streamflow during dry periods. A large discharge from an aquifer concentrated in a small area is a *spring*. Figure 4-5 illustrates a few of the many situations under which a spring may develop. Large springs are generally associated with fissures or caverns in the rocks. Springs associated with aquifers of large extent and moderate or low permeability usually flow at relatively constant rates. Springs receiving their flow from small or highly pervious aquifers may fluctuate widely in discharge and sometimes dry up during droughts.

GROUNDWATER HYDRAULICS

4-7 Movement of groundwater Except in large caverns and fissures, groundwater flow is almost exclusively laminar. Hagen (1839) and Poiseuille (1846) showed that the velocity of flow in capillary tubes is proportional to the slope S of the energy gradient. Darcy (1856) confirmed the

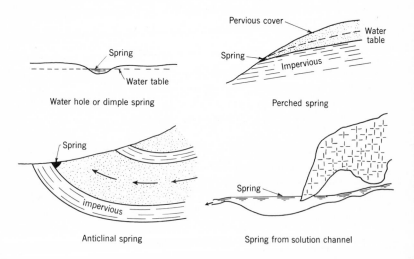

FIG. 4-5 *Typical springs.*

applicability of this principle to flow in uniform sands, and the resulting equation

$$V = KS \qquad (4\text{-}1)$$

is commonly called *Darcy's law*. The velocity V is an apparent one, that is,

$$V = \frac{Q}{A} \qquad (4\text{-}2)$$

where Q is the flow rate (volume per unit time) through a cross-sectional area A of aquifer. The term K in Darcy's law is referred to as the *saturated hydraulic conductivity* or *coefficient of permeability*. It has the same units as V (ft/day or m/day) since the slope S (ft/ft) is dimensionless. Since velocity heads are negligible in groundwater flow, S is also the slope of the water table or the piezometric surface.

The actual velocity varies from point to point throughout the medium. On the average, the actual velocity at which water is moving through an aquifer is given by

$$V_{act} = \frac{Q}{A_{act}} = \frac{Q}{pA} = \frac{V}{p} \qquad (4\text{-}3)$$

where p is the porosity of the medium expressed as a decimal. The V_{act} represents, on the average, the velocity at which a tracer would move through a permeable medium. In the case of aquifers made up of very fine particles, the p of Eq. (4-3) should be replaced by the effective porosity αp where α is the fractional part of pore space that is occupied by moving water. The term $1 - \alpha$ represents the fraction of pore space occupied by inactive water, i.e., water that does not contribute to the flow as it is held in the medium by molecular forces. For sands and gravels $\alpha \approx 1.0$.

The standard coefficient of permeability K_S is usually expressed in *Meinzer units*, the rate of flow in gallons per day through an area of 1 sq ft under a hydraulic gradient of unity (1 ft/ft). Since viscosity plays an important part in laminar flow, K_S is defined for laboatory determination at 60°F, a fairly representative temperature for groundwater. Permeability at temperatures other than 60°F varies inversely as the respective kinematic viscosities v. Hence

$$K = K_S \frac{v_{60}}{v_T} \qquad (4\text{-}4)$$

The transmissibility T, or the flow in gallons per day through a section of aquifer 1 ft wide under a hydraulic gradient of unity, is a substitute for K_S. Thus, the flow through an aquifer in gallons per day can be written as

$$Q(\text{gpd}) = K_S AS = TBS \qquad (4\text{-}5)$$

$$Q = VA = KSA$$

where B is the width of the aquifer in feet and S is the slope of the water table. Equation (4-5) is similar to the equation for the flow of electricity (Ohm's law) where S is analogous to the gradient of the voltage drop, $K_S A$ (or TB) to the conductance times the length of the circuit and Q to the amperage.

From Eq. (4-5) it is seen that the transmissibility of an aquifer may be expressed as

$$T = K_S(A/B) = K_S Y \qquad (4\text{-}6)$$

where Y is the thickness (depth) of the saturated zone within the aquifer.

The coefficients K, K_S, and T depend not only on the medium but also upon the fluid; a more rational concept of permeability would be to express it in terms that are independent of the fluid properties. The *intrinsic permeability* k of a medium can be defined as

$$k = Cd^2 \qquad (4\text{-}7)$$

where C is dimensionless and depends on the various properties of the medium such as porosity and particle shape and distribution; the term d is the mean particle diameter, and hence k is in effect a measure of the pore area.

The dimensions of k are L^2 or area. When expressed in square feet or square centimeters the numerical value of k is very small. The *darcy* has been adopted as the standard unit of intrinsic permeability. The darcy is defined through Darcy's law in metric units, and the equivalent conversions are as follows:

$$1 \text{ darcy} = 0.987 \times 10^{-8} \text{ sq cm}$$
$$1 \text{ darcy} = 1.062 \times 10^{-11} \text{ sq ft}$$

By dimensional analysis the relation between coefficient of permeability K and intrinsic permeability k can be shown to be

$$K = \frac{kg}{\nu} \qquad (4\text{-}8)$$

4-8 Permeability formulas For years many investigators[1] have tried to relate the factor of C of Eq. (4-7) to properties of the medium. Even for an ideal medium[2] of uniform-diameter spheres it is difficult to account for all possible variations of packing. The Fair and Hatch permeability for-

---

[1]G. M. Fair and L. P. Hatch, Fundamental Factors Governing the Streamline Flow of Water through Sands, *J. Am. Water Works Assoc.*, Vol. 25, pp. 1551–1565, 1933; H. E. Rose, An Investigation into the Laws of Flow of Fluids through Beds of Granular Materials, *Proc. Inst. Mech. Engrs.*, Vol. 153, pp. 141–148, 1945.

[2]J. B. Franzini, Porosity Factor for Case of Laminar Flow through Granular Media, *Trans. Am. Geophys. Union*, Vol. 32, pp. 443–446, 1951.

mula for laminar flow through permeable media is the most widely accepted relationship. It is expressed as

$$k = \frac{1}{m \left[\frac{(1-p)^2}{p^3} \left(\frac{\theta}{100} \Sigma \frac{P}{d_m} \right)^2 \right]}$$ (4-9)

where p is the porosity and m a packing factor found by experiment to have a value of 5. The term θ is a particle shape factor varying from 6.0 for spherical particles to 7.7 for angular grains. P is the percentage of particles on a weight basis held between each pair of adjacent sieves and d_m is the geometric mean $(d_1 d_2)^{1/2}$ opening of the pair.

4-9   Determination of permeability   Laboratory determinations of permeability are made with devices called *permeameters*. Many types of permeameters have been used, but all are similar in principle to that shown in Fig. 4-6. A sample of material is placed in a container, and the rate of discharge through the material under a known head is measured. A disadvantage of laboratory measurement of permeability is that the test sample is small and possibly not representative of average conditions in the aquifer. In many cases, it is difficult to obtain undisturbed samples which represent the true conditions in the aquifer. To avoid wall effect,[1] the permeameter diameter should be at least 40 times the mean particle diameter. To avoid

---

[1] J. B. Franzini, Permeameter Wall Effect, *Trans. Am. Geophys. Union*, Vol. 37, pp. 735–737, 1956.

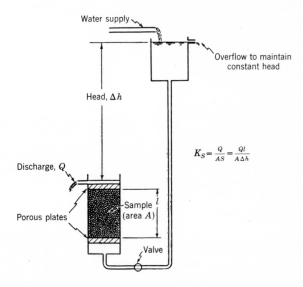

Water supply

Overflow to maintain constant head

Head, Δh

$K_S = \frac{Q}{AS} = \frac{Ql}{A \Delta h}$

Discharge, Q

Sample (area A) l

Porous plates

Valve

FIG. 4-6   *Simple constant-head upward-flow permeameter.*

difficulties from air bubbles, the water should be deaerated, and the medium should be saturated before test. With proper care good results can be obtained.

Field measurements of permeability are usually made by conducting pumping tests on wells. The well is pumped at a uniform rate, and the drawdown is measured. Permeability or transmissibility can be computed by methods outlined in Sec. 4-10. Another method of estimating the permeability of an aquifer is to introduce a tracer into a well and determine its time of arrival at a downstream well. Because of dispersion, the tracer concentration will vary with time as it passes the downstream well. Also the downstream sampling point cannot be far from the point of tracer application as the tracer concentration decreases rapidly with distance. The time of arrival at the downstream well can be assumed to be coincident with the centroid of the tracer-concentration versus time curve. Having thus determined the mean value of the actual velocity, the flow rate can then be determined from Eq. (4-3) if the porosity of the aquifer is known.

Various tracers such as common salt, dyes, and radioactive materials have been used successfully in groundwater studies,[1] particularly investigations of pollution. Some dyes and radioactive materials are unsuited for use in aquifers containing clay fractions because of base exchange and absorption phenomena. Tracer methods give only a rough evaluation of velocity and permeability. They are, however, useful in tracing the path of groundwater flow. For example, dye may be introduced into a cesspool or septic tank suspected of being a source of pollution for a well. If the dye subsequently appears in the well, the suspicion is confirmed.

4-10 Hydraulics of wells If a well penetrates an extensive homogeneous isotropic aquifer in which the water table is initially horizontal (Fig. 4-7), a circular depression in the water table must develop when the well is pumped since no flow could take place without a gradient toward the well. This depression is called a *cone of depression*, and the drop in water level Z is called the *drawdown*. Let the original thickness of aquifer penetrated by the well be Y and $Y - Z = y$. At any distance x from the well the flow toward the well in gallons per day is

$$Q = 2\pi x y K_S \frac{dy}{dx} \tag{4-10}$$

where $2\pi x y$ is the area of the cylinder through which flow occurs, K_S is the

---

[1] W. J. Kaufman and G. T. Orlob, Measuring Groundwater Movement with Radioactive and Chemical Tracers, *J. Am. Water Works Assoc.*, Vol. 48, pp. 559–572, 1956; E. Halevy and A. Nir, The Determination of Aquifer Parameters with the Aid of Radioactive Tracers, *J. Geophys. Res.*, Vol. 67, pp. 2403–2409, 1962.

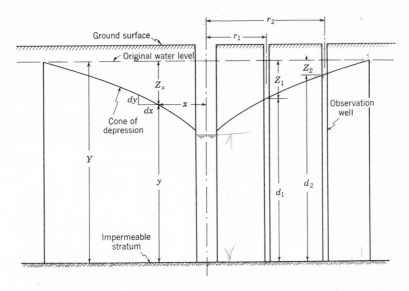

FIG. 4-7   *Definition sketch for a well-discharge equation for water-table conditions.*

permeability, and dy/dx is the slope of the water table. Note that the use of the vertical cylinder as a measure of the cross section of flow assumes that the streamlines are horizontal. If observation wells are located at distances r_1 and r_2 from the pumped well, and the values of y at these wells are d_1 and d_2, respectively, integration of Eq. (4-10) with respect to x from r_1 to r_2 and with respect to y from d_1 to d_2 gives

$$Q = \frac{\pi K_S(d_2{}^2 - d_1{}^2)}{\log_e(r_2/r_1)} = \frac{1.36 K_S(d_2{}^2 - d_1{}^2)}{\log_{10}(r_2/r_1)} \tag{4-11}$$

The foregoing analysis may be further simplified by noting that

$$d_2{}^2 - d_1{}^2 = (d_2 + d_1)(d_2 - d_1)$$

and that if Z is small compared with Y, $d_2 + d_1$ is approximately $2Y$. Moreover, since $T = K_S Y$,

$$Q = \frac{2.72 K_S Y(d_2 - d_1)}{\log_{10}(r_2/r_1)} = \frac{2.72 T(d_2 - d_1)}{\log_{10}(r_2/r_1)} \tag{4-12}$$

The analysis just outlined was originally proposed by Dupuit in 1863 and later modified by Thiem.[1] Equation (4-12) is also applicable to artesian wells, with the various terms defined as shown in Fig. 4-8.

---

[1] G. Thiem, "Hydrologische Methoden," J. M. Gebhardt, Leipzig, 1906.

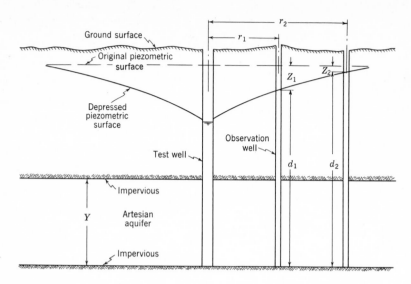

FIG. 4-8   Definition sketch for a well-discharge equation for artesian condition.

Equations (4-11) and (4-12) are severely limited in application by the many assumptions that enter their development. In order that the streamlines be approximately horizontal, the well should completely penetrate the aquifer (Fig. 4-7), and the drawdown in the well must be small compared with the thickness of the aquifer. Complete penetration of the aquifer by the well is rarely encountered in the field. However Eqs. (4-11) and (4-12) give reasonably good results for partial penetration if $r > 1.5Y$; special analysis[1] is required if $r < 1.5Y$.

In the development of Eqs. (4-11) and (4-12) it is assumed that equilibrium conditions exist although many years may be required for true equilibrium to develop. Pumping tests at Grand Island, Nebraska,[2] have shown that these equations can be used with reasonable accuracy even when the water table has an initial slope, provided that d_1 and d_2 be taken from wells lying in a straight line through the well being tested and in the direction of the initial slope of the water table or piezometric surface (Fig. 4-9) and that d_1 (and similarly d_2) be taken as the average of values at a distance r_1 both upslope and downslope of the test well. The distance r_1 should be great enough to extend beyond the immediate distortion of the streamlines

[1] D. Kirkham, Exact Theory of Flow into Partially Penetrating Well, *J. Geophys. Res.*, Vol. 64, pp. 1317–1327, 1959.

[2] L. K. Wenzel, The Thiem Method for Determining the Permeability of Water-bearing Materials, *U.S. Geol. Surv. Water Supply Paper* 679-A, pp. 1–57, 1936.

near the well, and all factors should be measured in a section of water table which has reached approximate equilibrium.

It is usually not practical to continue pumping tests in a new well until conditions even approximate equilibrium. As long as the cone of depression about a pumped well is enlarging, some water is withdrawn from storage in the unwatered portion of the aquifer. If pumping is continued at a constant rate, the drawdown in the well must increase slowly. Theis[1] first presented an analysis of well flow for the case of a homogeneous and isotropic aquifer which took into account the effect of time and the storage characteristics of the aquifer. His formula is

[1] C. V. Theis, The Relation between the Lowering of the Piezometric Surface and the Rate and Duration of Discharge of a Well Using Ground-water Storage, *Trans. Am. Geophys. Union*, Vol. 16, pp. 519–524, 1935.

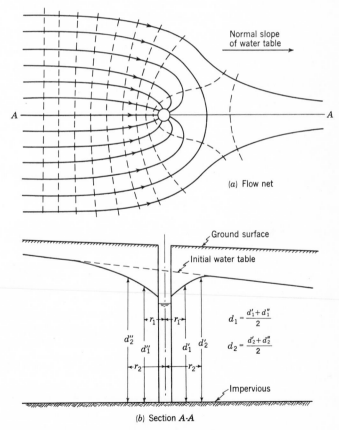

Normal slope of water table

(a) Flow net

Ground surface

Initial water table

$$d_1 = \frac{d_1' + d_1''}{2}$$

$$d_2 = \frac{d_2' + d_2''}{2}$$

Impervious

(b) Section A-A

FIG. 4-9   *Application of Eq. (4-12) to case of initially sloping water table.*

$$Z_r = \frac{Q}{4\pi T} \int_u^\infty \frac{e^{-u}}{u}\,du \tag{4-13}$$

where Z_r is the drawdown (ft) in an observation well at a distance r (ft) from the pumped well. The pumping rate Q and transmissibility T must be expressed in similar units, usually gallons per day and gallons per day per foot respectively. The term u is dimensionless and is given by

$$u = \frac{7.48 r^2 s_c}{4Tt} = \frac{1.87 r^2 s_c}{Tt} \tag{4-14}$$

In Eq. (4-14) t is the time in days since pumping began, and s_c, the *storage constant* of the aquifer, is dimensionless. The 1.87 factor has the dimensions of gallons per cubic foot, hence the equation is dimensionally correct. Physically s_c is the volume of water removed from a column of the aquifer 1 ft square when the water table or piezometric surface is lowered 1 ft. The integral in Eq. (4-13) is commonly written as $W(u)$ and is read as "the well function of u." It is not directly integrable but may be evaluated by the series

$$W(u) = -0.5772 - \log_e u + u - \frac{u^2}{2\cdot 2!} + \frac{u^3}{3\cdot 3!} \cdots \tag{4-15}$$

Values of $W(u)$ are given in Table 4-2, which is adapted from a more complete table by Wenzel.[1]

The Theis method is most conveniently solved by a graphical technique. The first step is the plotting of a "type curve" of u vs. $W(u)$ on logarithmic paper (Fig. 4-10). From Eq. (4-14)

$$\frac{r^2}{t} = \frac{T}{1.87 s_c} u \tag{4-16}$$

If the pumping rate is constant, it is evident from Eq. (4-13) that Z is equal to a constant times $W(u)$. Hence a curve of r^2/t vs. Z should be similar to a curve of u vs. $W(u)$. Using values of r, t, and Z from field observations, such a curve is plotted on logarithmic paper to the same scale as the curve of u vs. $W(u)$. The two curves are then superimposed with their coordinate axes parallel, and the coordinates of a common point are read from the curves. These coordinates are used in Eqs. (4-13) and (4-16) to solve for T and s_c.

---

[1] L. K. Wenzel, Methods for Determining the Permeability of Water-bearing Materials, *U.S. Geol. Surv. Water Supply Paper* 887, p. 88, 1942.

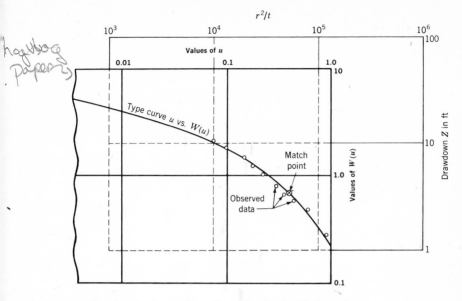

FIG. 4-10   *Graphical solution of a well problem by the Theis method.*

ILLUSTRATIVE EXAMPLE 4-1   A 12-in.-diameter well is pumped at a uniform rate of 500 gpm while observations of drawdown are made in a well 100 ft distant. Values of t and Z observed together with values of r^2/t are given below. Find T and s_c for the aquifer, and estimate the drawdown in the observation well at the end of 1 yr of pumping.

t, hr	1	2	3	4	5	6	8	10	12	18	24
Z, ft	0.6	1.4	2.4	2.9	3.3	4.0	5.2	6.2	7.5	9.1	10.5
$\dfrac{r^2}{t}, \dfrac{\text{sq ft}}{\text{day}} \times 10^{-5}$	2.4	1.2	0.8	0.6	0.5	0.4	0.3	0.24	0.2	0.13	0.1

The relations between r^2/t and Z and between u and $W(u)$ are plotted on separate sheets and superimposed (Fig. 4-10). The coordinates of the match points on the two curves are

Type curve: $u = 0.40$ $W(u) = 0.70$

Data curve: $Z = 3.4$ ft $\dfrac{r^2}{t} = 5.3 \times 10^4$ sq ft/day

Substituting these values in Eqs. (4-13) and (4-16) and noting that Q in gallons per day is $500 \times 1440 = 7.2 \times 10^5$,

$$T = \frac{QW(u)}{4\pi Z} = \frac{7.2 \times 10^5 \times 0.7}{12.56 \times 3.4} = 11,800 \text{ gpd/ft}$$

$$s_c = \frac{uT}{1.87r^2/t} = \frac{0.40 \times 11,800}{1.87 \times 5.3 \times 10^4} = 0.0478$$

At the end of 1 yr [from Eq. (4-14)],

$$u = \frac{1.87 \times 10^4 \times 0.0478}{11,800 \times 365} = 2.07 \times 10^{-4}$$

From Table 4-2, $W(u) = 7.91$, and substituting in Eq. (4-13),

$$Z = \frac{7.2 \times 10^5 \times 7.91}{12.56 \times 11,800} = 38.5 \text{ ft}$$

When u is small, the terms of Eq. (4-15) following $\log_e u$ become small and may be neglected. From Eq. (4-14) it is evident that u becomes small when t is large. For fairly large values of t the Theis formula may be modified to[1]

$$T = \frac{2.3Q}{4\pi \, \Delta Z} \log_{10} \frac{t_2}{t_1} \tag{4-17}$$

TABLE 4-2   Values of W(u) for Various Values of u (After Wenzel)

u	1.0	2.0	3.0	4.0	5.0	6.0	7.0	8.0	9.0
$\times 1$	0.219	0.049	0.013	0.0038	0.00114	0.00036	0.00012	0.000038	0.000012
$\times 10^{-1}$	1.82	1.22	0.91	0.70	0.56	0.45	0.37	0.31	0.26
$\times 10^{-2}$	4.04	3.35	2.96	2.68	2.48	2.30	2.15	2.03	1.92
$\times 10^{-3}$	6.33	5.64	5.23	4.95	4.73	4.54	4.39	4.26	4.14
$\times 10^{-4}$	8.63	7.94	7.53	7.25	7.02	6.84	6.69	6.55	6.44
$\times 10^{-5}$	10.95	10.24	9.84	9.55	9.33	9.14	8.99	8.86	8.74
$\times 10^{-6}$	13.24	12.55	12.14	11.85	11.63	11.45	11.29	11.16	11.04
$\times 10^{-7}$	15.54	14.85	14.44	14.15	13.93	13.75	13.60	13.46	13.34
$\times 10^{-8}$	17.84	17.15	16.74	16.46	16.23	16.05	15.90	15.76	15.65
$\times 10^{-9}$	20.15	19.45	19.05	18.76	18.54	18.35	18.20	18.07	17.95
$\times 10^{-10}$	22.45	21.76	21.35	21.06	20.84	20.66	20.50	20.37	20.25
$\times 10^{-11}$	24.75	24.06	23.65	23.36	23.14	22.96	22.81	22.67	22.55
$\times 10^{-12}$	27.05	26.36	25.95	25.66	25.44	25.26	25.11	24.97	24.86
$\times 10^{-13}$	29.36	28.66	28.26	27.97	27.75	27.56	27.41	27.28	27.16
$\times 10^{-14}$	31.66	30.97	30.56	30.27	30.05	29.87	29.71	29.58	29.46
$\times 10^{-15}$	33.96	33.27	32.86	32.58	32.35	32.17	32.02	31.88	31.76

---

[1] C. E. Jacob, Drawdown Test to Determine Effective Radius of Artesian Well, Trans. ASCE, Vol. 112, pp. 1047–1070, 1947.

where ΔZ is the change in drawdown between times t_1 and t_2. The simplest solution of the modified formula is found in plotting the drawdown Z on an arithmetic scale against time t on a logarithmic scale (Fig. 4-11). If ΔZ is taken as the change in drawdown over one log cycle, then $\log_{10}(t_2/t_1) = 1$, and T can be computed from Eq. (4-17). When $Z = 0$, it can be shown that

$$S_c = \frac{0.3Tt_0}{r^2} \tag{4-18}$$

where t_0 is the intercept (in days) obtained if the straight-line portion of the curve of Fig. 4-11 is extended to $Z = 0$.

ILLUSTRATIVE EXAMPLE 4-2   Using the modified Theis method, find the transmissibility and storage constant for the data of Illustrative Example 4-1.

The time-drawdown curve for these data is plotted in Fig. 4-11. Between $t = 3$ hr and $t = 30$ hr, $\Delta Z = 11.0$ ft. Hence,

$$T = \frac{2.3 \times 7.2 \times 10^5}{12.56 \times 11} = 12,000 \text{ gpd/ft}$$

and with $t_0 = 2.7$ hr $= 0.112$ day,

$$S_c = \frac{0.3 \times 12,000 \times 0.112}{10^4} = 0.0402$$

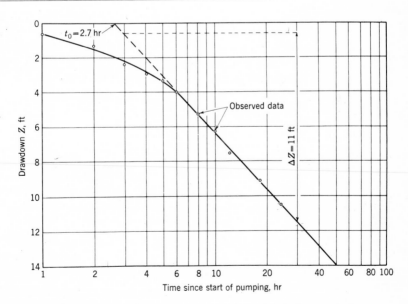

FIG. 4-11   Graphical solution of a well problem by the modified Theis method.

Since $u = $ constant times (r^2/t), for large values of t, Eqs. (4-17) and (4-18) can be written as:

$$T = \frac{2.3Q}{4\pi \, \Delta Z} \log_{10} \frac{r_1^2}{r_2^2} \tag{4-19}$$

and

$$s_c = \frac{0.3Tt}{r_0^2} \tag{4-20}$$

Thus, the transmissibility and storage constant of an aquifer can be estimated by observing the drawdowns at several observation wells at a particular instant in addition to observing the drawdown at a single well over a long period of time.

In the Theis approach parallel streamlines are assumed, i.e., small drawdown and full penetration of the well. It is also assumed that the water withdrawn from the well comes solely from storage; hence, aquifer recharge is neglected. These conditions are fairly well satisfied in artesian aquifers, but the method should be used with caution particularly when dealing with thin water-table aquifers of low permeability. A similar analysis can be applied to the recovery data from a well after pumping has stopped.[1]

Computation of unsteady flow in aquifers using the Theis approach can be facilitated by application of digital computers,[2] when the scope of the required computations is of such magnitude that considerable time and engineering cost can be saved.

4-11   Boundary effects   The cone of depression in the water table surrounding a well is rarely symmetric. Nonhomogeneity of the aquifer and the interference of one well with another disturb the symmetric drawdown assumed in the preceding sections. Where cones of depression overlap, the drawdown at a point is the sum of the drawdowns caused by the individual wells. When wells are located too close together, the flow from the wells is impaired because the increased drawdowns decrease the energy gradients toward the wells.

A stream or body of surface water in the vicinity of a well will influence the drawdown. The effect can be determined by the method of images devised by Lord Kelvin in his study of electrostatic theory.

Figure 4-12 shows an aquifer with a positive boundary in the form of an intersecting surface stream. The gradient from the stream to the well causes influent seepage from the stream. The cone of depression of the well will

[1] David K. Todd: "Groundwater Hydrology," p. 97, Wiley, New York, 1959.

[2] Joseph Foley, Computer Applications in Groundwater Hydrology, *J. Irrigation and Drainage Div., ASCE*, pp. 83–89, September, 1960.

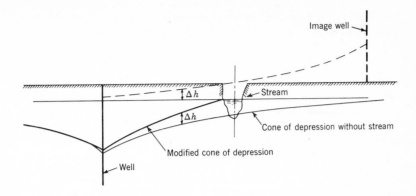

FIG. 4-12   Image well simulating the effect of seepage from a stream on water levels adjacent to a pumped well.

coincide with the water surface in the stream. A rigorous analysis would require that the channel be the full depth of the aquifer to avoid vertical flow components. However, no serious error is introduced if this condition is not satisfied, providing the stream is not too close to the well. An image well is an imaginary well introduced to create a flow pattern compatible to the physical boundary conditions of the aquifer. In Fig. 4-12 an image well is located on the opposite side of the stream and at the same distance from it as the real well. The image well is assumed to be a recharge well, i.e., one that adds water to the aquifer. Its discharge is assumed to be the same as that of the real well, hence its cone of depression is also the same but inverted. The resultant cone of depression of the real and image wells can be made to coincide with the water surface of the stream. The water table between the well and the stream is raised considerably from what it had been without the stream.

The effects of negative boundaries, i.e., faults and similar geologic discontinuities, can be analyzed by similar methods. For example, if a well is located near the bounds of an alluvial valley where solid rock forms a cutoff wall, hydraulic similarity is achieved by an image well similar to the pumping well. More complicated boundary problems require careful selection of multiple image wells. Relaxation methods and electrical and membrane analogies have been used successfully in the solution of boundary problems.

WELLS

4-12   Construction of wells   The simplest type of well is the *dug well* consisting of a pit dug to the groundwater level. A masonry lining is often used to support the excavation. Because of the difficulty of digging below the water table, dug wells do not penetrate to a sufficient depth to produce a high yield. Moreover, if the water table falls during a dry spell or period of heavy draft at nearby wells, a shallow dug well may easily go dry. This type of well is not often used for more than a single farm supply.

Bored wells are constructed by drilling in unconsolidated material with a large auger. The method is most commonly used for shallow wells in diameters up to 12 in. Such wells are useful as test holes or for temporary water supplies.

Driven wells up to 3 in. in diameter and 60 ft deep may be constructed in unconsolidated materials by use of *well points*. A well point is a section of perforated pipe with its lower end pointed for driving into soil. It may be driven with a maul or power ram. Additional sections of plain pipe are connected to the well point by threaded couplings until the required depth is reached. Because of limitations on size and depth, driven wells are not ordinarily adapted for large water-supply projects unless employed in large numbers. They are useful in prospecting for water, for home water supplies, or for temporary supplies. Batteries of well points are often used to dewater excavations. A series of well points are driven along a trench (Fig. 4-13), and the cones of depression created when these wells are pumped lower the water table below the trench bottom. The well points are connected to a manifold pipe, and a single pump is used for several wells. In some cases the well points are jetted into place by discharging a high-velocity jet from the tip of the well point.

Large-diameter shallow wells in unconsolidated material may be constructed by the *California stovepipe* method. Steel cylinders with wall thick-

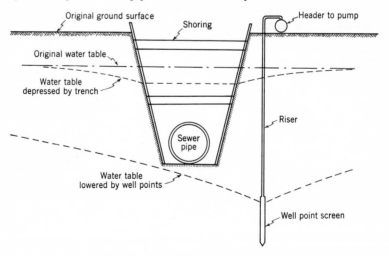

FIG. 4-13   *Simple well-point installation for dewatering a sewer trench.*

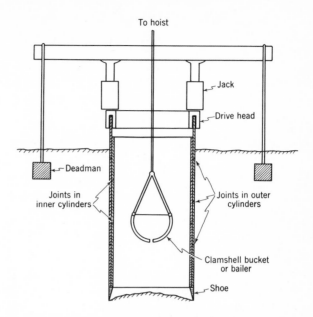

FIG. 4-14   Construction of a well by the California stovepipe method.

ness of 0.1 to 0.16 in. and 2 to 4 ft long are jacked into the soil (Fig. 4-14). Cylinders of two diameters are used such that the smaller will fit snugly within the larger. Alternate sections of large and small casing are joined with a 50 percent overlap so that the finished casing is double thickness. The two casings are often locked together by denting them with a pick. As the casing is jacked down, the earth within it is removed by use of a bailer or clamshell bucket. The stovepipe method has been used on wells 6 to 40 in. in diameter and up to 200 ft deep.

Probably the most common method of well drilling in rock is the use of the *cable-tool rig*.[1] A heavy bit suspended on a cable is raised and lowered in the well, crushing the material at the bottom. Sufficient water is introduced so that the crushed material can be removed at intervals with a bailer consisting of a hollow tube with a flap valve on the lower end. The valve permits entry of the water and crushed material from the bottom but prevents it from draining out as the bailer is raised from the well. Cable-tool wells have been drilled in diameters up to 12 in. and to depths as great as 5000 ft. In unconsolidated soils, casing is driven as the hole is sunk by attaching drive clamps to the drill stem.

[1] R. W. Gordon, "Water Well Drilling with Cable Tools," Bucyrus-Erie Co., South Milwaukee, Wis., 1958.

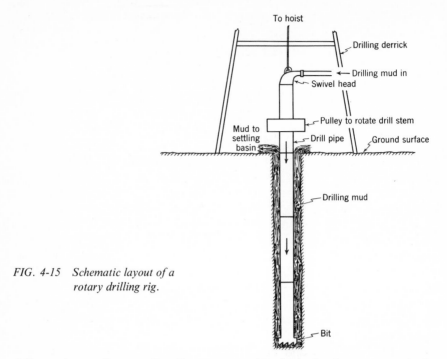

To hoist

Drilling derrick

← Drilling mud in
Swivel head

Pulley to rotate drill stem
Mud to
settling
basin
Drill pipe
Ground surface

Drilling mud

FIG. 4-15   Schematic layout of a
rotary drilling rig.

Bit

Large deep wells are ordinarily constructed by the *hydraulic-rotary* method. In this method a bit is rotated at the end of a string of pipe (Fig. 4-15). A mud slurry is continuously circulated through the drill shaft into the hole and back to the surface outside the drill pipe. The slurry serves to support the walls of the hole during the drilling and to bring the material loosened by the bit to the surface. This method has been used for wells as large as 60 in. in diameter and over 5000 ft deep. Oil wells more than 21,000 ft deep have been drilled by the rotary method.

Not all wells are vertical. The *kanats* of Persia are shafts driven horizontally or with a slight upward slope into a hillside until the water table is reached. Such wells are used because of the ease of construction and the possibility of transmitting water to a town at a lower elevation without pumping. Similar wells are used in the Hawaiian Islands to extract fresh water which is found in a relatively thin layer above salt groundwater.

Radial wells have found increasing use for large water supplies. A patented type of radial well known as a Ranney[1] collector consists of a caisson 13 ft in diameter driven into the aquifer to the required depth (Fig. 4-16).

---

[1] R. G. Kazmann, River Infiltration as a Source of Ground Water Supply, *Trans. ASCE*, Vol. 113, pp. 404–424, 1948.

Screens are driven radially from the caisson into the aquifer. The number, length, and placing of the screens are dictated by local conditions. The average length of screen is about 200 ft, and the resulting well has a screen area much larger than would be possible with the conventional vertical well. Radial wells have been installed adjacent to streams, where they serve to increase percolation from the stream to the groundwater. By using natural filtration in this way, potable water may be withdrawn from badly polluted streams. A single collector of this type has produced as much as 25 mgd.

4-13   Well finishing   The drilling of a hole does not complete the construction of an efficient well. Unless a well is drilled into rock, it must be cased to prevent collapse of the hole and a section of perforated casing (or screen) provided to permit entry of water. Casing for cable-tool wells is commonly inserted as drilling progresses. When the well has reached the proper level, the tools are withdrawn, and a section of screen slightly smaller than the casing is lowered to the bottom of the well. A lead packing ring is used to seal the top of the screen to the bottom of the casing. Devices which can rip or shoot holes in a solid casing after it has been placed are available, but the size and spacing of perforations made in this manner are difficult to control and are less satisfactory than a properly constructed screen (Fig. 4-17). Stovepipe casing cannot be withdrawn to permit placing of the screen; hence the casing must be perforated in place, or a screen bailed down within the casing.

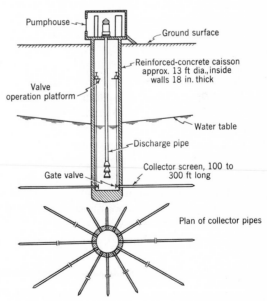

FIG. 4-16   Typical radial well of the Ranney type.

FIG. 4-17   Well screen in place showing the graded formation resulting from well development. (E. E. Johnson, Inc.)

Well screens are an important item in well design. Manufactured screens are commonly of bráss or bronze, but monel and other alloys are used in special cases to avoid corrosion. The size of the screen openings is selected on the basis of the grain-size distribution in the aquifer. The openings are often determined so as to permit 50 to 70 percent of the particles in the aquifer near the well to pass the screen. The screen area should be sufficient to keep the entrance velocity below 0.5 ft/sec to minimize head loss at entrance and to prevent excessive movement of the material in the aquifer. An increase in well diameter increases yield slightly. If r_1 of Eq. (4-11) is taken as the radius of a well and r_2 is 1000 ft, an increase in well diameter from 1 to 2 ft will increase discharge only about 10 percent.

After the screen is placed, the well is *developed* by pumping at a high rate or surging with a plunger.[1] This agitates the material around the screen and permits the finer particles to enter the well, from which they are removed by

[1] "Groundwater and Wells," Edward E. Johnson, Inc., St. Paul, Minn., 1966.

pumping or bailing. This permits freer flow to the well and increases the yield for a given drawdown. If the aquifer is of uniformly fine material, surging is of little benefit. In this case an artificial gravel envelope may be placed by drilling a larger hole than is actually required and placing an inner casing concentric with the outer casing (Fig. 4-18). Gravel is then fed down the annular space between the two casings as the outer casing is lifted so that the space surrounding the screen is filled with gravel.

The importance of careful construction and development of wells cannot be overemphasized. Inadequate screen area or improper development results in excessive head loss at entrance and an increased pumping lift. Poor alignment of the hole or a damaged casing may make it difficult to insert the usual deep-well pump and require the use of the less efficient air lift (Chap. 12). Inadequate seals, split casings, or perforations at the wrong level may permit pollution of the well and aquifer or leakage of water into other strata. Any of these defects may make it necessary to abandon the well or to undertake costly repairs.

4-14   Well sanitation   An important advantage of groundwater as a source of domestic supply is its comparative freedom from bacterial pollution. Groundwater which flows in large underground channels may transmit pollution for considerable distances, but water which percolates through

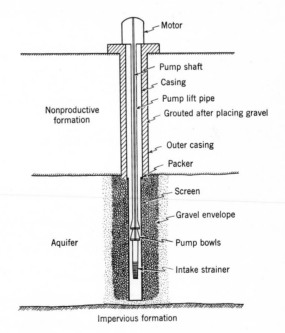

FIG. 4-18   *Gravel-packed well installation.*

fine-grained materials is usually cleared of bacterial pollution in short distances (20 to 100 ft). Polluted surface water may enter the groundwater around the top of the well casing or through the annular space between the casing and the wall of the hole. A suitable seal should be provided at the top of the casing, and the space around the casing may be grouted. Surface water should not be permitted to collect around the top of the well. Abandoned wells should be sealed with clay, concrete, or other filler to avoid contamination of the aquifer. Even though a well is built for irrigation or industrial use, sanitary precautions are advisable to avoid contamination of adjacent wells.

A final step in the construction of a well for domestic use is chlorination to eliminate any contamination introduced during construction. Chlorination is accomplished by filling the well with a solution of chlorine (50 ppm) and allowing it to stand for about 2 hr. It is then rinsed out with fresh water pumped in at the bottom of the well.

4-15 Maintenance of wells A properly constructed well requires little maintenance unless it is pumped at excessive rates. Excessive pumping may result in movement of the fine materials in the aquifer, with the possibility of clogging near the screen. Sand entering the well as a result of high entrance velocities may damage the pump. Highly mineralized waters can cause an incrustation on the well screen by the deposition of the dissolved minerals. The decrease in pressure caused by increased velocity near the screen may reduce the ability of the water to hold dissolved salts, particularly calcium carbonate, in solution. Thus incrustation may be accelerated by high entrance velocity as a result of inadequate screen area or excessive pumping. Incrustation can sometimes be relieved by surging the well with a plunger and causing alternating flow back and forth through the screen. Severe cases have been treated with hydrochloric acid, which is allowed to stand in the well for several hours. Bacterial clogging of wells is sometimes corrected by chlorination. A series of small explosive charges exploded in sequence will cause surges of gas, which may force water violently in and out of the well screen, as well as vibrations, which help to shake the incrustation loose. After treatment the well should be pumped vigorously to remove loosened deposits.

Very little can be done about corrosion of screen or casing. Leakage resulting from a corroded casing may sometimes be checked by grouting around the casing. If the well is large enough, a new casing can be inserted within the old casing. If the well is in rock, a damaged casing can sometimes be withdrawn and replaced, but if the well is in unconsolidated material, it may have to be abandoned.

4-16 Location of groundwater supplies Because of the cost of well drilling, it is desirable to have some assurance that a well will reach a satisfactory aquifer. It may be possible to predict the depth and productivity of an aquifer from conditions in other wells in the vicinity. Large projects will justify a more elaborate exploration by a competent geologist. Subsurface exploration is often done with small-diameter test holes from which samples of the soil and rock may be obtained and tested for permeability and specific yield. Pumping tests may also be conducted on these test wells to determine the transmissibility and storage constant for the aquifer.

Seismic surveys[1] are conducted by firing a charge of explosive near the ground surface and timing the travel of the resulting shock waves to a series of geophones remote from the shot point. The velocity of the shock wave depends on the type of formation and the presence of water. From the differences in the indicated velocities to the several geophones it may be possible to estimate the depth to the water table or to the interface between formations. *Resistivity surveys* make use of the fact that the depth of penetration of current between two electrodes in the soil surface increases as the electrode spacing increases. It is possible to estimate the relative resistivity of formations at different depths by measuring the current flow with various electrode spacings. Since water increases the conductivity of soil or rock, the presence of groundwater may be indicated by a decrease in resistivity. Both seismic and resistivity surveys should be made and interpreted by persons fully trained in the work. Neither method specifically locates groundwater but merely indicates discontinuities which may bound an aquifer. With a few test holes as control points, large areas may be surveyed quite rapidly by seismic or resistivity methods.

Geophysical methods are also useful in logging finished wells. Electrical logs include measurement of resistivity between a pair of electrodes lowered into the uncased well and the measurement of the self-potential (existing potential field) in the well. These data are useful in relating strata penetrated by one well with the same strata in another well. Resistivity data also give an indication of the chemical quality of the groundwater since dissolved salts reduce the resistivity of the water. Electrical logs in oil wells are often useful in studies of groundwater.

YIELD OF GROUNDWATER

4-17 Safe yield Withdrawal of water from the ground at rates greater than those at which it is replenished results in lowering the water table and an increase in pumping cost. In coastal areas an overdraft may reverse the normal seaward gradient of the water table and permit salt water

---

[1] C. A. Bays, Prospecting for Groundwater—Geophysical Methods, *J. Am. Water Works Assoc.*, Vol. 42, pp. 947–956, 1950.

to move inland and contaminate the aquifer. An aquifer undisturbed by pumping is in approximate equilibrium. Water is added by natural recharge and removed by natural discharge. In years of abundant water the water table rises, and in years of drought the water level declines, but rates of recharge and discharge tend to remain in approximate balance. When a well is put into operation, a new set of conditions is created. Some water may be removed from storage in the aquifer, or *mined* in the same sense that other minerals are mined. The depression in the water table caused by the well may induce increased recharge or may decrease the natural discharge. The concept of *safe yield* has been used for years to express the quantity of groundwater which can be withdrawn without impairing the aquifer as a water source, causing contamination, or creating economic problems from a severely increased pumping lift. Actually, safe yield cannot be defined in truly practical and general terms. The location of wells with respect to areas of recharge and discharge, the character of the aquifer, the potential sources of pollution, and many other factors are involved in estimates of the maximum feasible withdrawal from an aquifer. A number of closely spaced wells will cause much more rapid decline of local water levels than the same number of wells more widely dispersed.

Determination of safe yield is a complex problem in hydrology, geology, and economics for which each aquifer requires a unique solution. The general type cases are:

1. Aquifers in which safe yield is limited by the availability of water for recharge.
2. Aquifers in which safe yield is limited by the transmissibility of the aquifer.
3. Aquifers in which safe yield is limited by potential contamination.

The first case is commonly encountered in arid regions. The groundwater may be visualized as a large reservoir which is drawn down to supply water needs during periods of low recharge. Lowering of the water table during dry periods is not evidence that the safe yield has been exceeded, but a continuing decline during periods of abundant water warns of excessive withdrawals. The safe withdrawal from such a groundwater reservoir is equal to the annual recharge less the unavoidable natural discharge. Thus,

$$\text{Safe yield} = P - R - E_{\text{act}} - G_o \tag{4-21}$$

where P and E_{act} are the mean annual precipitation and evapotranspiration from the area tributary to the aquifer, R is the mean annual runoff from the tributary area, and G_o is the net mean annual subsurface discharge from the aquifer.

The transmissivity of aquifers may be so low that although adequate water is available for recharge, this water does not move toward the wells fast enough to permit its full utilization. Lowering the water table may increase the gradient from the recharge area and permit greater flow to the wells. The safe yield of such an aquifer is determined not by the availability of water but by the rate at which water can be delivered to the well. This problem is sometimes referred to as a *pipeline problem*, since it is analogous to a city supplied by a large reservoir but with an inadequate pipeline.

Where contamination of the groundwater is possible, the layout of the well field, the rates of use, and the types of wells must be planned in such a way that conditions permitting contamination cannot develop. Thus all three cases offer several possible values of safe yield depending upon the physical situation and the methods used to collect the groundwater. Hence, safe yield is essentially an academic concept which can be given quantitative significance only when all controlling conditions are specifically defined.

4-18   Artificial recharge   If the rate of recharge of a reservoir-type aquifer is increased, the safe yield is also increased. If an aquifer of low transmissibility can be recharged close to the point of withdrawal, the safe yield may also be increased. There are several advantages in storing water underground.[1] The cost of recharge may be considerably less than the cost of equivalent surface reservoirs. Furthermore, the aquifer serves as a distribution system and eliminates the need for surface pipelines or canals. The reduction in first cost may offset the cost of pumping. Water stored in surface reservoirs is subject to evaporation and to pollution, which may be avoided by underground storage. Even more important may be the fact that suitable sites for surface reservoirs may not be available. The groundwater can therefore be viewed as a reservoir to be operated alone or in conjunction with surface storage.

Artificial groundwater recharge may be accomplished by *induced infiltration*, *spreading*, and *recharge wells*. Induced infiltration is accomplished by increasing the water-table gradient from a source of recharge. This may be done by placing wells close to a stream or lake. Induced infiltration has been used along large rivers such as the Ohio to develop municipal and industrial supplies. Radial wells are often used for this purpose because of their large capacity. In one case[2] a two-level collector was employed to induce infiltration into an aquifer near a stream. Some of the water was then recharged into a lower aquifer for storage until periods of low streamflow (Fig. 4-19).

[1] O. E. Meinzer, General Principles of Artificial Ground-water Recharge, *Econ. Geol.*, Vol. 41, pp. 191–201, May, 1946.

[2] R. G. Kazmann, The Utilization of Induced Stream Infiltration and Natural Aquifer Storage at Canton, Ohio, *Econ. Geol.*, Vol. 44, pp. 514–524, September–October, 1949.

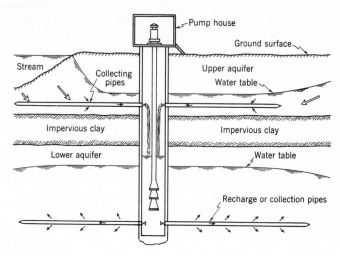

*FIG. 4-19   Two-level radial well using infiltrated river water to recharge a lower aquifer.
(After Kazmann)*

Water spreading involves diversion of surface water over permeable ground, where it may infiltrate to the groundwater. Shallow ditches or low earth dikes may be used to divert the occasional flows from small arroyos over adjacent flatlands. In areas where the main route of recharge is through the beds of the river channels, surface reservoirs may be used to store flood flows in excess of the percolation capacity of the channel. These waters may be released for percolation when the natural streamflow is low. A major problem in any percolation area is that of maintaining the percolation rate at a high level. Scarifying the area at intervals has sometimes been helpful. Vegetation is also reported to increase percolation but, of course, with an increase in transpiration losses. Water containing sediment should be avoided as it may seal the spreading area. Recharge rates are generally less than 5 ft/day though rates as high as 75 ft/day have been reported in the recharge pits at Peoria, Illinois.[1] The U.S. Agricultural Research Service[2] has been engaged in an extensive research program studying factors that affect recharge rates, and they have found that cotton-gin trash when mixed with soil will effectively increase rates. Certain chemicals which tend to cause aggregation of the soil particles show promise as additives to increase recharge. Bermuda grass provides a rugged ground cover that can withstand prolonged flooding while maintaining good recharge rates.

A proposed spreading area should be explored with test holes to assure

[1] M. Suter, The Peoria Recharge Pit: Its Development and Results, *J. Irrigation and Drainage Div.*, *ASCE*, Vol. 82, November, 1956.

[2] L. Schiff, The Status of Water Spreading for Groundwater Replenishment, *Trans. Am. Geophys. Union*, Vol. 36, pp. 1009–1020, 1955.

that subsurface conditions favor the spreading. If a stratum of low permeability is found, recharge through wells may be desirable. The physical details of recharge wells are essentially the same as for producing wells. In fact, producing wells are sometimes used for recharge during an offseason when the water is not required for some use. Water for recharge into wells should be free of suspended matter, which may clog the screen, or bacteria, which can form bacterial slimes. Water may be fed into the well by gravity or may be pumped under pressure to increase the recharge rate if subsurface conditions will permit. Recharge wells permit the water to be injected into the aquifer where it is most needed and may be particularly advantageous in dealing with pipeline-type aquifers. The recharge capacity of wells is, however, often quite low.

The temperature and chemical quality of the recharge water should be studied to determine the conditions which will result in the groundwater. Use of treated sewage for recharge has been proposed. However, sewage normally contains relatively large amounts of dissolved salts, especially sodium chloride, nitrates, and boron (from soaps). The effect of such compounds on the groundwater should be carefully considered. Since wells inject water directly into the aquifer, sewage for recharge through wells should generally be bacteriologically pure. Surface water is often warmer than the groundwater, and recharge may raise the temperature of the groundwater to the detriment of its use for cooling purposes.

4-19 Saltwater intrusion   A natural equilibrium between fresh and salt groundwater develops along coastlines. The specific gravity of seawater is about 1.025, and the fresh water floats on the seawater. Hydrostatic equilibrium would require a freshwater column about 1.025 times as high as a saltwater column, i.e., 1 ft of fresh water would exist above sea level for each 40 ft below sea level (Fig. 4-20a). Conditions of hydrostatic equilibrium do not occur, however, because of the hydraulic gradient imposed by the sloping water table. Magnification (Fig. 4-20b) of the interfaces near sea level shows that fresh water is flowing out of the freshwater aquifer through a seepage face and across a portion of the ocean bottom into the ocean. Thus the true shape of the interface is governed by hydrodynamic balance of the fresh and salt waters.

The 1/40 ratio between the water-table elevation and the depth to the fresh water–salt water interface immediately below applies quite accurately to two-dimensional flow, i.e., flow at right angles to the shoreline. In the case of wells, however, because of the three-dimensional aspect of the flow and the need for hydrodynamic balance, the 1/40 ratio does not hold. There are numerous examples where wells have been pumped with drawdown levels below sea level without inducing the entry of salt water into the well. However, in extreme cases of overpumping, saltwater entry may occur as shown in Fig. 4-20c.

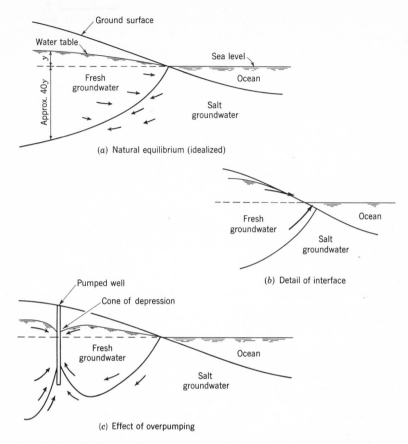

FIG. 4-20   *Fresh and salt groundwater near a coastline.*

Horizontal collectors or radial wells which operate with a small drawdown and need not extend below sea level can be used to avoid pumping salt water. This technique is widely used for the development of water supply on islands where the freshwater lens is surrounded by salt water. Overdraft of the fresh groundwater may reduce the seaward gradient and permit the salt water to advance inland. Recharge wells along the coast have been tried as a means of maintaining adequate fresh water to avoid saltwater intrusion. An outstanding example of this is the installation of the Los Angeles County Flood Control District near Manhattan Beach, California,[1] where 94 injection wells located over a nine-mile stretch on a line roughly parallel to the

[1]"West Coast Basin Barrier Project, a Los Angeles County Flood Control District Report on the Control of Seawater Intrusion," Los Angeles Flood Control District, June 1970.

coast are used to restrict saltwater intrusion. To prevent clogging from bacterial slimes and thus maintain satisfactory recharge rates, the water is chlorinated to about 1.5 mg/l. Some of the recharge water is wasted to the sea as part of the cost of saltwater control. A similar problem is encountered in inland areas where salt water may remain in sedimentary rocks formed beneath the ocean or may have formed by solution of salts from the surrounding rocks. Under these conditions pumping in adjacent freshwater aquifers must be limited to amounts which will not permit intrusion of the mineralized water.

BIBLIOGRAPHY

Butler, Stanley S.: "Engineering Hydrology," Prentice-Hall, Englewood Cliffs, N.J., 1957.

Davis, Stanley N., and Roger J. M. DeWeist: "Hydrogeology," Wiley, New York, 1966.

Ferris, John G.: Ground Water, chap. 7 in C. O. Wisler and E. F. Brater, "Hydrology," 2d ed., Wiley, New York, 1959.

"Groundwater and Wells," Edward E. Johnson, Inc., St. Paul, Minn., 1966.

Hantush, M. S.: Hydraulics of Wells, in Ven Te Chow (ed.), *Advances in Hydroscience*, Vol. I, Academic Press, New York, 1964.

Hubbert, M. K.: The Theory of Ground-water Motion, *J. Geol.*, Vol. 48, pp. 785–944, 1940.

"Isotopes in Hydrology," International Atomic Energy Agency, Vienna, 1967.

Jacob, C. E.: Flow of Ground Water, chap. 5 in Hunter Rouse (ed.), "Engineering Hydraulics," pp. 321–386, Wiley, New York, 1950.

Leggette, R. M.: Modern Methods of Developing Ground-water Supplies, *Proc. ASCE*, Vol. 81, Separate No. 627, 1955.

Linsley, R. K., M. A. Kohler, and J. L. H. Paulhus: "Hydrology for Engineers," McGraw-Hill, New York, 1958.

"Manual on Ground Water Basin Management," *ASCE Manual* 40, American Society of Civil Engineers, New York, 1961.

Meinzer, O. E.: Ground Water, chap. 10 in O. E. Meinzer (ed.), "Hydrology," Vol. IX, Physics of the Earth Series, McGraw-Hill, New York, 1942 (reprinted Dover, New York, 1949).

Thomas, Harold E.: "The Conservation of Ground Water," McGraw-Hill, New York, 1951.

Todd, David K.: "Ground Water Hydrology," Wiley, New York, 1959.

Walton, William C.: "Groundwater Resource Evaluation," McGraw-Hill, New York, 1970.

PROBLEMS

4-1 An undisturbed soil sample has an oven-dry weight of 712.63 g. After saturation with kerosene its weight is 815.41 g. The saturated sample is then immersed in kerosene and displaces 340.27 g. What is the porosity of the soil sample?

4-2 Well B is 1040 ft southeast of well A, and well C is 2700 ft west of well B. The static levels in the three wells are A, 1139 ft; B, 1120 ft; and C, 1127 ft. Find the slope of the water table and the direction of flow.

4-3 If the aquifer of Prob. 4-2 has a coefficient of permeability of 500 gpd/sq ft and a porosity of 18 percent, what is the actual velocity of flow in the aquifer, assuming all water is moving?

4-4 The map shows the water-table contours in a region where there are no wells in operation. Elevations are in feet. The horizontal scale of the map is shown.

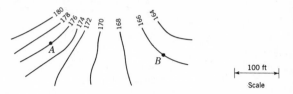

A and *B* represent two nonpumping observation wells. Dye injected into well *A* appears at well *B* in 70 days. Several tests indicate that the soil below the water table has a porosity of 28 percent. On the basis of these data determine the permeability of the aquifer. Express answer in gallons per day per square foot.

4-5 If the laboratory coefficient of permeability of a sample of soil is 350 gpd/sq ft, what would be the permeability of the same material at a temperature of 45°F?

4-6 If the permeability given in Prob. 4-3 is at 60°F, what would be the velocity of flow at 70°F?

4-7 Determine the specific permeability of Ottawa 20–30 sand if its porosity is 0.35. What is its coefficient of permeability when 80°F water flows through it? Sizes of sieve openings are: No. 20, 0.84 mm; No. 30, 0.59 mm.

4-8 Two uniform sands are to be mixed together in various proportions. The mean diameter of the sands are sand *A*, 0.08 mm; sand *B*, 0.40 mm. Plot a curve showing the relation between intrinsic permeability and percentage by weight of sand *A*. Assume 34 percent porosity for all mixtures.

4-9 A loosely packed sand with porosity 38 percent has a coefficient of permeability of 500 gpd/sq ft. What would be its coefficient of permeability if its porosity were reduced to 33.5 percent? Assume laminar flow.

4-10 In a field test it was observed that a time of 4 hr was required for a tracer to travel from one observation well to another. The wells are 80 ft apart, and the difference in their water-surface elevations is 1.50 ft. Samples of the aquifer between the wells indicate that the porosity is about 14 percent. Compute the coefficient of permeability of the aquifer assuming it is homogeneous.

4-11 An artesian well that penetrates a homogeneous aquifer of uniform depth has been pumped at a constant rate for many days and steady-state conditions have been achieved. Under these conditions dye travels from observation well *A* to observation well *B* in 40 hr. How long will it take the dye to travel from observation well *B* to observation well *C* if *A*, *B*, and *C* are all located on the same radial line? Distances from the pumped well are as follows: *A*, 60 ft; *B*, 40 ft; and *C*, 25 ft.

4-12 A well 12 in. in diameter penetrates 108 ft below the static water table. After a long period of pumping at a rate of 350 gpm, the drawdown in wells 57 and 148 ft from the pumped well is found to be 12 and 7.4 ft respectively. What is the transmissibility of the aquifer? What is the drawdown in the pumped well?

4-13 A 30-cm-diameter well penetrates vertically through an aquifer to an impermeable stratum which is located 18.0 m below the static water table. After a long period of pumping at a rate of 1.0 cu m/min, the drawdown in test holes 14 and 40 m from the pumped well is found to be 2.62 and 1.50 m respectively. What is the coefficient of permeability of the aquifer? Express in meters per day. What is its transmissibility? Express in cubic meters per day per meter. What is the drawdown in the pumped well?

4-14 A gravity-type well of diameter 12 in. produces 150 gpm with a drawdown of 10 ft.

When the drawdown is 50 ft the flow is 500 gpm. Find the discharge from the well if the drawdown is 20 ft. The aquifer is 150 ft thick. Note that when $d_2 = 150$ ft, $r_2 \approx cQ$, i.e., the radius of influence of the cone of depression is roughly proportional to the flow rate.

4-15 A 12-in.-diameter well produces 50 gpm when the drawdown is 6 ft. This well penetrates an aquifer 105 ft thick. Find the flow from this well for a drawdown of 6 ft if its diameter were: (a) 8 in., (b) 16 in. Assume that the radius of influence of the cone of depression is 2500 ft in all cases.

4-16 A 10-ft-high retaining wall holds back a homogeneous sandy soil. The wall rests on impermeable solid rock. For drainage purposes the wall is provided with 4-in.-diameter drains through the wall spaced every 6 ft along the bottom of the wall. An observation well 40 ft from the wall shows solid rock at an elevation of 86 ft, water table at an elevation of 100 ft, and ground surface at 105 ft. Another observation well 55 ft from the wall indicates solid rock at elevation 82 ft, water table at 102 ft, and ground surface at 109 ft. The two wells are on a line at right angles to the wall. Where this line intersects the wall the bottom of the wall is at an elevation of 92 ft. If the flow out of each drain is 10 gal every 4 min, compute the coefficient of permeability of the sandy soil behind the wall.

4-17 Tabulated below are data on an observation well 50 ft from an 18-in. well which is pumped for test at 250 gpm. Find the transmissibility and storage constant for the aquifer. What will be the drawdown in the observation well at the end of 6 months (180 days)? What will be the drawdown in the pumped well at the end of 6 months? Solve by the Theis method.

Time, min	2	3	5	7	9	12	15	20	40	60	90
Drawdown, ft	4.0	6.2	8.5	10.0	12.1	13.7	14.9	17.0	21.7	23.1	26.0

4-18 Tabulated below are data for an observation well 150 ft from a well pumped at a rate of 350 gpm. Find the transmissibility and storage constant by the modified Theis method.

Time, hr	0.4	1.8	2.7	5.4	9.0	18.0	54.0
Drawdown, ft	0.3	0.9	1.22	1.81	2.35	2.90	4.07

4-19 A well 300 ft deep is proposed in an aquifer having a transmissibility of 12,000 gpd per foot width of aquifer and a coefficient of storage of 0.006. The static water level is expected to be 100 ft below the ground surface, the well is to be pumped at 600 gpm, and the well diameter is to be 12 in. What will be the pumping lift from this well at the end of 1 yr? At the end of 2 yr?

4-20 After pumping a new well for 2 hr at 350 gpm, the drawdowns given below were noted in a number of nearby observation wells. What are the transmissibility and the storage constant for this aquifer? Note that these data supply a number of pairs of values of Z and r^2/t for use in the Theis method.

Well	A	B	C	D	E	F	G
Distance, ft	833	278	416	85	44	163	29
Drawdown, ft	0.7	4.3	2.5	10.0	12.7	6.4	14.9

4-21 A 12-in. well is in an aquifer with transmissibility of 15,000 gpd per foot width of aquifer and a storage constant of 0.009. What pumping rate can be adopted so that the drawdown will not exceed 35 ft within the next 2 yr?

4-22 Using the data of Illustrative Example 4-1, make the necessary calculations, and plot the drawdown curves for $t = 1$ day, 30 days, 365 days, 3650 days.

4-23 A 10-in.-diameter well is drilled and pumped at a constant rate of 40 gpm. After 6 hr of pumping the drawdown in the well is 3.00 ft and at 48 hr the drawdown is 3.52 ft. Compute the transmissibility of the aquifer. Find s_c. If the well penetrates an artesian aquifer whose average thickness is 28 ft determine the coefficient of permeability of the aquifer. At what time will the drawdown be 4.00 ft?

4-24 A man starts pumping an 18-in.-diameter well (Fig. 4-8) at a constant rate. In the first 60 days of pumping the drawdown is 10 ft. What additional drawdown might be expected if the pumping were to be continued for another 60 days at the same rate? Assume the well penetrates an artesian aquifer (Fig. 4-8) 140 ft thick. The aquifer consists of sandstone whose properties are given in Table 4-1.

4-25 Develop an expression for the steady-state relation between recharge rate and the parameters of Eq. (4-11) for the case where an unconfined aquifer is subject to a uniform recharge from rainfall.

Chapter 5

Probability concepts in design

All projects are designed for the future and the designer is uncertain as to the precise conditions to which the works will be subjected. The structural designer knows the intended loads for his structure, but he has no assurance that these loads will not be exceeded. He does not know what wind or earthquake loads may be exerted on the structure. He counters this uncertainty by making reasonable assumptions and allowing a generous factor of safety. The water-resources engineer is less certain of the flows which will affect his project. The hydrologic uncertainties are by no means the only ones in hydraulic design—future water requirements, benefits, and costs are all uncertain to some degree—but a serious error in the estimates of the expected hydrology can have devastating effects on the economy of the entire project.

Since the exact sequence of stream flow for future years cannot be predicted, something must be said about the probable variations in flow so that the design can be completed on the basis of a calculated risk. This chapter discusses the methods for estimating the probability of hydrologic events. The utilization of these probabilities in design is discussed in subsequent chapters.

5-1   The frequency series   Column 2 of Table 5-1 lists the *annual floods* (the highest flood each year) for the Susquehanna River at Harrisburg, Pennsylvania, in order of magnitude. This is a *data series* or *frequency series*. By grouping these data in *class intervals* (in this case of 25,000 cfs) the information may be presented graphically as a *frequency histogram* (Fig. 5-1). The histogram gives a good picture of the distribution of flood magnitude, but the engineer is more often interested in the number of times a flood of a given magnitude is equaled or exceeded in a stated period. This is illustrated by the integrated histogram (Fig. 5-2), or a plot of the total number of floods

TABLE 5-1   Annual Flood Data for the Susquehanna River at Harrisburg, Pennsylvania (1874–1949)

Year	Peak flow X, 1000 cfs	m	t_p, yr	$X - \bar{X}$	$(X - \bar{X})^2$	$\log X$	$(\log X - \overline{\log X})$	$(\log X - \overline{\log X})^2$	$(\log X - \overline{\log X})^3$
1936	740	1	77.	452	204,304	2.869	0.437	0.191	0.0835
1889	707	2	38.5	419	175,561	2.849	0.417	0.174	0.0725
1894	575	3	25.7	287	82,369	2.760	0.328	0.108	0.0353
1946	494	4	19.3	206	42,436	2.694	0.262	0.069	0.0180
1902	449	5	15.4	161	25,921	2.652	0.220	0.048	0.0106
1901	445	6	12.8	157	24,649	2.648	0.216	0.047	0.0101
1886	440	7	11.0	152	23,104	2.643	0.211	0.045	0.0094
1878	419	8	9.6	131	17,161	2.622	0.190	0.036	0.0069
1940	418	9	8.6	130	16,900	2.621	0.189	0.036	0.0068
1943	412	10	7.7	124	15,376	2.615	0.183	0.033	0.0061
1880	411	11	7.0	123	15,129	2.614	0.182	0.033	0.0060
1920	404	12	6.4	116	13,456	2.606	0.174	0.030	0.0053
1891	387	13	5.9	99	9,801	2.588	0.156	0.024	0.0038
1913	378	14	5.5	90	8,100	2.577	0.145	0.021	0.0030
1925	363	15	5.1	75	5,625	2.560	0.128	0.016	0.0021
1875	357	16	4.8	69	4,761	2.553	0.121	0.015	0.0018
1916	356	17	4.5	68	4,624	2.551	0.119	0.014	0.0017
1914	347	18	4.2	59	3,481	2.540	0.108	0.012	0.0013
1884	332	19	4.0	44	1,936	2.521	0.089	0.008	0.0007
1910	330	20	3.8	42	1,764	2.519	0.087	0.008	0.0007
1924	314	21	3.6	26	676	2.497	0.065	0.004	0.0003
1893	308	22	3.5	20	400	2.489	0.057	0.003	0.0002
1948	308	23	3.3	20	400	2.489	0.057	0.003	0.0002
1904	298	24	3.2	10	100	2.474	0.042	0.002	0.0001
1905	292	25	3.0	4	16	2.465	0.033	0.001	0.0000

1942	290	26	2.9	2	4	2.462	0.030	0.001	0.0000
1908	288	27	2.8	0	0	2.459	0.027	0.001	0.0000
1909	287	28	2.7	−1	1	2.458	0.026	0.001	0.0000
1898	283	29	2.6	−5	25	2.452	0.020	0.000	0.0000
1926	282	30	2.5	−6	36	2.450	0.018	0.000	0.0000
1915	278	31	2.4	−10	100	2.444	0.012	0.000	0.0000
1918	272	32	2.4	−16	256	2.435	0.003	0.000	0.0000
1921	269	33	2.3	−19	361	2.430	−0.002	0.000	0.0000
1933	269	34	2.2	−19	361	2.430	−0.002	0.000	0.0000
1919	268	35	2.2	−20	400	2.428	−0.004	0.000	0.0000
1903	266	36	2.1	−22	484	2.425	−0.007	0.000	0.0000
1896	260	37	2.1	−28	784	2.415	−0.017	0.000	0.0000
1892	256	38	2.0	−32	1,024	2.408	−0.024	0.001	0.0000
1945	252	39	1.9	−36	1,296	2.401	−0.031	0.001	0.0000
1927	247	40	1.9	−41	1,681	2.393	−0.039	0.002	−0.0001
1928	247	41	1.9	−41	1,681	2.393	−0.039	0.002	−0.0001
1879	245	42	1.8	−43	1,849	2.389	−0.043	0.002	−0.0001
1932	245	43	1.8	−43	1,849	2.389	−0.043	0.002	−0.0001
1941	244	44	1.7	−44	1,936	2.387	−0.045	0.002	−0.0001
1923	244	45	1.7	−44	1,936	2.387	−0.045	0.002	−0.0001
1935	242	46	1.7	−46	2,116	2.384	−0.048	0.002	−0.0001
1912	238	47	1.6	−50	2,500	2.377	−0.055	0.003	−0.0002
1885	238	48	1.6	−50	2,500	2.377	−0.055	0.003	−0.0002
1890	233	49	1.6	−55	3,025	2.367	−0.065	0.004	−0.0003
1929	233	50	1.52	−55	3,025	2.367	−0.065	0.004	−0.0003
1877	232	51	1.49	−56	3,136	2.366	−0.066	0.004	−0.0003
1882	232	52	1.46	−56	3,136	2.366	−0.066	0.004	−0.0003
1937	231	53	1.43	−57	3,249	2.364	−0.068	0.005	−0.0003
1895	229	54	1.41	−59	3,481	2.360	−0.072	0.005	−0.0004
1899	221	55	1.38	−67	4,489	2.344	−0.088	0.008	−0.0007

TABLE 5-1 Annual Flood Data for the Susquehanna River at Harrisburg, Pennsylvania (1874–1949) (Continued)

Year	Peak flow X, 1000 cfs	m	t_p, yr	$X - \bar{X}$	$(X - \bar{X})^2$	log X	$(\log X - \overline{\log X})$	$(\log X - \overline{\log X})^2$	$(\log X - \overline{\log X})^3$
1949	220	56	1.36	-68	4,624	2.342	-0.090	0.008	-0.0007
1888	219	57	1.33	-69	4,761	2.340	-0.092	0.008	-0.0008
1900	215	58	1.31	-73	5,329	2.332	-0.100	0.010	-0.0010
1947	214	59	1.29	-74	5,476	2.330	-0.102	0.010	-0.0011
1944	212	60	1.27	-76	5,776	2.326	-0.106	0.011	-0.0012
1907	210	61	1.25	-78	6,084	2.322	-0.110	0.012	-0.0013
1939	210	62	1.23	-78	6,084	2.322	-0.110	0.012	-0.0013
1883	206	63	1.21	-82	6,724	2.314	-0.118	0.014	-0.0016
1887	206	64	1.19	-82	6,724	2.314	-0.118	0.014	-0.0016
1876	199	65	1.17	-89	7,921	2.299	-0.133	0.018	-0.0024
1917	197	66	1.15	-91	8,281	2.294	-0.138	0.019	-0.0026
1922	187	67	1.13	-101	10,201	2.272	-0.160	0.026	-0.0041
1897	180	68	1.12	-108	11,664	2.255	-0.177	0.031	-0.0056
1938	178	69	1.10	-110	12,100	2.250	-0.182	0.033	-0.0060
1874	175	70	1.08	-113	12,769	2.243	-0.189	0.036	-0.0068
1930	166	71	1.07	-122	14,884	2.220	-0.212	0.045	-0.0095
1881	166	72	1.05	-122	14,884	2.220	-0.212	0.045	-0.0095
1906	164	73	1.04	-124	15,376	2.215	-0.217	0.047	-0.0102
1911	162	74	1.03	-126	15,876	2.210	-0.222	0.049	-0.0109
1931	145	75	1.01	-143	20,449	2.161	-0.271	0.073	-0.0199
1934	141	76	1.00	-147	21,609	2.149	-0.283	0.080	-0.0227
Sum	21,877	...	...	= -11	962,367	184.853	...	1.641	0.1619
Mean	287.8	...				2.432			

$\dfrac{21,877}{76}$

$\dfrac{184.853}{76}$

above the lower limit of any class interval. With a long period of record and small-class intervals the curve of Fig. 5-2 would be a smooth ogive.

Statistical analysis requires that all data in a series be gathered under similar conditions. The construction of reservoirs, levees, bypasses, or other

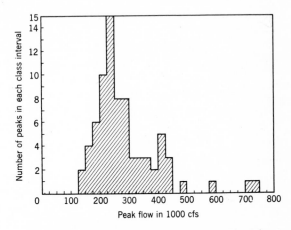

FIG. 5-1   *Frequency histogram of flood peaks on the Susquehanna River at Harrisburg,*

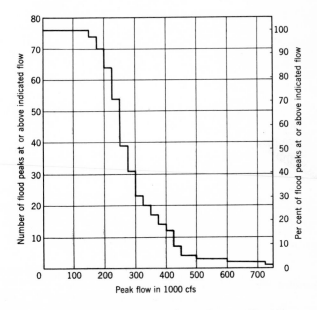

FIG. 5-2   *Integrated histogram of flood peaks for the Susquehanna River at Harrisburg, Pennsylvania.*

works which might alter flood flows on a stream results in a nonhomogeneous series. If the change caused by the works is large, an analysis may be limited to the period before or after the change, depending upon the purpose of the study. A study of natural flood conditions would be based on data collected prior to the change, while a forecast of future conditions would utilize data gathered subsequent to the construction. An alternative would be to adjust the data from one period to conform to the conditions existing during the other portion of the record. Isolated flood events distorted by unusual occurrences such as a failure of a dam or levee should be omitted from the series or adjusted to conform to the remainder of the data.

5-2   Recurrence interval   The recurrence interval is defined as the average interval in years between the occurrence of a flood of specified magnitude and an equal or larger flood. The mth largest flood in a data series has been equaled or exceeded m times in the period of record, N yr, and the best estimate of its recurrence interval t_p is

$$t_p = \frac{N + 1}{m} \tag{5-1}$$

Several other formulas have been suggested for the computation of recurrence interval or return period. The disagreement between the various formulas is limited to the larger floods, where m is small. If m equals 5 or more, the computed plotting positions by all methods are almost identical.

If an event has a true recurrence interval of t_p yr, then the probability P that it will be equaled or exceeded in any one year is

$$P = \frac{1}{t_p} \tag{5-2}$$

Since the only possibilities are that the event will or will not occur in any year, the probability that it will not occur in a given year is $1 - P$. From the principles of probability, the probability J that at least one event which equals or exceeds the t_p-yr event will occur in any series of N yr is

$$J = 1 - (1 - P)^N \tag{5-3}$$

Table 5-2, which has been computed from Eq. (5-3), shows that there are 4 chances in 10 that the 100-yr flood (or greater) will occur in any 50-yr period and even a 22 percent probability that the 200-yr flood (or greater) might occur in the 50 yr. On the other hand, there are 36 chances in 100 that the 50-yr flood will not occur in any 50-yr period. Thus, there is no assurance that the recurrence intervals computed for the larger floods of a series by Eq. (5-1) are even approximately correct. However, for $m = 5$ or more the computed values of t_p are reasonably reliable.

TABLE 5-2   *Probability That an Event of Given Recurrence Interval Will be Equaled or Exceeded during Periods of Various Lengths*

t_p, yr \ Period, yr	1	5	10	27	50	100	200	500
					Probability J			
1	1.0	1.0	1.0	1.0	1.0	1.0	1.0	1.0
2	0.5	0.97	0.999	*	*	*	*	*
5	0.2	0.67	0.89	0.996	*	*	*	*
10	0.1	0.41	0.65	0.93	0.995	*	*	*
50	0.02	0.10	0.18	0.40	0.64	0.87	0.98	*
100	0.01	0.05	0.10	0.22	0.40	0.63	0.87	0.993
200	0.005	0.02	0.05	0.12	0.22	0.39	0.63	0.92

*In these cases J can never be exactly 1, but for all practical purposes its value may be taken as unity.

Table 5-2 illustrates also that there can be no inference in the use of the phrase "N-yr flood" that this flood will be equaled or exceeded exactly once in every period of N yr. All that is meant is that in a long period of record, say 10,000 yr, there will be $10,000/N$ floods equal to or greater than the N-yr flood. All such floods might occur in consecutive years, but this is not very probable.

If the design flood for a particular project is to have a recurrence interval much shorter than the period of record, its value may be easily determined by plotting peak flows vs. t_p as computed from Eq. (5-1) and sketching a curve through the plotted points (Fig. 5-3). Because of inaccuracies in the plotted positions of the larger floods, a line sketched to conform to these floods may depart substantially from the location of the true frequency curve. The principles of statistics must be employed to construct a frequency curve which conforms to the available data.

5-3   Statistical methods for estimating the frequency of rare events With an extremely long period of record it would be possible to use a smaller class interval and Fig. 5-1 might approach a smooth frequency distribution such as Fig. 5-4. The ordinates of Fig. 5-4 are probability density and the abscissas are the magnitudes of the floods. The area under the curve is unity. The area under the curve above any magnitude X_1 is the probability that X_1 will be equaled or exceeded in any year.

Many kinds of events conform to one of several standard frequency distributions which have been studied at length and the equation of the distribution well established. The probability of such events can be determined quite easily. Only a very large number of samples (i.e., a long record length) will permit accurate definition of a distribution, and no streamflow

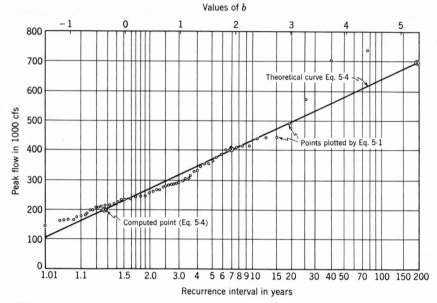

FIG. 5-3   *Frequency curve of annual floods for the Susquehanna River at Harrisburg,*
Pennsylvania (1874–1949).

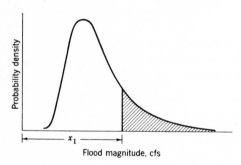

FIG. 5-4   *Idealized flood-frequency distribution.*

records are long enough to positively establish the appropriate distribution.
It is known that X must be greater than zero and that future floods will
exceed those which have been observed. One cannot even be sure that the
distribution should be the same for all streams.

Several distributions have been suggested[1] as appropriate for streamflow,

[1] H. A. Foster, Theoretical Frequency Curves, *Trans. ASCE*, Vol. 87, pp. 142–173, 1924;
Allen Hazen, "Flood Flows," Wiley, New York, 1930; L. R. Beard, Statistical Analysis in
Hydrology, *Trans. ASCE*, Vol. 103, pp. 1110–1160, 1943.

but there is no real proof of their adequacy. Fisher and Tippett[1] showed that if one selected the largest event from each of many large samples, the distribution of these extreme values was independent of the original distribution and conformed to a limiting function. Gumbel[2] suggested that this distribution of extreme values was appropriate for flood analysis since the annual flood could be assumed to be the largest of a sample of 365 possible values each year. Based on the argument that the distribution of floods is unlimited, i.e., that there is no physical limit to the maximum flood, he proposed that the probability P of the occurrence of a value equal to or greater than any value X be expressed as

$$P = 1 - e^{-e^{-b}} \qquad (5\text{-}4)$$

where e is the base of Napierian logarithms and b is given by

$$b = \frac{1}{0.7797\sigma}(X - \bar{X} + 0.45\sigma) \qquad (5\text{-}5)$$

In Eq. (5-5), X is the flood magnitude with the probability P, $\bar{X}$ is the arithmetic average of all floods in the series, and σ is the standard deviation of the series computed from

$$\sigma = \sqrt{\frac{\Sigma(X - \bar{X})^2}{N - 1}} \qquad (5\text{-}6)$$

where N is the number of items in the series (the number of years of record). The probability P is related to the recurrence interval t_p by Eq. (5-2). Values of b corresponding to various return periods are given in the Appendix.

ILLUSTRATIVE EXAMPLE 5-1   Using the data of Table 5-1, find the theoretical recurrence interval for a flood flow of 700,000 cfs using the Gumbel approach.

From the table, $\bar{X} = 287.8$, and $\sigma = (962,367/75)^{0.5} = 113.3$. When $X = 700$,

$$b = \frac{1}{0.7797 \times 113.3}[700 - 288 + 0.45(113.3)] = 5.24$$

The recurrence interval for $X = 700$ is, from Eqs. (5-2) and (5-4),

$$t_p = \frac{1}{1 - e^{-e^{-5.27}}} = 192 \text{ yr}$$

By the same method $t_p = 1.28$ yr when $X = 200$ and 6.89 yr when $X = 400$. These points are shown on Fig. 5-3 by the large circles.

---

[1] R. A. Fisher and L. H. C. Tippett, Limiting Forms of the Frequency Distribution of the Largest or Smallest Member of a Sample, *Proc. Cambridge Phil. Soc.*, Vol. 24, pp. 180–190, 1928.
[2] E. J. Gumbel, Floods Estimated by the Probability Method, *Eng. News-Record*, Vol. 134, pp. 833–837, 1945.

The plotting paper used for Fig. 5-3 is constructed by laying out on a linear scale of b the corresponding values of $t_p = 1/P$ from Eq. (5-4). Thus the computed line will be a straight line, and it is sufficient to calculate the return period corresponding to two flood flows and to draw the straight line defined by these points. A third point is a convenient check.

In 1967, the U.S. Water Resources Council[1] adopted the log Pearson Type III distribution (of which the log normal is a special case) as a standard for use by federal agencies. The purpose was to achieve standardization of procedures. The recommended procedure[2] is to convert the series to logarithms, compute the mean, standard deviation, and skew coefficient g which is

$$g = \frac{N\Sigma(\log X - \overline{\log X})^3}{(N-1)(N-2)(\sigma_{\log X})^3} \tag{5-7}$$

The values of X for various return periods are computed from

$$\log X = \overline{\log X} + K\sigma_{\log X} \tag{5-8}$$

where K is selected from tables (Appendix) for the computed value of g and the desired return period. Log-normal probability paper should be used for graphical display of the curves. A straight line will result only if $g = 0$.

ILLUSTRATIVE EXAMPLE 5-2   Using the data of Table 5-1, find the magnitude of the 10-yr and 100-yr floods using the log Pearson Type III distribution.

From the table $\overline{\log X} = 2.432$, $\sigma_{\log X} = (1.641/75)^{0.5} = 0.148$ and $g = (76 \times 0.1619)/[75 \times 74 \times (0.148)^3] = 0.682$. Obtaining K values from the table in the Appendix we get

$$10\text{-yr flood: } \log X = 2.432 + (1.332)(0.148) = 2.628$$
$$X_{10} = 429 \text{ cfs}$$
$$100\text{-yr flood: } \log X = 2.432 + (2.810)(0.148) = 2.848$$
$$X_{100} = 705 \text{ cfs}$$

Comparing these results with the Gumbel plot of Fig. 5-3 we note the following:

	Gumbel	log Pearson Type III
10-yr flood	435 cfs	429 cfs
100-yr flood	640 cfs	705 cfs

---

[1] A Uniform Technique for Determining Flood Flow Frequencies, *U.S. Water Resources Council Hydrology Comm. Bull.* 15, December, 1967.

[2] Subcommittee on Hydrology, Methods of Flow Frequency Analysis, Interagency Comm. on Water Resources, *Bull.* 13, U.S. Government Printing Office, Washington, D.C., April 1966.

Figure 5-3 is typical of frequency plots in that the computed line conforms well to most of the data but diverges from the three or four largest values. The discussion of recurrence interval in Sec. 5-2 points out that these higher points may be incorrectly plotted. Hence, to force a distribution to conform to these points may only perpetuate the error. Until much longer records are available, there is no absolute proof of the adequacy with which a theoretical distribution fits the actual distribution of floods. There are logical grounds for arguing that no single theoretical distribution can be expected to fit all streams.

5-4   Partial duration series   The annual flood series provides the best data for an analysis to determine the design flood for a spillway or a culvert. The annual series is sometimes criticized on the basis that the second highest flood in some years will exceed annual floods which are included in the series. The *partial duration series*, which consists of all floods above some arbitrary base value of flow, is sometimes suggested as a substitute. The two series give very nearly the same recurrence intervals for the larger floods, but the partial duration series will show higher flows for the shorter recurrence intervals. The partial duration series should not be extrapolated by methods such as outlined in Sec. 5-3. Where the lesser floods are of special interest and particularly where a recurrence interval less than 1 yr is desired, the partial series should be used by simply plotting flood peaks vs. t_p and sketching a curve by eye.

5-5   Flood frequency at points without streamflow records   Few projects are built at the exact spot where a streamflow record has been obtained. Many projects are built on streams for which no record at all exists. Three alternative methods have been used to estimate flood frequency in the absence of streamflow data. The first method is the use of flood formulas (Sec. 5-6). A second possibility is to utilize an analysis of rainfall frequency and compute the flood flows from the rainfall by procedures outlined in Chap. 3. The third alternative is to estimate the probable flood frequency at the ungaged point from data at nearby gaging stations.

Frequency curves for two gaging stations can be identical only when the two basins are quite similar. The basins should have geometric similarity in terms of area, shape, slope, and topography; hydrologic similarity in terms of rainfall, snowfall, soils, and valley storage; and geologic similarity with regard to those items which affect groundwater flow. The upper group of frequency curves in Fig. 5-5 are for six small basins in northern Minnesota,[1] while the lower group represents six basins of the same general size in the Puyallup River basin,[2] which drains from Mt. Rainier in western Washing-

[1] C. H. Prior, Magnitude and Frequency of Floods in Minnesota, *Minn. Dep. Conserv. Bull.* 1, November, 1949.

[2] G. L. Bodhaine and W. H. Robinson, Floods in Western Washington, *U.S. Geol. Surv. Circ.* 191, 1952.

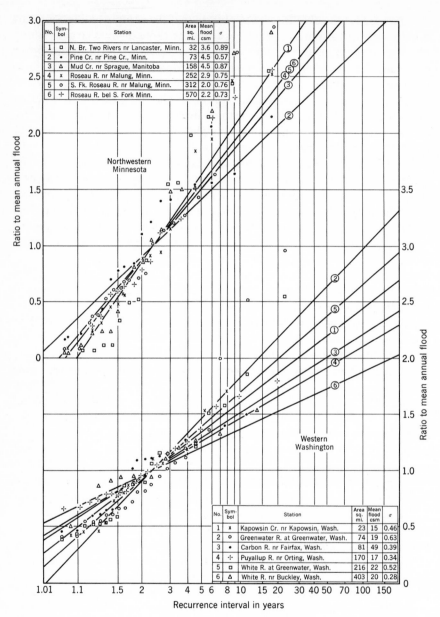

FIG. 5-5   Comparison of flood-frequency curves for adjacent basins.

ton. Although the stations in each group are close together, a considerable divergence of the curves is evident. Note that flood magnitudes are expressed in terms of the ratio to the mean annual flood. This ratio eliminates some of the differences caused by differences in basin size and by rainfall variations between basins. The characteristics of Eq. (5-4) are such that the mean annual flood has a recurrence interval of 2.33 yr.

A substantial error might result from the use of one of these frequency curves on any other basin. The agreement between the Minnesota basins is somewhat better than for those in Washington. This is indicative of the influence of topography on streamflow. The Minnesota basins are in a region of slight relief, and the differences may be caused largely by differences in valley storage characteristics between basins. The Washington basins are in a mountainous area, and their flood flows are materially influenced by orographic rainfall variations and by the varying role of snow in the hydrology of the basin. Some differences in both groups may be due to chance variations, which can be important when the length of record is short (in this case, 18 yr).

Estimates of flood frequency by comparison with other streams should be based on one or two streams selected because they most nearly resemble the project stream in all important features. Frequency curves for these streams can then be prepared, with flood magnitudes expressed as a ratio to the mean annual flood. The final step is an estimate of the mean annual flood on the project basin by comparison with other streamflow records in the vicinity. Some idea of the variation in mean annual flow is given by the tables of Fig. 5-5. The mean annual flood is stated in cubic feet per second per square mile to reduce variations resulting from differing drainage areas. Even a short streamflow record is helpful in comparing the flow of one basin with another, and a gaging station should be provided near the project site at an early date in the design.

Computer simulation offers a new and powerful tool for study of ungaged basins. By simulation an entire frequency series may be generated and a frequency curve constructed. The physical characteristics and the rainfall record of the watershed can thus be utilized in a rational way rather than by comparison with some other basin.

5-6   Flood formulas   If the highest floods of record at a group of gaging stations within a limited area are plotted on a graph such as Fig. 5-6, where peak flow Q_p in cubic feet per second per square mile of drainage area is plotted against drainage area A_d, a few of the higher floods seem to define an upper-limit line, or *enveloping curve* (solid line). On logarithmic paper the slope of the line is the exponent n, and the intercept where area is unity is the coefficient c in an equation of the form

$$Q_p = cA_d^n \tag{5-9}$$

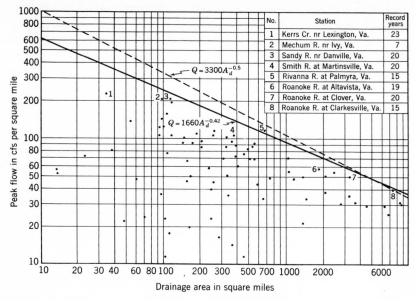

FIG. 5-6   *Record floods on coastal streams of Virginia prior to* 1949.

The exponent n is often taken as -0.5, which indicates that flood peaks vary inversely with the square root of drainage area (dashed line, Fig. 5-6). It can be seen from the figure that the slope of the line and the intercept are rather poorly defined by the data.

Most flood formulas assume that flood-peak magnitude is a function of drainage area, and many formulas include such factors as basin shape and mean annual precipitation in an attempt to minimize the variations in the coefficient c. Early engineers were forced to use equations like Eq. (5-9) because they had no other data to work with. Such a simple formula cannot possibly describe the complex phenomena involved in a flood. Moreover, such formulas offer no clue to the probability of the flow computed from them. With relatively abundant data and far better understanding of hydrology, the modern engineer has no excuse for utilizing empirical formulas. Since critical decisions concerning a project are often made on the basis of preliminary studies, approximations are not even justified in such studies.

5-7   Rainfall frequency   Frequency analysis of rainfall is complicated by the fact that one may be concerned with rainfall of various durations and over various areas. Most analyses are for rainfall of various durations at a single station. Each duration provides data which are analyzed for frequency and combined into a family of curves (Fig. 5-7). Recording rain gages provide the only satisfactory data for frequency analysis for rainfalls of short duration.

Figure 5-7 provides information on the frequency of rainfall on the small area of a single rain gage. Assuming that the gage is at the point of highest intensity in a storm, the average rainfall over an area surrounding the gage decreases as the area increases, but the nature of this decrease is not well established. Few gage networks are sufficiently dense to permit a study of the intensity variations within very small areas. Figure 5-8 shows the variation of average rainfall with area.[1] Variations in rainfall resulting from orographic influences in an area are not accounted for by Fig. 5-8.

The best method for estimating the frequency of average rainfall over large areas is to compute the average rainfall on the area during important storms and make a frequency analysis of these average rainfall values. It is incorrect to average the individual N-yr values at a number of rainfall stations to determine the N-yr average value over an area. Unless the several

---

[1] Rainfall Intensity-Frequency Regime, Part I: Ohio Valley, Part II; Southeastern United States, *U.S. Weather Bur. Tech. Paper* 29, 1957, 1958.

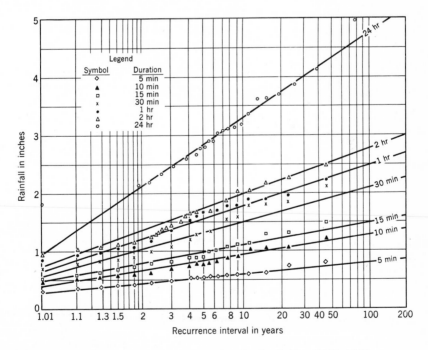

FIG. 5-7   *Rainfall-duration-frequency curves for Cleveland, Ohio* (1902–1947).

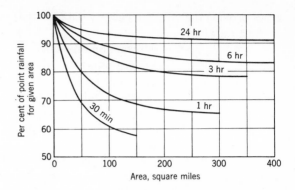

FIG. 5-8   *Reduction of point rainfall for application to basins of specific size. (National Weather Service)*

events can occur simultaneously, the average value has a return period greater than that of the individual values (see Sec. 5-9).

5-8   Rainfall maps and formulas   Figure 5-9 is one of a series of rainfall-intensity maps[1] based on analysis of all available recording gage records. The maps cover a range of durations from 30 min to 24 hr and frequencies from 1 to 100 yr. The maps are reliable in areas of negligible relief, but may be inaccurate in mountainous areas. Actually only the 1- and 24-hr maps are constructed from observed data, the others being interpolated by relationships presented in the report.

The relation between rainfall intensity i and duration t_R has often been expressed by formulas such as

$$i = \frac{c}{t_R + b} \tag{5-10}$$

and

$$i = \frac{k}{t_R{}^n} \tag{5-11}$$

where the constants c, b, k, and n are regional characteristics. Bernard[2] found that in the eastern United States n averaged about 0.75 and k about 0.30.

It is always preferable to utilize actual rainfall data in a frequency analysis instead of using maps or formulas. Sometimes, however, it is convenient

---

[1] Rainfall Frequency Atlas for the United States, *U.S. Weather Bur. Tech. Paper* 40, May, 1961.

[2] M. M. Bernard, Formulas for Rainfall Intensities of Long Duration, *Trans. ASCE*, Vol. 96, pp. 592–629, 1932.

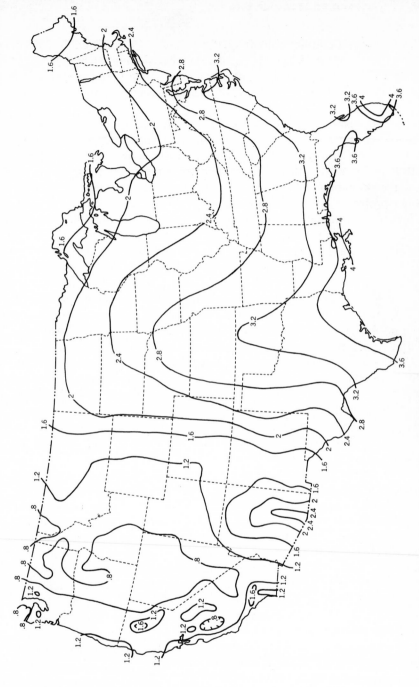

FIG. 5-9 One-hour rainfall to be expected once on the average in 25 yr. (National Weather Service)

to express the rainfall-intensity frequency relations for a station in terms of a formula such as Eq. (5-11). The constants in the equation can be determined by plotting intensity versus duration on logarithmic paper. If these data plot as a straight line, k is the intensity where t_R is unity, and n is the slope of the line.

5-9   Joint frequency   If two events are entirely independent (unrelated in cause), and their probabilities of occurrence are P_1 and P_2, respectively, the probability that they will occur at the same time is $P_1 P_2$. Since both P_1 and P_2 are less than 1, the probability of their joint occurrence is less than the probability of either event independently. Numerous hydraulic design problems involve the simultaneous occurrence of different events. It has been common practice to transform a design rainfall of known frequency to streamflow by the methods of Chap. 3 and to assume that the resulting flow has the same frequency as the rainfall. Since the transformation involves some assumption as to antecedent conditions it is a problem in joint probability. Storm rainfall appears to be substantially independent of antecedent conditions. Thus, unless the selected runoff coefficient or infiltration index has a probability of 1.0 of being the appropriate value, the resulting design flood has a return period different than that of the design rainfall—usually greater. This procedure should never be used for design.

The spillway design flood for a dam depends on the assumed flood flows into the reservoir and the available storage space in the reservoir at the time these flows occur. The worst possible condition is that the reservoir be full when the assumed inflow occurs. However, if the probability of the reservoir being full at the time the design flood occurs is less than one, the computed spillway design flood has a greater recurrence interval than the assumed reservoir inflow.

Because most hydrologic events are not strictly independent, it is usually necessary to solve problems of joint frequency by direct analysis rather than by use of the simple product rule.

Figure 5-10 illustrates a joint frequency analysis applied to the problem of the simultaneous flooding of two streams above their junction. Curves A and B of the figure are the separate frequency curves of the two streams. Curve C is a frequency curve for the sum of the two flows computed on the assumption of complete dependence, i.e., events of equal probability are assumed to occur simultaneously and the total flow below the junction is the sum of the flows on the two streams having the same return periods. Curve D assumes complete independence so that the probability of any two flows occurring simultaneously is the product of the probability that they will occur independently. Curve E is obtained by adding the flows which actually occurred at the same time and performing a conventional frequency analysis on the sums. It is evident that the flows on the two streams are partially but not wholly dependent.

Instead of attempting to combine probabilities, it is often more satisfactory to work with synthetic record. Using rainfall-runoff relations and unit hydrographs, one may synthesize[1] the flood peaks for years without stream-flow record and subject the computed peaks to frequency analysis. Computer simulation can serve the same purpose.

5-10 Probable maximum floods Since about 1940, the spillways of many major dams have been designed to discharge the *probable maximum flood*. The magnitude of this flood is determined by meteorological estimate[2] of the physical limit of rainfall over the drainage basin. This rainfall is used to compute the probable maximum flood flow by the methods outlined in Chap. 3. Figure 5-11 shows the estimated probable maximum rainfall[3] for a duration of 6 hr and an area of 10 sq mi. Values for other areas and durations can be determined from the reference. Although these values may seem

[1] J. L. H. Paulhus and J. F. Miller, Flood Frequencies Derived from Rainfall Data, *J. Hydraulics Div., ASCE*, December, 1957.

[2] M. M. Bernard, The Primary Role of Meteorology in Flood Flow Estimating, *Trans. ASCE*, Vol. 109, pp. 311–382, 1944.

[3] Generalized Estimates of Maximum Possible Precipitation in the United States, *U.S. Weather Bur. Hydrometeorol. Rept.* 23, June, 1947.

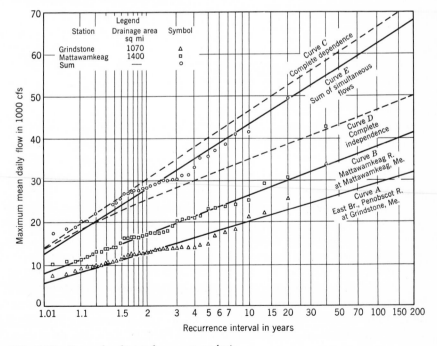

FIG. 5-10 Example of joint-frequency analysis.

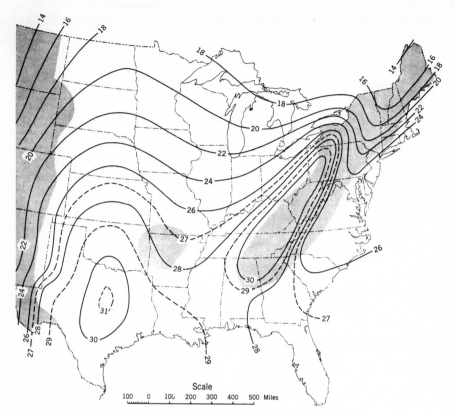

FIG. 5-11 *Probable maximum precipitation in inches for areas of 10 sq mi and duration of 6 hr.* (*National Weather Service; U.S. Corps of Engineers*)

large, several storms have exceeded 80 percent of the estimated probable maximum values in their region of occurrence. The consequences of project failure must be very serious to justify design against the probable maximum flood. Such a condition exists where the failure of a dam would result in heavy loss of life downstream.

5-11 Drought Drought is often defined in terms of a fixed period of time with less than some minimum amount of rainfall. Even when applied to a specified area and crop, such a definition is far from adequate since the critical time and rainfall depend on the stage of crop development, initial moisture content of the soil, temperature and wind during the drought period, and other factors. If drought is defined in terms of inadequate rainfall for crop production, most of the western United States has a drought every year, since rainless summers are common in much of the West. Since this is a normal occurrence, provision has been made to store or divert water from streams for crops and other needs. Under such conditions drought is defined in terms of inadequate streamflow. If irrigation is accomplished by direct diversion of water without storage, a single winter of low precipitation may cause a water shortage. If storage reservoirs are designed to carry water over from one year to the next, a single year of low runoff may not be critical.

In general terms a drought is a lack of water for some purpose. More specific definitions are possible only when local conditions are specified. The following sections deal with methods for expressing the probability of low streamflow. Methods of reservoir design are covered in Chap. 7.

5-12   Duration Curves   The natural streamflow characteristics of a river are frequently summarized in a *flow-duration curve*. Such a curve (Fig. 5-12) shows the percent of time that flow is equal to or less than various rates during the period of study. The same data may also be plotted to show the percent of time various flows are equaled or exceeded (Fig. 5-13). A duration curve is constructed by counting the number of days, months, or years with flow in various class intervals. As the length of the time unit increases, the range of the curve decreases. The selection of the time unit depends on the purpose of the curve. If a project for diversion without storage is under study, the time unit should be the day so that absolute minimum flows will

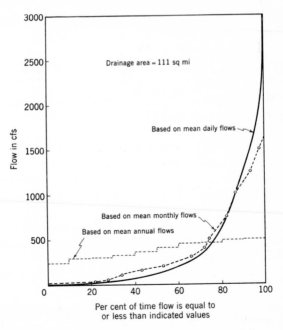

FIG. 5-12   *Flow-duration curves for Cherry Creek near Hetch Hetchy, California*
(1941–1950).

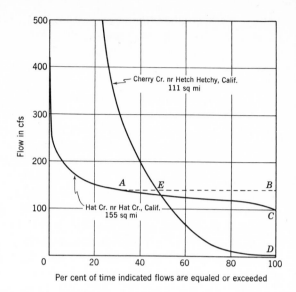

FIG. 5-13   Comparison of flow duration curves for two streams.

be indicated. For reservoir design, the month or year may be sufficient, depending upon reservoir size in relation to inflow.

The main defect of the flow-duration curve as a design tool is that it does not present flow in natural sequence. It is not possible to tell whether the lowest flows occurred in consecutive periods or were scattered throughout the record. Duration curves are most useful for preliminary studies and for comparisons between streams. Figure 5-13 compares the Cherry Creek duration curve with one for a stream with much more stable flow characteristics. Cherry Creek offers no chance of successful development without provision for storage to provide water during periods of low natural flow. Hat Creek could, however, provide at least 100 cfs for direct diversion. Storage would be required on both streams to meet a demand of 140 cfs, but the volume required for Hat Creek (*ABC*) is much less than for Cherry Creek (*EBD*). Cherry Creek produces considerably more runoff than Hat Creek and with proper storage facilities could provide a much higher yield. The exact storage requirements are dependent on the actual sequence of flow (Chap. 7) and cannot be accurately estimated from duration curves.

5-13   Drought frequency   If drought can be defined in specific terms for a particular project, drought frequency can be analyzed in the same manner as flood frequency. It is also possible to prepare generalized frequency curves of low flow (Fig. 5-14). These curves were constructed by determining the minimum flow each year during periods of various lengths, and the data for each period length were plotted as a frequency curve.[1] Because of the wide range of flow values, a logarithmic scale was chosen to permit more accurate plotting of the low flows. Curves of this type can serve many purposes in connection with design. For example, a small water-supply project requiring 0.3 cfs (195,000 gpd) might pump directly from the stream

---

[1] Data for this figure from W. P. Cross and E. E. Webber, Ohio Stream-flow Characteristics, *Ohio Dep. Nat. Resources Bull.* 13, Part 2, table 1, December, 1950. See also J. B. Stall and J. C. Neill, Partial Duration Series for Low-flow Analysis, *J. Geophys. Res.*, Vol. 66, pp. 4219–4225, December, 1961, for description of procedure for assembling data for figures such as Fig. 5-14.

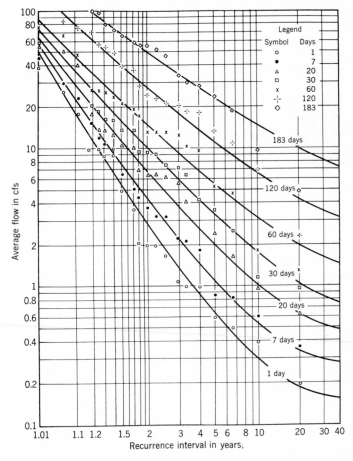

FIG. 5-14   *Frequency of minimum flows for Yellow Creek near Hammondsville, Ohio (1915–1935).*

of Fig. 5-14. Once in 10 yr, flow would be inadequate to meet the demand, but if storage were provided for about one-half of the daily requirement (100,000 gal), a shortage might occur only every 30 yr.

The degree of treatment required by wastewater is often dependent on the amount of water available for dilution of the treated wastewater when it is discharged into a stream (Chap. 19). The critical flow is not necessarily the absolute minimum, since an economic analysis may show a substantial saving in plant costs if moderate nuisance (without danger to health) is tolerable at infrequent intervals. Figure 5-14 provides the data necessary for the analysis of such a problem.

Drought frequency in terms of rainfall might be presented in a diagram similar to Fig. 5-14. Such a diagram might show the frequency of various minimum amounts of precipitation in periods of various lengths. Since precipitation is not continuous as is streamflow, zero totals will be common for the shorter periods. Such a chart might be useful in connection with studies for supplemental irrigation but for this purpose should be constructed only for the growing season in the locality under study.

5-14 Synthetic streamflow It is often important to know something of the probability of floods or droughts more severe than anything observed on a stream. The methods discussed in Sec. 5-3 provide some estimate of the probability of extreme floods. Because of the difficulty of defining a drought (Sec. 5-11) and because of the few cases of long-period drought in a short record, the procedures of Sec. 5-13 are really not adequate for defining recurrence intervals equal to or greater than the period of record.

On the assumption that streamflow is essentially a random variable, it is possible to develop a synthetic flow record by statistical methods. Such a record can be of any desired length and may well include flow sequences far more critical than any in the observed record. Thomas and Fiering[1] suggest a possible approach for the Clearwater River, Idaho, where a serial correlation exists between successive months of flow because of the seasonal snowmelt. In this case a random series may be generated by the equation

$$q_{i+1} = \bar{q}_{j+1} + b_j(q_i - \bar{q}_j) + t_i\sigma_{j+1}(1 - r_j^2)^{1/2} \tag{5-12}$$

where q_i and q_{i+1} are the flows in the ith and $(i + 1)$st months from the start of the synthetic sequence, $\bar{q}_j$ and $\bar{q}_{j+1}$ are the mean monthly flows in the jth and $(j + 1)$st month of the annual cycle, and b_j is the regression coefficient for estimating flows in the $(j + i)$st month from the flows in the jth month. The first two terms on the right-hand side of Eq. (5-12) represent the simple

[1] H. A. Thomas, Jr., and Myron Fiering, Mathematical Synthesis of Streamflow Sequences for the Analysis of River Basins by Simulation, chap. 12 in Arthur Maass and others, "Design of Water-resource Systems," Harvard, Cambridge, Mass., 1962.

serial correlation between months as determined from the historic data. In the final term, t_i is a random normal deviate with a zero mean and unit variance, σ_{j+1} is the standard deviation of the flows for the $(j + 1)$st month, and r_j is the correlation coefficient between flows in the jth and $(j + 1)$st months. By introducing the random deviate t, the final term of the equation imparts a random variation to the flows, but this variation is constrained by the known characteristics of the record, σ and r.

Stochastic analysis should be used to generate a number of synthetic flow traces of length equal to the expected useful life of the project under study. Each trace can then be analyzed to determine the required storage for the project demand (Sec. 7-4). If demand is expected to grow over time the analysis of each trace should assume a variable demand. Approximately one thousand traces should be generated and analyzed. The required storages for all of the traces will form a probability distribution conforming to the extreme-value theory. A design storage can be selected from this distribution with a specified probability that it will prove adequate during the project life.[1]

The procedure does not allow for time trends or cycles if any exist. Present evidence suggests that no significant trends can be expected within reasonable design periods and that the apparent cyclic variations in weather may be entirely random. The procedure also assumes that the statistical characteristics of the streamflow regime are adequately defined by the observed record—an assumption which should be reasonable if the observed record is reasonably long (approx. 50 yr) but which may not be valid if the record is quite short.

Stochastic process might be employed to generate a synthetic record of rainfall, which could be transformed to streamflow by an appropriate method. Such a procedure could be employed if a streamflow record were too short to provide a basis for stochastic generation or if more detail were desired in the streamflow than could be obtained from monthly flows. Pattison[2] demonstrated the feasibility of using a Markov process to generate a sequence of hourly rainfall data, and Franz[3] employed multivariate normal analysis to generate compatible hourly data at several rainfall stations.

[1] Stephen J. Burges, Use of Stochastic Hydrology to Determine Storage Requirements for Single-Purpose Storage Reservoirs—A Critical Analysis, *Engineering-Economic Planning Rept.* 34, Department of Civil Engineering, Stanford University, September, 1970.

[2] Allan Pattison, Synthesis of Rainfall Data, *Tech. Rept.* 40, *Department of Civil Engineering*, Stanford University, July, 1964.

[3] Delbert Franz, Hourly Rainfall Synthesis for a Network of Stations, *Tech. Rept.* 126, *Department of Civil Engineering*, Stanford University, March, 1970.

BIBLIOGRAPHY

Chow, Ven Te (ed.): "Handbook of Applied Hydrology," chap. 8, McGraw-Hill, New York, 1964.

Dixon, W. J., and F. J. Massey, Jr.: "Introduction to Statistical Analysis," 2d ed., McGraw-Hill, New York, 1957.

Drake, Alvin W.: "Fundamentals of Applied Probability Theory," McGraw-Hill, New York, 1967.

Fiering, Myron: "Streamflow Synthesis," Harvard, Cambridge, Mass., 1967.

Hufschmidt, M. M., and M. Fiering: "Simulation Techniques for the Design of Water Resources Systems," Harvard, Cambridge, Mass., 1966.

Jarvis, C. S., and others: Floods in the United States; Magnitude and Frequency, *U.S. Geol. Surv. Water Supply Paper* 771, 1936.

Linsley, R. K., M. A. Kohler, and J. L. H. Paulhus: "Applied Hydrology," chap. 20, McGraw-Hill, New York, 1949.

Linsley, R. K., M. A. Kohler, and J. L. H. Paulhus: "Hydrology for Engineers," chap. 11, McGraw-Hill, New York, 1958.

PROBLEMS

5-1 A rainfall station has been in operation for 3 yr. During this time there has been only one rainy day in November. On the basis of these data what can one say about the probability of having a rainy day at this station in November?

5-2 Analysis of an annual flood series covering the period 1920 to 1969 at Station A on a certain river shows that the 100-yr flood has a magnitude of 400,000 cfs, and the 10-yr flood a magnitude of 270,000 cfs. Assuming that the flood peaks are distributed according to the theory of extreme values, answer the following questions:

(a) What is the probability of having a flood as great as or greater than 350,000 cfs next year?
(b) What is the magnitude of the flood having a recurrence interval of 20 yr?
(c) What is the probability of having at least one 100-yr flood in the next 8 yr?
(d) Find $\bar{X}$, the mean of the annual floods.
(e) Find σ, the standard deviation of the annual floods.

5-3 Determine $\bar{X}$, σ, and b for the data of Table 5-1 using the years 1874 to 1911. Compute the magnitude of the 10-yr and 100-yr floods. Repeat using data from Table 5-1 for the years 1912 to 1949.

5-4 Determine $\overline{\log X}$, $\sigma_{\log X}$, and g for the data of Table 5-1 using the years 1874 to 1911. Compute the magnitude of the 10-yr and 100-yr floods. Repeat using data from Table 5-1 for the years 1912 to 1949.

5-5 Listed on page 145 are flood flows on the South Fork of the Bear River.

(a) Arrange the "annual floods" in order of magnitude as in Table 5-1. Tabulate the values of m, t_p, $(X - \bar{X})$, and $(X - \bar{X})^2$.
(b) Plot a frequency histogram using class intervals of 4000 cfs.
(c) Plot an integrated histogram of annual flood peaks.
(d) Plot a frequency curve similar to Fig. 5-3 using values of t_p as determined by Eq. (5-1). Draw your best estimate of the curve by eye.
(e) Compute and plot the position of the theoretical frequency curve using Gumbel's method. What is the magnitude of the 100-yr flood peak?
(f) Plot a frequency curve for the partial duration series. What differences are there between the line defined by the partial series and that for the annual flood series?

Date	Flow cfs	Date	Flow cfs
June 3, 1944	16,200	June 13, 1955	13,500
May 1, 1945	6,500	May 29, 1956	13,200
Jan. 14, 1946	19,800	May 20, 1957	17,000
June 4,	20,300	June 8, 1958	12,800
June 24,	14,600	May 25, 1959	15,500
Jan. 1, 1947	14,700	May 31,	15,900
Jan. 31, 1948	20,500	June 5,	19,900
June 13,	15,800	May 16, 1960	13,600
June 18,	16,400	May 9, 1961	3,900
May 30, 1949	12,400	May 29, 1962	11,300
May 23, 1950	7,200	May 25, 1963	11,600
Jan. 25, 1951	59,700	Feb. 18, 1964	19,200
May 30,	14,900	May 17,	14,000
June 1, 1952	18,300	May 15, 1965	6,800
June 9,	15,500	June 16, 1966	14,700
Jan. 17, 1953	45,400	May 28, 1967	8,400
May 5,	15,500	Dec. 28, 1968	20,100
June 9,	16,300	May 18, 1969	12,800
June 17,	15,900	June 15, 1970	14,400
Feb. 22, 1954	17,900		

5-6 Using the annual peaks of Prob. 5-5 compute the magnitude of the 10-yr and 100-yr floods assuming a log Pearson Type III distribution.

5-7 Obtain data on annual flood peaks from *U.S. Geological Survey Water Supply Papers* 1671–1689 (or other source) for a stream-gaging station selected by your instructor and plot a frequency curve similar to Fig. 5-3. Draw your best estimate of the curve by eye. What effect would floods which have occurred subsequent to the publication of your source have on the position of the curve you have sketched?

5-8 Compute the position of the theoretical frequency curve for the data of Prob. 5-7, using Gumbel's method. What differences are there between the theoretical curve and the one drawn by eye? What is the percent error in flow at a recurrence interval of 5 yr? At 100 yr?

5-9 Obtain from the latest *U.S. Geological Survey Water Supply Paper* the maximum flows of record for a group of basins designated by your instructor. Using these data, plot a chart similar to Fig. 5-6, and draw in the enveloping curve. Find the equation for the enveloping curve you have drawn. What is the approximate return period for the data you have used? Do stations with the longer records appear to control the position of the enveloping curve? Why? What differences are there in the curves as drawn by the various members of your class?

5-10 What is the probability of a flood equal to or greater than the 20-yr flood during the next 3 yr?

5-11 What is the probability that a flood equal to or greater than the 10-yr flood will not occur in any 10-yr period?

5-12 (a) What is the probability of having a flood equal to or greater than the 5-yr flood next year?

(b) What is the probability of having at least one flood equal to or greater than the 5-yr flood during the next 3 yr?

5-13   The analysis of annual flood peaks on a stream with 60 yr of record indicates that the mean value of the annual peaks is 80,000 cfs. The standard deviation of the peaks is 12,000 cfs.

(*a*) Find the probability of having a flood next year of magnitude equal to or greater than 112,000 cfs.

(*b*) What is the probability that at least one flood of such magnitude will occur during the next 25 yr?

(*c*) What is the probability that two floods of such magnitude will occur during the next 25 yr?

(*d*) What is the magnitude of a flood with a recurrence interval of 20 yr?

Assume distribution of extreme values is applicable.

5-14   The mean of the annual flood peaks on a given river is 32,100 cfs and the standard deviation of these flood peaks is 6000 cfs. What is the probability of a flood equalling or exceeding 40,000 cfs within the next 5 yr on this river?

5-15   A frequency analysis of Dry Creek using the Gumbel assumptions gives the following results:

Recurrence interval	Peak flow
2 yrs	30,000 cfs
40 yrs	70,000 cfs

(*a*) What peak flow will have a return period of 200 yr?

(*b*) What is the probability that a flow of 50,000 cfs or greater will not occur in the next 25 yr?

(*c*) A dam will be built on this creek. Flood damage will result only if the reservoir of the proposed dam is full and a flood peak of 60,000 cfs or more occurs. The reservoir will be full 20 percent of the time. What is the return period of flood damage if reservoir storage and flood peak probabilities are independent?

(*d*) During construction of the dam a diversion tunnel is to be used. Find a design flow for this tunnel such that there is a 20 percent probability of surcharge above the top of the tunnel entrance during a 2-yr construction period.

5-16   The analysis of annual flood peaks on a stream with 60 yr of record indicates that the mean value of the annual peaks is 190,000 cfs. The standard deviation of the peaks is 32,000 cfs. Assuming the annual peak floods on this stream conform to the theoretical distribution of extreme values,

(*a*) Find the probability of having a flood next year of magnitude equal to or greater than 300,000 cfs.

(*b*) What is the magnitude of a flood with a recurrence interval of 5 yr?

(*c*) What is the probability of having a flood equal to or greater than 300,000 cfs during the next 100 yr?

5-17   The annual precipitations at a given point during the 85-yr period of record are distributed as follows:

Less than 6.00 in.	6
Between 6.00 in. and 9.99 in.	14
Between 10.00 in. and 13.99 in.	20
Between 14.00 in. and 17.99 in.	35
Over 18.00 in.	10

(*a*) On the basis of these data what is the most likely probability of having a precipitation in excess of 18 in. in any one year?

(b) What is the most likely probability of having three successive years in which the precipitation exceeds 18 in./yr?

(c) What is the most likely probability of having less than 8 in. of precipitation next year? Give your best estimate on the basis of the given data.

5-18 In a certain tropical area where the precipitation consists of short storms more or less uniformly distributed throughout the year, there is a 60 percent chance of rain on any one day. What is the probability that there will be no rain on any two successive days? What is the probability that it will rain on both days, that it will rain on only one of the two days?

5-19 Obtain data on excessive rainfalls at a station selected by your instructor from the *U.S. Weather Bureau Bulletin* W (1930) or other source. Plot intensity-frequency curves for durations of 5, 15, 30, 60, and 120 min. Compare your results with Fig. 5-9.

5-20 Fit an equation of the form of Eq. (5-10) or (5-11) to the data of Prob. 5-19. If you use Eq. (5-11), how do your constants agree with those given by Bernard?

5-21 Make an analysis of the frequency of the average of the daily rainfalls at two stations about 20 mi apart. Draw the individual frequency curves for the two stations, and compute the frequency of average values on the assumptions of complete dependence and complete independence. Does the degree of dependence between the two stations appear to be greater at low or high rainfalls? Why?

5-22 Make an analysis of the frequency of low flows for periods of 1, 5, 10, and 30 days at some streamflow station designated by your instructor. Use a period of record of about 10 yr for the analysis.

5-23 Construct a duration curve, using daily streamflow values, for a station designated by your instructor. Use a period of about 3 yr for the curve. Superimpose on this curve one constructed on the basis of monthly average flows for the same period. Why is the range of variation of the latter curve less than that for daily data? If the members of your class used different periods for their duration curves, what differences resulted?

5-24 Refer to Fig. 5-14. How often on the average might one expect the flow over the driest 20-day period to be equal to or less than 80 acre-ft?

5-25 Refer to Fig. 5-7. How often on the average might one expect a storm of intensity 2.3 in./hr or more with a duration of 30 min to occur in Cleveland, Ohio?

Chapter **6**

Water law

In regions where water supply is inadequate to meet the needs of potential users, water is a commodity of considerable value. A system of laws has been developed to determine who has the right to water when shortages occur. Water rights play an important role in determining the availability of water in some parts of the country. Water law is a complex subject, and it is not the purpose of this chapter to prepare the reader to argue a water-law case in court but rather to provide a basic understanding of some of the legal problems which may be encountered in water-resources engineering. Water law can also play a major role in the economic aspects of water development since limitations on who may develop water often control how it is developed and utilized.

6-1 Riparian rights The doctrine of riparian rights was taken from French civil law by two American jurists, Story and Kent. During the early nineteenth century, English courts adopted the doctrine as part of their common law. Subsequently those American jurisdictions which adopted the English common law also accepted the riparian doctrine. Some form of riparianism was brought to Mexico by the Spaniards, and Texas courts hold that riparian doctrine has existed in Texas since Mexican days.

Under the concept of riparian rights the owner of land adjacent to a stream (*riparian land*) is entitled to receive the full natural flow of the stream without change in quality or quantity. The riparian owner is protected against the diversion of waters upstream from his property or from the diversion of excess floodwaters toward his property. In other words, no upstream owner may materially lessen or increase the natural flow of a stream to the disadvantage of a downstream owner.

The riparian doctrine has a serious defect in modern society—it does not provide for actual use of water by the riparian owners for irrigation or other purposes. Consequently, the riparian concept has been modified to permit reasonable use of water. Reasonable use allows riparian owners to divert and use streamflow in *reasonable amounts for beneficial purposes*. In regions of ample flow this permits riparian owners to use all the water they need, but if the flow is inadequate for all owners, the available water must be divided on some equitable basis. However, an upstream proprietor may always use as much water as he needs for domestic use and for watering domestic stock. Such use is considered an *ordinary* or *natural* use. Irrigation or watering of commercial herds of stock is an artificial use and not entitled to preference. Reasonableness of use is usually determined by such factors as area, character of the land, importance of the use, and possible injury to other riparian owners. No priority of right can exist between riparian owners, i.e., all riparian owners have equal rights to their reasonable share of water, and no owner can exercise his rights to the detriment of other owners. When riparian rights are transferred, the new owner must adhere to the conditions governing the original owner.

A riparian right inheres in the land and is not affected by use or lack of use. It can be voided by due process of law, as by exercise of eminent domain by a governmental unit. Riparian rights can be lost by upstream adverse use which ripens into a prescriptive right at the end of the period specified under the statute of limitations. If riparian property is sold, the right is automatically transferred to the new owner. If a parcel of riparian land is divided, any section not adjacent to the stream loses its riparian status unless the right is specifically preserved in the conveyance. Commonly, riparian status lost by division of land is not recoverable even though a new owner combines the land into a single parcel. Riparian rights do not attach to land outside of the stream basin, even though this land is contiguous to riparian land in the basin. Thus riparian owners cannot transport water from one basin to another. Riparian rights attach to all natural watercourses and all water in these channels from natural sources. Natural lakes have the same status as streams. Riparian rights do not inhere in artificial channels such as canals or drainage ditches unless by long existence and use these channels have developed characteristics of natural watercourses.

6-2   Appropriative rights   The idea of appropriation of water without regard to riparian rights was brought to the New World by the Spaniards, who adopted it from Roman civil law. The system, although not highly developed, influenced water laws of Arizona, New Mexico, and Colorado. The custom of appropriating water by diverting it and putting it to use was practiced by the Mormons from the time of their earliest settlement in Utah. The doctrine of *prior appropriation for beneficial use* was most profoundly influenced by developments in the mining camps of the Sierra Nevada in California. During the gold rush of 1849, mining claims were taken by the process of posting a notice of intention to utilize the land. Since water was essential to hydraulicking and placer mining, the miners used a similar procedure to lay claim to water.

Water was appropriated in the mining areas by posting a notice of intent at the point of diversion, filing a copy with the local recorder, and proceeding to construct the facilities and put the water to use. In most Western states a much more elaborate procedure is now prescribed by law. The outstanding feature of the doctrine of appropriation is the concept of "first in time, first in right." The right of the earliest appropriator is superior to any other claim, and further appropriation is possible only if water in excess of earlier claims is available. During water shortages the available supply is *not* apportioned among all users. Instead, those claimants with the earliest priority are entitled to their full share, and those with later priorities may have to do without.

Under an exclusive system of appropriative rights, all water in natural watercourses is subject to appropriation. An appropriator may store water in reservoirs for use during periods of shortage, but the amount stored is limited by the terms of the storage appropriation. In many states, appropriations for direct use and for storage are kept separate, although both appropriations may be granted at the same time by the same instrument. Wastewater, seepage, and releases from flood-mitigation or hydroelectric storage reservoirs discharged with no intent to recapture may be appropriated, but the appropriator cannot insist on the continuance of such flow by persons with prior rights. Reservoir releases intended for downstream use belong to the person storing the water or those to whom it is consigned and may not be appropriated.

6-3   Comparison of riparian and appropriation doctrines   The appropriation doctrine provides for acquiring rights to use of water by diverting it and putting it to beneficial use in accord with procedures set forth in state statutes or acknowledged by the courts. Appropriated water may be used on lands away from the stream as well as on lands adjoining the stream. The earliest appropriator in point of time has the exclusive right to use of water to the extent of his appropriation without diminution of quantity or deterioration of quality whenever the water is naturally available. Each subsequent appropriator has like priority over all appropriations later in time than his own. Appropriations are for a definite quantity of water and are valid only as long as the right is exercised. Appropriations may be made only for beneficial and reasonable uses.

The riparian doctrine gives to the owner of land adjacent to a stream the right to use water from that stream. This use is generally limited to riparian land but may be for any beneficial purpose. Use of water for irrigation must be reasonable in relation to the reasonable requirements of the owners of other lands riparian to the same watercourse. No riparian user acquires priority over other riparian users by virtue of the time he began his use. The riparian right is proportionate, not exclusive. Riparian rights are not expressed in specific quantities of water unless they have been apportioned by a court decree. Such a decree is based on conditions existing at the time of the hearing and is subject to change by the court if the conditions change.

6-4 Development of water law in the Western states The doctrines of appropriation and riparian right are quite different, and considerable confusion is to be expected where both doctrines are jointly recognized. The land of the West was largely the property of the United States government until it was transferred to other ownership by various acts. The Homestead Act of 1862 and the Pacific Railway Act of 1864 made public lands available for private patent under prescribed terms. Since the federal government was riparian owner of these lands prior to the issuance of the patents, the new owners generally claimed riparian rights on any streams adjacent to their property. Since the water of these streams was, in many cases, already appropriated, a conflict immediately arose. In the act of 1866, Congress recognized the existence of appropriative rights to water on public lands and provided that in claims which are recognized "by the local customs, laws, and decisions of courts, the possessors and owners of such vested rights shall be maintained and protected in same." The act of 1870 provided that all patents granted or preemptions or homesteads allowed should be subject to rights acquired under or recognized by the act of 1866. The Desert Land Act of 1877 provided specifically for appropriation of water for irrigation of desert lands in the Western states except Nebraska, Kansas, Oklahoma, and Texas. Although the act applied specifically to desert lands, the United States Supreme Court has held that it was the intent of Congress that the act should apply to all public lands in the states to which the legislation applied. This interpretation established the policy that all nonnavigable waters on the public lands were separate from the land and subject to appropriation under state laws.[1]

Although, by legislation of 1866, 1870, and 1877, Congress placed the administration of water rights under the states, it did not surrender the right of the federal government to make reservation of water for specific purposes. However, a prior appropriation under state law is good as against a subsequent federal reservation. The acts of 1866 and 1870 recognized water

---

[1] The Pelton Dam decision upholding the right of the Federal Power Commission to license a power plant which had been denied a license by the Oregon Fish and Game Commission; the Fallon case in which the U.S. Navy was not required to obtain a permit for a well under Nevada law; and the Fallbrook case in which the federal government attempted to void all rights on the Santa Marguerita River in California and to preempt the water for the use of a military installation cast some shadow on the strength of state water rights. Legislative and court action to establish the validity of state water rights has been pressed during the 1960's, but the uncertainties of federal-state relations in respect to water still remain.

rights vested under state law without limitation as to the navigability of the stream. Appropriation of water from navigable streams is subject, however, to the dominant easement of the public for navigation. Thus water may be diverted from a navigable stream so long as the diversion does not impair its navigability. Diversion of water within the public lands does not require permission from the federal government, but a right-of-way for a ditch or reservoir must be obtained from the agency controlling the particular land. If the diversion is for power, a license from the Federal Power Commission is necessary.

6-5 Water codes of the Western states Riparian law has been formulated largely by the courts, but the appropriation doctrine rests on statute law in most states. These states have water codes which govern the acquisition and control of water rights. In many states the riparian doctrine has been completely replaced by appropriation, while in other states the two concepts exist together. In those states where the riparian concept is still recognized, the water code may contain some reference to riparian rights, chiefly by way of limitation. The complications resulting from the joint adherence to two basically different ideas prevent any detailed discussion of the various state water codes in this text. In many states water law is still in the process of development through court decision and legislative action.[1]

In those states which still recognize the riparian doctrine the tendency of both law and court decision is toward restriction of riparian right to reasonable use. The riparian owner is granted a prior right to water to the extent of his reasonable requirements, but all water in excess of this amount is subject to appropriation. In many instances appropriators have been held to have acquired prescriptive rights superior to those of the riparian owners.

The typical complete water code consists of three parts: appropriation of water, adjudication of water rights, and administration of water rights and distribution of water. The various water codes differ greatly in detail and extent of coverage, but many of them contain similar provisions on fundamental matters. Some of the items with respect to appropriation of water are as follows:

1. *Method of appropriation.* The intending appropriator is usually required to file an application for a permit with a state agency such as the state engineer. The filing of this application is advertised, and if any interested parties object, a hearing is held before a permit is issued.

---

[1] A series of books presenting detailed discussion of each state water code has been prepared by Wells A. Hutchins. Books may be obtained from respective state agencies responsible for water rights.

2. *Conditions of fulfillment and forfeiture.*   A time limit is set within which construction of works must be commenced and the water put to use (often 6 months or a year). This limit is subject to extension for good cause. Nonuse of water for a specified period (usually 3 to 5 yr) constitutes forfeiture of the right.

3. *Preference for use.*   In most states domestic and municipal use have the first preference for water. The usual order of preference is then for irrigation, power and mining, wildlife, and recreation, although there is some variation in the codes. These preferences are exercised in several ways. If conflicting applications for appropriation are filed at the same time, approval is given to the one having the highest preference. In time of water shortage the highest preferences are entitled to water, but prior rights cannot be taken by the holders of junior rights without just compensation. Water rights may also be condemned in favor of a higher preference use. Some states permit municipalities to reserve water for future use without the time limits set for most appropriators.

State water codes usually give the state engineer or water-rights board the authority to grant or deny applications for water rights. If the water is available, and the application fulfills the statutory requirements, a permit must be issued. If there is a question as to the availability of water, the decision of the state engineer or water-rights board can be appealed in court. The burden of proof is usually on the claimant to demonstrate that sufficient water is available to satisfy his appropriation without detriment to the prior rights of the stream.

In addition to setting forth the methods of appropriation of water and adjudication of water rights, water codes often specify the procedures to be followed for the administration of water rights. Three methods are in common use: (1) distribution under the direction of commissioners appointed by a court as a result of litigation, (2) distribution by water masters appointed by the state and reporting to the state agency which administers the water code, and (3) distribution by water masters appointed by voluntary agreement of the interested water users. It is the function of the water master or commissioners to secure the measurements of flow necessary to the computation of the quantity of water to which each user is entitled; to provide proper facilities for the measurement of the quantities actually delivered to each user; and to see that the head gates of the various users are set and locked at all times at the proper opening to provide each user with his legal share of the available water.

6-6   Water law in the Eastern states   The Eastern states have not yet been faced with the problems of water rights which are present in the more arid West. Hence, few of the Eastern states have statutes which can be considered to be state water codes. In general, water law in the Eastern states conforms to the riparian doctrine as interpreted by the courts when specific issues are brought to litigation.[1]

6-7 Groundwater law   Under the common law, rights to groundwater are inherent in the overlying property, and the owner of this property is free to remove and use the water as he wishes. Like riparian law, this concept is satisfactory in areas with more than ample water, but if groundwater supply is inadequate to meet all needs, difficulties may be expected. Under the common law, early court decisions held that diversion of water from under a neighbor's property or lowering of the water table by excessive pumping was not a proper cause for court action.

For some time the trend of court decisions respecting groundwater in arid areas has been toward a doctrine of *reasonable use*. Under this doctrine the overlying landowner retains his rights to water under his property, but he is not permitted to use more than he really needs or to export the water to points distant from the source. The California doctrine of *correlative rights* goes even further in stating not only that the use of water must be reasonable but that the priorities of all landowners are equal and if the supply is not sufficient for all demands, each owner is entitled to no more than an equitable portion of the available water.

Water law with respect to groundwater is notably less advanced than for surface water. This results from a general lack of understanding of the mechanics of groundwater movement and a lack of specific information on the physical features of groundwater basins, as well as the comparatively moderate use of groundwater during the nineteenth century.

Legally, groundwater is commonly divided into *underground streams* and *percolating waters*. In almost all states underground streams have been accorded the same legal status as surface streams. The exact nature of an underground stream has never been thoroughly defined, but the burden of proof rests on the claimant who asserts the existence of such a stream. Percolating waters have been described as "vagrant, wandering drops moved by gravity in any and every direction along the line of least resistance." Such waters are further supposed "not to contribute to the flow of any definite stream or body of surface or subterranean water." In eight[2] of the Western states percolating waters are subject to appropriation in the same manner as for surface water or underground streams. The remainder of the states follow the common-law rule, generally modified to require reasonable use.

The interrelation between surface water and groundwater creates another

---

[1] H. T. Critchlow, Irrigation Water Rights in the Humid Areas, *Trans. ASCE*, Vol. 118, pp. 509–516, 1953.

[2] Idaho, Nevada, New Mexico, Oklahoma, Oregon, Utah, Washington, and Wyoming.

legal problem. Groundwater may be tributary to a stream, or it may be derived from streamflow. In the first case the use of groundwater reduces streamflow, while in the second case extensive diversion of surface water may reduce the available groundwater supplies. In many states this condition is not recognized, and the two sources of water are treated quite independently. Several states hold that groundwater which contributes to streamflow is a part of the stream and subject to the rules governing surface water. Underflow[1] of a stream is also considered to be part of the surface stream in some states. In other states all rights to the use of water from interrelated sources are adjudicated jointly. It may be presumed that future statutory action will be in the direction of correlating all sources of water.

6-8 Federal government water rights The federal government derives its authority over streams from several sources. Federal authority over navigable waters is based on the Commerce clause of the United States Constitution,[2] which states that Congress has the power "To regulate commerce with foreign Nations, and among the several States, and with the Indian Tribes." This places under federal control all navigable streams. One definition of a navigable stream was prescribed by Congress in the Federal Power Act of 1920, which states that navigable streams are those which "either in their natural or improved condition notwithstanding interruptions between the navigable parts of such streams or waters by falls, shallows, or rapids compelling land carriage, are used or suitable for use for the transportation of persons or property in interstate or foreign commerce." There are other definitions for other purposes.

The rapid expansion of federal water-control activities in the 1930's led to numerous court interpretations defining navigability in more detail. Continuous use of a stream for navigation is not essential; and past use, as well as present use, establishes navigability. Navigation is established by a vessel of any type including rafting or floating of timber. Navigation need not be commercially important, private pleasure boats serving to demonstrate navigability. A river which is susceptible of improvements which would make it navigable is considered navigable. Numerous court decisions have also held that federal authority over navigable streams extends to their tributaries as well, since diversion of flow on the tributaries might destroy the navigability of the main stream.

Federal interest in flood mitigation as well as navigation has been based on the Commerce clause since floods are detrimental to navigation. Private utilities must be licensed by the Federal Power Commission before they may construct dams on navigable streams. Numerous decisions have emphasized

---

[1] Underflow is that water flowing in permeable materials immediately below the stream bed.
[2] U.S. Const. Art. 1, §8, ¶3.

the right of the federal government to develop hydroelectric power at projects constructed for navigation or flood mitigation. In the Ashwander case (1936) the Supreme Court ruled that Wilson Dam on the Tennessee River was legally constructed under the Commerce clause and that "The power of falling water was an inevitable incident of the construction of the dam. That water power came into the exclusive control of the Federal government . . . and the water power, the right to convert it into electric energy, and electric energy thus produced, constitute property belonging to the United States."

The federal government does not, under the Commerce clause, have to compensate abutting owners for any damage resulting from work it may do below the ordinary high-water mark of a navigable stream, since the government holds an easement for navigation on the bed of navigable streams. Such owner may not even claim compensation for loss of streamflow from a navigable stream by virtue of federal activity. Federal activity which results in flooding of private land adjacent to a stream does require fair compensation to the landowners.

The Commerce clause is not the sole basis of federal authority over streams. The Property clause of the Constitution[1] authorizes Congress to "dispose of and make all needful Rules and regulations respecting the Territory or other property belonging to the United States." The Reclamation Act of 1902, which is the basis of the Federal Reclamation program, was intended to develop public lands under the authorization of the Property clause. The Constitution[2] also gives Congress power to levy taxes and to appropriate funds for the "common Defense." Wilson Dam on the Tennessee River was built during World War I to produce nitrates for ammunition. The Constitution also delegates to the President with the approval of the Senate the authority to make treaties and specifies[3] that treaties "shall be the supreme Law of the Land; and the Judges in every State shall be bound thereby, any Thing in the Constitution or laws of any state to the Contrary notwithstanding." Treaties with Canada and Mexico concerning such international streams as the Rio Grande, Colorado River, and Columbia River have been made under this authority. Certain water rights have also been recognized by the government as a result of treaties with Indian tribes.

Finally the spending power of the government to "provide for the general welfare" constitutes authority under which it may control and develop the nation's rivers. In discussing this authority, the Supreme Court has stated,[4]

---

[1] U.S. Const. Art. 4, §3, ¶2.
[2] U.S. Const. Art. 1, §8, ¶1.
[3] U.S. Const. Art. 6, ¶2.
[4] *United States v. Gerlach Live Stock Co.*, 339 U.S. 738.

"Thus the power of Congress to promote the general welfare through large-scale projects for reclamation, irrigation, and other internal improvements, is now as clear and ample as its power to accomplish the same results indirectly through resort to strained interpretation of the power over navigation."

6-9   Interstate problems   With many streams crossing state boundaries, it is inevitable that disputes over water rights will arise between states. In general these disputes take the form of a complaint by the downstream state that it is not getting its fair share of the water of the stream. Such disputes fall under the jurisdiction of the Supreme Court, and this Court has developed a doctrine of *equitable apportionment* based on the facts of the controversy and without adherence to any particular formula. In a dispute between Kansas and Colorado (1907) Kansas claimed riparian rights to the water of the Arkansas River. The Court held that diversions in Colorado had not been detrimental to users in Kansas and refused to enjoin Colorado from use of Arkansas River water or to allocate the water between the states. In a subsequent suit between Colorado and Wyoming over waters of the Laramie River the Court held that since both states adhered to the appropriation doctrine within their own boundaries, this same doctrine was a fair basis of allocation between the states. Subsequent decisions have all tended in the same direction. The Court has refused to allocate waters which have not been and may not be used or to enjoin existing beneficial uses without proof of serious detriment to the plaintiff.

In *Arizona v. California* (1963), the Supreme Court announced a new method of apportioning interstate streams: by congressional action. The Court found that the Congress had authorized in the legislation providing for the construction of Hoover Dam the apportionment of the Lower Colorado River among Arizona, California, and Nevada, and that this authority had been exercised by the Secretary of the Interior when he executed water-delivery contracts for the water stored in the reservoir. This case has some yet undetermined long-range effects, for it could be the basis for congressional action transferring large quantities of water from one basin to another, as from the Columbia Basin to the Colorado Basin.

The Supreme Court has also urged the use of interstate compacts as a basis of agreement between states. The Constitution[1] provides that "No State shall, without the consent of Congress, . . . enter into any Agreement or Compact with another State." In 1911, however, Congress passed a law[2] giving blanket consent to interstate compacts "for the purpose of conserving the forests and water supply of the States," but it is thought

---

[1] U.S. Const. Art. 1, §10, ¶3.
[2] Act of Mar. 1, 1911, 1, 35 Stat. 961, 16 U.S.C. 552.

necessary for Congress to ratify a compact after it has been negotiated among the states. Numerous interstate compacts have evolved to govern the apportionment of flow of interstate rivers for irrigation and other uses and setting up machinery for the control of pollution on interstate rivers. A compact for flood mitigation exists for the Red River of the North, and an early compact (1785) covers navigation on the Potomac River.

An interstate compact governing the allocation of water normally represents a mutual agreement between the states specifying the amount (or a formula by which the amount is determined) to which each state is entitled. The compact may also include provision for a commission or water master to supervise the terms of the compact and to determine in specific cases (under established rules) the quantities of water to which each state is entitled. The distribution of each state's allotment of water within its own boundaries is a matter for the state to determine. It has been held that the apportionment of water under a compact is binding on all citizens of the states involved even though it contravenes existing rights within the states. Because of the federal interest in water problems, federal representatives commonly participate in the negotiations for interstate water compacts. The Delaware River Basin Compact is unique in that it is a compact among several states and the United States. This type of compact is referred to as a federal-interstate compact, and several more of its type are pending ratification. This type of compact presents special problems because of the membership of the United States in them.

6-10   State control of hydraulic projects   Most states exercise their general welfare or police powers in connection with hydraulic projects. Many states require approval by the state engineer of plans for dams above some minimum size and also inspect the dam while under construction. These requirements are mainly to assure the safety of persons who are downstream of the dam.

Many states regulate the discharge of wastewater into streams and lakes and the use of water for domestic purposes and for irrigation. Such control is usually exercised through the state department of health or through special water-pollution control boards. These controls are intended to prevent the spread of disease and to avoid nuisance through careless discharge of waste or inadequate sanitary precautions in a water-supply system. The provisions of state laws with respect to supervision of dams and pollution control are so varied that they cannot be discussed in detail here. Engineers engaged in the design of hydraulic projects should become acquainted with the applicable state regulations in each case.

6-11   Drainage law   Two basic rules of law are applied in drainage problems. Although several states follow a common doctrine, each has some modifications as a result of local usage or interpretation. Legal advice should always be sought in important cases. Some states[1] follow the Roman civil law which specifies that the owner of high land (*dominant owner*) is entitled to the advantage that this elevation gives him and may discharge his drainage water onto lower land through *natural depressions and channels* without obstruction by the lower, or *servient*, owner. A dominant owner may accelerate the flow of surface water by constructing ditches or by improving natural channels on his property, and he may install tile drains. He may not carry water across a drainage divide and discharge it on land which would not have received the water naturally, nor may he locate the outlet of his drainage system at a point other than the natural outlet of the area. A servient owner can do nothing to prevent natural drainage from entering his property from above. Statute law in many states permits the dominant owner to construct drains on the land of a servient owner after a simple eminent-domain proceeding and payment of all costs and damages. Many states also modify the rule of law for cities, by relieving the servient owner of many restrictions. Otherwise a large number of lawsuits might develop as a result of grading and developing city lots.

English common law employs the *common-enemy rule*.[2] The basic principle here is that water is a common enemy of all, and any landowner may protect himself from water flowing onto his land from a higher elevation. Under this rule, the dominant landowner cannot construct drainage works which result in damage to the property of a servient owner without first securing an easement. The servient owner is allowed to construct dikes or other works to prevent the flow of surface water onto his property.

Both doctrines of drainage law place the responsibility for damages on any person or organization altering the natural stream pattern of an area or creating an obstacle which blocks the flow of a natural stream. Common law confers no rights to control of navigable streams under state jurisdiction except by the construction of levees to keep the stream from overflowing one's land.

6-12 Environmental law Congress, state legislatures, and the courts have recently been active in protecting the environment. The Federal Water Pollution Control Act of 1965 requires the states to set stream-quality standards satisfactory to the Secretary of the Interior, in whose department is located the Federal Water Quality Control Administration.[3] As a result of this legislation, all states in the 1960's enacted new water

---

[1] The rule of Roman civil law is followed in Alabama, California, Georgia, Illinois, Iowa, Kentucky, Louisiana, Maryland, Michigan, North Carolina, Ohio, Pennsylvania, and Texas.

[2] The common-enemy rule is used in Arkansas, Connecticut, Indiana, Kansas, Maine, Massachusetts, Minnesota, Missouri, Nebraska, New Hampshire, New Jersey, New Mexico, New York, Oklahoma, South Carolina, Virginia, Washington, and Wisconsin.

[3] This responsibility was transferred to the Environmental Protection Agency in 1970.

160   WATER-RESOURCES ENGINEERING

pollution control statutes. In addition, the Congress enacted the Federal Environmental Policy Act of 1969, which established the Environmental Policy Council to advise the President on environmental matters and which directed that reports be made on all federal projects affecting the environment. Lastly, the *Scenic Hudson* case gave standing to sue to conservationists who oppose federally licensed projects. It seems certain that the environmental impact of water development projects will be the subject of increasing legislation and litigation in the future.

BIBLIOGRAPHY

Clark, R. E. (ed.): "Water and Water Rights," The Allen Smith Company, Indianapolis (a multivolume treatise with some volumes not yet in print).
Harding, S. T.: "Water Rights for Irrigation—Principles and Procedure for Engineers," Stanford, Stanford, Calif., 1936.
McGuinness, C. L.: Legal Control of the Use of Ground Water, *Water Works Eng.*, Vol. 98, p. 475, May 2, 1945.
Meyers, Charles J., and A. D. Tarlock, "Water Resource Management," Foundation Press, Mineola, N.Y., 1971.
Sax, Joseph L.: "Water Law, Planning and Policy," Bobbs-Merrill, Indianapolis, 1968.
"State Water Law in the Development of the West," National Resources Planning Board, 1943.
Tolman, C. F., and A. C. Stipp: Analysis of Legal Concepts of Subflow and Percolating Waters, *Trans. ASCE*, Vol. 106, pp. 882–933, 1941.
Trelease, Frank J.: "Water Law," West, St. Paul, Minn., 1967.
"Water Resources and the Law," University of Michigan Law School, Ann Arbor, 1958.
"Water Resources Law," Vol. 3, President's Water Resources Policy Commission, Washington, D.C., 1950'

PROBLEMS

6-1 From one of the references above or from your state water code, prepare a summary of the procedure for appropriating water, adjudicating water rights, and administering water rights with respect to surface water.

6-2 If your state has a groundwater code, determine the rules governing the use of groundwater, and prepare a brief summary of the most important items.

Chapter 7

Reservoirs

A water-supply, irrigation, or hydroelectric project drawing water directly from a stream may be unable to satisfy the demands of its consumers during extremely low flows. This stream, which may carry little or no water during portions of the year, often becomes a raging torrent after heavy rains and a hazard to all activities along its banks. A *storage*, or *conservation*, *reservoir* can retain such excess water from periods of high flow for use during periods of drought. In addition to conserving water for later use, the storage of floodwater may also reduce flood damage below the reservoir. Because of the varying rate of demand for water during the day, many cities find it necessary to provide *distribution reservoirs* within their water-supply system. Such reservoirs permit water-treatment or pumping plants to operate at a reasonably uniform rate and provide water from storage when the demand exceeds this rate. On farms or ranches, *stock tanks* or *farm ponds* may conserve the intermittent flow from small creeks for useful purposes.

Whatever the size of a reservoir or the ultimate use of the water, the main function of a reservoir is to stabilize the flow of water, either by regulating a *varying supply* in a natural stream or by satisfying a *varying demand* by the ultimate consumers. The general aspects of reservoir design are discussed in this chapter, while the special aspects pertinent to specific uses are covered more fully in Chaps. 14 to 21.

7-1 Physical characteristics of reservoirs Since the primary function of reservoirs is to provide storage, their most important physical characteristic is *storage capacity*. The capacity of a reservoir of regular shape can be computed with the formulas for the volumes of solids. Capacity of reservoirs on natural sites must usually be determined from topographic surveys. An *area-elevation curve* (Fig. 7-1) is constructed by planimetering

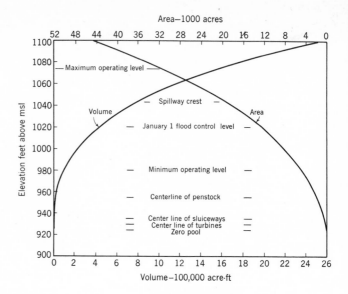

FIG. 7-1   Elevation-storage and elevation-area curves for Cherokee Reservoir on the Holston River, Tennessee. (Data from TVA Technical Report No. 7)

the area enclosed within each contour within the reservoir site. The integral of the area-elevation curve is the *elevation-storage*, or *capacity*, *curve* for the reservoir. The increment of storage between two elevations is usually computed by multiplying the average of the areas at the two elevations by the elevation difference. The summation of these increments below any elevation is the storage volume below that level. In the absence of adequate topographic maps cross sections of the reservoir are sometimes surveyed and the capacity computed from these vertical cross sections by use of the prismoidal formula.

   Normal pool level is the maximum elevation to which the reservoir surface will rise during ordinary operating conditions. For most reservoirs normal pool is determined by the elevation of the spillway crest or the top of the spillway gates. *Minimum pool level* is the lowest elevation to which the pool is to be drawn under normal conditions. This level may be fixed by the elevation of the lowest outlet in the dam or, in the case of hydroelectric reservoirs, by conditions of operating efficiency for the turbines. The storage volume between the minimum and normal pool levels is called the *useful storage*. Water held below minimum pool level is *dead storage*. In multi-purpose reservoirs the useful storage may be subdivided into *conservation storage* and *flood-mitigation storage* in accordance with the adopted plan of operation. During floods, discharge over the spillway may cause the water level to rise above normal pool level. This *surcharge storage* is normally uncontrolled, i.e., it exists only while a flood is occurring and cannot be retained for later use. Reservoir banks are usually permeable, and water enters the soil when the reservoir fills and drains out as the water level is lowered. This *bank storage* effectively increases the capacity of the reservoir above that indicated by the elevation-storage curve. The amount of bank storage depends on geologic conditions and may amount to several percent of the reservoir volume. The water in a natural stream channel occupies a variable volume of *valley storage* (Sec. 3-11). The net increase in storage capacity resulting from the construction of a reservoir is the total capacity less the natural valley storage. This distinction is of no importance for conservation reservoirs, but from the viewpoint of flood mitigation the effective storage in the reservoir is the useful storage plus the surcharge storage less the natural valley storage corresponding to the rate of inflow to the reservoir (Fig. 7-2).

The preceding discussion has assumed that the reservoir water surface is level. This is a reasonable assumption for most short, deep reservoirs. Actually, however, if flow is passing the dam, there must be some slope to the water surface to cause this flow. If the cross-sectional area of the reservoir is large compared with the rate of flow, the velocity will be small and the slope of the hydraulic grade line will be very flat. In relatively shallow and narrow reservoirs, the water surface at high flows may depart considerably from the horizontal (Fig. 7-3). The wedge-shaped element of storage above a horizontal is surcharge storage. The shape of the water-surface profile can be computed by using methods for nonuniform flow (Sec. 10-4). A different profile will exist for each combination of inflow rate and water-surface elevation at the dam. The computation of the water-surface profile is an important part of reservoir design since it provides information on the water

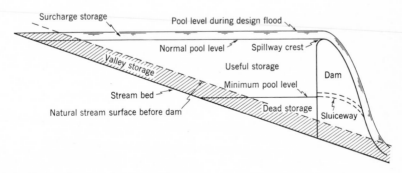

FIG. 7-2   *Zones of storage in a reservoir.*

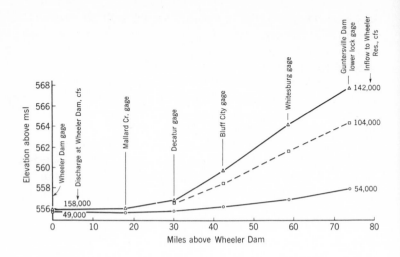

FIG. 7-3   *Profiles of the water surface in Wheeler Reservoir on the Tennessee River.*
(Data from TVA)

level at various points along the length of the reservoir from which the land
requirements for the reservoir can be determined. Acquisition of land or
flowage rights over the land is necessary before the reservoir can be built.
Docks, houses, storm drain outlets, roads, and bridges along the bank of the
reservoir must be located above the maximum water level expected in the
reservoir.

Storage in reservoirs subject to marked backwater effects cannot be related
to water-surface elevation alone as is done in Fig. 7-1. A second parameter
such as inflow rate or water-surface elevation on a gage near the upper end of
the reservoir must also be used. Storage volume under each profile can be
computed from cross sections by the methods used for earthwork computations. The resulting storage curves will look much like the routing curves
of Fig. 3-18.

7-2   Reservoir yield   Probably the most important aspect of storage-
reservoir design is an analysis of the relation between yield and capacity.
Yield is the amount of water which can be supplied from the reservoir in
a specified interval of time. The time interval may vary from a day for a
small distribution reservoir to a year or more for a large storage reservoir.
Yield is dependent upon inflow and will vary from year to year. The *safe*,
or *firm, yield* is the maximum quantity of water which can be guaranteed
during a critical dry period. In practice, the critical period is often taken as
the period of lowest natural flow on record for the stream. Hence, there is a

finite probability that a drier period may occur,[1] with a yield even less than the safe yield. Since firm yield can never be determined with certainty it is better to treat yield in probabilistic terms. The maximum possible yield equals the mean inflow less evaporation and seepage losses. If the flow were absolutely constant no reservoir would be required; but, as variability of the flow increases, the required reservoir capacity increases rapidly (Fig. 7-4).

Any reservoir with a yield less than the maximum will yield differing amounts in each period. The question is *Which value shall be used as the design yield?* A reservoir to supply municipal water should have a relatively low design yield so that the risk of a period with yield below the design value is small. By contrast, an irrigation system may tolerate 20 percent of the periods with yield below the nominal design value. Water available in excess of safe yield during periods of high flow is called *secondary yield.* Hydroelectric power developed from secondary water may be sold to large industries on a "when available" basis. Power commitments to domestic users must be on a firm basis and should not exceed the power which can be produced with the firm yield unless thermal power (steam or diesel) is available to support the hydroelectric power. The decision is an economic one based on costs and benefits for various levels of design.

7-3 Selection of distribution reservoir capacity for a given yield
Often a project design requires the determination of the reservoir capacity required to meet a specific demand. Examples are found in municipal water supply or in irrigation when it is desired to irrigate a specified area. Since the yield (outflow) is equal to the inflow plus or minus an increment of storage, the determination of the capacity to supply a given yield is based on the storage equation [Eq. (3-13)]. In the long run, outflow must equal inflow less waste and unavoidable losses. This is another way of saying that a reservoir does not make water but merely permits its redistribution with respect to time.

A simple problem involving the selection of distribution reservoir capacity is given in Illustrative Example 7-1. Here the required yield is based on an estimate of the maximum daily demand by the consumers. The inflow rate is fixed by a decision to pump at a uniform rate. The reservoir capacity must be sufficient to supply the demand at times when the demand exceeds the pumping rate. A similar solution would be used if a variable pumping rate were assumed.

ILLUSTRATIVE EXAMPLE 7-1   The water supply for a city is pumped from wells to a distribution reservoir. The estimated hourly water requirements for the maximum

---

[1] R. S. Gooch, Design Drouth Criteria, *J. Hydraulics Div., ASCE,* Vol. 92, pp. 233–242, March, 1966.

day are indicated below. If the pumps are to operate at a uniform rate, what distribution reservoir capacity is required?

Hour ending	Demand	Pumping rate	Required from reservoir
0100	0.52	1.00	0
0200	0.46	1.00	0
0300	0.43	1.00	0
0400	0.41	1.00	0
0500	0.43	1.00	0
0600	0.51	1.00	0
0700	0.61	1.00	0
0800	0.91	1.00	0
0900	1.18	1.00	0.18
1000	1.26	1.00	0.26
1100	1.28	1.00	0.28
1200	1.26	1.00	0.26
1300	1.24	1.00	0.24
1400	1.22	1.00	0.22
1500	1.22	1.00	0.22
1600	1.21	1.00	0.21
1700	1.23	1.00	0.23
1800	1.28	1.00	0.28
1900	1.35	1.00	0.35
2000	1.40	1.00	0.40
2100	1.39	1.00	0.39
2200	1.33	1.00	0.33
2300	1.17	1.00	0.17
2400	0.70	1.00	0
Total	24.00	24.00	4.02

Note: All flow rates in million gallons per hour.

The average pumping rate is determined by dividing the total pumped by 24. The required reservoir capacity is the sum of the hourly requirements from storage, or 4.02 million gallons.

7-4   Selection of capacity for a river reservoir   The determination of required capacity for a river reservoir follows a pattern not unlike that of Illustrative Example 7-1 but is generally more complex. The analysis is usually called an *operation study* and is essentially a simulation of the reservoir operation for a period of time in accord with an adopted set of rules. The operation study may be designed to define the optimum rules for operation, to select the most efficient installed capacity for the powerhouse, to establish the required sluiceway capacity for a flood control dam, or to achieve many other decisions required in the course of project planning.

An operation study may be made only for a period of extremely low flow which is selected as a critical period or it may extend over the entire period of observed or synthetic record (Chap. 5). In the first case the study can do little more than define the required capacity to overcome the selected drought, while in the latter case the study may determine the useful water (or power) production for each year of the record. The more complete study indicates the probability of water or power deficiencies of various magnitudes which is important in the economic planning and the integration of the project into a system.

An operation study may be performed with annual, monthly, daily, or even shorter-period data. Annual data generally yield relatively crude results because the sequence of flow during the year is usually quite important. For reservoirs which are relatively large compared with the inflows, a monthly study is usually adequate. If the reservoir is small, the sequence of flow within the month may become important, and daily data will be required.

Approximate analysis may be made graphically (Sec. 7-6), but in order to account for all important factors, a tabular solution is necessary. For lengthy analyses (including study of complex systems) the use of digital computers has many advantages.[1] By programming the operation on a computer, it is possible to make many trials with differing operating rules or changes in the assumed physical works.

Several preliminary steps are usually necessary before the data can be analyzed. Unless a streamflow record is available at the proposed reservoir site, the record from a station elsewhere on the stream or on a nearby stream must be adjusted to the dam site. Often, the available records are too short to include a really critical drought period, and the record may be extended by comparison with longer streamflow records in the vicinity, by hydrologic simulation, or by stochastic methods (Sec. 5-14).

After the streamflow at the dam site has been determined, an adjustment may be necessary for water which must be permitted to pass the reservoir to satisfy prior water rights. The construction of the reservoir also increases the exposed water-surface area above that of the natural stream and increases the evaporation loss. Since the evaporation from a free-water surface is nearly always in excess of the actual evapotranspiration from a land surface there is usually a net loss of water as a result of the reservoir construction. In an arid region the loss may be so great as to defeat the purpose of the reservoir. Assuming the inflow values represent flow at the dam site, the net

---

[1] D. F. Manzer and M. P. Barnett, Analysis by Simulation: Programming Techniques for a High-speed Digital Computer, chap. 9 in Arthur Maass and others, "Design of Water-resource Systems," Harvard, Cambridge, Mass., 1962; M. M. Hufschmidt and M. Fiering: "Simulation Techniques for the Design of Water-resources Systems," Harvard, Cambridge, Mass., 1966.

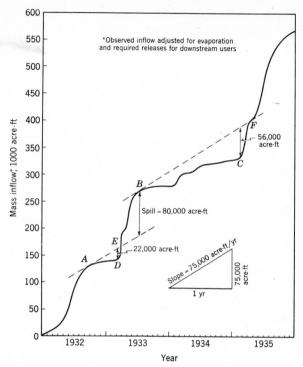

FIG. 7-4   *Use of a mass curve to determine the reservoir capacity required to produce a specified yield.*

evaporation loss as a result of the reservoir is

$$E_w - E_a \approx E_w - (P - q) \tag{7-1}$$

where E_w is the free-water evaporation, E_a is the actual evapotranspiration from the land area flooded by the reservoir, P is precipitation and q is the runoff from the area prior to flooding. Seepage from the reservoir may add still more to the loss resulting from reservoir construction. For preliminary studies it may be satisfactory to multiply the net loss per unit area by the area of the reservoir at average pool level to determine the volume of water involved. If the difference in area between maximum and minimum pool is large, the loss should be computed for each period on the basis of the estimated water-surface elevation for the period.

Illustrative Example 7-2 shows the computation of required capacity for a storage reservoir on a stream. The required storage is the sum of the monthly increments of demand in excess of streamflow.

ILLUSTRATIVE EXAMPLE 7-2   Given below are monthly inflows during a low-water period at the site of a proposed dam, the corresponding monthly evaporation and precipitation at a nearby station, and the estimated monthly demand for water. Prior water rights require the release of full natural flow or 100 acre-ft/month, whichever is least. Assume that 25 percent of the rainfall on the land area to be flooded by the reservoir has reached the stream in the past. Use a net increased pool area of 1000 acres. Find the required storage.

Month (1)	Flow, acre-ft (2)	Pan evap-ora-tion, in. (3)	Pre-cip-ita-tion, in. (4)	Demand, acre-ft (5)	Down-stream re-quire-ment, acre-ft (6)	Evap-ora-tion,* acre-ft (7)	Pre-cip-ita-tion,† acre-ft (8)	Ad-justed flow,‡ acre-ft (9)	Re-quired from stor-age,§ acre-ft (10)
January	2100	3.5	4.5	40	100	203	281	2078	0
February	4400	5.0	4.7	40	100	291	294	4303	0
March	30	5.8	0.5	80	30	340	31	−309	389
April	10	6.1	0.7	130	10	356	44	−312	442
May	5	5.4	0.2	140	5	315	12	−303	443
June	3	4.6	0	140	3	269	0	−269	409
July	1	3.0	0	130	1	175	0	−175	305
August	0	1.7	0	120	0	100	0	−100	220
September	0	0.8	0	80	0	47	0	−47	127
October	0	1.0	0.4	40	0	59	25	−34	74
November	0	1.3	0.8	30	0	76	50	−26	56
December	3	2.4	4.6	30	3	140	288	148	0
Total	6552	40.6	16.4	1000	252	2371	1025	4954	2465

*Col. 3 × $1000/12$ × 0.7.
†Col. 4 × $1000/12$ × 0.75.
‡Col. 2 − Col. 6 − Col. 7 + Col. 8.
§Col. 9 − Col. 5 whenever the result is negative.

7-5 Determination of yield from a given reservoir capacity In some cases the reservoir capacity is fixed by conditions at the site, and it is necessary to determine how much water this reservoir capacity will yield. The yield is equal to the sum of the usable storage in the reservoir and the usable inflow during the critical period. For the data of Illustrative Example 7-2 the available inflow during the critical months of March through November is − 1575 acre-ft. If the usable storage in the reservoir is 2000 acre-ft, the total water available during this period is 425 acre-ft, or 47 acre-ft/month.

7-6 Mass curves Mass curves permit a simple graphical inspection of the entire record or any portion of it for evaluation of yield. A *mass curve* (sometimes called a Rippl diagram) is a cumulative plotting of net reservoir inflow. Figure 7-4 is a mass curve for a 4-yr period. The slope of the mass curve at any time is a measure of the inflow rate at that time. Demand curves representing a uniform rate of demand are straight lines having a slope equal to the demand rate. Demand lines drawn tangent to the high points of the mass curve (A, B) represent rates of withdrawal from the reservoir. Assuming the reservoir to be full wherever a demand line intersects the mass curve, the maximum departure between the demand line and the mass curve represents the reservoir capacity required to satisfy the demand. The vertical distance between successive tangents represents water wasted over the spill-way. If the demand is not uniform, the demand line becomes a curve (actually a mass curve of demand) but the analysis is not changed. It is essential, how-ever, that the demand line for nonuniform demand coincide chronologically with the mass curve, i.e., June demand must coincide with June inflow, etc.

ILLUSTRATIVE EXAMPLE 7-3   What reservoir capacity is required to assure a yield of 75,000 acre-ft/yr for the inflows shown in Fig. 7-4?

Tangents to the mass curve at A and B have slopes equal to the demand of 75,000 acre-ft/yr. The maximum departure occurs at C and is 56,000 acre-ft. This is the required reservoir capacity. Such a reservoir would be full at A, depleted to 34,000 acre-ft of storage at D, and full again at E. Between E and B the reservoir would remain full, and all inflow in excess of the demand would be wasted downstream. At C the reservoir would be empty, and at F it would be full again. Note that in this case the storage must carry over 2 yr.

Mass curves may also be used to determine the yield which may be expected with a given reservoir capacity (Fig. 7-5). In this case tangents are drawn to the high points of the mass curve (A, B) in such a manner that their maximum departure from the mass curve does not exceed the specified reservoir capacity. The slopes of the resulting lines indicate the yields which can be attained in each year with the specified storage capacity. The slope of each demand line is the yield for the period. A demand line must intersect the mass curve when extended forward. If it does not, the reservoir will not refill.

ILLUSTRATIVE EXAMPLE 7-4   What yield will be available if a reservoir of 30,000 acre-ft capacity is provided at the site for which the mass curve of Fig. 7-5 applies?

The tangents to the mass curve of Fig. 7-5 are drawn so that their maximum depar-ture from the mass curve is 30,000 acre-ft. The tangent from B has the least slope, 60,000 acre-ft/yr, and this is the minimum yield. The tangent at A indicates a possible yield of 95,000 acre-ft in that year, but this demand could not be satisfied between points B and C without storage in excess of 30,000 acre-ft.

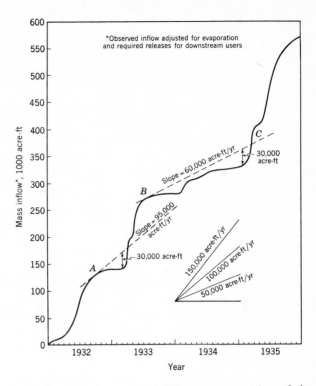

FIG. 7-5   *Use of a mass curve to determine the possible yield from a reservoir of specified capacity.*

7-7   Sediment transport by streams   Every stream carries some *suspended sediment* and moves larger solids along the stream bed as *bed load*. Since the specific gravity of soil materials is about 2.65, the particles of suspended sediment tend to settle to the channel bottom but upward currents in the turbulent flow counteract the gravitational settling. When sediment-laden water reaches a reservoir, the velocity and turbulence are greatly reduced. The larger suspended particles and most of the bed load are deposited as a *delta* at the head of the reservoir (Fig. 7-6). Smaller particles remain in suspension longer and are deposited farther down the reservoir, although the very smallest particles may remain in suspension for a long time and some may pass the dam with water discharged through sluiceways, turbines, or the spillway.

The suspended sediment load of streams is measured by sampling the water, filtering to remove the sediment, drying, and weighing the filtered material. Sediment load is expressed in *parts per million* (*ppm*), computed by

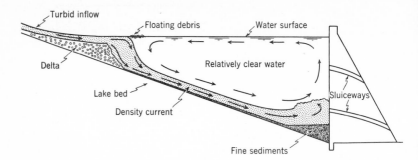

FIG. 7-6   *Schematic drawing of the sediment accumulation in a typical reservoir.*

dividing the weight of the sediment by the weight of sediment and water in the sample and multiplying the quotient by 10^6. The sample is usually collected in a pint bottle held in a sampler (Fig. 7-7) which is designed to avoid distortion of the streamlines of flow so as to collect a representative sample of the sediment-laden water.[1] Most of the available sediment-load data have been gathered since about 1938. Because of poorly designed samplers, many of the early data are of questionable accuracy.

No practical device for field measurement of bed load is now in use. According to Sheppard[2] bed load may vary from 0 to 100 percent of the suspended load. More commonly though, it lies in the 5 to 25 percent range. Einstein[3] has presented an equation for the calculation of bed-load movement on the basis of the size distribution of the bed material and the stream-flow rates.

The relation between suspended-sediment transport Q_s and streamflow Q is often represented by a logarithmic plot (Fig. 7-8) which may be expressed mathematically by an equation of the form

$$Q_s = kQ^n \qquad (7-2)$$

where n commonly varies between 2 and 3, and k, the intercept when Q is unity, is usually quite small. A sediment-rating curve such as Fig. 7-8 may be used to estimate suspended-sediment transport from the continuous record of streamflow in the same manner that the flow is estimated from the continuous stage record by use of a stage-discharge relation. The

---

[1] Measurement of the Sediment Discharge of Streams, *Rept.* 8, St. Paul District, Corps of Engineers, March, 1948.

[2] J. R. Sheppard, Methods and Their Suitability for Determining Total Sediment Quantities, *Proceedings of the Federal Inter-Agency Sedimentation Conference 1963*, U.S. Dept. of Agriculture, Washington, D.C., 1965.

[3] H. A. Einstein, The Bed-load Function for Sediment Transportation in Open-Channel Flow, *U.S. Dept. Agr. Tech. Bull.* 1026, September, 1950.

sediment rating is much less accurate than the corresponding streamflow-rating curve. Rates of erosion vary from storm to storm with variations in rainfall intensity, soil condition, and vegetal development. Sediment eroded from a basin during one storm may be deposited in the stream channel, to remain until a subsequent storm washes it downstream. Portions

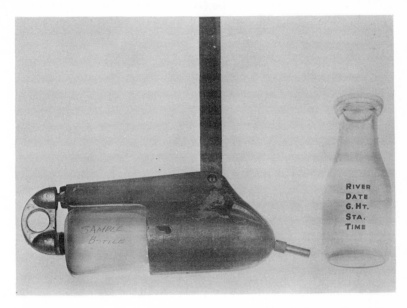

FIG. 7-7   Depth-integrating sediment sampler, model U.S. DH-48, for small streams.

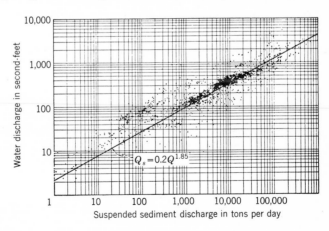

$$Q_s = 0.2 Q^{1.85}$$

FIG. 7-8   Sediment-rating curve for the Powder River at Arvada, Wyoming. (Leopold and Maddock)

of the drainage area may be more susceptible to erosion than others, and higher sediment loads may be expected when a storm centers over such areas. Thus, the rate of suspended-sediment transport and the rate of streamflow are rarely closely correlated. Despite these inaccuracies, the sediment rating provides a useful tool for estimates of suspended-sediment transport. The total sediment transport may be estimated by adding an arbitrary 10 to 20 percent to the suspended-sediment transport to allow for the bed-load contribution.

In the absence of suspended-sediment data, the total sediment transport of a stream may be estimated by comparison with similar watersheds whose sediment transports have been previously determined from suspended-sediment-load data or from studies of reservoir sediment accumulation. The total amount of sediment that passes any section of a stream is referred to as the *sediment yield* or sediment production. Rates of sediment production for typical watersheds in the United States are presented in Table 7-1. Mean annual sediment production rates generally range from 200 to 4000 tons per square mile.

   7-8   Reservoir sedimentation   The ultimate destiny of all reservoirs is to be filled with sediment. If the sediment inflow is large compared with the reservoir capacity, the useful life of the reservoir may be very short. A small water-supply reservoir on the Solomon River near Osborne, Kansas, filled with sediment during the first year after its completion. Reservoir planning must include consideration of the probable rate of sedimentation in order to determine whether the useful life of the proposed reservoir will be sufficient to warrant its construction.

Our knowledge of reservoir sedimentation rates (Table 7-1) is based on surveys to determine the rate of sediment accumulation[1] in reservoirs which have been in existence for many years. These surveys indicate the specific weight of the settled sediments and the percèntage of entering sediment which is deposited in the reservoir. These data are necessary in order to interpret the data on sediment load of streams in terms of reservoir sedimentation. The specific weight of settled sediments seems to vary with the age of the deposit and the character of the sediment. Specific weights (dry) of sediment samples from reservoirs range from about 40 to 110 lb/cu ft with an average of about 60 lb/cu ft for fresh sediments and 80 lb/cu ft for old sediments.[2] The percent of the inflowing sediment which is retained in a

---

   [1] J. M. Caldwell, Supersonic Sounding Instruments and Methods, *Trans. ASCE*, Vol. 117, pp. 44–58, 1952, and L. C. Gottschalk, Measurement of Sedimentation in Small Reservoirs, *Trans. ASCE*, Vol. 117, pp. 59–71, 1952.

   [2] D. C. Bondurant, Sedimentation Studies at Conchas Reservoir in New Mexico, *Trans. ASCE*, Vol. 116, pp. 1283–1295, 1951, and V. A. Koelzer and J. M. Lara, Densities and Compaction Rates of Deposited Sediment, *J. Hydraulics Div., ASCE Paper* 1603, pp. 1–15, April, 1958.

TABLE 7-1   Rates of Sediment Accumulation in Selected Reservoirs in the United States*

Name and location	Net drainage area, sq mi	Original capacity, acre-ft	Annual sediment production rate, tons/sq mi	Annual loss of storage, percent
Schoharie (Prattsville, N.Y.)	312.	63,812	217	0.07
Roxboro (Roxboro, N.C.)	7.52	531	447	0.69
Norris (Norris, Tenn.)	2,823.	2,045,300	450	0.05
Bloomington (Bloomington, Ill.)	60.3	6,678	514	0.50
Crab Orchard (Carbondale, Ill.)	160.	67,320	1,976	0.45
Abilene (Abilene, Tex.)	97.5	10,325	274	0.19
Dallas (Denton, Tex.)	1,157.	180,759	1,304	0.72
Mission (Horton, Kans.)	7.76	1,852	3,874	1.20
Morena (San Diego, Calif.)	109.	66,767	2,444	0.31
Roosevelt, (Globe, Ariz.)	5,760.	1,522,200	1,112	0.25
Mead (Boulder City, Nev.)	167,600.	31,250,000	877	0.33
Arrowrock (Boise, Idaho)	2,170.	279,250	173	0.09

*Extracted from Louis C. Gottschalk, "Reservoir Sedimentation," Chapter 17-I of Ven Te Chow (ed.), *Handbook of Applied Hydrology*, McGraw-Hill, 1964, pp. 17–28 to 17–29.

reservoir (*trap efficiency*) is a function of the ratio of reservoir capacity to total inflow. A small reservoir on a large stream passes most of its inflow so quickly that the finer sediments do not settle but are discharged downstream. A large reservoir, on the other hand, may retain water for several years and permit almost complete removal of suspended sediment. Figure 7-9 relates reservoir trap efficiency to the capacity-inflow ratio on the basis of data from surveys of existing reservoirs.[1] The trap efficiency of a reservoir

---

[1] G. M. Brune, Trap Efficiency of Reservoirs, *Trans. Am. Geophys. Union*, Vol. 34, pp. 407–418, June, 1953.

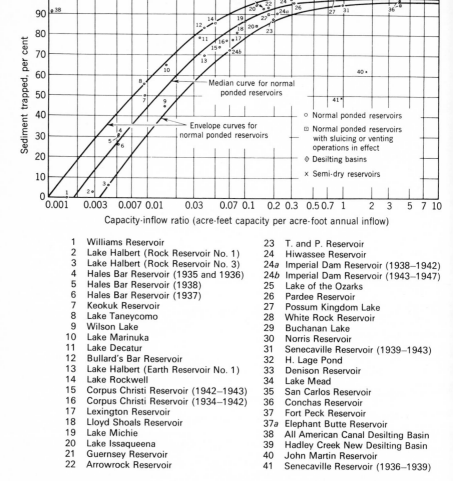

FIG. 7-9   Reservoir trap efficiency as a function of the capacity-inflow ratio. (*Brune*)

1	Williams Reservoir	23	T. and P. Reservoir
2	Lake Halbert (Rock Reservoir No. 1)	24	Hiwassee Reservoir
3	Lake Halbert (Rock Reservoir No. 3)	24a	Imperial Dam Reservoir (1938–1942)
4	Hales Bar Reservoir (1935 and 1936)	24b	Imperial Dam Reservoir (1943–1947)
5	Hales Bar Reservoir (1938)	25	Lake of the Ozarks
6	Hales Bar Reservoir (1937)	26	Pardee Reservoir
7	Keokuk Reservoir	27	Possum Kingdom Lake
8	Lake Taneycomo	28	White Rock Reservoir
9	Wilson Lake	29	Buchanan Lake
10	Lake Marinuka	30	Norris Reservoir
11	Lake Decatur	31	Senecaville Reservoir (1939–1943)
12	Bullard's Bar Reservoir	32	H. Lage Pond
13	Lake Halbert (Earth Reservoir No. 1)	33	Denison Reservoir
14	Lake Rockwell	34	Lake Mead
15	Corpus Christi Reservoir (1942–1943)	35	San Carlos Reservoir
16	Corpus Christi Reservoir (1934–1942)	36	Conchas Reservoir
17	Lexington Reservoir	37	Fort Peck Reservoir
18	Lloyd Shoals Reservoir	37a	Elephant Butte Reservoir
19	Lake Michie	38	All American Canal Desilting Basin
20	Lake Issaqueena	39	Hadley Creek New Desilting Basin
21	Guernsey Reservoir	40	John Martin Reservoir
22	Arrowrock Reservoir	41	Senecaville Reservoir (1936–1939)

decreases with age as the reservoir capacity is reduced by sediment accumulation. Thus complete filling of the reservoir may require a very long time, but actually the useful life of the reservoir is terminated when the capacity occupied by sediment is sufficient to prevent the reservoir from serving its intended purpose. Figure 7-9 may be used to estimate the amount of sediment which a reservoir will trap if the average annual sediment load of the stream is known. The volume occupied by this sediment can then be computed, using a reasonable value of specific weight for the deposited sediment.

The useful life may be computed by determining the total time required to fill the critical storage volume.

ILLUSTRATIVE EXAMPLE 7-5   Using Fig. 7-9, find the probable life of a reservoir with an initial capacity of 30,000 acre-ft if the average annual inflow is 60,000 acre-ft and the average annual sediment inflow is 200,000 tons. Assume a specific weight of 70 lb/cu ft for the sediment deposits. The useful life of the reservoir will terminate when 80 percent of its initial capacity is filled with sediment.

| Capacity, acre-ft | Capacity-inflow ratio | Trap efficiency | | Annual sediment trapped | | Increment volume, acre-ft | Years to fill |
		At indicated volume, percent	Average for increment, percent	Tons	Acre-ft*		
30,000	0.5	96.0					
24,000	0.4	95.5	95.7	191,400	126	6000	48
18,000	0.3	95.0	95.2	190,400	125	6000	48
12,000	0.2	93.0	94.0	188,000	123	6000	49
6,000	0.1	87.0	90.0	180,000	118	6000	51
Total	...	...	...	...	...	...	196

$$*1 \text{ acre-ft} = \frac{43,560 \times 70}{2000} = 1525 \text{ tons.}$$

7-9 Reservoir sedimentation control   The most common procedure for dealing with the sediment problem is to designate a portion of the reservoir capacity as *sediment storage*. This is a negative approach which in no way reduces the sediment accumulation but merely postpones the date when it becomes serious. Since sediment is deposited all through the reservoir, the allocation for sediment storage cannot be exclusively in the dead storage but must also include some otherwise useful storage. Figure 7-6 shows schematically the distribution of sediment within a reservoir, while Fig. 7-10 shows the relative disposition of sediment in several reservoirs and a tentative design curve suggested by the U.S. Bureau of Reclamation.[1]

Actually reservoir sedimentation cannot be prevented, but it may be retarded. One way of doing this is to select a site where the sediment inflow is naturally low. Some basins are more prolific sources of sediment than others because of soil type, land slopes, vegetal cover, and rainfall character-

[1] *U.S. Bur. Reclamation Manual*, Vol. 7, Part 9, chap. 9–4, April, 1948.

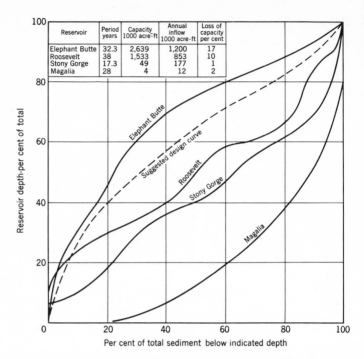

The following data table appears within the figure:

Reservoir	Period years	Capacity 1000 acre-ft	Annual inflow 1000 acre-ft	Loss of capacity per cent
Elephant Butte	32.3	2,639	1,200	17
Roosevelt	38	1,533	853	10
Stony Gorge	17.3	49	177	1
Magalia	28	4	12	2

FIG. 7-10   *Distribution of sediment in several reservoirs and a suggested design curve. (U.S. Bureau of Reclamation)*

istics. If an alternative site exists, such prolific sediment sources should be avoided. After a site has been selected, the reservoir capacity should be made large enough to create a useful life sufficient to warrant the construction. Although trap efficiency of large reservoirs is high, it does not increase linearly, and the useful life of a large reservoir is longer than that of a small reservoir if all other factors remain constant.

Some reduction in sediment inflow to a reservoir is possible by use of soil-conservation methods within the drainage basin. Terraces, strip cropping, contour plowing, and similar techniques retard overland flow and reduce erosion. Check dams in gullies retain some sediment and prevent it from entering the streams. Vegetal cover on the land reduces the impact force of raindrops and minimizes erosion. However, if a stream is denied its normal sediment load, it will tend to scour its bed or cave its banks. Consequently, stream-bank protection by revetment, vegetation, or other means is a necessary feature of a sediment-control plan. Conservation methods will never completely eliminate erosion and may be difficult to justify economically in some areas.

Sediment accumulation in reservoirs may be reduced by providing means for discharge of some sediment. Sluice gates at various levels will sometimes permit discharge of the finer sediments before they have time to settle to the bottom. In many reservoirs, a sediment-laden inflow may move through the pool as a *density current*, or layer of water with a density slightly different from that of the main body of reservoir water. The density difference may result from the sediment, dissolved minerals, or temperature. Because of the density difference, the water of the density current does not mix readily with the reservoir water and maintains its identity for a considerable time. Reservoir trap efficiency may be decreased from 2 to 10 percent if it is possible to vent such density currents through sluiceways.

Physical removal of sediment deposits is rarely feasible. Sluice gates near the base of the dam may permit flushing some sediment downstream, but the removal will not extend far upstream from the dam. An acre-foot of sediment contains 1613 cu yd, and, at the most favorable prices, removal by ordinary earth-moving methods would be expensive unless the excavated sediment has some sales value.

7-10   Wind setup and waves in reservoirs   Earth dams must have sufficient freeboard above maximum pool level so that waves cannot wash over the top of the dam. Waves in reservoirs may also damage shoreline structures and embankments adjacent to the water and interfere with navigation. Part of the design of any reservoir is an estimate of wind setup and wave height.

Wind setup is the tilting of the reservoir water surface caused by the movement of the surface water toward the leeward shore under the action of the wind. This current of surface water is a result of tangential stresses between the wind and the water and of differences in atmospheric pressure over the reservoir. The latter, however, is, typically, a smaller effect. As a consequence of wind setup, the reservoir water surface is above normal still-water level on the leeward side and below the still-water level on the windward side. This results in hydrostatic unbalance, and a return flow at some depth must occur. The water-surface slope which results is that necessary to sustain the return flow under conditions of bottom roughness and cross-sectional area of flow which exist. Wind setup is generally larger in shallow reservoirs with rough bottoms.

Wind setup may be estimated from

$$Z_s = \frac{V_w^2 F}{1400d} \tag{7-3}$$

where Z_s is the rise in feet above still-water level, V_w is the wind speed in miles per hour, F is the *fetch* or length of water surface over which the wind blows in miles, and d is the average depth of the lake along the fetch in feet.

Equation (7-3) is modified[1] from the original equation developed by Dutch engineers on the Zuider Zee. Additional information and techniques are given in other references.[2] Wind setup effects may be transferred around bends in a reservoir and the value of F used may be somewhat longer than the straight-line fetch.

When wind begins to blow over a smooth surface, small waves, called capillary waves, appear in response to the turbulent eddies in the wind stream. These waves grow in size and length as a result of the continuing push of the wind on the back of the waves and of the shearing or tangential force between the wind and the water. As the waves grow in size and length, their speed increases until they move at speeds approaching the speed of the wind. Because growth of a wave depends in part upon the difference between wind speed and wave speed, the growth rate approaches zero as the wave speed approaches the wind speed.

The duration of the wind and the time and direction from which it blows are important factors in the ultimate height of a wave. The variability of the wind and the amazingly complex and yet to be fully understood response of the water surface to the wind lead to a wave pattern that is a superposition of many waves. The pattern is often described by its energy distribution or spectrum. The growth of wind waves as a function of fetch, wind speed, and duration can be calculated from knowledge of the mechanism of wave generation and use of collected empirical results.[3] The duration of the wind and the fetch play an important role because a wave may not reach its ultimate height if the wave passes out of the region of high wind or strikes a shore during the growth process. The depth of water also plays a key role, tending to yield smaller and shorter waves in deep water.

Wave-height data gathered at two major reservoirs[4] confirm the theoretical and experimental data for ocean waves if a modified value of fetch is used. The derived equation is

$$z_w = 0.034 V_w^{1.06} F^{0.47} \tag{7-4}$$

where z_w is the average height (in ft) of the highest one-third of the waves and is called the *significant wave height*, V_w is the wind velocity (in mph) about 25 ft above the water surface, and F is the fetch in miles. The equation is

---

[1] T. Saville, Jr., E. W. McClendon, and A. L. Cochran, Freeboard Allowances for Waters in Inland Reservoirs, *J. Waterways and Harbors Div.*, *ASCE*, pp. 93–124, May, 1962.

[2] Shore Protection, Planning and Design, *Tech. Rept.* 3, 3d ed., U.S. Army Coastal Engineering Research Center, June, 1966.

[3] W. J. Pierson, Jr., and R. W. James, Practical Methods for Observing and Forecasting Ocean Waves, *U.S. Navy Hydrographic Office Pub.* 603, 1955 (reprinted 1960).

[4] T. Saville, Jr., E. W. McClendon, and A. L. Cochran, Freeboard Allowances for Waters in Inland Reservoirs, *J. Waterways and Harbors Div.*, *ASCE*, pp. 93–124, May, 1962.

shown graphically in Fig. 7-11 together with lines showing the minimum duration of wind required to develop the indicated wave height. Figure 7-12 shows the method of computing the effective fetch for a narrow reservoir.

Since the design must be made before the reservoir is complete, wind data over land must generally be used. Table 7-2 gives ratios of wind speed over land to those over water and may be used to correct observed wind to reservoir conditions. Waves are critical only when the reservoir is near maximum levels. Thus in selecting the critical wind speed for reservoirs subject to seasonal fluctuations, only winds which can occur during the season of maximum pool levels should be considered. The direction of the wind and the adopted fetch must also be the same.

The height of the significant wave is exceeded about 13 percent of the time. If a more conservative design is indicated, a higher wave height may be chosen. Table 7-3 gives ratios of z'/z_w for waves of lower exceedance.

When a wave strikes a land slope, it will *run up* the slope to a height above its open-water height. The amount of run-up depends on the surface. Figure 7-13 shows the results of small-scale experiments[1] on smooth slopes and

---

[1] T. Saville, Jr., Wave Run-up on Shore Structures, *Trans. ASCE*, Vol. 123, pp. 139–158, 1958; R. Y. Hudson, Laboratory Investigation of Rubble-mound Breakwaters, *Trans. ASCE*, Vol. 126, Part IV, pp. 492–541, 1962.

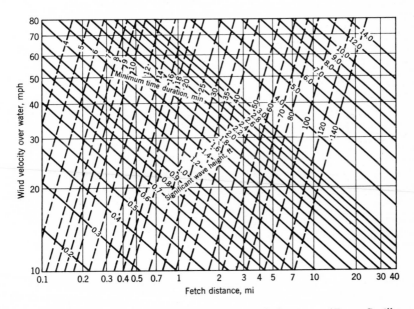

FIG. 7-11   *Significant wave heights and minimum wind durations.* (*From Saville, McClendon, and Cochran*)

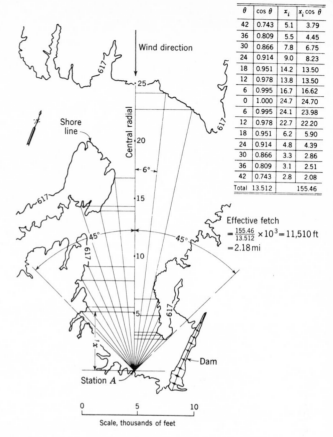

θ	$\cos \theta$	x_i	$x_i \cos \theta$
42	0.743	5.1	3.79
36	0.809	5.5	4.45
30	0.866	7.8	6.75
24	0.914	9.0	8.23
18	0.951	14.2	13.50
12	0.978	13.8	13.50
6	0.995	16.7	16.62
0	1.000	24.7	24.70
6	0.995	24.1	23.98
12	0.978	22.7	22.20
18	0.951	6.2	5.90
24	0.914	4.8	4.39
30	0.866	3.3	2.86
36	0.809	3.1	2.51
42	0.743	2.8	2.08
Total	13.512		155.46

Effective fetch
$$= \frac{155.46}{13.512} \times 10^3 = 11{,}510 \text{ ft}$$
$$= 2.18 \text{ mi}$$

FIG. 7-12   *Computation of effective fetch. (Modified from Saville, McClendon and Cochran)*

rubble mounds. Height of run-up z_r is shown as a ratio z_r/z_w and is dependent on the ratio of wave height to wavelength (wave steepness). Wavelength λ for deep-water waves may be computed from

$$\lambda = 5.12 t_w^2 \tag{7-5}$$

where the wave period t_w is given by

$$t_w = 0.46 V_w^{0.44} F^{0.28} \tag{7-6}$$

For shallow-water waves other length relations are appropriate.[1]

---

[1] Shore Protection, Planning and Design, *Tech. Rept.* 3, 3d ed., U.S. Army Coastal Engineering Research Center, June, 1966.

TABLE 7-2   *Relationship between Wind over Land and That over Water*
(*After Saville, McClendon, and Cochran*)

Fetch, mi	0.5	1	2	4	6	8
$\dfrac{V_{water}}{V_{land}}$	1.08	1.13	1.21	1.28	1.31	1.31

The curves for rubble mounds represent extremely permeable construction, and for more typical riprap on earth embankments the run-up may be somewhat higher, depending on both the permeability and the relative smoothness of the surface.

7-11   Reservoir clearance   The removal of trees and brush from a reservoir site is an expensive operation and is often difficult to justify on an economic basis. The main disadvantages resulting from leaving the vegetation in the reservoir are the possibilities that (1) trees will eventually float and

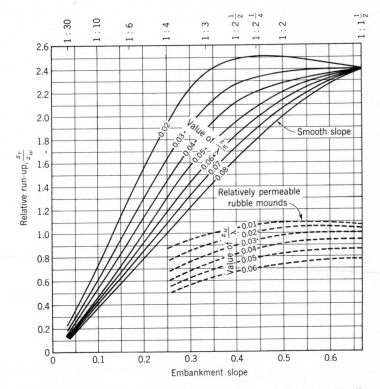

FIG. 7-13   *Wave run-up ratios vs. wave steepness and embankment slopes.* (*From Saville, McClendon, and Cochran*)

TABLE 7-3   *Percent of Waves Exceeding Various Wave Heights Greater than* z_w
(*After Saville, McClendon, and Cochran*)

z'/z_w	1.67	1.40	1.27	1.12	1.07	1.02
Percent of waves $> z'$	0.4	2	4	8	10	12

create a debris problem at the dam, (2) decay of organic material may create undesirable odors or tastes in water-supply reservoirs, and (3) trees projecting above the water surface may create an undesirable appearance and restrict the use of the reservoir for recreation.

Frequently all timber which would project above the water surface at minimum pool level is removed. This overcomes most of the problems cited above at some savings over the cost of complete clearance.

7-12   Reservoir leakage   Most reservoir banks are permeable, but the permeability is so low that leakage is of no importance. If the walls of the reservoir are of badly fractured rock, permeable volcanic material, or cavernous limestone, serious leakage may occur. This leakage may result not only in a loss of water but also in damage to property where the water returns to the surface. If leakage occurs through a few well-defined channels or within a small area of fractured rock, it may be possible to seal the area by pressure grouting.[1] If the area of leakage is large, the cost of grouting may be excessive.

7-13   Reservoir-site selection   It is virtually impossible to locate a reservoir site having completely ideal characteristics. General rules for choice of reservoir sites are:

1. A suitable dam site must exist. The cost of the dam is often a controlling factor in selection of a site.
2. The cost of real estate for the reservoir (including road, railroad, cemetery, and dwelling relocation) must not be excessive.
3. The reservoir site must have adequate capacity.
4. A deep reservoir is preferable to a shallow one because of lower land costs per unit of capacity, less evaporation loss, and less likelihood of weed growth.
5. Tributary areas which are unusually productive of sediment should be avoided if possible.
6. The quality of the stored water must be satisfactory for its intended use.

---

[1] L. A. Schmidt, Jr., Reservoir Leakage Stopped at Outlets, *Eng. News-Record*, Vol. 134, pp. 137–139, Jan. 11, 1945; Foundation Treatment, *J. Am. Water Works* Assoc., Vol. 38, pp. 1170–1178, October, 1946.

BIBLIOGRAPHY

American Society of Civil Engineers, Committee on Sedimentation: Sediment Transportation Mechanics, *J. Hydraulics Div., ASCE,* Vol. 88, pp. 77–128, July, 1962.
Brown, C. B.: Sediment Transportation, chap. XII in Hunter Rouse (ed.), "Engineering Hydraulics," Wiley, New York, 1950.
Gottschalk, L. C.: Reservoir Sedimentation, chap. 17-I in V. T. Chow (ed.), "Handbook of Applied Hydrology," McGraw-Hill, New York, 1964.
Koelzer, Victor A.: Reservoir Hydraulics, sec. 4 in C. V. Davis and K. E. Sorenson (eds.), "Handbook of Applied Hydraulics," 3d ed., McGraw-Hill, New York, 1969.
Linsley, R. K., M. A. Kohler, and J. L. H. Paulhus: "Applied Hydrology," chaps. 13, 16, 20, 21, McGraw-Hill, New York, 1949.
Linsley, R. K., M. A. Kohler, and J. L. H. Paulhus: "Hydrology for Engineers," chaps. 12, 13, McGraw-Hill, New York, 1958.
Stall, J. B., and J. C. Neill: Calculated Risks of Impounding Reservoir Yield, *J. Hydraulics Div., ASCE,* Vol. 89, pp. 25-34, January, 1963.
Techniques of Forecasting Wind Waves and Swell, *U.S. Navy Hydrographic Office Pub.* 604, 1951.
Thomas, N. O., and G. E. Harbeck: Reservoirs in the United States, *U.S. Geol. Surv. Water Supply Paper* 1360-A, 1956.
Trask, P. D.: "Applied Sedimentation," Wiley, New York, 1950.

PROBLEMS

7-1 What reservoir capacity is required for the demand rates of Illustrative Example 7-1 if pumping is to be limited to the day shift (8 A.M. to 5 P.M.)? What pump capacity is required?

7-2 The mass curve of available water during the critical dry period at a given storage reservoir is as shown.

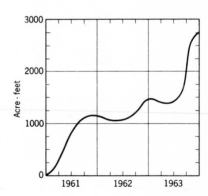

(*a*) What continuous constant yield (in acre-ft/yr) is possible with a reservoir having a storage capacity of 600 acre-ft?
(*b*) What storage capacity (in acre-ft) is required for a constant yield rate of 120 gal/min?

7-3 The flows into and out of a reservoir are as follows:

Time	Inflow, cfs	Outflow, cfs
1000	20	30
1200	26	26
1400	42	20
1600	54	12
1800	48	8
2000	36	8

At 10 A.M. there are 4.0 acre-ft of water in the reservoir. How much water is in the reservoir at 5 P.M.?

7-4 What reservoir capacity is required to produce a yield (at uniform rate) of 60 acre-ft/yr for a site where the monthly flows during a critical flow period are as tabulated below?

October	18	October	5	October	15
November	22	November	6	November	17
December	17	December	6	December	25
January	26	January	5	January	47
February	15	February	3	February	16
March	32	March	2	March	18
April	8	April	1	April	7
May	3	May	0	May	4
June	0	June	0	June	0
July	0	July	0	July	1
August	0	August	0	August	3
September	0	September	7	September	4

Note: These monthly flows are expressed in acre-feet. Neglect effect of evaporation and precipitation.

7-5 Given below are monthly flows, pan evaporation, rainfall, and demand rates for a proposed reservoir site. Prior water rights require the release of natural flow or 5 cfs, whichever is least. Assume that the average reservoir area is 1500 acres and that the runoff coefficient for the land which will be flooded is 0.4. What reservoir capacity is required? If the previous year was substantially similar to the one given in the table, is there sufficient water to meet the demand?

Month	Mean flow, cfs	Demand, acre-ft	Evaporation, in.	Rainfall, in.
January	107	1000	2.1	5.3
February	95	1500	2.7	3.0
March	109	2000	3.2	1.6
April	52	3000	3.6	1.0
May	20	3600	5.5	0.2
June	9	3600	7.9	0
July	3	3600	8.3	0
August	5	3500	8.0	0
September	25	3400	7.8	0
October	4	2000	6.1	0
November	95	1800	3.4	4.7
December	103	1000	2.1	6.3

7-6   A city engineer estimates the hourly demand for water on the maximum day as tabulated below. If pumping is to be at a uniform rate for the 24 hr, what pump capacity is required? What reservoir capacity?

Hour ending	Demand, gpm	Hour ending	Demand, gpm
0100	720	1300	910
0200	600	1400	935
0300	535	1500	935
0400	530	1600	920
0500	500	1700	925
0600	500	1800	940
0700	560	1900	940
0800	690	2000	970
0900	775	2100	985
1000	850	2200	1000
1100	870	2300	950
1200	900	2400	890

7-7   Tabulated below are monthly flows for a period of low runoff on a small stream, corresponding monthly rainfall, and the normal monthly pan evaporation. Find the safe yield, assuming a constant demand rate and a 5000-acre-ft reservoir. What is the maximum possible yield from this stream for the period given, and what reservoir capacity would be required to sustain this yield? Assume the average water-surface area of the reservoir to be 400 acres, and the runoff coefficient of the flooded land to be 0.2.

Month	Flow, acre-ft				Rainfall, in.				Normal pan evaporation, ft
	1932	1933	1934	1935	1932	1933	1934	1935	
January	2030	1045	62	1820	4.5	9.1	1.3	9.1	0.20
February	4460	26	300	18	4.7	1.2	5.5	1.3	0.29
March	0	340	13	1630	0.5	3.4	0	5.9	0.42
April	0	6	8	3680	0.7	0.2	0.7	5.8	0.48
May	8	4	6	23	0.2	1.9	0.6	0	0.51
June	0	1	0	8	0	0	0.6	0	0.45
July	0	0	0	1	0	0	0	0	0.38
August	0	0	0	0	0	0	0	0	0.25
September	0	0	0	0	0	0	1.0	0.2	0.14
October	0	0	0	0	0.4	1.9	1.0	0.8	0.07
November	0	0	0	0	0.8	0	5.1	0.5	0.08
December	3	0	1	8	4.6	6.9	3.7	3.5	0.11

7-8 Plot a mass curve for a period of 20 yr for a stream selected by your instructor. Use monthly flow data from *U.S. Geological Survey Water Supply Papers*. What is the safe yield of this stream if a reservoir with a capacity equal to the annual runoff volume were provided? How many years during the 20-yr period would have produced an annual yield not more than 25 percent greater than the safe yield? Fifty percent greater than the safe yield? Twice the safe yield? Would it be safe to plan a water-use project on the basis of a yield greater than the safe yield?

7-9 Using the data of Prob. 7-8, find the safe yield with a reservoir capacity of one-half the mean annual runoff of the stream. Estimate the probable change in this yield if the mass curve were corrected for rainfall on and evaporation from the reservoir surface. Assume a reasonable reservoir area.

7-10 Plot a sediment rating such as Fig. 7-8, and estimate the equation of the curve. *U.S. Geological Survey Water Supply Paper* 1163 contains sediment data and on page 15 gives references to earlier publications.

7-11 The water-year precipitations at a certain Weather Bureau station were as follows:

28.13	36.45	26.73	27.98	30.18
31.62	34.72	27.12	26.12	32.66
27.45	29.16	30.06	28.35	36.81
34.12	26.81	38.55	31.17	33.22
31.06	28.62	30.16	30.25	24.16
32.77	33.15	31.74	22.06	27.60
39.01	23.64	32.42	28.48	29.15
35.20	29.12	35.21	25.26	33.12
29.17	36.20	34.65	39.22	29.62
32.24	33.13	29.17	28.43	26.45
25.96	30.62	36.71	32.17	32.53
24.87	30.54	34.36	27.45	35.64
33.72	27.42	23.18	31.28	31.52

The relation between water-year precipitation and runoff from a watershed in the vicinity of the rain gage is given by this curve:

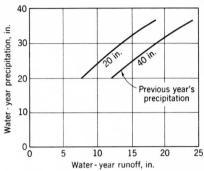

Water - year runoff, in.

(a) Develop a 50-yr record of water-year runoff on the assumption that the given 65-yr record of precipitation is a representative one. Do this by selecting precipitations at random from the entire set of data. Assume 20 in. of precipitation during the year preceding the start of record. How could this procedure be improved to account for the possibility of precipitation outside the bounds of the given data?
(b) Assuming the basin has an area of 7000 acres, plot a mass curve of runoff for the 50 yr of record.
(c) Determine the mean annual flow.
(d) Look at the most critical dry periods of your mass curve and determine the storage required for yields of 3000, 2500, 2000, 1500, and 1000 acre-ft/yr. Neglect the effects of local inflow, precipitation, evaporation, seepage, downstream releases, etc. It is suggested that the instructor tabulate and discuss the results obtained by the students.

7-12 A reservoir is contemplated on a stream which has an average annual runoff of 400,000 acre-ft. Measurements indicate that the average sediment inflow is 280,000 tons/yr. Assuming that a cubic foot of settled sediment will dry out to a weight of 75 lb/cu ft, show a plot of probable reservoir capacity versus time if the original capacity of the reservoir is 20,000 acre-ft. Use the median curve of Fig. 7-9.

7-13 Repeat the preceding problem for the situation where the average sediment inflow is 2,800,000 tons/yr.

7-14 What is the average concentration of suspended sediment for the inflows of the two preceding problems? Express answers in parts per million (ppm) on a weight basis.

7-15 A small reservoir (5000 acre-ft capacity) is proposed on the Red River. The average annual sediment load is estimated at 1200 tons/sq mi, drainage area is 850 sq mi. If the average annual runoff from this basin is 1.2 in., what is the most probable life of the reservoir to the point where it is 80 percent full of sediment? Assume 1500 tons of sediment occupies 1 acre-ft. If 25 percent of the incoming sediment could be vented through the sluiceways, what would be the probable life of the reservoir?

7-16 A reservoir has a fetch of 5 mi and the estimated variation of wind speed with duration as tabulated below. What is the significant wave height to be expected on this reservoir? Show also the significant wave height for each pair of data.

Wind duration, hr	0.2	0.5	1.0	1.5	2.0
Wind speed, mph	51	38	30	27	21

7-17 Analysis of westerly winds at a given point reveals the following relation between wind speed and duration:

Wind duration, min.	15	30	60	90	120
Wind speed, mph	61	52	43	41	40

Determine from Fig. 7-11 the significant wave heights for westerly winds with fetches of 1, 2, 3, 5, and 7 mi.

7-18 Repeat the preceding problem with these wind data:

Wind duration, min.	15	30	60	90	120
Wind speed, mph	78	63	45	36	30

7-19 What wind setup may be expected on a reservoir with a fetch of 8 mi, average depth of 20 ft, and critical wind speed (land station) of 37 mph? What wind setup would occur if the depth were 200 ft?

7-20 A reservoir has an effective fetch for waves of 9 mi and for setup of 23 mi. Average depth is 120 ft. The critical wind velocity (land station) is 47 mph. What freeboard allowance should be made for an 8 percent exceedance if the upstream face of the dam is smooth? How much less should the allowance be for an upstream facing of riprap? Slope of upstream face of dam is 1:3.

7-21 For a reservoir in your vicinity selected by the instructor, carry out a complete setup and wave analysis, securing wind data from the National Weather Service and physical data on the reservoir from the operating agency.

Chapter 8

Dams

The first dam for which there are reliable records was built on the Nile River sometime before 4000 B.C. It was used to divert the Nile and provide a site for the ancient city of Memphis. This dam is no longer in existence. The oldest dam still in use is the Almanza Dam in Spain, which was constructed in the sixteenth century. With the passage of time, materials and methods of construction have improved, making possible the erection of such large structures as Hoover Dam on the Colorado River and Grand Coulee on the Columbia River. The failure of a dam may cause serious loss of life and property; consequently, the design and maintenance of dams are commonly under governmental surveillance. In the United States over 28,000 dams are under the control of state authorities.[1]

8-1　Types of dams　Dams are classified on the basis of the type and materials of construction, as *gravity*, *arch*, *buttress*, and *earth-fill* (Fig. 8-1). The first three types are usually constructed of concrete. A gravity dam depends on its own weight for stability and is usually straight in plan although sometimes slightly curved. Arch dams transmit most of the horizontal thrust of the water behind them to the abutments by arch action and have thinner cross sections than comparable gravity dams. Arch dams can be used only in narrow canyons where the walls are capable of withstanding the thrust produced by the arch action. The simplest of the many types of buttress dams is the slab type, which consists of sloping flat slabs supported at intervals by buttresses. Earthfill dams are embankments of rock or earth with some provision for controlling seepage by means of an impermeable core or upstream blanket. More than one type of dam may be included in a single structure. Curved dams may combine both gravity and arch action to

---

[1] "Supervision of Dams by State Authorities," Subcommittee of the U.S. Committee on Large Dams, New York, July, 1966.

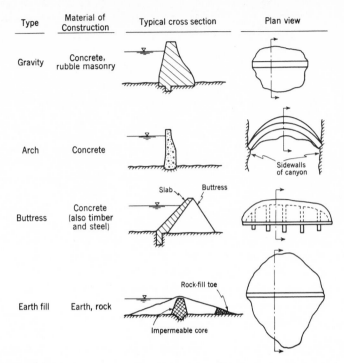

Type	Material of Construction	Typical cross section	Plan view
Gravity	Concrete, rubble masonry		
Arch	Concrete		Sidewalls of canyon
Buttress	Concrete (also timber and steel)	Slab Buttress	
Earth fill	Earth, rock	Rock-fill toe Impermeable core	

FIG. 8-1   Basic types of dams.

achieve stability. Long dams often have a concrete river section containing spillway and sluice gates and earth or rock-fill wing dams for the remainder of their length.

The selection of the best type of dam for a given site is a problem in both engineering feasibility and cost. Feasibility is governed by topography, geology, and climate. For example, because concrete spalls when subjected to alternate freezing and thawing, arch and buttress dams with thin concrete sections are sometimes avoided in areas subject to extreme cold. The relative cost of the various types of dams depends mainly on the availability of construction materials near the site and the accessibility of transportation facilities.

The height of a dam is defined as the difference in elevation between the roadway, or spillway crest, and the lowest part of the excavated foundation.[1] However, figures quoted for heights of dams are often determined in other ways. Frequently the height is taken as the net height above the old river bed.

---

[1] "Glossary: Water and Sewage Control Engineering," American Society of Civil Engineers, 1949.

8-2   **Forces on dams**   A dam must be relatively impervious to water and capable of resisting the forces acting on it. The most important of these forces are *gravity* (*weight of dam*), *hydrostatic pressure, uplift, ice pressure, and earthquake forces*. These forces are transmitted to the foundation and abutments of the dam, which react against the dam with an equal and opposite force, the *foundation reaction*. The effect of hydrostatic pressure caused by sediment deposits in the reservoir and of dynamic forces caused by water flowing over the dam may require consideration in special cases.

The weight of a dam is the product of its volume and the specific weight of the material. The line of action of this force passes through the center of area of the cross section. Hydrostatic forces may act on both the upstream and downstream faces of the dam. The horizontal component H_h of the hydrostatic force is the force on a vertical projection of the face of the dam and for unit width of dam is

$$H_h = \frac{\gamma h^2}{2} \tag{8-1}$$

where γ is the specific weight of water and h is the depth of water. The line of action of this force is $h/3$ above the base of the dam. The vertical component of the hydrostatic force is equal to the weight of water vertically above the face of the dam and passes through the center of gravity of this volume of water.

Water under pressure inevitably finds its way between the dam and its foundation and creates uplift pressures. The magnitude of the uplift force depends on the character of the foundation and the construction methods. It is often assumed that the uplift pressure varies linearly from full hydrostatic pressure at the upstream face (*heel*) to full tailwater pressure at the downstream face (*toe*). For this assumption the uplift force U is

$$U = \gamma \frac{h_1 + h_2}{2} t \tag{8-2}$$

where t is the base thickness of the dam and h_1 and h_2 are the water depths at the heel and toe of the dam, respectively. The uplift force will act through the center of area of the pressure trapezoid (Fig. 8-4).

Actual measurements on dams (Fig. 8-2) indicate that the uplift force is much less than that given by Eq. (8-2). Various assumptions have been made regarding the distribution of uplift pressures.[1] The U.S. Bureau of Reclamation sometimes assumes that the uplift pressure on gravity dams varies linearly from two-thirds of full uplift at the heel to zero at the toe.

---

[1] Uplift in Masonry Dams, Final Report of the Subcommittee on Uplift in Masonry Dams, *Trans. ASCE*, Vol. 117, pp. 1218–1252, 1952.

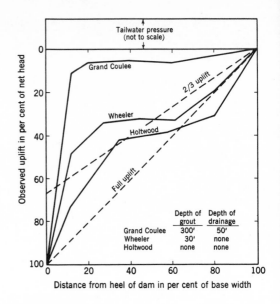

FIG. 8-2   Maximum observed uplift pressures under some existing gravity dams. (*Data from the ASCE Committee on Uplift in Masonry Dams*)

Drains are usually provided near the heel of the dam to permit the escape of seepage water and relieve uplift.

An ice sheet subjected to a temperature increase will expand and exert a thrust against the upstream face of a dam. Many dams in the northern United States have been designed for ice loads as large as 60,000 lb/lin ft applied at maximum winter-water level. Such forces are too high because the plastic properties of ice prevent such large stresses from developing except during very rapid temperature changes. The curves of Fig. 8-3 show probable ice thrusts[1] for various ice thicknesses and rates of temperature change. These curves assume no lateral restraint of the ice sheet and for complete restraint should be multiplied by 1.58 (assuming Poisson's ratio = 0.365). Solar radiation accelerates the heating of the ice and hence produces larger thrusts. More detailed information on ice pressure is available in the literature.[2]

When an earthquake shakes the earth upon which a dam rests, the resulting inertia force equals the product of the mass of the dam and the acceleration caused by the earthquake. In regions of known earthquake activity, horizontal accelerations of as high as 0.5 are assumed to act on the dam. Vertical motion may also occur during an earthquake, with a resultant

[1] E. Rose, Thrust Exerted by Expanding Ice Sheet, *Trans. ASCE*, Vol. 112, pp. 871–900, 1947.
[2] Ice Pressure against Dams—A Symposium, *Trans. ASCE*, Vol. 119, pp. 2–42, 1954.

vertical inertia force which acts momentarily to change the effective weight of the dam.

In addition to the inertia forces acting on the dam, earthquakes also cause oscillatory increases and decreases in the hydrostatic pressure on the face of the dam. Von Karman[1] suggested that this force be computed from

$$E_w = 0.555k\gamma h^2 \qquad (8\text{-}3)$$

where k is the ratio of the acceleration caused by the earthquake to that of gravity. The force E_w acts at a distance $4h/3\pi$ above the bottom of the reservoir. Zangar[2] also proposed formulas for computing this force which give results comparable to the Von Karman approach.

The material underlying a dam must be capable of withstanding the foundation pressures exerted on it under both moist and dry conditions. Most failures of dams have been caused by failure of the underlying strata. The Austin Dam on the Colorado River in Texas collapsed in 1900 because large cavities had been dissolved in its limestone foundation. The St. Francis Dam, a concrete gravity structure in California 205 ft high and 700 ft long, failed in 1928 soon after completion because conglomerate in one abutment became weakened after exposure to moisture from the reservoir. The failure of the 200-ft-high Malpasset arch dam in southern France (1959) has been attributed to the presence of a clay seam in the rock at one of the abutments.[3]

[1] T. von Karman, Discussion of Pressures on a Dam during Earthquakes, *Trans. ASCE*, Vol. 98, pp. 434–436, 1933.
[2] C. N. Zangar, Hydrodynamic Pressures on Dams Due to Horizontal Earthquake Effects, *Eng. Monograph* 11, Bureau of Reclamation, May, 1952.
[3] Malpasset Dam on French Riviera Fails, *Civil Eng.* (*N.Y.*), p. 91, January, 1960.

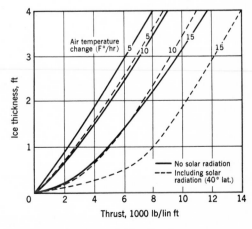

FIG. 8-3   Effect of change in air temperature on ice thrust. (After Rose)

Though an extensive geologic investigation had been made before the construction of the dam, the clay seam had gone undetected. A carefully conducted geologic investigation should be one of the first steps in the design of a dam. The geologic investigation should include inspection of the rock outcrops and extensive underground exploration by means of test holes, core drilling, and geophysical methods. Some of the test holes should be large enough for a man to enter and inspect thoroughly.

GRAVITY DAMS

Prior to the middle of the nineteenth century, dams were designed by rule of thumb with little concern for the principles of mechanics of materials, and, as a result, they were usually much more massive than necessary. Pioneers in the field of improved design practice were the French engineer DeSazilly, the English scientist Rankine, and in this country Wegmann. Some of the larger gravity dams in the United States are listed in Table 8-1.

TABLE 8-1   Some Large Gravity Dams in the United States

Name	State	Volume, 1000 cu yd	Height, ft	Length, ft	Top width,* ft	Base thickness, ft	Plan	Reservoir capacity, 1000 acre-ft	Year completed
Grand Coulee	Wash.	9737	550	4173	30	400	Straight	9517	1942
Shasta	Calif.	6230	602	3460	30	568	Curved	4389	1945
Fontana	N.C.	2700	470	1750	30	374	Straight	1600	1945
Friant	Calif.	2030	319	3488	20	267	Straight	520	1942
Norfork	Ark.	1500	244	2620	34	200	Straight	1983	1944
Wilson	Ala.	1240	137	4860	15	101	Straight	490	1926
Tygart	W. Va.	1200	232	1880	...	195	Straight	327	1937
Norris	Tenn.	1195	265	1570	20	203	Straight	2710	1935
Marshall Ford	Tex.	1774	278	2422	20	215	Straight	1934	1942
New Croton	N.Y.	855	238	710	22	185	Straight	92	1907
Pardee	Calif.	640	358	1320	16	246	Curved	222	1929
Elephant Butte	N. Mex.	605	301	1674	18	214	Straight	2219	1916
Morris	Calif.	450	328	756	20	280	Curved	37	1934
Wachusett	Mass.	267	228	971	22	187	Straight	193	1906
Medina	Tex.	205	178	1580	25	128	Straight	254	1913
Cross River	N.Y.	161	170	772	23	116	Straight	33	1908
San Mateo	Calif.	157	154	680	25	176	Curved	54	1889

*Exclusive of overhang for roadway, operating bridges, etc.

A simplified approach to the analysis of gravity dams based on elastic behavior of concrete is presented in the following sections. This approach is adequate for the design of small dams. It is also useful for preliminary design of large dams, though final design is generally accomplished through application of more sophisticated methods involving finite-element procedures[1] in which the digital computer plays an important role.

8-3   Structural stability of gravity dams   Figure 8-4 is a simplified free-body diagram of the cross section of a gravity dam. The forces shown are the weight of the dam W, the horizontal components of hydrostatic force H_h, the vertical components of hydrostatic force H_v, uplift U, ice pressure F_i, the increased hydrostatic pressure caused by earthquakes E_w, and the inertia force caused by the earthquake on the dam itself E_d. The vectorial resultant of these forces is equal and opposite to R, the equilibrant, which is the effective force of the foundation on the base of the dam. A gravity dam may fail by sliding along a horizontal plane, by rotation about the toe, or by failure of the material. Failure may occur at the foundation plane or at any higher level in the dam. Sliding (or shear failure) will occur when the net horizontal force above any plane in the dam exceeds the shear resistance developed at that level. It is good construction practice to step the foundation of a dam to increase resistance to sliding. Overturning and excessive compressive stress can be avoided by selecting a cross section of proper size and shape. Typical

---

[1] O. C. Zienkiewicz, The Finite-Element Method, sec. 10 in C. V. Davis and K. E. Sorenson (eds.), "Handbook of Applied Hydraulics," McGraw-Hill, New York, 1969.

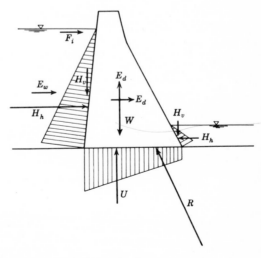

FIG. 8-4   Free-body diagram of the cross section of a gravity dam.

working stresses employed in the design of concrete dams are about 600 psi for compression and 0 psi for tension. Tensile stresses are avoided by keeping the resultant of all forces within the middle third of the base.

8-4   Analysis of gravity dams   Preliminary analysis of a gravity dam is made by isolating a typical cross section of unit width. This section is assumed to act independently of adjoining sections, although this assumption disregards beam action in the dam as a whole. Structural analysis of a section proceeds step by step from top to bottom and must consider both reservoir-full and reservoir-empty conditions.

ILLUSTRATIVE EXAMPLE 8-1   Figure 8-5a shows a typical cross section near the center of a gravity dam. Analyze the block B_1 for stability. Assume zero freeboard.

The forces acting on block B_1 are shown in Fig. 8-5b. For the condition of empty reservoir the hydrostatic forces are zero, and, neglecting earthquake and ice forces, the only active force is the weight W acting at the center of gravity. Taking the specific weight of concrete as 150 lb/cu ft, the weight of a unit width of block B_1 is

$$W = 150 \left(\frac{20 \times 1}{2} + 20 \times 10 + \frac{20 \times 6}{2} \right) = 40{,}500 \text{ lb}$$

which acts at

$$\bar{x} = \frac{(10 \times 16.33) + (200 \times 11) + (60 \times 4)}{10 + 200 + 60} = 9.65 \text{ ft} \qquad \text{from } t_1$$

The reaction R from the underlying block will be equal and opposite to W. Since the line of action of W is vertical and cuts the base, block B_1 is statically stable.

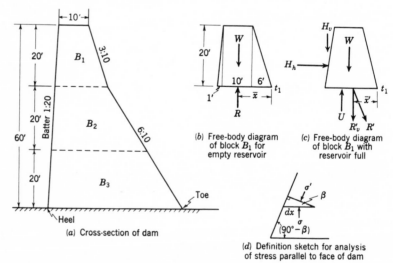

(b) Free-body diagram of block B_1 for empty reservoir

(c) Free-body diagram of block B_1 with reservoir full

(a) Cross-section of dam

(d) Definition sketch for analysis of stress parallel to face of dam

FIG. 8-5   *Analysis of a gravity dam.*

Assuming a linear distribution of stress, the direct normal stress in the concrete is a combination of the stress caused by the central load and that caused by the moment which arises from the eccentricity of the load. In this case the eccentricity e is

$$e = 9.65 - 8.50 = 1.15 \text{ ft}$$

and the normal stresses at the extreme fibers are

$$\sigma = \frac{P}{A} \pm \frac{Mc}{I} = \left[\frac{40{,}500}{17 \times 1} \pm \frac{40{,}500 \times 1.15 \times 8.5}{(1 \times 17^3)/12} \right] \frac{1}{144}$$

whence σ at the heel is 23.2 psi and at the toe is 9.8 psi. The stresses parallel to the face of the dam may be found by writing the equation for equilibrium of vertical forces, using the notation of Fig. 8-5d.

$$\sigma \, dx = (\sigma' \cos \beta)(dx \cos \beta)$$

from which

$$\sigma' = \frac{\sigma}{\cos^2 \beta}$$

At the heel

$$\sigma' = \frac{23.2}{(0.9988)^2} = 23.2 \text{ psi}$$

while at the toe

$$\sigma' = \frac{9.8}{(0.958)^2} = 10.6 \text{ psi}$$

For reservoir-empty conditions the assumed design of block B_1 is adequate. Note that the assumption of a linear stress distribution implies perfect elasticity of the concrete. Though concrete is not perfectly elastic this approach provides a convenient mode of analysis.

For the reservoir-full condition (Fig. 8-5c) the components of hydrostatic pressure on a unit width of block B_1 are

$$H_h = \tfrac{1}{2}(62.5)(20)^2 = 12{,}500 \text{ lb} \qquad \text{and} \qquad H_v = 62.5(^{20}\!/_2) = 625 \text{ lb}$$

Since water might seep through joints in the concrete and exert hydrostatic pressure, it is customary to assume[1] that uplift pressure acts at any level in the dam. The uplift force is

$$U = \frac{(62.5 \times 20) + 0}{2}(17 \times 1) = 10{,}625 \text{ lb}$$

The lines of action of H_h, H_v, and U are all at the one-third points of their respective force triangles. The tendency for the block to overturn is found by taking moments about t_1. The sum of the overturning moments is

---

[1] There are indications that this assumption is overly conservative. See Roy W. Carlson, Permeability, Pore Pressure, and Uplift in Gravity Dams, *Trans. ASCE*, Vol. 122, pp. 587–613, 1957.

$$12{,}500 \times {}^{20}\!\!/_3 + 10{,}625 \times 17 \times (\tfrac{2}{3}) = 203{,}800 \text{ ft-lb}$$

The forces W and H_v create a righting moment of

$$40{,}500 \times 9.65 + 625 \times 16.67 = 401{,}200 \text{ ft-lb}$$

The factor of safety against overturning is the ratio of the righting moment to the overturning moment, or

$$\text{Factor of safety (overturning)} = \frac{401{,}200}{203{,}800} = 1.97$$

The force H_h tends to cause block B_1 to slide with respect to block B_2. If it is assumed that the two blocks are monolithic, the average shearing stress along the plane between the blocks is

$$\tau = \frac{12{,}500}{17 \times 144} = 5.1 \text{ psi}$$

which is well below the 250 psi commonly taken as a working stress for concrete in shear. If it is assumed that there is no bond between the blocks, the friction force that could develop along the contact plane is $F_f = \mu(W + H_v - U)$. Taking the coefficient of friction $\mu = 0.65$, $F_f = 19{,}800$ lb, and the factor of safety against sliding is

$$\text{Factor of safety (sliding)} = \frac{F_f}{H_h} = \frac{19{,}800}{12{,}500} = 1.58$$

To compute the compressive stress at the base of the block, the forces W, H_h, H_v, and U can be combined into a resultant R' which has a vertical component

$$R'_v = 40{,}500 + 625 - 10{,}625 = 30{,}500 \text{ lb}$$

which cuts the base at a distance

$$\bar{x}' = \frac{401{,}200 - 203{,}800}{30{,}500} = 6.48 \text{ ft from } t_1$$

The eccentricity $e' = 8.50 - 6.48 = 2.02$ ft and the normal stresses are 3.6 psi at the heel and 21.3 psi at the toe.

The calculations of Illustrative Example 8-1 indicate that the assumed dimensions of the upper block are adequate. The next step would be similar calculations for blocks B_1 and B_2 as a unit. After that the entire dam should be analyzed in the same fashion. The final step is an analysis of the adequacy of the foundation. It is assumed that the direct normal stresses in the concrete at the base of the dam are transferred to the foundation. Typical working stresses for various foundation materials are given in Table 8-2.

In actual practice a cross section is first assumed and checked for stability as outlined above. Factors of safety against overturning of about 2 and against sliding of 1 to 1.5 are usually considered acceptable. If the computed factors of safety fall below these limits, the section should be modified. The

TABLE 8-2   *Allowable Compressive Stresses for Foundation Materials*

Material	Allowable stress, psi
Granite	600–1000
Limestone	400– 800
Sandstone	400– 600
Gravel	40– 80
Sand	20– 60
Firm clay	50
Soft clay	15

dam of Fig. 8-5*a* with a base width about two-thirds its height has a fairly representative cross section. However, the downstream face should be straight or smoothly curved rather than a series of intersecting planes where sharp corners and discontinuities give rise to stress concentrations. The elevation of the crest of the dam is found by adding the expected maximum depth of flow over the spillway (Chap. 9) and the anticipated wave heights to the adopted normal pool level. Top widths of gravity dams vary from about 0.15 times the height for low dams to the width necessary for a roadway or gate operating facilities on high dams. A top width of 20 ft is ample for most dams unless special roadways are desired.

The analysis above was applied to a section near the center of a dam. Conditions near the ends of the dam will be quite different since the height need not be as great. Calculated stresses in Morris Dam (curved in plan) are presented in Fig. 8-6. It should be noted that the maximum pressures at

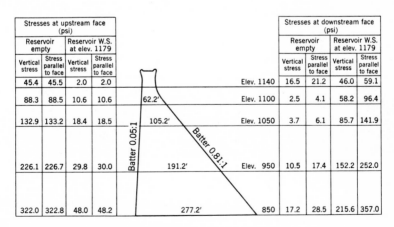

Stresses at upstream face (psi)								Stresses at downstream face (psi)			
Reservoir empty		Reservoir W.S. at elev. 1179						Reservoir empty		Reservoir W.S. at elev. 1179	
Vertical stress	Stress parallel to face	Vertical stress	Stress parallel to face					Vertical stress	Stress parallel to face	Vertical stress	Stress parallel to face
45.4	45.5	2.0	2.0				Elev. 1140	16.5	21.2	46.0	59.1
88.3	88.5	10.6	10.6		62.2'		Elev. 1100	2.5	4.1	58.2	96.4
132.9	133.2	18.4	18.5		105.2'		Elev. 1050	3.7	6.1	85.7	141.9
226.1	226.7	29.8	30.0		191.2'		Elev. 950	10.5	17.4	152.2	252.0
322.0	322.8	48.0	48.2		277.2'		850	17.2	28.5	215.6	357.0

FIG. 8-6   *Calculated stresses in Morris Dam. (Courtesy of the Water Department of Pasadena, California)*

the heel occur when the reservoir is empty, while maximum pressures at the toe develop with full reservoir. Low gravity dams may be built on unconsolidated foundation material as long as the foundation pressure is within reasonable limits. Since no shear resistance can be expected, the sliding factor of safety for this case should be about 2.5.

8-5   Construction of gravity dams   Before construction work in a river channel can be started, the streamflow must be diverted. In two-stage construction the flow is diverted to one side of the channel by a cofferdam (Fig. 8-7b) while work proceeds on the other side. After work on the lower portion of one side of the dam is complete, flow is diverted through outlets in this portion or may even be permitted to overtop the completed portion while work proceeds in the other half of the channel. If geologic and topographic conditions are favorable, a tunnel or diversion channel may be used to convey the entire flow around the dam site. A tunnel is particularly advantageous if it will serve some useful purpose after completion of the dam. Four 50-ft circular concrete-lined tunnels were used for diversion at Hoover Dam and later converted to outlet works. A diversion channel or tunnel should be capable of carrying a flow selected by frequency analysis as a reasonable risk in view of the hazards on each particular job. It is advantageous to schedule construction of the lower portion of a dam during normal low-flow periods to minimize the diversion problem.

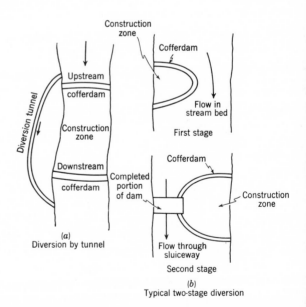

FIG. 8-7   *Typical streamflow-diversion procedures for construction of dams.*

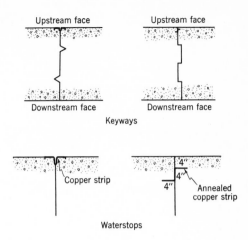

FIG. 8-8   Keyways and waterstops for dams.

The foundation must be excavated to solid rock before any concrete is poured. After excavation, cavities or faults in the underlying strata are sealed with concrete or grout. Frequently a *grout curtain* is placed near the heel of the dam to reduce seepage and uplift. A grout of cement and water sometimes mixed with a small amount of fine sand is forced under pressure into holes drilled into the rock. Grouting at pressures up to about 40 psi may be done before concrete is placed for the dam, but high-pressure grouting (200 psi) is done from permanent galleries in the dam after the dam is complete so that the weight of the dam can resist the grouting pressures.

Concrete for the dam is usually placed in blocks[1] depending on the dimensions of the dam, with a maximum width of about 50 ft on large dams. Maximum height of a single pour is usually about 5 ft. Sections are poured alternately so that each block is permitted to stand several days before another one is poured next to it or on top of it (Fig. 8-13). After individual sections are poured, they are sprinkled with water and otherwise protected from the drying effect of the air. After the forms are removed, the lateral surfaces of each section are painted with an asphaltic emulsion to prevent adherence to adjoining sections and to form construction joints to reduce cracking of the concrete. *Keyways* (Fig. 8-8) are provided between sections to carry the shear from one section to the adjacent one and make the dam act as a monolith. Metal *water stops* (Fig. 8-8) are also placed in the vertical construction joints near the upstream face to prevent leakage. Inspection galleries to permit access to the interior of the dam are formed as the concrete

---

[1] Recent Technical Advances in Constructing Large Dams, *Travaux, Monthly Review*, Paris, May, 1961.

is placed. These galleries may be necessary for grouting operations, for operation and maintenance of gates and valves, and as intercepting drains for water which seeps into the dam.

When concrete sets, a great deal of heat is liberated, and the temperature of the mass is raised. As the concrete cools, it shrinks and cracks may develop. To avoid cracks, special low-heat cement may be used. Very lean mixes are also used for the interior of the dam. Two sacks of cement per cubic yard of concrete is not uncommon. In addition, the materials which go into the concrete may be cooled before mixing. For best results the temperature of the concrete mix should be between 50 and 80°F. Occasionally, further cooling is accomplished by circulating cold water through pipes embedded in the concrete, although this is expensive and is generally used only on large gravity dams.

ARCH DAMS

An arch dam[1] is curved in plan and carries most of the water load horizontally to the abutments by arch action. The thrust thus developed makes it essential that the sidewalls of the canyon be capable of resisting the arch forces. The first arch dam in the United States was Bear Valley Dam, built in southern California in 1884 and now no longer in use. Like many early arch dams Bear Valley Dam was of rubble masonry, but practically all arch dams constructed in recent years are of concrete. Relatively few arch dams have failed, in comparison with the more numerous failures of other types of dams. Pertinent data on some typical arch dams in the United States are given in Table 8-3.

8-6   General considerations   Structural analysis of arch dams is complex, and the computations are lengthy. In principle an arch dam is visualized as consisting of a series of horizontal arches transmitting thrust to the abutments or a series of vertical cantilevers fixed at the foundation (Fig. 8-9). The horizontal component of the water load is resisted jointly by the

---

[1] Symposium on Arch Dams, *Trans. ASCE*, Vol. 125, pp. 909–993, 1960.

FIG. 8-9   Structural elements of an arch dam.

(a) Series of horizontal arches

(b) Series of vertical cantilevers

TABLE 8-3   Typical Arch Dams in the United States

Name	State	Height, ft	Length, ft	Top width, ft	Base thick-ness, ft	Vol-ume, 1000 cu yd	Radius	Reser-voir capac-ity, 1000 acre-ft	Year com-pleted
Hoover	Ariz.-Nev.	726	1244	45	660	3251	Constant	31,141	1936
Glen Canyon	Ariz.	700	1550	25	340	4830	Constant	28,040	1962
Hungry Horse	Mont.	520	2115	39	330	2900	Variable	3,500	1952
Diablo	Wash.	426	1180	16	140	350	Variable	90	1930
Owyhee	Oreg.	417	833	30	265	488	Constant	1,120	1932
Pacoima	Calif.	372	640	10	100	225	Variable	6	1928
Arrowrock	Idaho	350	1150	16	223	578	Constant	286	1915
Parker	Ariz.-Calif.	320	856	39	100	260	Constant	717	1938
Horse Mesa	Ariz.	305	784	8	43	147	Variable	245	1927
Seminoe	Wyo.	295	530	17	88	173	Constant	1,026	1939
Shannon	Wash.	263	493	20	135	132	Constant	132	1926
Big Tujunga No. 1	Calif.	253	800	8	73	79	Variable	6	1931
Cheeseman*	Colo.	236	710	18	176	103	Constant	80	1904
Calderwood	Tenn.	230	814	25	48	400	Variable	34	1930
Pelton	Oreg.	204	583	8	26	39	Variable	37	1957
Gibson	Mont.	199	960	15	87	161	Constant	105	1929
East Canyon Cr.	Utah	193	273	5	26	16	Variable	28	1916
Safford	Ariz.	110	180	2	5	1	Variable	0.2	1929
Las Vegas	N. Mex.	50	210	4	16	3	Constant	2	1911

*Rubble masonry.

arch and cantilever action. The distribution of the load between the arches and the cantilevers is usually determined by the *trial-load method*,[1] which begins with an assumption as to the load distribution. Near the bottom of the dam most of the load is carried by the cantilevers, while near the top the arches take more of the load. After assuming a division of the load, the resulting deflections of the arches and the cantilevers are computed. The deflection of the arch at any point should equal the deflection of the cantilever at the same point. If computed deflections are not equal, new loads are assumed until a distribution is found which produces equal arch and cantilever deflections at all points. Stresses in the dam and foundation can then be computed on the basis of this load distribution.

[1] M. D. Copen and L. R. Scrivner, Arch Dam Design: State of the Art, *J. Power Div.*, *ASCE*, Vol. 96, No. PO1, *Proc. Paper* 7021, pp. 93–108, January, 1970.

There are two main types of arch dams, *constant-center* and *variable-center* (Fig. 8-10). The constant-center arch dam, also known as the constant-radius dam, usually has a vertical upstream face, although some batter may be provided near the base of large dams. Intrados curves are usually, but not always, concentric with extrados curves. The variable-center arch dam, also known as the variable-radius or constant-angle dam, is one with decreasing extrados radii from top to bottom so that the included angle is nearly constant to secure maximum arch efficiency at all elevations. This design often results in an overhang of the upstream face near the abutments and sometimes of the downstream face near the crown of the arch. The variable-center dam is best adapted to V-shaped canyons since arch action can be depended upon at all elevations. The constant-center dam is sometimes preferred for U-shaped canyons as cantilever action will carry a large portion of the load at the lower levels. The formwork for a constant-center dam is much simpler to construct, but the increased arch efficiency of the variable-center dam usually results in a saving of concrete.

8-7   Design of arch dams   The same forces which act on gravity dams also act on arch dams, but their relative importance is different.

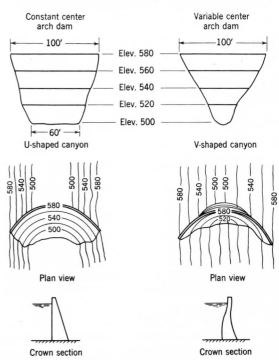

FIG. 8-10   Types of arch dams.

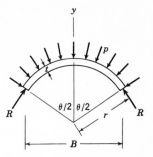

FIG. 8-11   Free-body diagram of an arch rib.

Because of the narrow base width of arch dams, uplift pressures are less important than for gravity dams. However, internal stress caused by ice pressure and temperature changes may become quite important in arch-dam design.

The simplest approach to arch analysis is to assume that the horizontal water load is carried by arch action alone. Most early arch dams were designed on this basis. Figure 8-11 represents a free-body diagram of the forces in the horizontal plane acting on an arch rib of Fig. 8-9. Since the intensity of hydrostatic pressure is $p = \gamma h$, the total downstream component of hydrostatic force on a rib of unit height is

$$H_h = \gamma h 2r \sin \frac{\theta}{2} \qquad (8\text{-}4)$$

This force is balanced by the upstream component of the abutment reaction $R_y = 2R \sin \theta/2$. Since $\Sigma F_y = 0$,

$$2R \sin \frac{\theta}{2} = 2\gamma h r \sin \frac{\theta}{2} \qquad (8\text{-}5)$$

or

$$R = \gamma h r \qquad (8\text{-}6)$$

If the thickness t of the arch rib is small as compared with r, there is little difference between the average and maximum compressive stress in the rib and $\sigma \approx R/t$. The required thickness of the rib is

$$t = \frac{\gamma h r}{\sigma_w} \qquad (8\text{-}7)$$

where σ_w is the allowable working stress for concrete in compression. This indicates that the thickness of the ribs should increase linearly with distance below the water surface and that for a given water pressure the required thickness is proportional to the radius of curvature.

The volume of concrete required for a single arch rib (Fig. 8-11) across a canyon of width B is

$$V = rA\theta \tag{8-8}$$

where A is the cross-sectional area of the rib and θ is the central angle in radians. Since t is proportional to r, $A = kr$ and

$$V = kr^2\theta \tag{8-9}$$

From trigonometry $r = B/2 \sin (\theta/2)$, and

$$V = k \left(\frac{B}{2 \sin \dfrac{\theta}{2}} \right)^2 \theta \tag{8-10}$$

Differentiating Eq. (8-10) with respect to θ and equating to zero gives $\theta = 133°34'$ for a rib of minimum volume. This is the reason why a constant-angle dam can be designed to require less concrete than a constant-center dam. In practice the central angles of arch dams vary from 100 to 140°. The base width of arch dams is usually between 0.1 and 0.5 the height; however in Europe many ultrathin arch dams have been built. The 260-ft-high Tolla Dam,[1] for example, has a maximum thickness of only 8 ft. The dimensions of Pacoima Dam, one of the largest variable-radius arch dams in the United States, are given in Fig. 8-12.

Rigorous analysis of an arch dam involves many factors not considered in the preceding approximate analysis.[2] The cantilevers are actually trapezoidal in cross section, and their deflection includes that due to shearing

[1] Andre Coyne, Carlo Semenza, Tore Nilsson, Henri Gicot, and Calvin Davis, World Progress in Dams, *Eng. News-Record*, Sept. 11, 1958.

[2] W. A. Perkins, Analysis of Arch Dam of Variable Thickness, *Trans. ASCE*, Vol. 118, pp. 725–766, 1953.

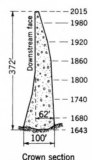

Elevation	Radius-feet		Thickness-feet	
	Extrados	Intrados	Crown	Abut.
2015	335.8	325.4	10.4	10.4
1980	318.3	297.3	21.0	21.0
1920	281.6	232.8	40.0	43.4
1860	251.7	152.0	52.0	72.0
1800	234.5	111.3	59.7	78.0
1740	212.0	79.0	72.0	82.0
1680	171.3	64.0	90.0	
1643			100.0	

Crown section

FIG. 8-12   Dimensions of Pacoima Dam.

FIG. 8-13   Construction of Alder Dam. (City of Tacoma, Washington)

action as well as bending. Deflection of arch ribs is caused mainly by the
water load, but it is also greatly affected by temperature changes. Shrinkage
and plastic flow of the concrete must also be considered. Yielding of the
foundation or abutments affects the structural behavior of arch dams. If
the foundation yields relatively more than the abutments, cantilever action
is suppressed, while if the boundary conditions are reversed, arch action
plays a lesser role.

8-8   Construction of arch dams   The foundation of an arch dam
must be stripped to solid rock and the abutments should be stripped and
excavated at approximately right angles to the line of thrust to prevent
sliding of the dam. Seams and pockets in the foundation and abutment are
grouted in the usual manner. Since the cross section of an arch dam is
relatively thin, care must be taken in the mixing, pouring, and curing of the
concrete in order to secure adequate resistance to seepage and weathering.
Concrete is placed in a manner similar to that for gravity dams (Fig. 8-13),
usually in 10-ft lifts, although 20-ft lifts are not uncommon at the upper
levels, where the section is quite thin. A layer of mortar is usually placed
between lifts to ensure better bond. Small arch dams are provided with
only radial construction joints, while large arch dams have circumferential
joints as well. All joints must have keyways, and water stops must be provided to prevent leakage. To minimize temperature stresses, the closing

section of the dam is poured only after the heat of setting in the other sections is largely dissipated.

BUTTRESS DAMS

A buttress dam consists of a sloping membrane which transmits the water load to a series of buttresses at right angles to the axis of the dam. There are several types of buttress dams, the most important ones being the *flat-slab* (Fig. 8-14) and the *multiple-arch* (Fig. 8-15). These differ in that the water-supporting member in one case is a series of flat reinforced-concrete slabs, while in the other it is a series of arches which permit wider spacing of buttresses.

Buttress dams usually require only one-third to one-half as much concrete as gravity dams of similar height but are not necessarily less expensive because of the increased formwork and reinforcing steel involved. Since a buttress dam is less massive than a gravity dam, the foundation pressures are less and a buttress dam may be used on foundations which are too weak to support a gravity dam. If the foundation material is permeable, a cutoff wall extending to rock may be desirable. The upstream faces of buttress dams usually slope at about 45°, and with a full reservoir a large vertical component of hydrostatic force is exerted on the dam. This assists in stabilizing the dam against sliding and overturning. The first reinforced-concrete slab and buttress dam was built in this country by Nils Ambursen in 1903, and this type of dam is often called an *Ambursen dam*.

The height of a buttress dam can be increased by extending the buttresses and slabs. Consequently buttress dams are often used where a future increase in reservoir capacity is contemplated. Powerhouses and water-treatment plants have been placed between the buttresses of dams with some saving in cost of construction. Some typical buttress dams are listed in Table 8-4.

8-9   Forces on buttress dams   Buttress dams are subjected to the same forces as gravity and arch dams. Because of the slope of the upstream face, ice pressures are not usually important as the ice sheet tends to slide up

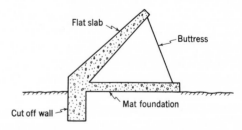

FIG. 8-14   Cross section of a typical flat-slab buttress dam.

FIG. 8-15   Bartlett Dam, Arizona, a multiple-arch dam. (U.S. Bureau of Reclamation)

the dam. Uplift pressures are relieved by the gaps between the buttresses. The total uplift forces are usually quite small and can generally be neglected except when a mat foundation is used.

8-10   Flat-slab buttress dams   A cross section of a typical flat-slab and buttress dam is shown in Fig. 8-14. Buttress spacing varies with height of dam from about 15 ft for dams under 50 ft high to 50 ft for dams more than 150 ft high. Closely spaced buttresses can be less massive, and the slabs can be thinner, but more formwork is required. The best buttress spacing is that which gives minimum overall cost. Actually, a curve of cost vs. buttress spacing is usually quite flat over a wide range of spacings. Concrete beams or diaphragms, placed as stiffeners between adjacent buttresses, or concrete braces may be used to resist buckling of the buttresses. For wide buttress spacings such devices are uneconomical, and hollow buttresses are sometimes used to increase the effective buttress width.

A feature of the flat-slab dam is its articulation, i.e., the slab is not rigidly attached to the buttresses (Fig. 8-16). The joint between the slab and buttress is filled with asphaltic putty or some flexible joint compound. This permits each slab to act independently, and minor settlement of the foundation will not seriously harm the structure. Buttresses are usually haunched

TABLE 8-4   Typical Buttress Dams in the United States

Name	State	Height, ft	Up-stream slope, deg	Length, ft	Vol-ume, 1000 cu yd	Buttress spacing, ft	Reser-voir capac-ity, 1000 acre-ft	Year com-pleted
Flat slab (Ambursen type)								
Possum Kingdom	Tex.	189	50	1640	325	40	990	1940
La Prele	Wyo.	148	40	330	23	18	24	1910
Stony Gorge	Calif.	139	45	868	38	18	50	1928
Bristol	N.H.	100	45	525	...	15	*	1924
Tiger Creek	Calif.	100	45	470	14	18	1	1932
Multiple-arch type								
Bartlett	Ariz.	287	48	800	153	60	182	1938
Big Dalton	Calif.	180	48	480	45	60	1	1928
Palmdale	Calif.	175	45	648	25	24	6	1924
Pensacola	Okla.	155	48	4284	230	84	540	1940
Hamilton	Tex.	154	48	8377	...	35	1000	1938
Mt. Dell	Utah	150	50	560	...	35	*	1917
Hodges	Calif.	136	45	558	18	24	37	1917
Lake Lure	N.C.	122	45	585	40	41	*	1927

*Less than 5000 acre-ft.

where they join the slab. Flat-slab and buttress dams are particularly adapted to wide valleys where a long dam is required and foundation materials are of inferior strength. By placing the buttresses on spread footings, the foundation pressures can be much reduced. Flat-slab dams have been built on materials ranging from fine sand to solid rock. The maximum practical height is necessarily less on poor foundations.

8-11   Design of flat-slab and buttress dams   The slab, or water-supporting member, is designed by assuming that it consists of a series of parallel beams which act independently of one another (Fig. 8-17). Since slabs are not rigidly connected to the buttresses, they are designed as simply supported beams by standard methods of reinforced-concrete design. Beam thickness and amount of reinforcement required increase with depth below the water surface, since each beam is designed to withstand the component of water load normal to it.

Buttress design is based on simplifying assumptions, as a rigorous analysis is quite difficult. A buttress is usually assumed to consist of a system of

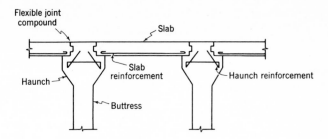

FIG. 8-16   Method of joining slab to buttress.

independent columns (Fig. 8-17). The load on each column is a combination structure load and water load. The columns are assumed to be curved so as to avoid eccentric loading. Buttresses are usually reinforced.

After a trial design of slab and buttresses, foundation pressures are calculated and buttress footings are designed. In some cases the base of the buttress will give adequate bearing area, while in other instances a spread footing or mat foundation may be required. Spread footings and mats are amply reinforced with steel to improve their effectiveness in distributing the load. On very poor foundations a continuous slab may be provided underneath the entire length of the dam. As a final step in design, the stability of the entire structure against sliding and overturning is investigated (Sec. 8-4). If the computed factors of safety are not sufficiently high, the slope of the upstream face may be flattened.

8-12   Multiple-arch dams   The multiple-arch dam is more rigid than the flat-slab type and consequently requires a better foundation. Circular arches of uniform thickness are most economical for short spans (30 or 40 ft),

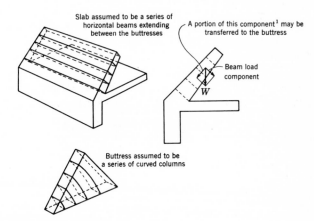

FIG. 8-17   Structural elements of a flat-slab buttress dam.

while greater economy may be secured by using an arch of variable thickness for longer spans. The central angle of the arches is usually between 150 and 180°. A central angle of 180° eliminates horizontal thrusts between adjoining arches and simplifies buttress design. Arches for a multiple-arch dam are designed in the same manner as a single-arch dam, but cantilever action is commonly ignored. The multiple-arch design is most economical for high dams, where the savings in concrete and steel are sufficient to offset the increased cost of forms.

8-13   Miscellaneous types of buttress dams   In addition to the slab or arch-deck buttress dams, a few *massive buttress* dams (Fig. 8-18) have been built. In this type the water-supporting member is formed by enlarging the upstream end of the buttress. In a few instances a series of columns or a truss has been used to support a girder on which a deck slab rests. Two steel deck dams, Ash Fork Dam, Arizona (1898), and Redridge Dam, Michigan (1901), have been built in the United States. These dams have deck plates which are slightly curved, placed with their concave side to the water, and supported by struts. Regular painting seems to have prevented serious corrosion of these steel dams. Still another type of buttress dam consists of a timber deck sloping at an angle of about 30° with the horizontal and supported on A-frame bents 5 to 15 ft apart.

8-14   Construction of buttress dams   Removal of overburden down to a suitable foundation and excavation of a trench for the cutoff wall are the first steps in the construction of buttress dams. Great care must be taken in the construction of forms, handling of concrete, and placing of reinforcing steel in order to develop fully the strength and watertightness of the thin sections used in buttress dams. Deck and buttresses are placed in lifts of 12 ft or more, the buttress construction being kept well in advance of the deck. Keyways are required in all construction joints. Since buttress dams require much less concrete than comparable gravity dams, the time for construction is usually less and the problem of water diversion somewhat simplified.

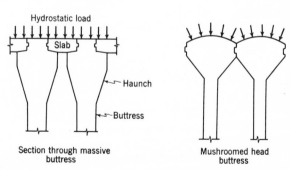

FIG. 8-18   Sections of typical massive-buttress dams.

EARTH DAMS

Earth dams utilize natural materials with a minimum of processing and may be built with primitive equipment under conditions where any other construction material would be impracticable. It is not surprising that the earliest known dams were of earth. Modern developments in earth-moving equipment have resulted in decreased cost for earth moving as compared with an increase in cost of concrete as a result of increased wage and material costs. Earth dams are now competitive in cost with masonry in all sizes. The design of earth levees and dikes (Chap. 20) follows the same principles as for earth dams. Some of the larger earth and rock-fill dams in the United States are listed in Table 8-5.

Unlike high-arch and gravity dams, which require a sound rock foundation, earth dams are readily adapted to earth foundations. They are logical choices for many sites where foundation conditions make concrete dams unsatisfactory. It should not be assumed that the construction of earth dams is a simple operation and that their design requires little more than rule-of-thumb criteria. Numerous failures of poorly designed earth embankments make it apparent that earth dams require as much engineering skill in their conception and construction as any other type of dam. In no other type of dam are the construction and design procedures so interdependent; continuous field observations of deformations and pore water pressures are often made during the construction period to evaluate the initial design. Modifications of design based on these observations are not uncommon for large earth dams.

8-15 Types of earth dams The *simple embankment* is essentially homogeneous throughout, although a blanket of relatively impervious material may be placed on the upstream face. Levees are often simple embankments, but large dams are rarely constructed in this manner. *Zoned embankments*[1] (Fig. 8-19) usually have a central zone of selected soil material, to form a relatively *impermeable core*, a *transition zone* along both faces of the core to prevent piping through cracks which may form in the core, and outer zones of more pervious material for stability. This construction is widely used in earth dams and is selected whenever suitable materials are available. Clay, even though highly impermeable, may not make the best core if it shrinks and swells too much. The most satisfactory cores are of clay mixed with sand and fine gravel. *Diaphragm-type dams* have a thin central section of concrete, steel, or timber which serves as a water barrier, while the surrounding earth or rock fill provides stability. Thin concrete sections are easily cracked by differential earth loads, and it is difficult to

---

[1] Karl Terzaghi and Thomas M. Leps, Design and Performance of Vermilion Dam, *Trans. ASCE*, Vol. 125, pp. 63–100, 1960.

TABLE 8-5   Large Earth and Rock-fill Dams in the United States

Name	State	Volume, 1000 cu yd	Height, ft	Length, ft	Top width, ft	Reservoir capacity, 1000 acre-ft	Year completed
Earth-fill							
Fort Peck	Mont.	100,000	270	21,000	100	19,000	1940
Oroville	Calif.	77,500	770	5,140	80	3,484	1968
Garrison	N.D.	68,000	210	12,000	60	23,000	1954
Kingsley	Nebr.	28,000	165	16,000	25	2,200	1941
Navajo	N. Mex.	26,250	408	3,800	30	1,709	1965
El Capitan	Calif.	2,680	240	1,200	20	118	1934
McKay	Oreg.	2,338	165	2,700	23	74	1927
Davis Bridge	Vt.	1,900	200	1,250	15	116	1924
Cobble Mt.	Mass.	1,800	246	800	50	812	1936
Alcova	Wyo.	1,600	182	763	40	190	1936
Rock-fill							
Salt Springs	Calif.	3,000	325	1,300	15	130	1931
Nantahala	N.C.	2,260	260	1,040	30	124	1942
Dix River	Ky.	1,750	250	910	20	300	1924
El Vado*	N. Mex.	607	212	1,300	20	198	1935
Morena	Calif.	335	167	505	16	66	1912
Cucharas	Colo.	195	125	550	20	50	1911
Composite rock-fill and earth							
Trinity	Calif.	33,000	537	2,450	40	2,500	1963
San Gabriel No. 1	Calif.	10,620	375	1,670	40	44	1938
Anderson Ranch	Idaho.	9,653	456	1,356	40	500	1948
Stevens	Wash.	2,109	425	700	50	130	1942
Inland	Ala.	1,600	190	1,100	30	68	1938

*Has steel-plate facing on the upstream slope.

form a perfectly watertight barrier of timber or steel. In addition, the diaphragm must be tied into bedrock or a very impermeable material if excessive underseepage is to be avoided. Rock-fill dams are built of coarse rock, which provides structural stability, with a concrete membrane on the upstream slope as a water barrier.

8-16   Methods of construction *Hydraulic-fill dams* are constructed by using water for transporting the material to its final position in the dam. The material is discharged from pipes along the outside edges of the fill (Fig. 8-20), and the coarse materials are deposited soon after discharge,

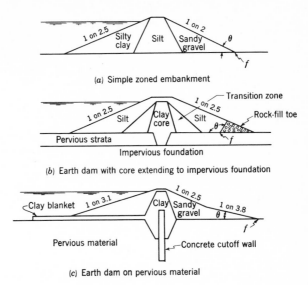

(a) Simple zoned embankment

(b) Earth dam with core extending to impervious foundation

(c) Earth dam on pervious material

FIG. 8-19   *Cross sections of typical earth dams.*

while the fines are carried into the central pool. The result is a zoned embankment with a relatively impermeable core. This type of construction originated in the western United States as a result of experience with hydraulic mining. Recent improvements in equipment for transporting and compacting dry fill have tended to discourage use of the hydraulic method. Hydraulic fill is best suited for placing well-graded soils containing a considerable amount of coarse sand and gravel. An adequate water supply is necessary, although some of the water may be recirculated. Since the fill is saturated when placed, high pore pressures develop in the core material and the embankment must be designed for stability under these pressures. Because of the slow drainage of the water from the core, considerable settlement is to be expected over a long period. The embankment is especially susceptible to earthquake damage until drainage is reasonably complete.

Semihydraulic-fill dams are constructed by dumping the fill material from trucks into its approximate position in the dam and sluicing the fines into a core. Like the hydraulic-fill method this procedure requires careful control

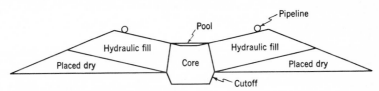

FIG. 8-20   *Construction of a hydraulic-fill dam.*

to assure a satisfactory embankment. Dumping soil in a pool of water to produce a puddled core often washes the fines from the material and leaves a more permeable core than might have been obtained from proper dry compaction. Numerous failures of hydraulic fill dams have occurred during construction.

Rolled-fill dams are constructed by placing selected materials in thin layers (6 to 18 in.) and compacting them with a heavy roller. Some compaction may be obtained by proper routing of trucks and other construction equipment. Usually, however, special equipment is used for compacting the fill. Both sheepsfoot rollers and heavy pneumatic-tired rollers are used singly or in combination. Ordinary road rollers have been used successfully on small projects. In any case the material should be placed at a moisture content near that for optimum density. Coarse gravels are not suited for compaction by rolling, but vibrating equipment can be used.

8-17   Design of earth dams   The design of an earth dam consists in developing a fill of sufficiently low permeability for the intended purpose out of available materials at minimum cost. Borrow for the fill must usually be close to the site because of the high cost of long truck hauls. Since quantity of fill varies approximately with the square of the height, earth-fill dams of great height are rare.

The structural design of an earth dam is a problem in soil mechanics involving assurance of stability of the fill and foundation and sufficient control of the water flow and seepage pressures. There is little harm in seepage through a flood-control dam if the stability of the embankment is not impaired, but a conservation dam must be as watertight as possible. It is difficult to analyze the probable behavior of natural fill materials in a zoned embankment with the assurance attained in the design of masonry structures. Present design practice is to conform with existing dams of similar characteristics with analytical checks on the adequacy of the dam under special conditions. Empirical rules are frequently employed for the preliminary design of large dams and for the final design of small dams. Fill quantities in dams under 15 ft high are so small that a considerable excess factor of safety can be provided at low cost.

8-18   Height of dam   The required height of an earth dam is the distance from the foundation to the water surface in the reservoir when the spillway is discharging at design capacity, plus a freeboard allowance for wind setup, waves, frost action and earthquake motions. Recent studies of earth-dam failures indicate that 40 percent resulted from overtopping of the dam because of insufficient freeboard or inadequate spillway capacity. Means of estimating wind setup, wave height, and wave run-up are discussed in Sec. 7-10.

Frost in the upper portion of a dam may cause heaving and cracking of the soil with dangerous seepage. An additional freeboard allowance up to a maximum of about 5 ft should be provided for dams in areas subject to low temperatures.

Earth materials consolidate under load, but this consolidation is not instantaneous. Consolidation results from a reduction in void space accompanied by compression or extrusion of air and water from the voids. In coarse gravels the void openings are large enough to permit rapid escape of confined water and air, and full consolidation may occur before an embankment is finished. In fine-grained soils consolidation is less rapid, and it may be necessary to provide additional height of fill so that, after settlement, the embankment will be the desired height. The allowance for consolidation may be determined by laboratory tests and observation of the settlement during construction. The usual consolidation allowance is between 2 and 5 percent of the total height of the dam. Dewatering of the foundation material is sometimes used to accelerate consolidation.

Parapet walls 2 or 3 ft high are sometimes provided on the upstream side of the crest of an earth dam. Such walls are considered only as an additional safety factor, but they may be constructed strongly enough to be considered as an element of freeboard. This latter practice is economical only on dams exceeding about 30 ft in height.

8-19   Top width   The top width of an earth dam should be sufficient to keep the phreatic line, or upper surface of seepage, within the dam when the reservoir is full. Top width should also be sufficient to withstand earthquake shock and wave action. Top widths of low dams may be governed by secondary requirements such as minimum roadway widths. Various empirical expressions have been suggested for determination of top width. For example, the U.S. Bureau of Reclamation has proposed

$$B = \frac{H_d}{5} + 10 \tag{8-11}$$

where B is top width in feet and H_d is height of dam in feet. A minimum width of 10 ft is usually required for maintenance.

8-20   Seepage   No earth dam can be considered impervious, and some seepage through the dam and its foundation must be expected. If the rate of pressure drop resulting from seepage exceeds the resistance of a soil particle to motion, that particle will tend to move. This results in *piping*, the removal of finer particles usually from the region just downstream of the toe of the embankment.

Seepage through earth dams may be reduced by the use of a very broad base, by the placing of an impervious blanket on the upstream face, by use of a clay core, or by a diaphragm of timber, steel, or concrete. Seepage through permeable foundation materials may be reduced by increasing the length of seepage with a relatively impervious blanket extending upstream from the dam (Fig. 8-19c) or by one or more pile, concrete, or clay cores extending into the foundation and connecting to the core of the dam. A grout curtain formed by forcing cement grout down through closely spaced drill holes is an effective means of checking leakage through fractured rock. In any case drains are normally provided near the toe of the dam to permit the free escape of seepage water which passes any barriers provided.

Drains usually consist of a *rock toe* (Fig. 8-19b) or a *drainage blanket* (Fig. 8-21a) of coarse material in which seepage water collects and moves to a point where it can be safely discharged. To prevent movement of fine material from the dam into the drain, the drain material is graded from relatively fine on the periphery of the drain to coarse near the center. A drainage blanket extends farther under the dam than a rock toe does. If a pervious foundation layer is overlain by a less pervious layer, relief wells or a drain trench through the upper layer may be necessary to permit escape of seepage water. If this is not done, water may boil up near the toe of the dam unless the weight of the overlying material is sufficient to resist the upward pressure.

Amount of seepage is estimated from a *flow net*,[1] which consists of two groups of curves, equipotential lines (lines of equal energy) and streamlines of flow which are normal to the equipotential lines. It is convenient to draw only a limited number of flow lines and equipotential lines such that the rate of flow between each pair of flow lines is equal and the energy drop between successive potential lines is the same. The distance between the flow lines is made equal to the distance between the potential lines, thus forming a series of squares. Where the flow lines are curved, the squares will be distorted, but they will be more nearly perfect as the number of lines is increased, approaching true squares as they become infinitesimal in size. Some typical flow nets are shown in Fig. 8-21.

The pattern of the flow net may be determined in several ways. A sand model of the embankment may be constructed and the flow lines traced by dye. Electrical analog employ a conductor (tray of electrolyte or special conducting paper) in which equipotential lines between electrodes representing the upstream and downstream water surfaces are traced. This method rests on the similarity between Ohm's law and Darcy's law (Sec. 4-7). Various mathematical and graphical procedures and computer solutions by numerical analysis are available.

---

[1] In this text it will be assumed that all soils are homogeneous and isotropic, i.e., that permeability is independent of location and direction of flow. In many soils the permeability is substantially greater in the horizontal direction than in the vertical. Flow nets for such cases are discussed in most soil-mechanics textbooks.

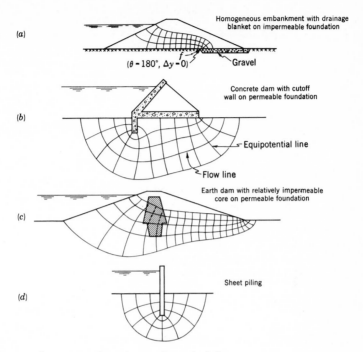

(a)   Homogeneous embankment with drainage blanket on impermeable foundation

$(\theta = 180°, \Delta y = 0)$ Gravel

(b)   Concrete dam with cutoff wall on permeable foundation

Equipotential line

Flow line

(c)   Earth dam with relatively impermeable core on permeable foundation

(d)   Sheet piling

FIG. 8-21   *Flow nets for seepage through or under typical dams.*

In simple cases flow nets are usually constructed[1] by freehand sketching, with gradual adjustment and correction until the flow lines and equipotential lines intersect at right angles. The exterior flow lines are the seepage line and any impermeable boundary in the dam or foundation. The *seepage line* (also *line of saturation*, or *phreatic line*) in an earth dam is the line above which there is no hydrostatic pressure. Casagrande[2] has suggested an approximate method for drawing the seepage line for an earth dam on an impervious foundation, based on the assumption that it is a parabola with a focus at f (Fig. 8-22), which is the intersection of the downstream equipotential line and the streamline along an impermeable boundary. The parabola also intersects the water surface at about 0.3 of the horizontal distance from the face of the dam to the upstream toe of the embankment, that is, $AB = 0.3BC$. The directrix of the parabola is a distance $AD = Af$ from A, and every point on the parabola is equidistant from the focus and the directrix. The upstream terminus of the seepage line is at B, the intersection

[1] H. R. Vallentine, "Applied Hydrodynamics," 2d ed., Plenum Press, New York, 1967.

[2] Arthur Casagrande, Seepage through Dams, *J. New Engl. Water Works Assoc.*, Vol. 51, pp. 131–172, June, 1937.

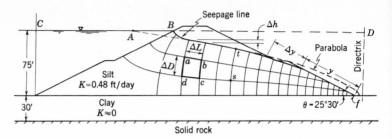

FIG. 8-22   Location of the seepage line and construction of a flow net for an earth dam.

of the reservoir water surface and the upstream face of the dam. The terminus of the seepage line at the downstream face (rock toe or blanket) is below the point where the parabola intersects the face of the dam by the distance Δy, which is defined by the equation

$$\Delta y = (y + \Delta y)\frac{180 - \theta}{400} \qquad (8\text{-}12)$$

where y, Δy, and θ are defined in Figs. 8-19, 8-21, and 8-22.

The seepage rate can be computed with Darcy's law [Eq. (4-1)]. Applying the principle of continuity between each pair of flow lines, it is evident that the velocity must vary inversely with the spacing. Assuming the cross section of Fig. 8-22 to have unit width, the flow through square $abcd$ is $\Delta q = K\Delta D(\Delta h/\Delta L)$. Since $\Delta D = \Delta L$ and $\Delta h = h/N$, where N is the number of increments into which the potential drop is divided, it follows that $\Delta q = Kh/N$. The total flow through a unit width of the dam is therefore

$$q = N'\Delta q = \frac{N'Kh}{N} \qquad (8\text{-}13)$$

where N' is the number of spaces between the flow lines.

8-21   Pore pressure   A soil mass is composed of solid particles and void spaces filled with water and air. The air-water mixture in the voids creates a pore pressure which reduces the contact pressure between soil particles. On consolidation of the soil, excess moisture leaks out and the intergranular pressure between soil particles gradually increases.

Immediately after the construction of an earth dam sizable pore pressures may be present. These are gradually dissipated as the soil moisture is redistributed, but as the reservoir is filled, water enters the pores of the dam, and a new pattern of pore pressure develops. Under steady-state seepage conditions the pore-pressure head at any point in a dam is equal to the hydrostatic head due to water in the reservoir less the head loss in seepage through the dam to that point. The pore-pressure distribution under steady-state seepage conditions can be found from the flow net. For example, the pore-pressure head at point s (Fig. 8-22) is equal to the difference in elevation between s and t, the point where the line of equipotential through s intersects the seepage line. This elevation difference corresponds to the height that water would rise in a piezometer tube with opening at s.

8-22 Slope stability The usual failure of an earth embankment consists in the sliding of a large mass of soil along a curved surface. Several methods for checking the stability of a fill can be found in the literature of soil mechanics. Only the simplest form of the method of slices will be described here. This method assumes a condition of plane strain with failure along a cylindrical surface (Fig. 8-23).

The location of the center of the failure arc is assumed, the earth mass is divided into a number of vertical segments, and the weight W of each segment is calculated. The forces W_1, W_2, etc., are assumed to act through the center of mass of the respective segments. The forces acting on the sliding mass as a whole as well as the forces acting on each slice must satisfy equilibrium. In its most approximate form, the method sets the forces acting on the sides of each slice equal to zero. The moment tending to rotate the soil mass about point O is

$$M = \Sigma Wx \tag{8-14}$$

where W is the weight and x is the moment arm of the individual segments. Tangential shear stresses acting along the failure arc can create a resisting moment M_r,

$$M_r = \Sigma s_s(\Delta L)r \tag{8-15}$$

where s_s is the shear strength of the soil, ΔL the length of the failure arc for a segment, and r the radius of the failure arc. The shearing strength s_s is given by Coulomb's equation

$$s_s = c + \sigma \tan \phi \tag{8-16}$$

where c is the cohesion, ϕ the angle of internal friction, and σ the effective contact pressure between soil particles along the failure arc. Generally,

$$\bar{\sigma} = \frac{W \cos \theta}{\Delta L} - u_w \tag{8-17}$$

where θ is as indicated on Fig. 8-23, and u_w is the pore pressure. In sands for steady seepage conditions, u_w can readily be determined from the flow net. In clays, u_w is difficult to estimate because the very slow movement of the water through the interstices rarely permits the existence of steady-state seepage conditions. The properties of a soil are commonly determined from laboratory triaxial compression tests. Cohesion is a function of the soil material and its moisture content, and for clays it ranges from about 100 to 1200 psf. Cohesion of sand is negligible. The angle of internal friction is usually between 5 and 20° for clay, while for sand it is about 30°.

Calculations for the stability of an earth dam by the method described above are presented in Fig. 8-23. Several possible failure arcs should be tested to find the one with the minimum factor of safety to assure that the dam and its foundation are stable. The factor of safety along an assumed arc is the ratio of the clockwise moments divided by the counterclockwise moments.

Segment	W, 1000 lb	x, ft	$M = Wx$, 1000 ft-lb	$W \cos \theta$, 1000 lb	ΔL, ft	$\dfrac{W \cos \theta}{\Delta L}$, 1000 psf	u_w,[*] 1000 psf	$\bar{\sigma}$, 1000 psf	$\bar{\sigma} \tan \phi$, psf	$s_s = c + \bar{\sigma} \tan \phi$, psf	$M_r = s_s(\Delta L)r$, 1000 ft-lb
1	73	94	6850	45	58	0.78	0.60	0.18	90	490	3560
2	92	74	6800	71	29	2.44	2.00	0.44	210	610	2210
3	168	46	7730	156	36	4.33	†	4.33	910	1960	8800
4	143	16	2290	141	33	4.27	†	4.27	900	1950	8050
5	86	14	−1200	85	33	2.57	†	2.57	540	1590	6560
6	25	40	−1000	23	34	0.68	†	0.68	140	1190	5050
Total	...	...	21,470	...	...	....	....	....	...	....	34,230

[*]From the flow net (Fig. 8-22) it is seen that the pore-pressure head along the assumed-failure arc beneath segment W_2 varies from 30 to 34 ft. Hence,

$$u_w = 32 \times \frac{14.7}{33.9} \times 144 = 2000 \text{ psf}$$

†Pore pressure in foundation is accounted for by the c and ϕ for the clay as determined in triaxial test.

Free-body diagram of segment:

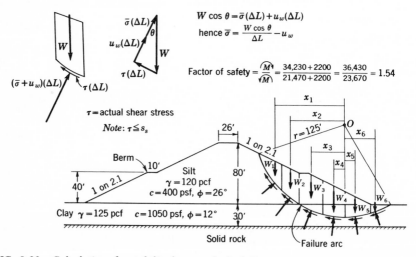

$$W \cos \theta = \bar{\sigma}(\Delta L) + u_w(\Delta L)$$

$$\text{hence } \bar{\sigma} = \frac{W \cos \theta}{\Delta L} - u_w$$

$$\text{Factor of safety} = \frac{\Sigma M}{\Sigma M} = \frac{34,230 + 2200}{21,470 + 2200} = \frac{36,430}{23,670} = 1.54$$

$\tau = $ actual shear stress

Note: $\tau \leq s_s$

FIG. 8-23   Calculations for stability by a method of slices.

The above method generally gives conservative estimates for the computed factor of safety. More nearly correct and elaborate procedures[1] include side forces on the individual slices and also noncircular sliding surfaces. These methods are programmed for use with digital computers; the critical sliding surface in each situation is automatically found.

Side slopes of earth dams depend upon the available embankment materials and vary from as flat as 1 vertical on 7 horizontal to a maximum of about 1 on 2. To assure stability, the base of a dam can be broadened. The total volume of fill in a dam can be reduced by placing coarse material in the outer shell so that its weight prevents failure of the core, while at the same time free drainage from the shell prevents formation of serious pore pressures.

With a full reservoir the critical region in an earth dam is near the downstream face. If the line of saturation or upper limit of free water in the dam intersects the downstream face, a critical situation may develop. This condition can be avoided by providing a sufficiently wide embankment so that the head loss is great enough to bring the line of saturation out beneath the downstream toe of the dam. A facing of coarse material on the downstream slope or a rock filter at the toe is usually a more economical solution. Rapid drawdown of a reservoir after it has been filled may lead to a similar condition at the upstream face of the dam if the embankment does not drain readily. In addition the drawdown removes the water above the upstream slope of the dam which contributed by its weight to the stability of the earth mass. A rock facing or a limit on the rate of drawdown may be employed to prevent damage of this sort.

ILLUSTRATIVE EXAMPLE 8-2   A dam 80 ft high is to be constructed from a silt which, when compacted at optimum moisture, has a specific weight $\gamma = 120$ lb/cu ft, cohesion $c = 400$ psf, angle of internal friction $\phi = 26°$, and a coefficient of permeability of 0.48 ft/day. A cross section of the proposed design is shown in Fig. 8-23. The top width as calculated with Eq. (8-11) is 26 ft, and a 10-ft berm will be located at midheight on both faces. Side slopes of 1 on 2.1 are proposed for both faces. The overall base width is 382 ft. Check this proposal for seepage and embankment stability.

The flow net for this cross section is shown in Fig. 8-22. Total seepage is $q = 3/20 \times 0.48 \times 75 = 5.4$ cu ft/day per foot of width. This seepage rate is not excessive. The computations for embankment stability are shown in Fig. 8-23. Allowance is made for pore pressures in the embankment. The factor of safety for the slip circle shown is $36,430/23,670 = 1.54$, which is adequate. Other circles should be checked.

8-23   Foundation analysis   Foundation stresses and deformations under earth dams may be critical when the foundation material consists of

---

[1] Karl Terzaghi and Ralph B. Peck, "Soil Mechanics in Engineering Practice," 2d ed., pp. 232–260, Wiley, New York, 1967.

normally consolidated soft clay or silt. A failure by spreading may occur as the dam tends to squeeze the material above the bedrock out from under the dam. The foundation conditions should be investigated by soil mechanics theories of bearing capacity and settlements. In general, stability will improve with time as the soft foundation material consolidates under the embankment load.

Special deformation and stability analyses are required for dams constructed in earthquake regions.[1] The seismic resistance of an earth dam is high because of its ductile behavior and large energy absorption. In regions with earthquakes it is especially important to provide a zoned earth dam with side-filter transition zones between the outer shell and the dam core to prevent piping through cracks and promote the subsequent healing of cracks suffered during an earthquake. Failures may also be caused by strong earthquake vibrations due to liquefaction of loose sand strata in the dam foundation.[2]

8-24 Slope protection The upstream slope of an earth dam should be protected against wave action by a cover of riprap or concrete. When available, a 3-ft layer of dumped rock is usually most economical. The rock should be sound and not subject to rapid weathering and should be placed over a filter layer of graded gravel at least 12 in. thick. Hand-placed riprap requires a lesser thickness and may be more economical if suitable rock is limited in quantity. A filter layer of gravel to prevent the washing of fines from the dam into the riprap is required. Concrete slabs are often employed for facing the upstream slope of earth dams but must be carefully constructed. The usual failure results from the washing of embankment material through the joints until the slab is only partially supported and cracks under its own weight. A filter layer of graded gravel is therefore desirable under the slab. Weep holes are also required to permit escape of water when the reservoir is drawn down. In two recent instances[3] it has been necessary to replace concrete paving on earth dams. Upstream-slope protection should extend from above the upper limit of wave action to a *berm,* or horizontal shelf, in the fill a few feet below the lowest anticipated pool level.

The downstream slope of an earth dam may be subject to erosion from rainfall. On dams having a rock shell this is no problem, but earth slopes

[1] Nathan M. Newmark, Effects of Earthquakes on Dams and Embankments, *Geotechnique,* Vol. 15, No. 2, pp. 139–160, 1965.

[2] H. Bolton Seed, Landslides During Earthquakes Due to Liquefaction, *Proc. ASCE,* Vol. 94, No. SM5, pp. 1055–1122, 1968.

[3] Kingsley Dam Gets a New Stone Facing, *Eng. News-Record,* Vol. 136, pp. 117–119, Jan. 10, 1946; New Upstream Slope Protection Added on Santee-Cooper Earthfill Dams, *Eng. News-Record,* Vol. 140, pp. 86–88, Mar. 18, 1948.

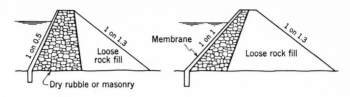

FIG. 8-24   Sections of typical rock-fill dams.

should be planted to grass as soon as possible after completion. Since the erosive action of water increases as the slope length increases, berms should be placed at about 50-ft intervals of elevation to intercept rainwater and discharge it safely.

8-25   Rock-fill dams   Rock-fill dams[1] have characteristics midway between gravity dams and earth dams. The rock-fill dam has two basic structural components—an impervious membrane and an embankment which supports the membrane (Fig. 8-24). The embankment usually consists of an upstream section of dry rubble or masonry and a downstream section of loose rock fill. The rock used should be capable of resisting erosion and strong enough to withstand loads of high intensity even when wet. The rock fill (downstream section of the embankment) supports the remainder of the embankment, the membrane, and the water load. Since load is transmitted through the fill by rock-to-rock contact, a dense fill with well-graded rocks is best, but sand and gravel in small quantities do no harm as long as rock-to-rock contact is maintained and drainage is adequate. Rocks may vary from small stones to boulders 10 or more ft in diameter.

The upstream dry-rubble section transfers the load from the membrane to the rock fill. Large stones of regular shape are used to give a flat surface on which the membrane can be placed. The impervious membrane is most commonly constructed of concrete. The membrane is sometimes poured as a monolith without expansion joints but with ample steel reinforcement in both horizontal and vertical directions. Most membranes have expansion joints at intervals of about 30 ft with asphaltic joint filler to minimize leakage. Slab thickness is usually between 6 and 18 in., with greater thickness often used near the base of the dam. Rock-fill dams are subject to considerable settlement, which may result in cracking of the membrane. This is perhaps the greatest weakness of rock-fill dams, although in many instances leakage has been controlled by periodic repair of the membrane. Salt Springs Dam (325 ft high) settled more than 2 ft in 10 yr.

A rock-fill dam of good design and careful construction has high resistance to earthquakes because of its flexible character. Low rock-fill dams may

---

[1] Symposium on Rock-fill Dams, *Power Div. ASCE*, August, 1958.

have upstream face slopes of 1 vertical on $\frac{1}{2}$ horizontal, but dams over 200 ft high usually have face slopes of 1 on 1.3, the natural angle of repose of rock fill. Downstream slopes of all rock-fill dams should be about 1 on 1.3. Since the face slopes are quite steep, much less material is required for a rock-fill dam than for an earth dam. Because of the narrow base width and the possibility of high seepage, foundation requirements for rock-fill dams are more rigid than for earth dams. Rock-fill dams are generally cheaper than concrete dams and can be constructed more rapidly if the proper material is available.

MISCELLANEOUS TYPES OF DAMS

8-26 Timber-crib dams A timber-crib dam is made of timber bolted into cribs and filled with rock. This type of dam usually leaks considerably, and its resistance against sliding is reduced by buoyant forces which decrease the effective weight of the dam. The life span of a timber dam varies from 10 to 40 yr depending upon climatic conditions and the type of timber used. Cedar, redwood, and cypress are the most durable timbers. A sloping timber deck on the upstream face serves to decrease seepage and increase stability. Where there is an adequate supply of timber, crib dams are quite inexpensive and generally satisfactory for heights not exceeding about 20 ft.

8-27 Cofferdams Cofferdams are temporary structures used to exclude water from an area to permit construction operations. A cofferdam must be relatively low in cost but as watertight as practicable. The simplest kind of cofferdam is a single-wall sheet-pile type, sometimes strengthened by an earth fill on both sides. A single-wall sheet-pile cofferdam can be used only in very shallow water. For depths over about 5 ft, cellular cofferdams are commonly used. These consist of circular sheet-pile cells filled with earth. A mixture of sand and clay is the best filler since clay itself tends to wash out through the gaps between the piles while sand is too permeable. Timber-crib dams are sometimes used for cofferdams, as are simple earth embankments. The simple embankment often requires too much room because of the large base width necessary for an embankment of any great height.

BIBLIOGRAPHY

Davis, C. V., and K. E. Sorenson (eds.): "Handbook of Applied Hydraulics," 3d ed., secs. 8–19, McGraw-Hill, New York, 1969.
"Design of Small Dams," U.S. Bureau of Reclamation, 1960.
Justin, J. D., J. Hinds, and W. P. Creager: "Engineering for Dams," Vols. 2 and 3, Wiley, New York, 1945.

Leliavsky, Serge: "Design of Dams for Percolation and Erosion," Chapman and Hall, London, 1965.
Mermel, T. W.: "Register of Dams in the United States," McGraw-Hill, New York, 1958.
Outland, Charles F.: "Man-made Disaster, the Story of St. Francis Dam," Arthur Clark Co., Glendale, Calif., 1963.
"Reclamation Project Data," U.S. Bureau of Reclamation, 1961.
Scott, R. F., and J. J. Schoustra: "Soil Mechanics and Engineering," McGraw-Hill, New York, 1968.
Sherard, J. L., R. J. Woodward, S. F. Gizienski, and W. A. Clevenger: "Earth-Rock Dams: Engineering Problems of Design and Construction," Wiley, New York, 1963.
Taylor, Donald W.: "Fundamentals of Soil Mechanics," Wiley, New York, 1948.
Walters, R. C. S.: "Dam Geology," Butterworth, London, 1962.
Wegmann, Edward: "Design and Construction of Dams," 8th ed., Wiley, New York, 1927.

PROBLEMS

8-1 For some site selected by your instructor, suggest the most appropriate type of dam. Outline your reasons.

8-2 Design the cross section of a gravity dam to retain a maximum water level 40 ft above its base. Use a maximum wave height of 3.5 ft as freeboard. Use a top width of 4 ft, and assume concrete at 150 lb/cu ft. Consider only water loads and weight of the dam. Compute the maximum normal stress on the foundation, assuming full uplift.

8-3 How would the addition of earthquake loads resulting from an earthquake acceleration of $0.1g$ affect the stability of the dam designed in Prob. 8-2?

8-4 Complete the analysis of the gravity dam used in Illustrative Example 8-1.

8-5 Recalculate the factors of safety against sliding and overturning for block B_1 of Illustrative Example 8-1, including ice thrust for a temperature change of 15°F/hr without solar radiation and ice thickness of 2 ft (Fig. 8-3), and an earthquake acceleration of $0.1g$. Consider both horizontal and vertical earthquake accelerations and the added thrust of the water behind the block. Redesign the section so that it is stable, and compute the maximum normal stresses in the concrete.

8-6 Analysis shows that the resultant reactive force on a cross section of gravity dam is 140,000 lb/ft. This force is inclined at an angle of 20° from the vertical and cuts the 30-ft base at a point 12 ft from the toe of the dam. Determine the maximum foundation pressure under these conditions, expressing the answer in pounds per square inch.

8-7 A 20-ft-high gravity dam with a vertical upstream face has a top width of 8 ft. What must be its base width if the resultant of the active forces is to cut the third point of the base when the water depth is 18 ft? Assume full uplift pressure, but neglect ice loads and earthquake forces. Verify the analytical result graphically by adding force vectors together. Use 1 in. = 10 ft, 1 in. = 20,000 lb.

8-8 On the basis of "arch-rib" analysis, design an arch dam 380 ft high to span a 600-ft-wide U-shaped canyon. Use 800 psi as the allowable compressive stress in concrete. Compare the result with the Pacoima Dam cross section (Fig. 8-12).

8-9 Design a flat slab and buttress dam to meet the conditions of Prob. 8-2. Use a buttress spacing of 15 ft and buttress thickness of 2 ft; 45° slope for the upstream face; and working stress in steel of 18,000 psi and concrete of 800 psi. Assume the foundation does not require spread footings. Compare the volume of concrete in this dam with that for the gravity dam of Prob. 8-2.

8-10 Determine the factor of safety against sliding and overturning and the maximum normal pressure under the buttress of a slab and buttress dam 20 ft high. Buttresses are 2 ft thick on 15-ft centers with a 25-ft base normal to the dam. Slab is 9 in. thick and on a 45° slope. Maximum water level is 18 ft above the base of the dam. Use a coefficient of friction at the foundation of 0.65 and a weight of concrete of 150 lb/cu ft. Ignore earthquake or ice forces.

8-11   Determine the required thickness of concrete and amount of reinforcing steel required at the top, mid-height, and bottom of the slab of a buttress dam 30 ft high and retaining a maximum depth of water of 26 ft. Buttresses are 3 ft thick on 20-ft centers; face slope of dam is 45°. Working stress for steel = 18,000 psi; concrete = 800 psi.

8-12   Draw the flow net for an earth dam 40 ft high with a maximum water level of 35 ft above the base. Top width is 20 ft, and side slopes are 2.25 horizontal to 1 vertical. The foundation material is impervious. A drainage blanket 4 ft thick extends from the toe to a point 60 ft upstream of the toe. If the permeability of the embankment material is 0.015 ft/min, what is the seepage rate per foot of dam?

8-13   Draw the flow net for the dam of Prob. 8-12 with the drainage blanket omitted. Compute the seepage rate.

8-14   Repeat the preceding problem for the case where the dam has a centrally located, completely impermeable core 12 ft high, top width of 15 ft, and side slopes of 1 on 1.

8-15   A small concrete gravity dam with a base width of 8 ft rests on a sandy silt stratum 15 ft thick which overlies impermeable rock. If the coefficient of permeability of the sandy silt is 0.05 ft/min, compute the seepage under the dam when the water depth behind the dam is 10 ft. Plot the lines of equal pore pressure in the soil beneath the dam.

8-16   From the flow net of Prob. 8-13, sketch lines of equal pore pressure, for pore pressures of 3, 6, and 9 psi.

8-17   Refer to the earth dam of Fig. 8-23. Check this dam for stability using a failure arc of radius 80 ft with center 80 ft vertically above a point 60 ft upstream of the toe. Consider the effect of pore pressures along failure arc.

8-18   An earth dam with homogeneous embankment 100 ft high and a top width of 30 ft, side slopes of 1 on 2, rests on an impervious foundation and is designed to retain water to a depth of 90 ft. The embankment is a silty material whose properties are: $c = 500$ psf, $\phi = 15°$, $\gamma = 110$ pcf, and $K = 0.035$ ft/day. Sketch the flow net, compute the seepage rate, and plot lines of equal pore pressure. Check the embankment for stability under the following conditions:

(a) Steady-state seepage with failure arc of radius 100 ft centered at a point 100 ft vertically above the toe of the dam.

(b) Steady-state seepage with failure arc of radius 100 ft centered at a point 100 ft vertically above the heel of the dam.

(c) Water in reservoir completely drawn down so fast that pore pressures have not had time to dissipate. Failure arc as indicated in (b).

8-19   Draw the tentative cross section for an earth dam to retain 60 ft of water above stream bed. Allow 3 to 5 ft for wave conditions. Material available includes a limited supply of decomposed granite with a very low permeability when compacted to optimum specific weight of 125 lb/cu ft, unlimited quantities of river-run gravel varying from coarse sand to 8 in. in diameter, and limited quantities of granite which will be obtained from the spillway excavation. Disregard foundation design.

8-20   Using a failure arc with its center 40 ft vertically above the center of the berm of Fig. 8-23 (on the downstream slope) and with a radius of 100 ft, compute the factor of safety against embankment failure. Consider the effect of pore pressures in the embankment. Also determine the factor of safety if the embankment had been completely free of pore pressure.

Chapter 9

Spillways, gates,
and outlet works

Some provision must be made in the design of almost every dam to permit the discharge of water downstream. A spillway is necessary to discharge floods and prevent the dam from being damaged. Gates on the spillway crest, together with sluiceways, permit the operator to control the release of water downstream for various purposes. In some cases facilities to regulate the flow in canals or pipelines leading from the reservoir are also necessary. For each of these functions a variety of devices, each with its special characteristics, are available for selection.

SPILLWAYS

A spillway is the safety valve for a dam. It must have the capacity to discharge major floods without damage to the dam or any appurtenant structures, at the same time keeping the reservoir level below some predetermined maximum level. A spillway may be controlled or uncontrolled; a controlled spillway is provided with crest gates or other facilities so that the outflow rate can be adjusted.

    9-1 Spillway design capacity The required capacity (maximum outflow rate through the spillway) depends on the spillway design flood (inflow hydrograph to the reservoir), the normal discharge capacity of the outlet works, and the available storage. The selection of the spillway design flood is related to the degree of protection that ought to be provided to the dam which, in turn, depends on the type of dam, its location, and consequences of failure of the dam. A high dam storing a large volume of water located upstream of an inhabited area should have a much higher degree of protection than a low dam storing a small quantity of water whose down-stream reach is uninhabited. The probable maximum flood (Sec. 5-10) is commonl used for the former while a smaller flood is suitable for the latter.

Overflow spillways An overflow spillway is a section of dam designed to permit water to pass over its crest. Overflow spillways are widely used on gravity, arch, and buttress dams. Some earth dams have a concrete gravity section designed to serve as a spillway. The design of the spillway for low dams is not usually critical, and a variety of simple crest patterns are used. In the case of large dams it is important that the overflowing water be guided smoothly over the crest with a minimum of turbulence. If the over-flowing water breaks contact with the spillway surface, a vacuum will form at the point of separation and cavitation may occur. Cavitation plus the vibration from the alternate making and breaking of contact between the water and the face of the dam may result in serious structural damage.

Cavities filled with vapor, air, and other gases will form in a liquid when-ever the absolute pressure of the liquid is close to the vapor pressure. This phenomenon, *cavitation*,[1] is likely to occur where high velocities cause reduced pressures. Such conditions may arise if the walls of a passage are so sharply curved as to cause separation of flow from the boundary. The cavity, on moving downstream, may enter a region where the absolute pressure is much higher. This causes the vapor in the cavity to condense and return to liquid with a resulting implosion, or collapse, of the cavity. When the cavity collapses, extremely high pressures result.[2] Some of the implosive activity will occur at the surfaces of the passage and in the crevices and pores of the boundary material. Under a continual bombardment of these im-plosions, the surface undergoes fatigue failure and small particles are broken away, giving the surface a spongy appearance. This damaging action of cavitation is called pitting.

The ideal spillway would take the form of the underside of the nappe of a sharp-crested weir when the flow rate corresponds to the maximum design capacity of the spillway (Fig. 9-1a). Figure 9-1c shows an *ogee weir* which closely approximates the ideal.[3] More exact profiles may be found in more extensive treatments of the subject.[4] The reverse curve on the downstream face of the spillway should be smooth and gradual. A radius

[1] Cavitation in Hydraulic Structures—A Symposium, *Trans. ASCE*, Vol. 112, pp. 1–124, 1947.

[2] In tests at the Stanford University Fluid Mechanics Laboratory pressures as high as 350,000 psi were measured in the collapse of a cavity.

[3] Hydraulic Models as an Aid to the Development of Design Criteria, *U.S. Waterways Expt. Sta. Bull.* 37, Vicksburg, Miss., June, 1951.

[4] Studies of Crests for Overfall Dams, Boulder Canyon Project, *U.S. Bur. Reclamation Bull.* 3, Part VI, 1938.

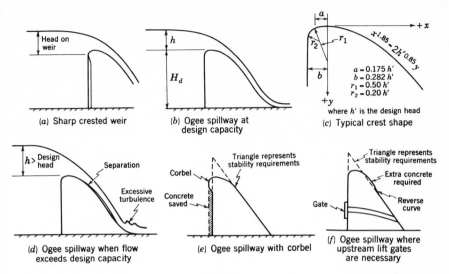

FIG. 9-1   Characteristics of an ogee spillway.

of about one-fourth of the spillway height has proved satisfactory. Structural design of an ogee spillway is essentially the same as the design of a concrete gravity section. The pressure exerted on the crest of the spillway by the flowing water and the drag forces caused by fluid friction are usually small in comparison with the other forces acting on the section. The change in momentum of the flow in the vicinity of the reverse curve may, however, create a force which must be considered. The requirements of the ogee shape usually necessitate a thicker section than the adjacent nonoverflow sections. A saving of concrete can be effected by providing a projecting corbel on the upstream face of the spillway section (Fig. 9-1e). If the dam is to have gates on the upstream face to control the flow in outlet conduits through the dam, a corbel will interfere with gate operation and a solution such as that shown in Fig. 9-1f may be used.

The discharge of an overflow spillway is given by the weir equation

$$Q = C_w L h^{3/2} \qquad (9\text{-}1)$$

in which Q is the discharge in cubic feet per second, C_w is a coefficient, L is the length of the crest in feet, and h is the head on the spillway (vertical distance in ft from the crest of the spillway to the reservoir level). The coefficient C_w varies from about 3.0 to 4.0 for ogee spillways, with the higher values applicable to the standard crest of Fig. 9-1c. The coefficient for any spillway also varies with the head (Fig. 9-2). Experimental models are often used to determine spillway coefficients. End contractions on a

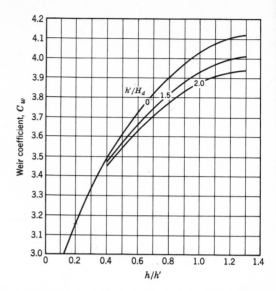

FIG. 9-2   *Variation of discharge coefficient with head for an ogee-spillway crest such as shown in Fig. 9-1c.*

spillway reduce the effective length below the actual length L. Square-cornered piers disturb the flow considerably and reduce the effective length by the width of the piers plus about $0.2h$ for each pier (Fig. 9-3). Streamlining the piers or flaring the spillway entrance minimizes the flow disturbance. If the cross-sectional area of the reservoir just upstream from the spillway is less than five times the area of flow over the spillway, the approach velocity will increase the discharge a noticeable amount. For such conditions the effect of approach velocity can be accounted for by the equation

$$ Q = C_w L \left(h + \frac{V_0^2}{2g} \right)^{3/2} \tag{9-2} $$

where V_0 is the approach velocity in feet per second.

9-3   Chute spillways   Water flows over the crest of a chute spillway into a steep-sloped open channel which is called a chute, or trough (Fig. 9-4). The channel is usually constructed of reinforced-concrete slabs 10 to 20 in. thick. Such a structure is relatively light and is well adapted to earth or rock-fill dams when topographic conditions make it necessary to place the spillway on the dam. A chute spillway may be constructed around the end of any type of dam when topographic conditions permit, and such a location is preferred for earth dams to prevent possible damage to the embankment.

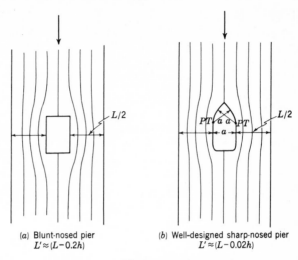

(a) Blunt-nosed pier
$L' \approx (L - 0.2h)$

(b) Well-designed sharp-nosed pier
$L' \approx (L - 0.02h)$

L' - Effective length of spillway crest
h - Head on spillway

FIG. 9-3   Effective length of spillway crest.

FIG. 9-4   The chute spillway and powerhouse at Anderson Ranch Dam, Idaho.
(U.S. Bureau of Reclamation)

The chute is sometimes of constant width but is usually narrowed for economy and then widened near the end to reduce discharge velocity. If the grade of the chute can conform to topography, excavation will be minimized. However, it is desirable that the slope be steep enough to maintain flow below critical depth in order to avoid unstable flow conditions. Vertical curves should be gradual and designed to avoid separation of the flow from the channel bottom.

Expansion joints are usually required in chute spillways at intervals of about 30 ft. If water penetrates under the slab, it may cause troublesome uplift. The expansion joints should, therefore, be as watertight as possible, and drains under the spillway pavement are highly desirable. The drains may be rock-filled trenches, clay tile, or perforated steel pipe. The slabs of a chute spillway should be keyed together in such a manner that the upstream end of a slab cannot rise above the block next upstream (Fig. 9-5).

9-4 Side-channel spillways A side-channel spillway (Fig. 9-6) is one in which the flow, after passing over the crest, is carried away in a channel running parallel to the crest. The crest is usually a concrete gravity section, but it may consist of pavement laid on an earth embankment or the natural ground surface. This type of spillway is used in narrow canyons where sufficient crest length is not available for overflow or chute spillways.

The hydraulic theory of flow in a side-channel spillway will not be presented here. Figure 9-7 shows a sketch of the flow in such a channel. Analysis of flow in the side channel is made by application of the momentum principle in the direction of flow.[1] Residual energy of the water after passing the spillway crest is ignored in the design of the side channel. In fact, a weir, or sill, is often placed across the channel to create a stilling basin beneath the spillway crest to dissipate this energy.

After passing through the side channel, the water is ordinarily carried away through a chute or tunnel. There are many spillways which change

---

[1] Julian Hinds, Side-channel Spillways, *Trans. ASCE*, Vol. 89, pp. 881–939, 1926.

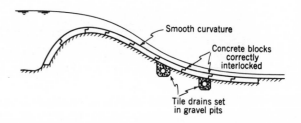

FIG. 9-5   *Details of a typical chute spillway.*

FIG. 9-6   The side-channel spillway on the Arizona side of Hoover Dam, looking upstream. Drum gates (Sec. 9-14) on crest are in open position. (U.S. Bureau of Reclamation)

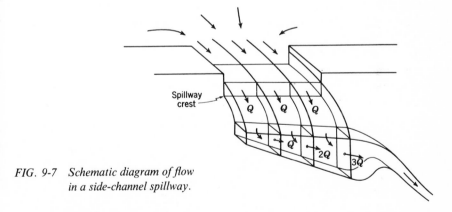

FIG. 9-7   Schematic diagram of flow in a side-channel spillway.

direction immediately after the crest and whose characteristics are intermediate between the chute and the side channel.

9-5 Shaft spillways In a shaft spillway the water drops through a vertical shaft to a horizontal conduit which conveys the water past the dam (Fig. 9-8). A shaft spillway can often be used where there is inadequate

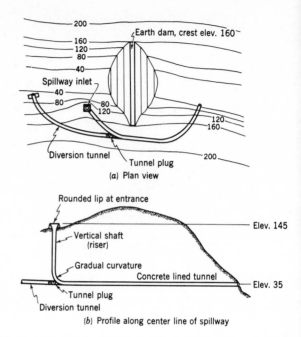

FIG. 9-8   *Typical shaft spillway.*

space for other types of spillways. It is generally considered undesirable to carry a spillway over or through an earth dam. If topography prevents the use of a chute or side-channel spillway around the end of the dam, a shaft spillway through the foundation material may be a good alternative.

For low dams where the shaft height is small, no special inlet design is necessary, but on large projects a flared inlet, referred to as a *morning glory*, is often used. Small shaft spillways may be constructed entirely of metal or concrete pipe or clay tile. The vertical shaft of large structures is usually of reinforced concrete, while the horizontal conduit is tunneled in rock. Frequently the diversion tunnel is planned so that it may be used for the spillway outlet. In some favorable situations the vertical shaft has been driven in rock.

There are three possible conditions of flow in a shaft spillway (Fig. 9-9). At low heads the outlet conduit flows partly full, the perimeter of the inlet serves as a weir, and the discharge of the spillway varies as $h_1^{3/2}$. As the head is increased, water rises in the shaft, and the outlet may flow partly full (weir flow) or full (orifice flow). When the shaft is completely filled with water, and the inlet is submerged, the discharge becomes approximately proportional to $h_2^{1/2}$ (pipe flow), where h_2 is the total head on the outlet. In this third stage an increase in h_2 results in only a very slight increase in discharge. This, in effect, places a limit on the capacity of a shaft spillway; hence, such spillways must be conservatively designed.

An abrupt transition between the shaft and outlet conduit may result in cavitation,[1] hence a smooth transition is preferred in large structures. Hydraulic analysis of shaft spillways is difficult, and model tests are often employed. Models must be used with caution, for the air pressure in the model is not reduced to model scale. An undesirable feature of shaft spillways is the hazard of clogging with debris. Trash racks, floating booms, or other types of protection are necessary to prevent debris from entering the inlet.

9-6 Siphon spillways If a large capacity is not necessary, and space is limited, the siphon spillway (Fig. 9-10) may be a practical selection. Siphon spillways have the advantage that they can automatically maintain water-surface elevation within very close limits. At low flows, the siphon spillway operates like an overflow spillway with its crest at C. At higher flows, after the siphon has primed, discharge is given by $C_d A \sqrt{2gh}$, where C_d is a coefficient of discharge which is usually about 0.9. If the outlet of the siphon is not submerged (Fig. 9-10a), the head h is the vertical distance from the water surface in the reservoir to the end of the siphon barrel. When the outlet is submerged, h is the difference in elevation between the head-

---

[1] J. N. Bradley, Morning-glory Shaft Spillways: Prototype Behavior, *Trans. ASCE*, Vol. 121, pp. 312–333, 1956.

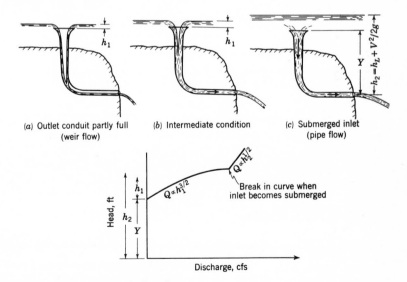

FIG. 9-9   *Flow conditions in a shaft spillway.*

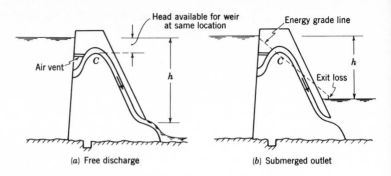

FIG. 9-10   Cross section of a siphon spillway.

water and tailwater (Fig. 9-10b). If air is prevented from entering the outlet end of the siphon, flow through the siphon will entrain and remove the
air at the crown and prime the siphon. Entrance of air can be prevented by
deflecting the flow across the barrel in such a way as to seal it off or by sub-
merging the outlet end of the barrel. Siphon action will continue until the
water level in the reservoir drops to the elevation of the upper lip of the
siphon entrance, unless a vent is provided at a higher level. A siphon may
be designed so that variations in upstream water level are small with respect
to total head, and thus the discharge is nearly always at capacity when the
siphon is primed. This makes the siphon spillway particularly advantageous
in disposing of sudden surges of water such as may occur in canals and fore-
bays when the outlet gates are closed rapidly.

As soon as a siphon is primed, a vacuum forms at the crown. In order
to prevent cavitation, it is wise to limit this vacuum to three-fourths atmospheric pressure. In other words, at sea level the vertical distance from the
crown of the siphon down to the hydraulic grade line should not exceed
25 ft. At higher elevations the limiting distance from crown to grade line is
still smaller. If the entrance to the siphon remains submerged to a depth
of a few feet, there is little likelihood of clogging from debris or ice but trash
racks may be a wise precaution. One disadvantage of the siphon spillway
is the relatively high cost of forming the barrel, but if the siphon can be
made from pipe, the cost may not be high.

9-7   Service spillways and emergency spillways   On many projects a
single spillway serves to discharge all rates of outflow. In some instances,
however, it is economic to have more than one spillway—a *service spillway*
to convey frequently occurring outflow rates and one or more *emergency
spillways* that are used only rarely during extreme floods. Often a saddle or
low point on natural ground at the periphery of the reservoir will serve as
the emergency spillway. In other instances an engineered structure is used.

An interesting example of a "service-emergency spillway structure" is shown in Fig. 9-11 in which the side-channel spillway with concrete discharge chute is designed to handle the outflow from the 50-yr flood. Larger flows pass over the backside of the side-channel spillway (secondary weir) and are conveyed to the river through a natural depression.

9-8 Dynamic forces on spillways Newton's second law of motion states that force equals the time rate of change of momentum. The resultant of the forces on an element of water is

$$\Sigma \vec{F} = \rho Q \, \vec{\Delta V} \qquad (9\text{-}3)$$

where ρ is the density of the water[1] in slugs per cubic foot, Q is the flow rate, and ΔV is the change in velocity. Since Eq. (9-3) is vectorial, it may be expressed as

$$\Sigma F_x = \rho Q(V_{2_x} - V_{1_x}) \quad \text{and} \quad \Sigma F_y = \rho Q(V_{2_y} - V_{1_y}) \qquad (9\text{-}4)$$

where the subscripts x and y represent any convenient system of coordinates. Equation (9-4) may be used to find the dynamic forces exerted by water on spillways, deflectors, turbine blades, pipe bends, etc. The forces F_x and F_y

---

[1] For water, $\rho = 62.4/32.2 = 1.94$ slugs/cu ft.

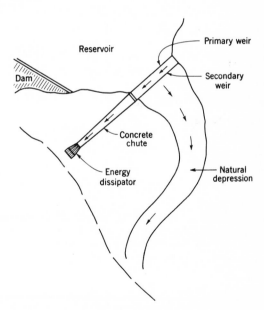

FIG. 9-11   Spillway for Auld Valley Project of Metropolitan Water District of Southern California.

are those acting on a significant free body of fluid and include gravity forces, hydrostatic pressures, and the reaction of any object in contact with the water.

ILLUSTRATIVE EXAMPLE 9-1   Given the ogee spillway of Fig. 9-12 with $C_w = 3.8$, find the total force of the water on the curved section AB.

$$\frac{Q}{L} = 3.8 \times (5)^{3/2} = 42.5 \text{ cfs/ft} \quad\longleftarrow Q = C_w L H^{3/2}$$

Assuming no loss in energy and neglecting approach velocity,

$$45 = 6 + d_A \cos 60° + \frac{V_A{}^2}{2g} = d_B + \frac{V_B{}^2}{2g}$$

Substituting $V_A = 42.5/d_A$ and $V_B = 42.5/d_B$ and solving by trial, $d_A = 0.85$ ft, $V_A = 50.0$ ft/sec, $d_B = 0.80$ ft, and $V_B = 53.4$ ft/sec. In the free-body diagram (Fig. 9-12) F_H and F_V represent components of force on the water by the curved section AB of the spillway. The hydrostatic forces F_1 and F_2 are

$$F_1 = 62.4 \times (0.425 \times \cos 60°) \times 0.85 = 11.2 \text{ lb}$$

$$F_2 = 62.4 \times 0.400 \times 0.80 = 19.9 \text{ lb}$$

and the weight W of the water in the section AB is

$$W = 62.4 \times {}^{60}\!/_{360} \times 2\pi(12 \times 0.825) = 644 \text{ lb}$$

For a 1-ft length of spillway Eq. (9-4) is

$$(11.2 \times 0.5) - 19.9 + F_H = (1.94 \times 42.5)(53.4 - 25.0)$$

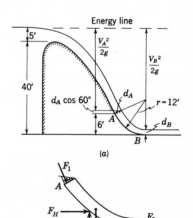

(a)

(b)

FIG. 9-12   Sketch for Illustrative
Example 9-1.

and F_H = 2354 lb/ft of spillway. The vertical component may be found from

$$F_V - 0.866F_1 - W = (1.94 \times 42.5)[0 - (-0.866 \times 50)]$$

and F_V = 4224 lb/ft of spillway. The plus values indicate that the assumed directions were correct. Thus, the force of the water on the section AB per foot of spillway length is 2354 lb to the left and 4224 lb down.

CREST GATES

Additional storage above the spillway crest can be made available by the installation of temporary or movable gates. Such an increase in reservoir level is permissible in the low-water season, when low flows may be permitted over the crest-control device. If a large flood occurs, full spillway capacity may be made available by removing the temporary barriers. These devices must be used with caution on the spillway of an earth dam, where operational failure of the gates may result in overtopping of the dam.

9-9   Flashboards   The usual flashboard installation consists of wooden panels supported by vertical pins placed on the crest of the spillway (Fig. 9-13a). Such installations are temporary and are designed to fail when the water surface in the reservoir reaches a predetermined level. A common design uses steel pipe or rod set loosely in sockets in the crest of the dam and designed to bend and release the flashboards at the desired water level. Temporary flashboards of this type have been used in heights up to 4 or 5 ft. Since temporary flashboards are lost each time the supports fail, permanent flashboards are more economical for large installations. Permanent flashboards usually consist of wooden panels which can be raised or lowered from an overhead cableway or bridge (Fig. 9-13b). The lower edge of the panels is placed in a seat or hinge on the spillway crest, and the panels are supported in the raised position by struts or by attaching the upper edge of the panel to the bridge.

9-10   Stop logs and needles   Stop logs consist of horizontal timbers spanning the space between grooved piers (Fig. 9-14a). The logs may be raised by hand or with a hoist. There is usually much leakage between the logs, and considerable time may be required for removing the logs if they become jammed in the slots. Stop logs are ordinarily used for small installations where the cost of more elaborate devices is not warranted or in situations where removal or replacement of the stop logs is expected only at very infrequent intervals.

Needles consist of timbers with their lower ends resting in a keyway on the spillway crest and their upper ends supported by a bridge (Fig. 9-14b). Needles are somewhat easier to remove than stop logs but are quite difficult

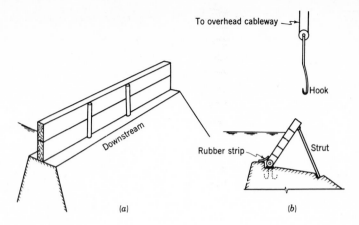

FIG. 9-13   *Two types of flashboards.*

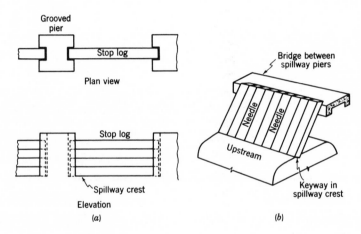

FIG. 9-14   *Stop-logs (a) and needles (b).*

to place in flowing water. Consequently they are used mainly for emergency bulkheads, where they need not be replaced until flow has stopped.

9-11   Vertical lift gates   Simple timber or steel gates which slide in vertical guides on piers on the crest of the dam are used for small installations. Their size is limited by the high friction force developed in the guides because of the hydrostatic force on the gate. By placing cylindrical rollers between the bearing surfaces of the gate and guides the frictional resistance can be much reduced. The *Stoney gate* has rollers that are independent of the gate or guides, thus eliminating axle friction. The independent roller train of the Stoney gate is difficult to design and build, and the development of low-friction roller bearings has led to the use of the *fixed-wheel gate,* which has wheels attached to the gate and riding in tracks on the downstream side of the gate guide. Fixed-wheel gates have been used in sizes up to 50 by 50 ft (Bonneville Dam). In the large sizes excessive headroom is required to lift the gate clear of the water surface, and large vertical lift gates are often built in two horizontal sections so that the upper portion may be lifted and removed from the guides before the lower portion is moved. This design also reduces the load on the hoisting mechanism. Discharge may occur over either one or both sections of the gate or over the spillway crest.

A gate 50 ft square may have to support a water load of over 2000 tons, and the gate itself may weigh 150 tons.[1] Design of such a gate and its operating mechanism is a structural and mechanical problem of considerable magnitude. Accurate alignment of the rollers and guides is necessary so that the gate will operate satisfactorily.

9-12 Radial gates The cross section of a *radial*, or *Tainter*, *gate* is shown in Fig. 9-15. The face of the gate is a cylindrical segment supported on a steel framework which is pivoted on trunnions set in the downstream portion of the piers on the spillway crest. Hoisting cables are attached to the gate and lead to winches on the platform above the gate. The winches are usually motor-driven, although hand power may be used for small gates. Each gate may have its independent hoisting mechanism, or a common unit may be moved from gate to gate. Flexible fabric or a rubber strip is used to form a water seal between the gates and the piers and spillway crest. Tainter gates vary in size from 3 to 35 ft in height and 6 to 60 ft in length. One of the largest Tainter-gate installations is at Clark Hill dam in Georgia, which has 23 Tainter gates 35 ft high and 60 ft long.

Radial gates have several advantages. Friction is concentrated at the pin and is usually much less than for sliding gates. Since the trunnion bears part of the load, the hoisting load is nearly constant for all gate openings and is much less than for vertical lift gates of the same size. Counterweights are sometimes required for large gates of either radial or vertical lift type.

9-13 Rolling gate A *rolling* (or *roller*) *gate* consists of a steel cylinder spanning between the piers. Each pier (Fig. 9-16) has an inclined rack which engages gear teeth encircling the ends of the cylinder. When a pull is exerted on the hoisting cable, the gate rolls up the rack. The lower portion of the gate consists of a cylindrical segment which makes contact with the spillway crest and increases the gate height. Rolling gates are well adapted to long spans of moderate height. A rolling gate 147 ft long and 21 ft high is installed on the Glommen River in Norway.

9-14 Drum gates Another type of gate adapted to long spans is

---

[1] F. L. Boissonault, Estimating Data for Reservoir Gates, *Trans. ASCE*, Vol. 113, pp. 992–1026, 1948.

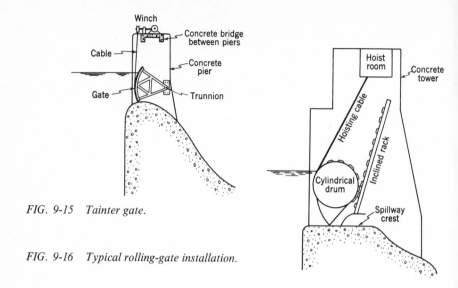

FIG. 9-15   Tainter gate.

FIG. 9-16   Typical rolling-gate installation.

the *drum gate*[1] (Fig. 9-17). This gate consists of a segment of a cylinder which, in the open or lowered position, fits in a recess in the top of the spillway. When water is admitted to the recess, the hollow drum gate is forced upward to the closed position. The type developed by the U.S. Bureau of Reclamation (Fig. 9-17a) is a completely enclosed gate, hinged at the upstream edge so that buoyant forces aid in its lifting. This type of gate is adapted to automatic operation and also conforms closely to the shape of the ogee crest when lowered. A second type (Fig. 9-17b) has no bottom plate and is raised by water pressure alone. Because of the large recess required by drum gates in the lowered position, they are not adapted to small dams.

9-15   Movable dams   A movable dam or weir is one that can be readily activated or deactivated. A *bear-trap gate* consists of two leaves of timber or steel hinged and sealed to the dam (Fig. 9-18). When water is admitted to the space under the leaves, they are forced upward. The downstream leaf is frequently hollow so that its buoyancy aids the lifting operation. This type of gate is adapted to low navigation dams, since at high-river stages the gate may be lowered so as not to interfere with navigation over it.

Another type of movable dam is the *rubber dam* which consists of a full channel-width, shallow steel container attached to the concrete channel bottom in which lies a collapsed nylon-rubber bladder. The dam is activated by pumping water into the rubber bladder thus inflating it to form a barrier

---

[1]Joseph N. Bradley, Rating Curves for Flow over Drum Gates, *Trans. ASCE*, Vol. 119, pp. 403–433, 1954.

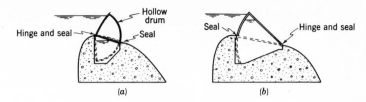

FIG. 9-17   Two types of drum gates.

that stands out above the channel bottom thus blocking the channel. The dam is deactivated by releasing the water from inside the bladder. In its inflated state the rubber dam may be strong enough for use as a temporary bridge for pedestrians and lightweight vehicles.[1] Other types of movable dams are discussed in Sec. 17-5.

9-16   Ice control at spillways   If ice forms on gates, they may become inoperable. Gates which are lowered to increase flow, such as bear-trap or drum gates, are advantageous in that ice is not likely to interfere with their operation. Moreover, ice and drift can pass over such gates more readily than where the flow is between the gate and the spillway crest. Some gate installations are provided with steam coils or electrical heating units to prevent ice formation. It is not essential that every gate be equipped with a heating unit, since if every third or fourth gate is kept from freezing, the others will be thawed by the flowing water. If compressed air is introduced into the water below the bottom of the gates, the rising air brings warmer water from lower levels to the surface, thus preventing formation of ice. This does not eliminate the formation of ice on those portions of the structure which are above the water.

OUTLET WORKS

The major portion of the storage volume in most reservoirs is below the spillway crest. Outlet works must be provided in order that water can be drawn from the reservoir as needed. This water may be discharged into the

---

[1] Movable Weir, 92 Feet Wide, Helps Control Netherlands Rivers and Canals, *Civil Eng. (N.Y.)*, Vol. 39, p. 91, November, 1969.

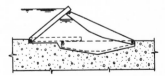

FIG. 9-18   Bear-trap gate.

channel below the dam or may be transported in pipes or canals to some distant point.

9-17 Sluiceways A sluiceway is a pipe or tunnel that passes through a dam or the hillside at one end of the dam and discharges into the stream below. Sluiceways for masonry dams generally pass through the dam, while those for earth or rock-fill dams are preferably placed outside the limits of the embankment. If a sluiceway must pass through an earth dam, projecting collars should be provided to reduce seepage along the outside of the conduit. A common rule is that the collars should increase the length of the seepage path by at least 25 percent, i.e., in Fig. 9-19, $2Nx > 0.25L$, where N is the number of collars and x is the projection of the collar.

The outlets of most dams consist of one or more sluiceways with their inlets at about minimum reservoir level. Large dams may have sluiceways at various levels. If the quantity of discharge is to vary considerably, a number of sluiceways is desirable. In most cases a single large-capacity sluiceway may be structurally unsatisfactory. Sluiceways may be circular or rectangular. The interior should be smooth and without projections or cavities which might induce separation of the flow from the boundary of the conduit and cause negative pressures and cavitation.[1] There have been several instances where excessive damage has resulted from cavitation in the vicinity of square-edged entrances (Fig. 9-20a) operating under high heads.

Considerable attention should be given to the design of the sluiceway entrance. A bellmouth entrance (Fig. 9-20b) is far superior to other types, and the extra cost is usually justified except for small projects under low head. Douma[2] suggested that the bellmouth should be elliptical, the equation for circular conduits being

---

[1] H. A. Thomas and E. P. Schuleen, Cavitation in Outlets of High Dams, *Trans. ASCE,* Vol. 107, pp. 421–493, 1942.

[2] J. H. Douma, Discussion of Cavitation in Outlets of High Dams, *Trans. ASCE,* Vol. 107, p. 481, 1942.

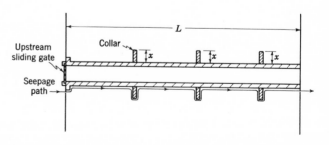

FIG. 9-19   Sluiceway collars for earth dams.

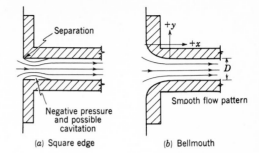

FIG. 9-20   Sluiceway entrances.

$$4x^2 + 44.4y^2 = D^2 \tag{9-5}$$

and that for rectangular conduits

$$x^2 + 10.4y^2 = D^2 \tag{9-6}$$

where x and y are defined in Fig. 9-20, and D is the diameter of a circular conduit or the width or height of a rectangular conduit, depending on whether the side or top and bottom curves are being considered.

9-18   Intakes   An intake structure is required at the entrance to a conduit through which water is withdrawn from a river or reservoir unless this entrance is an integral part of a dam or other structure. Intake structures vary from a simple concrete block supporting the end of a pipe to elaborate concrete intake towers, depending upon reservoir characteristics, climatic conditions, capacity requirements, and other factors. The primary function of the intake structure is to permit withdrawal of water from the reservoir over a predetermined range of pool levels and to protect the conduit from damage or clogging as a result of ice, trash, waves, and currents.

Intake towers are often used where there is a wide fluctuation of water level. Ordinarily they are provided with ports at various levels which may aid flow regulation and permit some selection of the quality of water to be withdrawn. If the ports are submerged at all levels, trouble from ice and floating debris is unlikely. It may be necessary to place the lowest ports far enough above the bottom of the reservoir so that sediment will not be drawn into them. A *wet intake tower* (Fig. 9-21) consists of a concrete shell filled with water to the level of the reservoir and has a vertical shaft inside connected to the withdrawal conduit. Gates are normally provided on the inside shaft to regulate flow. A *dry intake tower* has no water inside of it at close-off since the entry ports are connected directly to the withdrawal conduit. Each entry port is provided with a gate or valve. When the entry ports are closed, dry intake towers are subject to buoyant forces and hence must be of heavier construction than wet towers. An advantage of the dry tower is that water can be withdrawn from any selected level of the

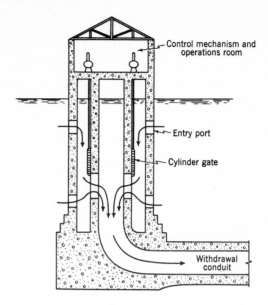

FIG. 9-21   Section through a wet intake tower.

reservoir. Intake towers should be located so as not to interfere with navigation and must be designed to withstand hydrostatic pressures and forces
from earthquakes, wind, waves, and ice.

A *submerged intake* (Fig. 9-22) consists of a rock-filled crib or concrete
block which supports the end of the withdrawal conduit. Because of their
low cost, submerged intakes are widely used on small projects. Because
they do not obstruct navigation, they are well-suited to use as water-supply
intakes from rivers. They are sometimes used for intakes to sluiceways of
earth dams with hydraulically operated gates for flow regulation. The
main disadvantage of the submerged intake is that it is not readily accessible
for repair of the gates. Submerged intakes in reservoirs and rivers must be
placed where they will not be buried by sediment.

9-19   Trash racks   The entrances to intakes and sluiceways should
be provided with trash racks to prevent entrance of debris. These racks are
usually made of steel bars spaced on 2- to 6-in. centers, depending upon the
maximum size of debris which can be permitted in the conduit. The velocity
through the trash rack should be kept low, preferably less than 2 ft/sec to
minimize head loss. This is sometimes accomplished by constructing the
rack in the form of a half cylinder. Debris which accumulates on the racks
is sometimes removed by hand when necessary, but where much debris is
expected, automatic power-driven rack rakes are preferable. Electrically
charged racks are sometimes used to discourage entry of fish into the

outlet conduit. In cold climates trash racks are often clogged by accumulations of *frazil ice* crystals from the flowing water. A steel rack which extends above the water surface may be slightly colder than the water, and the ice will adhere to the rack quite readily. Steam or electrical heating may be used to prevent ice formation on trash racks.[1]

9-20   Entrance gates   Most intakes and sluiceways are provided with some type of gate or valve at their entrance. On projects with head less than 100 ft, entrance gates may act as flow regulators, but with higher heads they are ordinarily used only as emergency gates to permit inspection and repair of the conduit.

The frictional resistance caused by hydrostatic pressure and the weight of the gate together determine the type of gate and hoisting mechanism required. Small gates on very low-head installations are often simple sliding gates operated by a hand-powered worm drive. Motor-driven slide gates will suffice for moderate-head projects, but for high-head installations or very large gates under low heads, roller gates are required. Slide gates usually have bronze bearing surfaces to minimize friction.

Because rubber water seals do not perform well under high heads, high-head entrance gates are designed differently from vertical-lift crest gates. High-head *tractor gates* (Fig. 9-23) consist of a gate leaf and crossbeam to which a wedge-shaped roller train is attached. As the gate is lowered into the closed position, its downward motion is halted when its bottom edge

---

[1] P. E. Gisiger, Safeguarding Hydro Plants against the Ice Menace, *Civil Eng.* (*N.Y.*), Vol. 17, pp. 24–27, January, 1947.

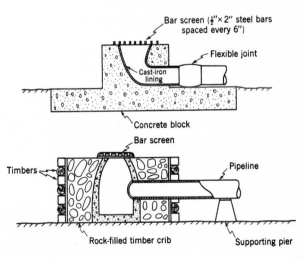

FIG. 9-22   *Two typical submerged intakes.*

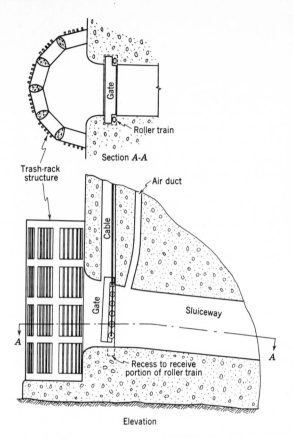

FIG. 9-23   *Details of a tractor gate installation.*

comes in contact with the bottom of the gate frame. The roller trains, moving in slots beside the gate, continue their downward movement and, because of their wedge shape, permit the gate leaf to move a small distance downstream. Finally hydrostatic pressure pushes the gate tightly into the gate frame to form a watertight seal. The first tractor gates were designed to close the 16- by 20-ft entrance to the penstocks at Norris Dam under a maximum head of 184 ft.

At tractor-gate installations a closed concrete structure resembling a chimney is sometimes built on the upstream face of the dam to protect the hoisting cables and gate assembly. Since these gates are normally open, an alternative to the protective structure is to store the gates above the water surface, but on high dams this lengthens the closure time considerably. The hoisting equipment is usually placed at the crest of the dam.

Large intake towers of the wet type are often provided with *cylinder gates* which can be used to close off the flow into the tower (Fig. 9-21). These gates consist of steel plates and supporting framework connected together in the form of a hollow cylinder which can be raised or lowered to uncover the ports. Cylinder gates may be mounted inside or outside of the shaft.

9-21   Interior gate valves   Interior gate valves[1] are located downstream from the conduit entrance. For sluiceways in a gravity dam the valve-operating mechanism is often in a gallery inside the dam. For heads under 75 ft, interior gate valves are often used to regulate flow, but for greater heads they are ordinarily used only in the fully open or fully closed position. Standard gate valves of the type used in water-distribution systems may be used for interior valves in circular conduits up to 4 ft in diameter under heads of less than 300 ft. Large interior gate valves consist of a rectangular gate sliding in a frame formed by two castings bolted together as shown in Fig. 9-24. A bonnet receives the gate when it is in the open position. The transition from rectangular to circular section, if required, should be gradual. Large gates of this type are provided with a hydraulic operating mechanism.

At heads over about 250 ft the discontinuities in the water passage created by the gate slots are likely to induce cavitation. To overcome this difficulty, *ring-follower gates* (Fig. 9-25) were devised. The unique feature of the ring-follower gate is that unbroken continuity of the water passage is provided by a ring attached below the gate leaf in such a way that the interior surface of the ring is aligned with the interior of the conduit when the gate is fully opened. A recess is required below the conduit to receive the ring when the gate is closed, and outlets must be provided for flushing sediment from this recess.

The difficulty of opening large interior gates under high heads caused the U.S. Bureau of Reclamation to develop *paradox gates*. These differ from the ring-follower type in that the paradox gate has a wedge-shaped roller system similar to those on tractor gates. Final closure of the paradox gate is not made until the gate leaf is aligned with the conduit. Hence there is no sliding between sealing and seating surfaces, which would otherwise cause considerable wear on these surfaces.

9-22   Butterfly valves   A butterfly valve consists of a circular gate pivoted about a shaft coincident with a diameter. This type of valve has been used in sizes of 2 in. to 27 ft. The small sizes may be used as regulating valves, but the larger ones are used only as guard gates. This type of valve is best suited for moderate heads, although they have been used under high heads. Careful design is necessary because severe structural loadings may result from the flowing water. One difficulty with butterfly valves is the problem of securing a truly watertight seal.

---

[1] Dow A. Buzzell, Trends in Hydraulic Gate Design, *Trans. ASCE*, Vol. 123, pp. 27–42, 1958.

FIG. 9-24   *Shop testing four 3-ft by 6.5-ft high-pressure slide gates for Willow Creek Dam of the Colorado—Big Thompson project. Total weight 125,000 lb. Hydraulic cylinders have 12.5-in. bore and are rated at 85,000-lb lift. (North West Marine Iron Works)*

9-23   High-head regulating valves   Slide gates are not suitable for flow regulation under high heads because of the hazard of cavitation at partial gate openings. On high-head installations, needle valves and tube valves are widely used for flow regulation. Needle valves are also used to form the jet for impulse turbines (Chap. 12). An *interior-differential needle valve* (Fig. 9-26) consists of three water-filled chambers in which the hydraulic pressure can be varied. Chambers A and C are interconnected so that the pressure in them is the same. The valve is opened by increasing the pressure in chamber B and releasing it from chambers A and C, thus forcing the needle to the left. To close the valve, chamber B is exhausted to the atmosphere while pressure is increased in chambers A and C. The outlet diameter of needle valves in current use ranges from about 4 in. to 105 in.

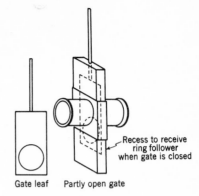

FIG. 9-25   Schematic diagram of a
ring-follower gate.

Recess to receive
ring follower
when gate is closed

Gate leaf   Partly open gate

Tube valves are opened and closed by mechanical means rather than by hydraulic pressure. A screw stem incased in oil is activated by a bevel gear to drive the valve cylinder (or tube) toward or away from the valve seat. Tube valves are generally shorter, lighter in weight, and more economical to build than needle valves. Since there are no hydraulic chambers, the close tolerances to prevent leakage between chambers are unnecessary. Because of cavitation effects, tube valves cannot be operated extensively at flows below about 35 percent of their capacity.

The most recent development in needle valves is the *hollow-jet valve*, in which the needle moves upstream to seat against a ring at the entrance to the valve chamber. The resulting jet is in the form of a hollow cylinder. The hollow-jet needle valve is easier to construct and operate than earlier needle valves and is virtually free of cavitation problems.

Needle valves and tube valves are usually placed at the downstream end of sluices and discharge directly into the atmosphere. The jets issuing from these valves possess a great deal of energy, which may cause considerable scour. A divergent-type valve, the Howell-Bunger valve (Fig. 9-27), which dissipates the energy of the jet over a large area may be used to prevent scour. This type of valve consists of a fixed, cylindrical body with a cone-shaped head at the downstream end which creates a flaring flow pattern.

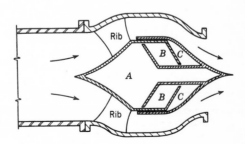

Rib

B C

A

B C

Rib

FIG. 9-26   Sectional view of a
needle valve.

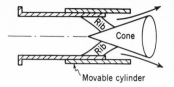

FIG. 9-27   Diagram of a Howell-Bunger valve.

The flow rate through the valve is regulated by the movable cylinder. These valves have been built with outlet diameters ranging from 24 to 96 in.

9-24   Gate installations   For sluiceways passing through earth or rock-fill dams (Fig. 9-28a) it is good practice to install a gate at the upstream end of the conduit so that the conduit is under pressure only when the gate is open. On low-head installations these gates may act as flow regulators, but for high-head projects they are used only as guard gates for regulating valves at the downstream end of the conduit. It is important that the downstream outlet be located so that the discharging water will not damage the downstream face of the dam.

High-pressure regulating valves (needle, tube, or Howell-Bunger) should always be provided with guard gates to permit inspection and repair. The guard gates may be located near the upstream face of the dam (Fig. 9-28b) or immediately upstream from the regulating valve (Fig. 9-28c), in which case a single operating chamber may serve both valves. On large projects, guard gates are often installed in tandem to provide a safeguard against one of them becoming inoperative (Fig. 9-28d). The installation pictured in Fig. 9-28d has the outlet end of the conduits curved downward so as to discharge the water tangentially to the downstream face of the dam. This design is used at Shasta Dam, where eighteen 102-in. valves are employed to regulate the flow in eighteen sluices located at three different levels. The several sluiceways serve as a precaution against one or more becoming inoperative and also permit regulation of outflow with the valves wide open.

To minimize negative pressure and cavitation, air inlets should be provided downstream from high-pressure gates and valves that do not discharge into the atmosphere. The cross-sectional area of the air inlet will usually be 0.5 to 1 percent that of the sluiceway.[1] The air-inlet manifold may be cast with the gate frame. As a further precaution, a lining of cast iron or steel is often provided in sluiceways where high negative pressures may develop.

The hoisting mechanism for a gate must have adequate capacity to handle the weight of the gate and other moving parts, the frictional resistance,

---

[1] W. H. Kohler and J. W. Ball, High-pressure Outlets, Gates and Valves, sec. 22 in C. V. Davis and K. E. Sorensen (eds.), "Handbooks of Applied Hydraulics," 3d ed., McGraw-Hill, New York, 1969.

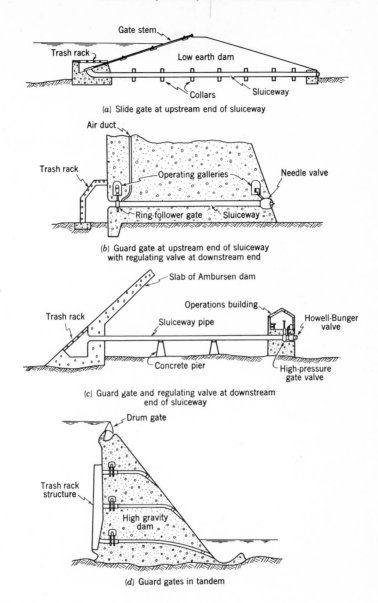

(a) Slide gate at upstream end of sluiceway

(b) Guard gate at upstream end of sluiceway
with regulating valve at downstream end

(c) Guard gate and regulating valve at downstream
end of sluiceway

(d) Guard gates in tandem

FIG. 9-28   *Some typical sluiceway arrangements showing location and type of gates.*

and the hydraulic forces tending to open or close the gate. These usually take the form of *hydraulic downpull* resulting from reduced pressure at the bottom of the gate when in the partially open position. The hydraulic downpull force may be quite large in some instances.[1]

9-25   Hydraulics of outlet works   The outlet works of a dam must be designed to discharge water at rates dictated by the purpose of the dam. Head losses encountered in outlet conduits include those caused by the trash racks, conduit entrance, conduit friction, gates and valves, transitions, and bends.

Trash-rack losses are approximately as indicated in Table 9-1. Head loss at the entrance to a conduit depends on the shape of the entrance and varies from $0.04h_v$ for a bellmouth entrance to $0.50h_v$ for a square-edged opening (Fig. 9-20), where h_v is the velocity head in the conduit just downstream from the entrance. Head losses caused by conduit friction may be calculated by standard pipe-friction formulas (Chap. 11).

The head loss through fully open gate and butterfly valves is about $0.2h_v$, but for ring-follower gates the head loss is assumed to be zero since there is no discontinuity in the walls of the conduit when the gate is open. Discharge curves for needle, tube, and Howell-Bunger valves may be secured from the manufacturer. Most needle valves[2] at full capacity have a coefficient C_d of about 0.7 in the equation

$$Q = C_d A \sqrt{2gh} \tag{9-7}$$

where A is the inlet area of the valve and h is the total head at the valve. Because of their simple construction, Howell-Bunger valves have coefficients of about 0.9.

9-26   Fishways   Salmon, steelhead trout, and other types of anadromous fish, having spent their adult years in salt water, instinctively migrate upstream for spawning. The fingerlings that are hatched generally head downstream during their first year. Dams built below the spawning area on streams inhabited by these types of fish must be provided with fish-passing facilities if it is desired to maintain these fish. Upstream passage of fish can be accomplished through the installation of fish ladders.[3] These structures provide a series of stepped pools having an elevation difference of about 1 ft on a slope of 1:10 to 1:15. The water flows from one pool to

[1] R. I. Murray and W. P. Simmons, Jr., Hydraulic Downpull Forces on Large Gates, *Water Resources Tech. Pub. Res. Rept.* 4, U.S. Bureau of Reclamation, Washington, D.C., 1966.

[2] Charles W. Thomas, Discharge Coefficients for Gates and Valves, *ASCE Proc. Paper* 746, July, 1955.

[3] R. Banys and K. R. Leonardson, Fishways at Dams, sec. 23 in C. V. Davis and K. E. Sorensen (eds.), "Handbook of Applied Hydraulics," 3d ed., McGraw-Hill, New York, 1969.

TABLE 9-1   Head Loss at Trash Rack

Velocity through rack, ft/sec	Head loss, ft
0.5	0.02
1.0	0.10
1.5	0.30
2.0	0.50

the other through a weir and orifice; the fish generally travel upstream through the orifices. Fish ladders are not economic for use on dams higher than 100 ft. In such instances fish are usually transported upstream by mechanical means using tramways or tank trucks.

The seaward-migrating fingerlings must be protected from turbines and spillways on high-head projects. If the head is less than 15 ft, the fish can pass through turbines and spillways with very low mortality. Various types of screens and louvers are used to divert the fish into a safe bypass which might take the form of a short length of pressure pipe through which the fish pass downstream, or the fish may be diverted into a pond from which they can be extracted for downstream transport by tank truck.

PROTECTION AGAINST SCOUR BELOW DAMS

Water flowing over a spillway or through a sluiceway is capable of causing severe erosion of the stream bed and banks below the dam. Consequently the dam and its appurtenant works must be so designed that harmful erosion is minimized.

9-27   General considerations of scour protection   The decision as to what degree of erosion protection should be provided immediately downstream from a dam depends largely on the amount of erosion expected and the damage which might result from this erosion. The time required to develop serious erosion depends not only on the character of the stream-bed material and the velocity distribution but also on the frequency with which scouring flows occur. Thus the results of model tests must be interpreted in the light of the expected flows at the dam site. If many years are expected to pass before serious erosion conditions develop, it may not be economical to provide expensive protection works in the initial construction.

Loose earth and rock are vulnerable to the erosive action of flowing water and may scour severely at velocities as low as 2 or 3 ft/sec. Solid rock is usually resistant to scour, although if the rock has marked bedding planes, it may not endure high-velocity flow. If the rock has a rough, jagged surface, cavitation may assist in the erosion. The general distribution of velocity and the resulting scour are depicted reasonably well in a *movable-bed* hydraulic model. Gravel, sand, or powdered coal may be used to simulate the river bed. The erosion patterns in the bed of the model indicate qualitatively the location and amount of erosion to be expected in the prototype. Ordinary models do not reproduce effects dependent on surface tension, such as air entrainment and spray formation, nor do they reproduce pressure effects such as cavitation. Special glass-enclosed models in which the pressure can be reduced below atmospheric are used to study cavitation.[1] The effectiveness of various energy-dissipating devices may also be estimated from trials on hydraulic models.

9-28   The hydraulic jump   The expression[2] for the hydraulic jump in a horizontal channel of rectangular cross section is

$$d_2 = -\frac{d_1}{2} \pm \sqrt{\frac{d_1{}^2}{4} + \frac{2V_1{}^2 d_1}{g}} \tag{9-8}$$

where d_1 and d_2 are the depths before and after the jump, respectively, and V_1 is the mean velocity in the water before the jump. The approximate depth of flow d_1 at the toe of a spillway (Fig. 9-29) may be found by applying the energy equation along a streamline between point A on the surface of the reservoir and point B at the toe of the spillway. Neglecting friction and velocity of approach, the energy equation is

$$H_d + h = d_1 + \frac{V_1{}^2}{2g} \tag{9-9}$$

The average velocity of flow V_1 at the toe of the spillway is $Q/d_1 B$. By substituting values of h and Q in the energy equation, the corresponding

---

[1] J. E. Warnock, Hydraulic Similitude, chap. 2 in H. Rouse (ed.), "Engineering Hydraulics," Wiley, New York, 1950.

[2] This expression is derived from the impulse-momentum principle. The derivation, as well as procedures for handling nonrectangular sections, may be found in most elementary fluid-mechanics textbooks.

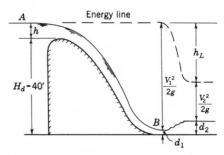

FIG. 9-29   *Sketch for Illustrative Example 9-2.*

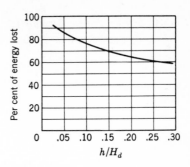

FIG. 9-30   Relation between percent of
energy loss and h/H_d for the
hydraulic jump below the
spillway of Fig. 9-29.

values of d_1 and V_1 can be found. The hydraulic-jump equation may then be used to find the depth after jump d_2. The energy dissipated in the jump is equal to the difference in specific energy before and after the jump. Figure 9-30 shows a plot of energy lost as a function of h/H_d. This shows that a fully developed hydraulic jump below an ogee spillway is particularly effective as an energy dissipator in situations where h is small compared with the height of the spillway.

ILLUSTRATIVE EXAMPLE 9-2   Find the energy loss in the jump at the foot of the spillway of Fig. 9-29 for values of h of 2 to 10 ft. Take C_w from Fig. 9-2 for a design h' = 10 ft. Assume tailwater depth = d_2 in all instances.

For h = 2 ft the discharge per foot of crest is

$$\frac{Q}{B} = 3.16 \times 2^{3/2} = 8.9 \text{ cfs/ft}$$

Hence, the energy equation gives

$$2 + 40 = d_1 + \frac{8.9^2}{64.4 d_1{}^2}$$

from which d_1 = 0.17 ft and V_1 = 8.9/0.17 = 51.9 ft/sec. The depth after jump is given by Eq. (9-8).

$$d_2 = -\frac{0.17}{2} \pm \sqrt{\frac{0.17^2}{4} + \frac{2 \times 51.9^2 \times 0.17}{32.2}} = 5.27 \text{ ft}$$

whence V_2 = 8.9/5.27 = 1.69 ft/sec. The specific energy is $d + V^2/2g$. Therefore,

$$\text{Energy before jump} = 0.17 + \frac{51.9^2}{64.4} = 42.0 \text{ ft-lb/lb}$$

$$\text{Energy after jump} = 5.27 + \frac{1.69^2}{64.4} = 5.3 \text{ ft-lb/lb}$$

The energy loss in the jump is 42.0 − 5.3 = 36.7 ft-lb, and the percent of energy loss $\Delta E/E_1$ is 36.7/42.0 = 87.4 percent. A summary of the computations for other depths over the spillway is given in Table 9-2.

TABLE 9-2   Energy Loss in Hydraulic Jump for Spillway of Fig. 9-29

h, ft	C_w	$\dfrac{Q}{B}$ cfs/ft	d_1, ft	d_2, ft	V_1, ft/sec	V_2, ft/sec	$\dfrac{V_1{}^2}{2g}$, ft	$\dfrac{V_2{}^2}{2g}$, ft	E_1, ft	E_2, ft	ΔE, ft	$\dfrac{\Delta E}{E_1}$
2	3.16	8.9	0.17	5.27	51.9	1.7	41.83	0.04	42.0	5.3	36.7	0.87
4	3.50	28.0	0.53	9.33	52.9	3.0	43.48	0.14	44.0	9.5	34.5	0.78
6	3.72	54.7	1.02	13.04	53.8	4.2	45.00	0.27	46.0	13.3	32.7	0.71
8	3.90	88.3	1.62	16.52	54.6	5.3	46.43	0.44	48.0	17.0	31.0	0.65
10	4.02	127.1	2.30	19.85	55.4	6.4	47.80	0.64	50.1	20.5	29.6	0.59

In order that a hydraulic jump may occur, the flow must be below critical depth (Sec. 10-3). This condition is satisfied in almost all cases where there is a potential scour hazard. The location and character of the jump depend on tailwater elevation (Fig. 9-31). For the jump to form at the toe of the dam, the tailwater elevation must coincide with the upper conjugate depth. If the tailwater depth is less than the upper conjugate depth, flow will con-

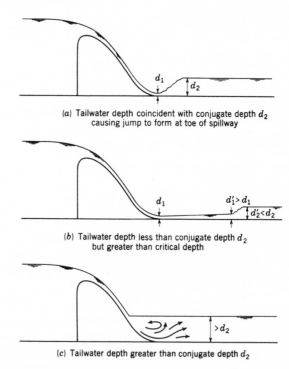

(a) Tailwater depth coincident with conjugate depth d_2 causing jump to form at toe of spillway

(b) Tailwater depth less than conjugate depth d_2 but greater than critical depth

(c) Tailwater depth greater than conjugate depth d_2

FIG. 9-31   Effect of tailwater depth on the character and location of a hydraulic jump.

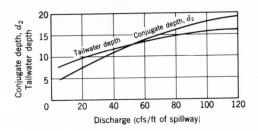

FIG. 9-32   Upper conjugate depth and tailwater depth vs. discharge for the spillway of Fig. 9-29.

tinue below critical depth for some distance downstream from the toe of the dam (Fig. 9-31*b*). Because of energy loss to friction, the depth of flow will gradually increase until a new depth d'_1 is reached which is conjugate with the tailwater depth. At that point a hydraulic jump will occur. If the tailwater depth is greater than the upper conjugate depth (Fig. 9-31*c*), the jump will occur upstream of the toe and it may be completely submerged.

The measures necessary to control erosion and dissipate the energy of the spillway flow are dependent on the relation between the upper conjugate depth d_2 and tailwater depth. Figure 9-32 shows the upper conjugate depth data from Table 9-2 plotted against the corresponding discharges. The tailwater rating curve is dependent on downstream channel conditions and may be determined from hydraulic computations (Chap. 10) or by measurements (Chap. 2). For any situation there are five possible cases based on the location of the upper conjugate depth curve with respect to the tailwater rating. These are:

1. Upper conjugate depth and tailwater coincide at all flows
2. Upper conjugate depth is always below tailwater
3. Upper conjugate depth is always above tailwater
4. Upper conjugate depth is above tailwater at low flows and below at high flows
5. Upper conjugate depth is below tailwater at low flows and above at high flows

9-29   Scour protection below overflow spillways   The best method of energy dissipation for any dam depends on which of the five cases noted in Sec. 9-28 exists.[1] Representative types of energy-dissipation works are illustrated in Fig. 9-33.

Class 1   If the upper conjugate depth and tailwater always coincide, a

[1] J. N. Bradley and A. J. Peterka, Six papers on the Hydraulic Design of Stilling Basins, *J. Hydraulics Div. ASCE*, Vol. 83, Papers 1401–1406, October, 1957; "Hydraulic Design of Stilling Basins and Energy Dissipators," *U.S. Bureau of Reclamation Eng. Monograph 25*, July, 1963.

hydraulic jump will form at the toe of the spillway at all flow rates. In this situation a simple concrete apron (Fig. 9-33a) should provide adequate protection. In order to assure the formation of a perfect jump, training walls[1] are usually required on the downstream face of the spillway and at the sides of the jump pool to confine the flowing stream. Class 1 conditions are rarely encountered, and scour protection for the other cases is usually achieved by modifying the flow conditions to create class 1 conditions as nearly as possible.

    Class 2   If the upper conjugate depth is always below the tailwater, the jump will be drowned out by the tailwater, and little energy will be dissipated. Water may continue at high velocity along the channel bottom

---

[1] D. B. Gumensky, Design of Side Walls in Chutes and Spillways, *Trans. ASCE*, Vol. 119, pp. 355-372, 1954.

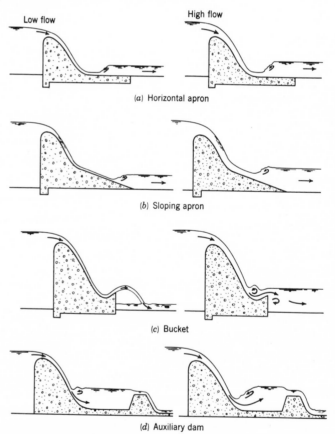

Low flow   High flow

(a) Horizontal apron

(b) Sloping apron

(c) Bucket

(d) Auxiliary dam

FIG. 9-33   *Typical scour-protection works.*

for a considerable distance. One method of solving the class 2 problem is to build a sloping apron (Fig. 9-33*b*) above stream-bed level. The slope of the apron can be such that proper conditions for a jump will occur somewhere on the apron at all flow rates. A second solution is to use an apron with an up-turned bucket (Fig. 9-33*c*) which deflects the overflow up through the tail-water. A roller will form downstream from the bucket, but it will tend to move scoured bed material toward the dam, thus preventing serious scour at the toe of the dam. Some of the scoured material may enter the bucket, where, under the action of a second roller, it may cause severe abrasion.[1] A dentated bucket lip (Fig. 9-34*a*) will permit escape of material caught in the bucket. If the tailwater is very low, the sheet of water will shoot up out of the bucket and fall harmlessly into the stream at some distance downstream from the bucket.

Class 3   When upper conjugate depth is always above the tailwater, a secondary dam may be built below the main dam to increase the tailwater height and cause a jump to form at the toe of the main dam (Fig. 9-33*d*). In order to get a good jump at all flow rates, it may be necessary to provide a sloping apron between the main and secondary dams. Another method of increasing tailwater depth is to excavate a pool, or *stilling basin*, downstream from the spillway. The bottom of the pool may be sloped to assure a jump at all discharges.

If the stream bed is of solid rock, an upturned bucket may provide satisfactory protection, as it will throw the overflow up and out so that it strikes the stream at a safe distance below the dam. If the spillway is low and the tailwater almost sufficient to form a jump, baffles or sills may be used to dissipate the energy. Baffles (Fig. 9-34) are unsuitable where very high velocities make cavitation possible. Furthermore, they must be designed to withstand impact from ice and floating debris. The location, shape, size, and spacing of baffles are best determined by model studies.

If the depth of flow on a spillway is decreased for a given discharge, the upper conjugate depth will be decreased. Hence a solution to class 3 conditions is to spread the flow over a greater width. This may be accomplished by providing a longer spillway crest or by constructing the lower portion of the spillway with a warped surface sloping downward from its longitudinal center line.

Class 4   This case occurs where the tailwater depth is not sufficient to cause a jump at low flows and is too great during high flows. A secondary dam, or sill, which will increase the tailwater depth sufficiently at low flow rates may be combined with a sloping apron to provide a satisfactory jump

[1] Kenneth Keener, Spillway Erosion at Grand Coulee, *Eng. News-Record*, Vol. 133, pp. 95–101, July 13, 1944.

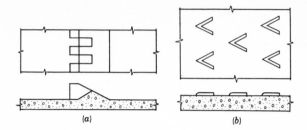

FIG. 9-34   Types of baffle structures.

at high discharge. If the water velocity is not too great, baffle piers may be used to break up the flow.

Class 5   This case requires that the tailwater depth be increased sufficiently to form a jump at high flows. This can usually be accomplished by a secondary dam or stilling pool.

9-30   Scour protection below other types of spillways and sluiceways The five hydraulic conditions discussed in Sec. 9-29 can occur below any type of spillway or sluiceway, and erosion control can be obtained by any of the methods described. In the case of chute, side-channel, and shaft spillways the point of discharge is often far enough from the dam so that protection is required only for the spillway itself. In most cases a stilling basin at the discharge end of the spillway or a bucket which throws the water into the stream at some distance from the end of the spillway will suffice. The *ski-jump spillway* which has been used extensively in Europe directs the water into a trajectory that partially disintegrates the jet and brings it into river-bed level a hundred or more feet downstream of the toe of the dam.

Discharge through sluiceways is usually small compared with that over spillways, and provision for scour protection is less important. Often the jet from a sluiceway is allowed to drop into a stilling basin. Another solution is to spread the jet by a jet deflector at the outlet or by use of a Howell-Bunger valve. The sluiceway design of Fig. 9-28d utilizes the energy-dissipation works for the main spillway.

On arch dams with overflow spillways the nappe usually springs free of the dam, and a deep stilling pool (plunge pool) may be used to absorb the energy of overflow.

BIBLIOGRAPHY

American Society of Civil Engineers, Committee on Hydraulic Structures: Bibliography on the Hydraulic Design of Spillways, *J. Hydraulics Div., ASCE*, Vol. 89, pp. 117–139, July, 1963.

Bradley, J. N., and A. J. Peterka: The Hydraulic Design of Stilling Basins, *ASCE Papers* 1401–1406, October, 1957.

Chow, Ven Te: "Open-channel Hydraulics," chaps. 14, 15, 16, McGraw-Hill, New York, 1959.

Creager, W. C., J. D. Justin, and Julian Hinds: "Engineering for Dams," 3 vols., Wiley, New York, 1945.

"Dams and Control Works," U.S. Bureau of Reclamation, Washington, D.C., 1954.

Davis, Calvin V., and K. E. Sorensen (eds.): "Handbook of Applied Hydraulics," 3d ed., chaps. 20, 21, 22, McGraw-Hill, New York, 1969.

"Design of Small Dams," U.S. Bureau of Reclamation, Washington, D.C., 1960.

Elevatorski, Edward A.: "Hydraulic Energy Dissipators," McGraw-Hill, New York, 1959.

Ippen, A. T.: Channel Transitions and Controls, chap. 8 in Hunter Rouse (ed.), "Engineering Hydraulics," pp. 496–588, Wiley, New York, 1950.

King, H. W., and E. F. Brater: "Handbook of Hydraulics," 5th ed., McGraw-Hill, New York, 1963.

"Nomenclature for Hydraulics," *Manual and Reports on Engineering Practice* 43, American Society of Civil Engineers, New York, 1962.

PROBLEMS

9-1 Draw the profile of a spillway conforming to the design shown in Fig. 9-1c. The spillway is to be 100 ft long, and the design flow is 4500 cfs. Assume that this is the spillway for the dam of Prob. 8-2, and compare the two sections. How much should a corbel project on this dam?

9-2 Water flows over an ogee spillway ($C_w = 3.8$) with a crest length of 20.0 ft under a head of 6.00 ft. Upstream of the spillway where the head is measured the flow cross section may be approximated by a rectangle of height 15 ft and width 40 ft. What percentage error is introduced in calculating the flow rate if the velocity of approach is omitted? Neglect the effect of spillway end contractions.

9-3 Compute the discharge over an ogee weir with $C_w = 3.5$ at a head of 5 ft. The weir length is 200 ft, the weir crest is 10 ft above the bottom of the approach channel, and the approach channel is 200 ft wide.

9-4 An ogee spillway 16 ft long is designed according to Figs. 9-1c and 9-2 to pass 420 cfs when the water-surface elevation upstream of the spillway is 23.0 ft. The reservoir bottom is horizontal and at elevation 0.0 ft upstream of the spillway. Find the flow when the water surface elevation upstream of the spillway is 21.5 ft. Assume there are no end contractions and neglect velocity of approach.

9-5 A dam has a siphon spillway whose cross section is 4 ft high and 10 ft wide. Tailwater elevation at design flow is 80 ft below the summit of the siphon, and the headwater elevation is 8 ft above the summit. Assuming a coefficient of discharge = 0.9, what is the capacity of the siphon? What head would be required on an ogee spillway ($C_w = 3.9$) 10 ft long to discharge this flow? What length of ogee weir would be required to discharge the same flow as the siphon with a head of 8 ft on the weir crest?

9-6 Find the force exerted by overflow water on section *AB* of Fig. 9-12 if the head is 10 ft and $C_w = 3.9$.

9-7 Water flows over the ogee section of a low diversion dam 5 ft high at a rate of 15 cfs/ft of spillway length. The spillway coefficient is 3.6 and the tailwater depth (Fig. 9-31b) is 3.0 ft. Find the net force in the horizontal direction exerted by the water on a 20-ft length of this spillway.

9-8 What is the total force exerted against the slab of a chute spillway if the slab is curved on a radius of 20 ft, has an included angle of 40°, and is tangent to the horizontal at its downstream end and to a straight chute at its upper end? The spillway is 30 ft wide, the discharge is 3000 cfs, and the water depth at the upstream edge of the slab is 2 ft.

9-9 A spillway has flashboards installed in panels 12 ft long. Each panel is supported by two pieces of $\frac{3}{4}$-in. pipe (1.05 in. OD, 0.824 in. ID) set in sockets on the spillway crest. The yield point of the steel in the pipes is 35,000 psi in tension. What height of boards could be kept in place without overflow over their top? What depth of overflow would cause failure if the flashboards were 1 ft high?

9-10 It is proposed to use flashboards on a small dam, but these boards must fail under a head of 3 ft above the spillway crest. If the boards are 2 ft high and each 8-ft panel is supported by two pipes, what pipe size (yield point 35,000 psi) should be selected?

9-11 Plot the profile of a circular conduit with bellmouth entrance designed in accordance with Eq. (9-5) with $D = 6$ ft.

9-12 The chute spillway of Prob. 9-8 terminates at an elevation of 20 ft above the channel into which it discharges. How far from the end of the chute will the water strike the channel?

9-13 A slide gate 8 ft square weighs 74,000 lb. If the coefficient of friction between the gate and the seat is 0.3, and the head on the gate is 125 ft, what lifting force is required to move this gate?

9-14 An hydraulic jump occurs in a rectangular channel of 10 ft wide. The water depth changes from 3.0 ft before jump to 5.2 ft after jump. Compute the flow rate.

9-15 Calculate the energy loss in a hydraulic jump at the toe of a 75-ft-high ogee spillway under a head of 12 ft and with $C_w = 4.0$. What is the horsepower loss?

9-16 For the situation depicted in Prob. 9-7 determine the head loss and the horsepower loss.

9-17 Water traveling at 25 ft/sec along a horizontal concrete apron at a depth of 0.25 ft impinges on an upturned concrete bucket of height 6 ft. The bucket diverts the water through an angle of 35°. Find the horizontal force exerted by the water on the bucket per foot of bucket width.

Chapter 10

Open channels

Two types of conduits are used to convey water, the open channel and the pressure conduit. The open channel may take the form of a canal, flume, tunnel, or partly filled pipe. Open channels are characterized by a free water surface, in contrast to pressure conduits, which always flow full.

HYDRAULICS OF OPEN-CHANNEL FLOW

It is presumed that the reader is familiar with fluid mechanics, and the following sections review only the salient features pertinent to problems encountered in the field of water resources. There are two distinct types of flow, one in which the individual fluid particles follow a straight-line path, and another in which the particles take random paths. The first type is known as *laminar flow* and is rarely encountered in hydraulic engineering practice, and consequently the following discussion will be limited to *turbulent-flow* conditions.

10-1   Uniform flow   Figure 10-1 illustrates conditions for uniform flow of water in an open channel. Writing the energy equation between sections A and B gives

$$z_A + d_A + \frac{V_A{}^2}{2g} = z_B + d_B + \frac{V_B{}^2}{2g} + h_L \tag{10-1}$$

where z is the elevation of the channel bottom above an arbitrary datum in feet, d is the depth of flow in feet, V is the average velocity in feet per second, and h_L is the head loss between A and B expressed in foot-pounds per pound of water flowing. In uniform flow $d_A = d_B$, and $V_A = V_B$; hence $h_L = z_A - z_B = SL$, where S is the slope of the energy grade line, and L is the distance between sections.

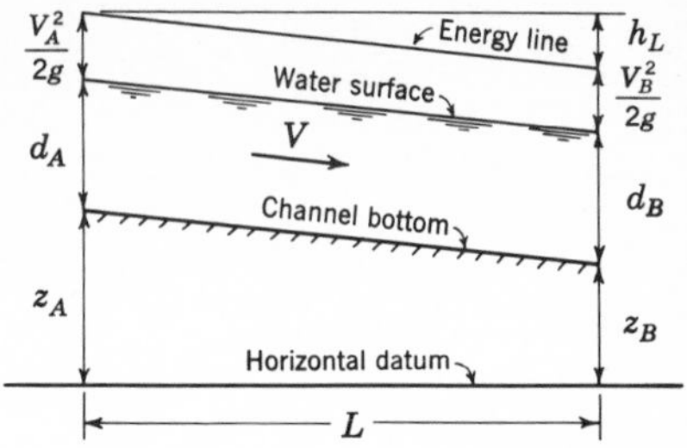

FIG. 10-1   Definition sketch for uniform flow in an open channel.

Several equations are used to calculate the rate of flow in an open channel. The Chézy equation, which can be derived by use of basic principles of fluid mechanics, is

$$V = C\sqrt{RS} \tag{10-2}$$

where V is the average velocity of flow in feet per second, C is a coefficient, R is the hydraulic radius (cross-sectional area divided by wetted perimeter), and S is the *slope of the energy grade line* (equal to the slope of the water surface and also the channel bottom in uniform flow). The Chézy coefficient is most frequently expressed by

$$C = \frac{1.49}{n} R^{1/6} \tag{10-3}$$

Combining Eqs. (10-2) and (10-3) results in what is commonly called the Manning equation,

$$V = \frac{1.49}{n} R^{2/3} S^{1/2} \tag{10-4}$$

which is applicable when the channel slope is less than about 0.10. Under such conditions the slant length of the channel does not differ materially from L.

The factor n in Eqs. (10-3) and (10-4) is the roughness coefficient of the channel (Table 10-1). If the roughness is not uniform across the channel width, an average value of n must be selected or the channel may be treated as two or more contiguous channels, each having its own value of n. The discharge in each subdivision of the channel is computed independently, and the separate values are added to obtain the total flow. Natural channels with overbank flood plains are often treated in this manner.

Various types of open-channel problems occur in engineering practice. Tables, nomographs (Fig. 10-2), slide rules, and other devices have been prepared to simplify calculations and facilitate solution of these problems. The nomograph of Fig. 10-2 is applicable when $n = 0.013$. It should be

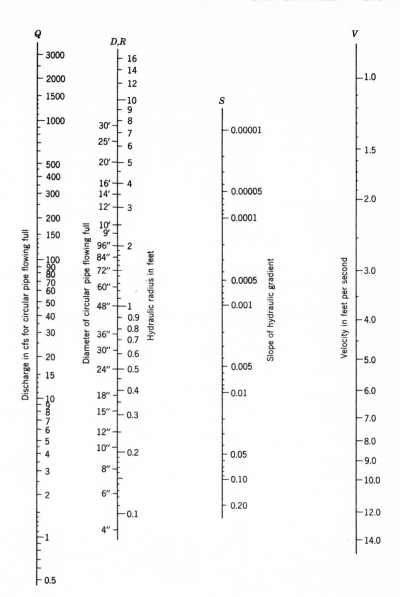

FIG. 10-2   Nomograph for solution of the Manning equation (n = 0.013).

TABLE 10-1   Values of the Roughness Coefficient n

Channel material	n
Plastic, glass, drawn tubing	0.009
Neat cement, smooth metal	0.010
Planed timber, asbestos pipe	0.011
Wrought iron, welded steel, canvas	0.012
Ordinary concrete, asphalted cast iron	0.013
Unplaned timber, vitrified clay	0.014
Cast-iron pipe	0.015
Riveted steel, brick	0.016
Rubble masonry	0.017
Smooth earth	0.018
Firm gravel	0.023
Corrugated metal pipe	0.022
Natural channels in good condition	0.025
Natural channels with stones and weeds	0.035
Very poor natural channels	0.060

noted that V and Q are proportional to $1/n$ and S proportional to n^2 so that values from the nomograph may be readily adjusted to any other value of n.

A situation often encountered in hydraulic engineering, particularly in the case of sewers, is that of a closed conduit flowing partly full. Under this condition the liquid surface is at atmospheric pressure, and the flow is the same as that in an open channel. It is often inconvenient to compute R and A for partially full sections, and it is simpler to calculate V or Q for the pipe flowing full and to adjust to partly full conditions by use of a chart[1] such as Fig. 10-3. When the depth of flow in a circular pipe increases above $0.8D$, the wetted perimeter increases more rapidly than the cross-sectional area because of the convergence of the pipe walls. Hence R, and consequently V, decreases. Maximum discharge occurs when $d = 0.94D$.

10-2   Normal depth   Normal depth d_n is the depth at which uniform flow will occur in an open channel. Normal depth may be determined by writing the Manning equation for discharge,

$$Q = \frac{1.49}{n} AR^{2/3} S^{1/2} \tag{10-5}$$

and substituting for A and R expressions involving d and other necessary dimensions of the channel cross section. The resulting equation requires

---

[1] The roughness coefficient n varies somewhat with depth of flow. This variation is reflected in Fig. 10-3. See "Design and Construction of Sanitary and Storm Sewers," *ASCE Manual of Practice* 37, 2d ed., pp. 87–95, American Society of Civil Engineers, New York, 1969.

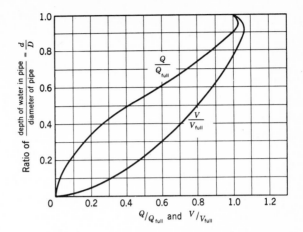

FIG. 10-3   Hydraulic characteristics of circular pipes flowing partly full.

a trial-and-error solution (see Illustrative Example 10-1) or the normal depth may be computed through the use of tables.[1]

10-3   Critical depth   The critical depth d_c for flow in an open channel is defined as that depth for which the specific energy (sum of depth and velocity head) is a minimum. It can be shown mathematically that critical depth occurs in a channel when

$$\frac{Q^2}{g} = \frac{A^3}{B} \tag{10-6}$$

where B is the surface width. On a mild slope $(d_n > d_c)$ flow is subcritical while on a steep slope $(d_n < d_c)$ flow is supercritical.

10-4   Nonuniform (varied) flow   Uniform flow rarely occurs in natural streams because of changes in depth, width, and slope along the channel. While the simplest design for man-made channels provides a uniform cross section and constant slope, this is not always feasible because of topographic conditions. Therefore, the engineer is often concerned with nonuniform flow in open channels. One method of solving problems of nonuniform flow utilizes a step-by-step procedure and the assumption that the head loss in a short reach of the channel is identical with that caused by uniform flow in a channel whose hydraulic radius and mean velocity are equal to the numerical averages of the respective quantities at the end points of the reach. Writing the energy equation for the conditions of Fig. 10-4, substituting $h_L = Sx$ and $z_A - z_B = S_0 x$, and solving for x gives

[1] Various tables for the solution of open-channel flow problems are available in the "Handbook of Hydraulics" by H. W. King and E. F. Brater, 5th ed., McGraw-Hill, New York, 1963.

$$x = \frac{(d_A + V_A^2/2g) - (d_B + V_B^2/2g)}{S - S_0} \qquad (10\text{-}7)$$

in which S is the slope of the energy line between the two sections and S_0 is the slope of the channel bottom.

The value of S may be determined by substituting the means of the hydraulic elements of the two sections into the Manning equation, which for this purpose may be transformed to

$$S = \frac{n^2 V_{AV}^2}{2.22 R_{AV}^{4/3}} \qquad (10\text{-}8)$$

Equation (10-7) is used by starting at some point of known depth and computing the distance x to a section of slightly different depth. A reasonably accurate determination of the water-surface profile is possible by this method if the increments of depth are small.

The shape of the water-surface profile in a given channel depends on the relationship between the actual depth d, critical depth d_c, and normal depth d_n. Figure 10-5 shows six common nonuniform-flow situations for mild (M) and steep (S) slopes. This figure is a useful guide in a preliminary analysis of many problems in nonuniform flow. In addition to the six cases shown, there are six other possible cases, two each for critical, horizontal, and adverse bottom slopes.[1]

10-5 Location of the hydraulic jump   Whenever flow changes from supercritical to subcritical, a hydraulic jump will occur. One problem facing the designer is to find the location of the hydraulic jump under various conditions of flow. Equation (9-8) permits computation of conjugate depths, i.e., depths before and after jump, for the case of a channel with rectangular cross section on a horizontal slope. For moderate slopes this equation gives good results. For channel slopes greater than 0.10 a modification[2] is required. Another method of determining conjugate depths in a rectangular channel

---

[1] Ven Te Chow, "Open-channel Hydraulics," pp. 227–232, McGraw-Hill, New York, 1959.
[2] C. E. Kindsvater, The Hydraulic Jump in Sloping Channels, *Trans. ASCE*, Vol. 109, pp. 1107–1154, 1944.

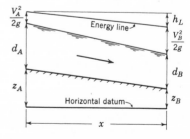

FIG. 10-4   *Definition sketch for nonuniform flow in an open channel.*

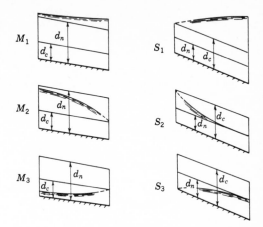

FIG. 10-5 *Typical water-surface profiles for some nonuniform flow situations.*

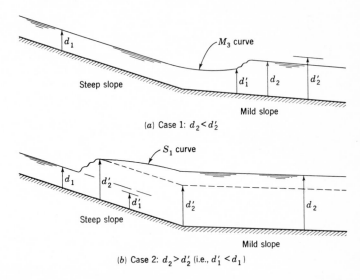

(a) Case 1: $d_2 < d_2'$

(b) Case 2: $d_2 > d_2'$ (i.e., $d_1' < d_1$)

FIG. 10-6 *Examples of location of a hydraulic jump. d_1 and d_2 are the normal depths on upper and lower slopes respectively; d_2' is conjugate to d_1; d_1' is conjugate to d_2.*

with moderate bottom slope for any flow rate q cfs per foot of width is to plot the function $(d^2/2 + q^2/gd)$ vs. depth d. For any value of the function within the range of the curve there will be two depths, which are conjugate. A separate curve must be plotted for each flow rate. Examples of the location of a hydraulic jump are shown in Fig. 10-6.

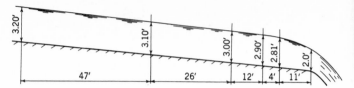

FIG. 10-7   Sketch of water-surface profile for Illustrative Example 10-1 (vertical scale exaggerated).

10-6   Free outfall   If the flow in a channel is subcritical, critical depth should theoretically occur at a free outfall (Fig. 10-7) since the specific energy is at a minimum at that point. However, curvature of the streamlines in the vicinity of the outfall alters the flow conditions and results in a depth at the brink considerably less than d_c. The depth at a free outfall has been observed experimentally to be about $0.7d_c$ for subcritical flow,[1] and under such conditions critical depth occurs about $4d_c$ upstream from the brink. In contrast, if the flow is supercritical, the depth at brink will be only slightly less than the normal depth, which is necessarily below the critical depth.

ILLUSTRATIVE EXAMPLE 10-1   A discharge of 160 cfs occurs in a rectangular open channel 6 ft wide with $S_0 = 0.002$ and $n = 0.012$. If the channel ends in a free outfall, calculate the depth at the brink, d_n, and d_c. Determine the shape of the water-surface profile for a distance of 100 ft upstream from the brink.

$$Q = \frac{1.49}{n} AR^{\frac{2}{3}} S^{\frac{1}{2}} = \frac{1.49}{0.012} \times 6d_n \times \left(\frac{6d_n}{6 + 2d_n}\right)^{\frac{2}{3}} \times 0.002^{\frac{1}{2}} = 160 \text{ cfs}$$

By trial $d_n = 3.5$ ft. Critical depth occurs when

$$\frac{Q^2}{g} = \frac{A^3}{B} \qquad \text{or} \qquad \frac{160^2}{32.2} = \frac{(6d_c)^3}{6}$$

from which $d_c = 2.81$ ft. Since $d_n > d_c$, the flow is subcritical and the water-surface profile is M_2 (Fig. 10-5). The depth at the outfall is approximately $0.7d_c = 2.0$ ft. Critical depth occurs at about $4d_c = 11$ ft upstream from the brink. Computations for the water-surface profile are shown in Table 10-2.

10-7   Hydraulic efficiency of channels   From Eq. (10-5) it is evident that, with uniform flow and fixed values of A, S, and n, the rate of flow will be a maximum when the hydraulic radius R is a maximum. The cross section having the highest hydraulic efficiency (maximum R for a given A) is the half circle. For a trapezoidal channel it can be shown that for maxi-

---

[1] Hunter Rouse, Discharge Characteristics in Free Outfall, *Civil Eng.* (*N.Y.*), Vol. 6, p. 257, April, 1936.

TABLE 10-2   Computation of Water-surface Profile for Illustrative Example 10-1

d, ft	A, sq ft	$B + 2d$, ft	R, ft	V, ft/sec	$\dfrac{V^2}{2g}$, ft	$d + \dfrac{V^2}{2g}$, ft	$\Delta\left(d + \dfrac{V^2}{2g}\right)$, ft	V_{avg}, ft/sec	R_{avg}, ft	S	$S - S_0$	x, ft	Σx,* ft
2.81	16.86	11.62	1.451	9.49	1.398	4.208							
							0.005	9.34	1.463	0.00341	0.00141	4	4
2.90	17.40	11.80	1.475	9.20	1.313	4.213							
							0.014	9.04	1.488	0.00312	0.00112	12	16
3.00	18.00	12.00	1.500	8.89	1.227	4.227							
							0.022	8.74	1.512	0.00284	0.00084	26	42
3.10	18.60	12.20	1.525	8.60	1.149	4.249							
							0.029	8.47	1.536	0.00262	0.00062	47	89
3.20	19.20	12.40	1.548	8.33	1.078	4.278							

*Summation x is measured from the point of critical depth 11 ft upstream from the brink.

mum channel efficiency $R = d/2$. The best trapezoidal section has the shape of a half hexagon, while the best rectangular section is one with depth equal to one-half the width.

10-8   Channel transitions   Special transition sections are often used to join conduits of differing size or shape in order to avoid undesirable flow conditions and minimize head loss. Transition design for supercritical flow is a complicated problem which will not be discussed in detail here. If flow is subcritical, a straight-line transition (Fig. 10-8a) with an angle of

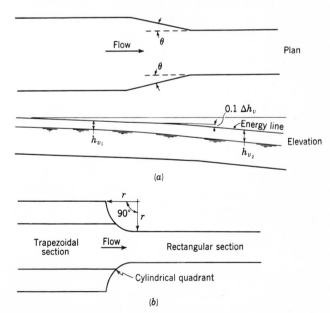

(a)

(b)

FIG. 10-8   Open channel transitions. (a) Simple straight-line contraction. (b) Cylinder-quadrant transition from trapezoidal to rectangular section.

about 12.5° is fairly satisfactory and will result in a head loss of about 0.1 Δh_v at a channel contraction and 0.2 Δh_v at an expansion, where Δh_v is the change in velocity head in the transition.

The cylinder-quadrant transition (Fig. 10-8b) is effective for a change from trapezoidal to rectangular cross section if the Froude number $(N_F = V/\sqrt{gd})$ in the downstream section is less than 0.5. At higher Froude numbers complex warped transitions are advisable. The main problem in the design of transitions for supercritical flow $(N_F > 1.0)$ is to avoid excess buildup of standing waves on the water surface. These waves may overtop the channel walls. No general rules for the design of a transition for supercritical flow are possible, and each case must be treated independently.[1]

10-9 Flow around bends When a body moves along a curved path at constant speed, it is acted upon by a force directed toward the center of curvature of the path. The centrifugal force acting on element $ABCD$ (Fig. 10-9) must be equal and opposite to the net pressure force. Assuming that the velocity V across the section is uniform and that $r \gg B$,

$$\frac{\gamma B(d_1 + d_2)/2}{g} \frac{V^2}{r} = \frac{\gamma d_2^2}{2} - \frac{\gamma d_1^2}{2} \qquad (10\text{-}9)$$

By algebraic transformation this expression becomes

$$d_2 - d_1 = \frac{V^2 B}{gr} \qquad (10\text{-}10)$$

It can be shown that Eq. (10-10) applies to any shape of cross section. If the effect of velocity distribution and variations in curvature across the stream are considered, the difference in water depths between the outer and inner banks may be as much as 20 percent more than that given by Eq.

---

[1] Arthur T. Ippen, Design of Channel Contractions, *Trans. ASCE*, Vol. 116, pp. 326–346, 1951; F. M. Henderson, "Open Channel Flow," chap. 7, Macmillan, New York, 1966.

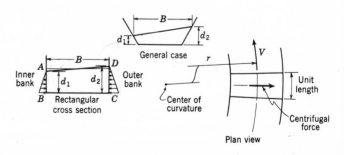

FIG. 10-9   Flow in an open-channel bend.

(10-10). If the actual velocity distribution across the stream is known, the width may be divided into sections and the difference in elevation computed for each section. The total difference in surface elevation across the stream is the sum of the differences for the individual sections.

In designing a channel bend, additional wall height must be provided on the outside of the bend. In addition to centrifugal effects, diagonal waves will occur in the region of the bend if the flow is supercritical. These waves, of height approximately equal to $V^2 B/gr$, may be suppressed by superelevating the channel bottom, using a spiral transition to introduce the superelevation gradually.[1] Diagonal sills which set up counterdisturbances have also been used to reduce wave heights.

10-10   Flow on steep slopes   When flow occurs on steep slopes, large amounts of air may be carried below the water surface in the highly turbulent flow. This *entrained air*[2] reduces the density of the fluid, resulting in an increase in volume called *bulking*. An increase in flow cross section of 25 percent as a result of bulking is not uncommon. Although not strictly applicable, the Manning equation is often used to design channels on steep slopes, and the cross sections thus determined are increased by an arbitrary bulking allowance to provide for air entrainment. Hall[3] has presented empirical data which permit use of a modified value of n in the Manning equation to allow for the effect of air entrainment.

MEASUREMENT OF FLOW IN OPEN CHANNELS

The flow in large natural channels is usually measured with a current meter, as described in Chap. 2. The flow in small streams and man-made channels may often be more conveniently measured by other methods. Among the most common methods are the use of weirs and venturi-type flumes.

10-11   Weirs   For weir formulas to give accurate values of discharge, the upstream face of the weir must be vertical and at right angles to the channel, and the crest of the weir must be horizontal. In addition atmospheric pressure should be maintained under the nappe, and the approach channel should be straight and unobstructed. The head h should be measured

[1] R. T. Knapp, Design of Channel Curves for Supercritical Flow, *Trans. ASCE*, Vol. 116, pp. 296–325, 1951. See also H. W. King and E. F. Brater, "Handbook of Hydraulics," 5th ed., pp. 9-16 to 9-20, McGraw-Hill, New York, 1963.

[2] L. G. Straub and O. P. Lamb, Studies of Air Entrainment in Open Channel Flows, *Trans. ASCE*, Vol. 121, pp. 30–41, 1956.

[3] L. S: Hall, Open Channel Flow at High Velocities, *Trans. ASCE*, Vol. 108, pp. 1393–1434, 1943.

far enough upstream from the weir to avoid the effects of curvature of the water surface near the weir.

The standard formula for discharge over a suppressed rectangular weir is

$$Q = C_w' \frac{2}{3} \sqrt{2g}\, L \left[\left(h + \frac{V_0^2}{2g} \right)^{\!3/2} - \left(\frac{V_0^2}{2g} \right)^{\!3/2} \right] \tag{10-11}$$

where C_w' is a coefficient characteristic of flow conditions over the weir, L is the length of the weir crest in feet, h is the head on the crest in feet, and V_0 is the velocity of flow in the channel just upstream from the weir. If the weir is contracted, an approximate correction for the effect of the contractions may be made by subtracting $0.1h$ from L for each contraction. It is convenient to simplify Eq. (10-11) to

$$Q = C_w L h^{3/2} \tag{10-12}$$

where C_w is a coefficient which, in addition to other factors, accounts for the velocity of approach. The characteristics of the flow pattern which affect C_w can be defined by h and H_d/h, where H_d is the height of the weir. Values of C_w for sharp-crested rectangular[1] weirs are given in Table 10-3.

If the discharge to be measured is quite small, a triangular, or V-notch, weir will usually prove more accurate than a rectangular weir. The discharge over a triangular weir is given by

$$Q = C_w'' \frac{8}{15} \sqrt{2g}\, h^{5/2} \tan \frac{\theta}{2} = 4.28\, C_w'' h^{5/2} \tan \frac{\theta}{2} \tag{10-13}$$

where θ is the vertex angle of the notch. The coefficient C_w'' has a value close to 0.58, varying slightly with head and notch angle.[2]

A Cipoletti weir is trapezoidal in shape, with side slopes of 4 vertical to 1 horizontal. The excess flow permitted by the flaring ends of the Cipoletti weir corresponds to the decrement of flow induced by the lateral contraction, and the discharge of a Cipoletti weir may be determined by use of Eq. (10-12) and Table 10-3. The Sutro, or proportional, weir (Fig. 10-10) is one for which the flow rate varies directly with head. This type is often used for irrigation diversions or for measuring very small flows.

Corrosion of the crest of a sharp-edged weir or damage by floating debris may alter the weir coefficient. Broad-crested weirs of timber or concrete

---

[1] Th. Rehbock, Wassermessung mit scharfkantigen Uberfallwehren, *Z. Ver. deut. Ing.*, Vol. 73, p. 817, 1929.

[2] F. W. Greve, Flow of Water through Circular, Parabolic, and Triangular Vertical Notch Weirs, *Purdue Univ., Eng. Bull., Res. Ser.* 40, 1932; A. T. Lenz, Viscosity and Surface-tension Effects on V-notch Weir Coefficients, *Trans. ASCE*, Vol. 108, pp. 759–802, 1943.

TABLE 10-3   Coefficient C_w for Rectangular Sharp-crested Weirs

H_d/h	Head h on weir, ft						
	0.2	0.4	0.6	0.8	1.0	2.0	5.0
0.5	4.18	4.13	4.12	4.11	4.11	4.10	4.10
1.0	3.75	3.71	3.69	3.68	3.68	3.67	3.67
2.0	3.53	3.49	3.48	3.47	3.46	3.46	3.45
10.0	3.36	3.32	3.30	3.30	3.29	3.29	3.28
∞	3.32	3.28	3.26	3.26	3.25	3.25	3.24

are often preferred because of their durability. If a broad-crested weir is to be used for measurement purposes, its shape must conform to one for which coefficients have been established by test. Equation (10-12) and the coefficients of Table 10-4 may be used to compute the flow over the broad-crested weirs shown in Fig. 10-11.

Equations (10-12) and (10-13) apply to free-flow conditions (Fig. 10-12a). When the water level downstream from a weir rises above the level of the weir crest, the weir crest is said to be *submerged* (Fig. 10-12b). Formulas have been developed for flow over submerged weirs,[1] but under such conditions accurate flow measurement is not possible because surface dis-

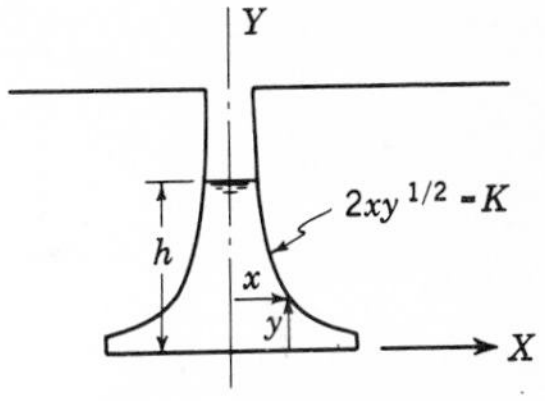

FIG. 10-10   Sutro weir.

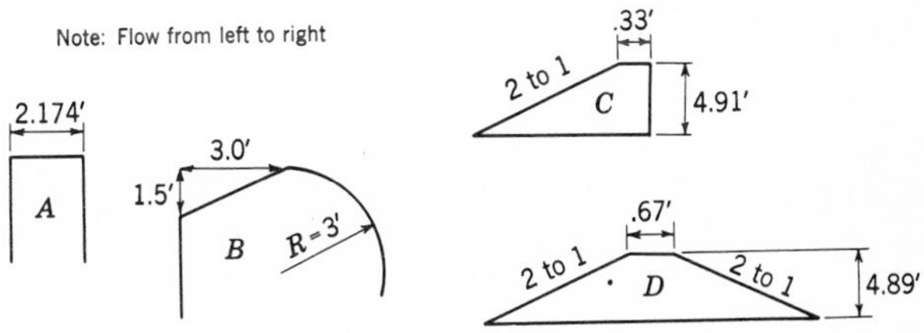

FIG. 10-11   Dimensions of broad crested weirs of Table 10-4.

---

[1] J. R. Villemonte, Submerged Weir Discharge Studies, *Eng. News-Record*, Vol. 139, pp. 54–57, Dec. 25, 1947.

TABLE 10-4   Coefficients* for Broad-crested Weirs of Fig. 10-11

Shape	Head h on weir, ft						
	0.5	1.0	1.5	2.0	3.0	4.0	5.0
A	2.70	2.64	2.64	2.70	2.89		
B	3.27	3.38	3.46	3.51	3.58	3.67	3.83
C	...	3.85	3.82	3.79	3.75	3.70	3.64
D	...	...	3.58	3.56	3.58	3.62	3.68

*From R. E. Horton, Weir Experiments, Coefficients, and Formulas, *U.S. Geol. Survey Water Supply Paper* 200, pp. 59–134, 1907.

turbances downstream from the weir make it difficult to measure the depth of submergence, h_s.

The head on a weir is usually measured in the field with a staff gage, though better results are obtainable with a hook or point gage in a stilling well. The gage reading corresponding to zero head should be determined accurately and checked from time to time. Inaccurate head determination will produce sizable error in flow measurement.[1]

ILLUSTRATIVE EXAMPLE 10-2   The head on a sharp-crested rectangular weir of height 2.0 ft and crest length 4.00 ft was incorrectly observed to be 0.38 ft when it was actually 0.40 ft. Determine the percentage error in the computed value of flow rate.

$$Q = C_w L h^{3/2}$$

Taking the partial differential of Q with respect to h,

$$\partial Q = C_w L (3/2) h^{1/2} \partial h$$

and $$\frac{\partial Q}{Q} = \frac{3}{2} \frac{\partial h}{h} = \frac{3}{2} \frac{0.02}{0.40} = 7.5 \text{ percent error}$$

10-12   Venturi flumes   If water contains suspended sediment, some will be deposited in the pool above a weir, resulting in a gradual change in the weir coefficient. Moreover, the use of a weir results in a relatively large head loss. Both these difficulties are at least partially overcome by use of venturi-type flumes. One of the most common of the venturi flumes is the *Parshall flume*[2] (Fig. 10-13).

---

[1] Charles W. Thomas, Common Errors in Measurement of Irrigation Water, *Trans. ASCE*, Vol. 124, pp. 319–340, 1959.

[2] R. L. Parshall, The Parshall Measuring Flume, *Colo. Agr. Exp. Sta. Bull.* 423, 1936.

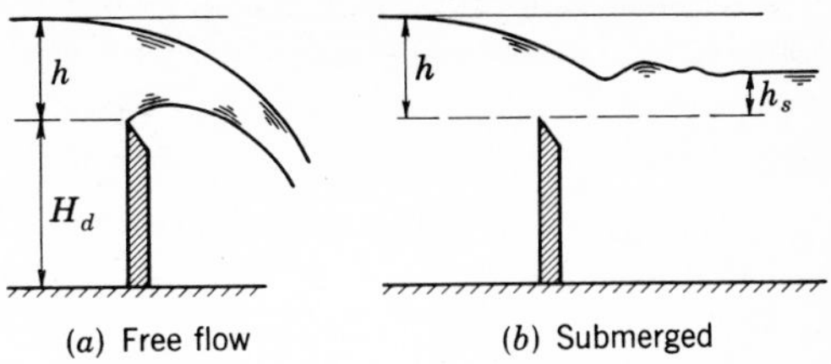

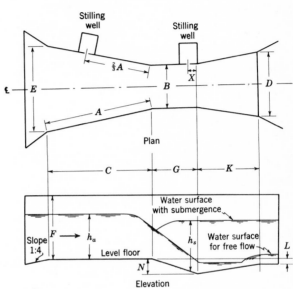

FIG. 10-12   *Weir flow.*

(a) Free flow   (b) Submerged

FIG. 10-13   *Parshall measuring flume.*

Flow through a Parshall flume usually occurs in a free-flow condition with critical depth at the crest and a hydraulic jump in the exit section. The discharge equation for Parshall flumes with throat widths from 1 ft through 8 ft under free-flow condition is

$$Q = 4Bh_a^{1.522B^{0.026}} \tag{10-14}$$

Under certain conditions of flow the hydraulic jump becomes submerged, in which case the flow rates given by Eq. (10-14) must be modified by subtracting a correction factor. Tables of free-flow discharge through Parshall flumes and correction factors for submerged flows are available in the literature.[1] For best results a Parshall flume ought to be installed in a straight section of channel where flow conditions are relatively uniform.

---

[1] Ven Te Chow, "Open-channel Hydraulics," pp. 72–81, McGraw-Hill, New York, 1959.

Another technique of measuring flow in an open channel makes use of a vane that is free to rotate on a horizontal axis at right angles to the channel axis. The angle through which the vane rotates is indicative of the velocity and hence the flow rate. By proper calibration in a channel section of specified shape and size, good results can be obtained with this device. The big advantage is that there is no head loss.

ILLUSTRATIVE EXAMPLE 10-3   Examine the flow conditions in a 10-ft wide open rectangular channel of rubble masonry with $n = 0.017$ when the flow rate is 400 cfs. The channel slope is 0.020, and an ogee weir 5.0 ft high with $C_w = 3.8$ is located at the downstream end of the channel (Fig. 10-14).

The normal depth of flow in the channel is found from

$$400 = \frac{1.49}{0.017} \times 10d_n \left(\frac{10d_n}{10 + 2d_n}\right)^{2/3} \times 0.020^{1/2}$$

from which, by trial, $d_n = 2.36$ ft. Critical depth in this channel is found from Eq. (10-6).

$$d_c = \sqrt[3]{\frac{Q^2}{B^2 g}} = \left(\frac{400^2}{10^2 \times 32.2}\right)^{1/3} = 3.7 \text{ ft}$$

Since $d_n < d_c$, the flow is supercritical (case S_1, Fig. 10-5). The head required at the weir to discharge the given flow is found from Eq. (9-2).

$$400 = 3.8 \times 10 \times \left\{ h + \frac{[400/(5 + h) \times 10]^2}{64.4} \right\}^{3/2}$$

from which, by trial, $h = 4.74$ ft. Hence the depth of water just upstream from the weir is 9.74 ft, which is greater than d_c. The flow at this point is subcritical, and a hydraulic jump must occur upstream. The depth d_2 after the jump is

$$d_2 = -\frac{2.36}{2} \pm \left[\frac{2.36^2}{4} + \frac{2(400/23.6)^2 \times 2.36}{32.2}\right]^{1/2} = 5.40 \text{ ft}$$

The distance from the weir to the jump is determined with Eq. (10-7).

$$d_A = 5.40 \text{ ft} \qquad V_A = \frac{400}{54} = 7.41 \text{ ft/sec} \qquad \frac{V_A^2}{2g} = 0.85 \text{ ft}$$

$$d_B = 9.74 \text{ ft} \qquad V_B = \frac{400}{97.4} = 4.11 \text{ ft/sec} \qquad \frac{V_B^2}{2g} = 0.26 \text{ ft}$$

$$V_{\text{avg}} = \frac{V_A + V_B}{2} = 5.76 \text{ ft/sec} \qquad R_{\text{avg}} = 2.95 \text{ ft}$$

$$S = \frac{(0.017 \times 5.76)^2}{2.21 \times 2.95^{4/3}} = 0.00102$$

$$x = \frac{5.40 + 0.85 - 9.74 - 0.26}{0.00102 - 0.02000} = 198 \text{ ft}$$

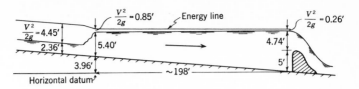

FIG. 10-14   *Sketch of flow conditions for Illustrative Example* 10-3.

The hydraulic and energy grade lines are shown in Fig. 10-14. It is interesting to note that in this example the water surface downstream of the jump slopes uphill in the direction of flow. The slope of the energy line, of course, is downhill. It has been assumed that the water-surface profile from the end of the jump to the weir is a straight line. Actually the surface profile is slightly curved, and greater accuracy could have been obtained if x had been computed in increments. Since computations of this type involve small differences between large numbers, it is important to carry as many significant figures as the data justify.

TYPES OF OPEN CHANNELS

10-13   Canals The problem of canal location is similar in many respects to highway location, but the solution may be more difficult since the slope of the canal bottom must be downgrade, and frequent changes in slope (and hence changes in section) should be avoided. Within the limitations of topography, the exact route of a canal is determined by the slopes that can be tolerated. Excessive slope may result in a velocity sufficient to cause erosion of the channel bottom or sides. The velocity at which scour will begin depends on the bed material and the shape of the channel cross section. Fine-grained soils generally scour at a lower velocity than coarse-grained soils, but this is not always the case for the presence of cementing material in the soil may greatly increase its resistance to scour. The bed material of a canal tends to consolidate with use and develop increased resistance to erosion. Water which carries abrasive material is more effective in eroding cohesive or consolidated materials. Table 10-5 lists approximate maximum permissible velocities in channels of various materials. A more sophisticated approach to the scour problem[1] involves comparing the boundary shear stress (tractive force per unit area) with the permissible unit tractive force. Through such an approach it is possible to achieve a *balanced design*, i.e., to determine the bottom width and side slope of the channel such that scour of bottom and sides is equally unlikely.

If the channel slope is too gradual the velocity may be so low that growth

---

[1] Ven Te Chow, "Open-channel Hydraulics," secs. 7-11 through 7-15, McGraw-Hill, New York, 1959.

TABLE 10-5   Maximum Permissible Velocities in Canals and Flumes

	Velocity, ft/sec	
Channel material	Clear water	Water with abrasive sediment
Fine sand	1.5	1.5
Silt loam	2.0	2.0
Fine gravel	2.5	3.5
Stiff clay	4.0	3.0
Coarse gravel	4.0	6.0
Shale, hardpan	6.0	5.0
Steel	*	8.0
Timber	20.0	10.0
Concrete	40.0	12.0

*Limited only by possible cavitation.

of aquatic plants will reduce the hydraulic efficiency of the channel. Moreover, suspended sediment in the water may be deposited. Consequently design velocities should be slightly less than the maximum permissible if topography permits.

Earth canals are generally trapezoidal, with side slopes determined by the stability of the bank material. The determination of stable slopes uses the procedures described for earth dams (Sec. 8-22). Table 10-6 lists typical side slopes for unlined canals in various materials.

Freeboard must be provided above the design water level as a precaution against accumulation of sediment in the canal, reduction in hydraulic efficiency by plant growth, wave action, settlement of the banks, and flow in excess of design quantities during storms. Economy in the cost of excavation and earthwork is achieved primarily by balancing cut and fill. It may, however, be advantageous to borrow or waste where the haul distance is great. On sidehill locations the canal may be quite deep in order to balance cut and fill. Typical canal sections are shown in Fig. 10-15.

TABLE 10-6   Typical Side Slopes for Unlined Canals

Bank material	Slopes (horizontal: vertical)
Cut in firm rock	$\frac{1}{4}$:1
Cut in fissured rock	$\frac{1}{2}$:1
Cut in firm soil	1:1
Cut or fill in gravelly loam	$1\frac{1}{2}$:1
Cut or fill in sandy soil	$2\frac{1}{2}$:1

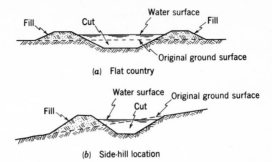

Water surface
Fill   Cut   Fill
Original ground surface
(a)   Flat country

Water surface   Original ground surface
Fill   Cut

(b)   Side-hill location

FIG. 10-15   Typical canal cross sections.

If the water has high value and the soil in which the canal is constructed is quite permeable, it may be economical to provide a canal lining to reduce seepage from the canal. The rate of seepage from unlined canals is influenced chiefly by the character of the soil and the location of the groundwater level. Table 10-7 indicates the order of magnitude of seepage from canals situated above the water table. Seepage rates[1] may be measured by (1) ponding, (2) inflow-outflow measurements, and (3) seepage meter determinations. In the ponding method temporary watertight bulkheads are used to isolate a reach of canal. Water is admitted, and its rate of disappearance less an allowance for evaporation loss is the seepage rate. The inflow-outflow method is not reliable because a small error in one of these measurements will produce a relatively large error in the computed seepage rate. The seepage meter consists of an inverted metal cup with a face area of 2 sq ft connected by tube to a flexible water bag that is initially filled with water. The cup is inserted face down into the canal bottom with the bag submerged in the canal water. A valve is opened permitting water to flow from the bag to the cup to satisfy seepage through the canal bottom. The

TABLE 10-7   Seepage Rates from Unlined Canals

Material	Seepage rate, cu ft/sq ft/day
Clay loam	0.25–0.75
Sandy loam	1.0 –1.5
Loose sandy soils	1.5 –2.0
Gravelly soils	3.0 –6.0

[1] August R. Robinson, Jr., and Carl Rohwer, Measurement of Canal Seepage, *Trans. ASCE,* Vol. 122, pp. 347–373, 1957.

rate of loss of water from the bag is indicative of the seepage rate. To determine an average seepage rate by this method, numerous tests at various points of the canal bottom are necessary.

Various types of linings[1] are used to reduce seepage losses from canals. Clay, asphalt, cement mortar, and reinforced concrete have been used effectively. An effective and inexpensive lining is a *buried membrane* constructed by spraying asphalt over the sides and bottom of the channel and then placing a protective cover of about 6 in. of soil. The presence of fine sediment in the water may help to make the canal self-sealing. For important canals a concrete lining is usually most satisfactory because of its permanency. Reinforced concrete is used for canal linings in thicknesses of 2 to 8 in. depending on the size and importance of the canal. Standard reinforcement is 0.5 percent in the longitudinal direction and 0.2 percent in the transverse direction. Watertight construction joints are required at regular intervals. Mortar linings for small canals are often placed by guniting over steel mesh or by use of movable forms. Special paving machines (Fig. 10-16) are used on large canals. For a concrete lining to be

---

[1] "Canal Linings and Methods of Reducing Costs," U.S. Bureau of Reclamation, Washington, D.C.

FIG. 10-16   Canal-lining machine in operation. (Courtesy Caterpillar Tractor Co.)

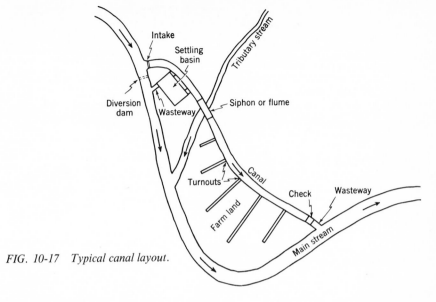

FIG. 10-17   Typical canal layout.

successful, the canal banks must be stable and well-drained. Uplift under the lining may cause serious damage when the canal is empty. Seepage loss from properly lined canals may be as low as 0.05 cu ft/sq ft/day. In addition to a reduction in seepage, lining may permit higher water velocities and smaller cross sections in the canal, with a resulting saving in cost.

10-14   Canal appurtenances   Numerous structures are necessary for the proper operation of canals. A general layout of a canal system is shown in Fig. 10-17. The usual *diversion structure* is an overflow dam built across a stream to maintain the water level above the floor of the intake structure. Diversion dams may be provided with sluiceways for flushing sediment from the pool above the dam or discharging water during low-flow periods. Canal *intakes* (Fig. 10-18) are ordinarily located a short distance upstream from the diversion structure and serve to regulate the flow into the canal. On small installations a simple slide gate may be sufficient, while more elaborate gates are required for large canals. *Fish screens*[1] are often provided at the intake to keep fish out of the canal. This is mainly a wildlife conservation measure, as the fish do little damage in the canals. The transition from the intake to the canal should be long enough to permit smooth adjustment of the flow.

If the overall change in elevation for a canal is large, it may be necessary to use *chutes* or *drops* to avoid excessive slope in the canal. A chute (Fig.

[1] R. Banys and K. R. Leonardson, Fishways at Dams, sec. 20 in C. V. Davis and K. E. Sorenson (eds.), "Handbook of Applied Hydraulics," McGraw-Hill, New York, 1969.

FIG. 10-18   *Typical intake structure for a small canal. (Armco Metal Products, Division of Armco Steel Corp.)*

10-19) ordinarily consists of an intake structure, a long inclined section, and an outlet designed to dissipate excessive kinetic energy. Because of the high velocity in the chute, the entire structure is usually concrete-lined. A drop is similar to a chute, but the change in elevation is effected in a shorter distance. In some cases a vertical drop is used.

If a canal meets an obstruction such as a highway embankment or hill extending above the canal grade, a culvert or tunnel may be used to carry water through the obstruction (Fig. 10-20a). If the canal must cross a depression such as a highway cut, stream, or gully, a *flume* (Fig. 10-20b) or a pipe may be used to convey the water. In some cases an *inverted siphon* (Fig. 10-20c and Sec. 11-27) may be used in preference to an elevated crossing. In all cases a suitable transition structure is required at the inlet and outlet of the special section. Many times the alternative of carrying the canal around the obstruction may be chosen if the annual cost of the additional length of canal is less than that of the comparable flume or inverted siphon.

Regulation of flow in the canal and the distribution of the water is facilitated by various structures. A *check* is a short, concrete-lined structure placed in the canal and provided with piers so that flashboards or gates can be used for flow regulation. The main purpose of the check is to raise the

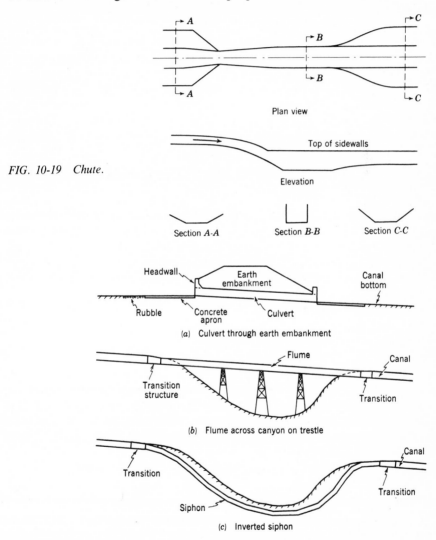

FIG. 10-19   Chute.

FIG. 10-20   *Arrangements for carrying a canal past an obstruction.*

upstream water level to permit diversion. *Turnouts* are usually pipes through the canal embankment for diversion of flow from the main canal to a smaller distribution canal. In some cases true siphons are used to carry the flow over the canal bank and thus avoid placing pipes through the embankment. The intake to a turnout is provided with a gate or stop logs so that it may be opened or closed as required. *Wasteways* are canals or pipes used to return excess water to the stream. They are necessarily provided with gates and are sometimes used for sluicing sediment from the canal.

If the use of water from a canal is suddenly discontinued, the channel may fill up and overflow unless the inflow is reduced at the intake. The wasteway mentioned above may provide some protection against such flooding, but automatic overflow spillways and siphon spillways are often provided along the canal to maintain a safe maximum water level. Excess water is discharged into a natural drainage course. Excess storm water may be prevented from entering a canal by use of *drainage culverts* under the canal or an *overchute* (flume passing over the canal). In either case, it is essential that the culvert or overchute have adequate capacity to pass the floodwaters that are expected.

Settling basins are sometimes constructed just below the point of diversion so that sediment can be collected and sluiced back to the river. A settling basin consists of a large, shallow basin through which the water passes with a low velocity (1 ft/sec or less). As much as 90 percent of the suspended sediment can be collected in a well-designed settling basin. *Sand traps* are sections of depressed canal bed located upstream from a check structure and provided with wasteways or sluiceways so that accumulated sediment can be sluiced from the canal.

10-15 Flumes A flume is a channel of wood, concrete, or metal which is usually supported aboveground. Flumes are used to convey water over terrane where construction of canals is difficult or expensive. They are often employed to carry a canal over a depression. The flume channel must be designed to carry its own weight and that of the water as a beam between supports, while the supporting piers or trestle must carry the flume and water load plus such wind and snow loads as may be appropriate. Wooden flumes are usually of rectangular cross section, but triangular and semicircular sections are also used. Various types of wood have been used in timber flumes, but redwood and cypress are superior to all others. The cross sections of several typical flumes are shown in Fig. 10-21.

The most permanent type of flume is a properly constructed concrete flume. However, care must be taken to avoid cracking as a result of unequal settlement of the supporting structure. Suitable contraction joints must be provided at each pier. Small concrete flumes are often constructed of pre-

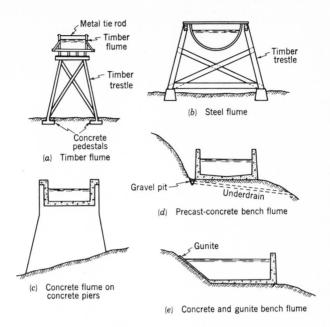

FIG. 10-21   Cross sections of some typical flumes.

cast sections supported on steel, timber, or concrete piers. Large flumes are usually cast in place with concrete piers and channel. Because of the thin sections which are exposed to the weather, concrete flumes are not well adapted to extremely cold climates. Construction and transportation problems may make concrete flumes expensive for use in rugged terrane where access is difficult.

The formation of ice in flumes or canals will reduce their capacity. In cold climates canals are sometimes patroled by boats to keep the ice broken up. Flumes and canals are sometimes covered to minimize evaporation, pollution, and freezing. The covering is placed well above the highest water level so that the flow will always be under open-channel conditions.

10-16 Tunnels Occasionally it is cheaper to convey water by tunnel through a hill than by flume or canal around the hill. Tunnels are usually of circular or horseshoe section (Fig. 10-22) to take advantage of arch action. If the tunnel material is unstable, a concrete or steel liner may be used to prevent collapse. Tunnels driven through stable rock need not be lined for structural reasons, but their interior surfaces are usually smoothed with gunite to improve hydraulic characteristics. The minimum economic diameter for hand-excavated tunnels is about 6 ft, while machine-excavated tunnels are usually at least 8 ft in diameter. Tunnels may flow full or partly

(a) Horseshoe tunnel   (b) Pressure tunnel   (c) Reinforced-concrete tunnel
                                    (unreinforced)

FIG. 10-22   Typical tunnel cross sections.

full. When flowing full, they act as pressure conduits and metal liners may be necessary to control bursting pressures. When flowing partially full, the tunnel acts as an open channel and flow in it (if uniform) is governed by the Manning equation.

BIBLIOGRAPHY

Brown, C. B.: Sediment Transportation, chap. 12 in Hunter Rouse (ed.), "Engineering Hydraulics," pp. 769–857, Wiley, New York, 1950.

Chow, Ven Te: "Open-channel Hydraulics," McGraw-Hill, New York, 1959.

Daugherty, R. L., and J. B. Franzini: "Fluid Mechanics with Engineering Applications," 6th ed., McGraw-Hill, New York, 1965.

Henderson, F. M.: "Open Channel Flow," Macmillan, New York, 1966.

Hinds, Julian: Canals, Flumes, Covered Conduits, Tunnels, and Pipe Lines, sec. 10 in C. V. Davis (ed.), "Handbook of Applied Hydraulics," 2d ed., pp. 415–445, McGraw-Hill, New York, 1952.

Howe, J. W.: Flow Measurement, chap. 3 in Hunter Rouse (ed.), "Engineering Hydraulics," pp. 177–212, Wiley, New York, 1950.

Ippen, A. T.: Channel Transitions and Controls, chap. 8 in Hunter Rouse (ed.), "Engineering Hydraulics," pp. 496–588, Wiley, New York, 1950.

King, H. W., and E. F. Brater: "Handbook of Hydraulics," 5th ed., McGraw-Hill, New York, 1963.

Leliavsky, Serge: "River and Canal Hydraulics," Chapman and Hall, London, 1965.

Posey, C. J.: Gradually Varied Channel Flow, chap. 9 in Hunter Rouse (ed.), "Engineering Hydraulics," pp. 589–643, Wiley, New York, 1950.

Scobey, F. C.: Flow of Water in Irrigation and Similar Canals, *U.S. Dept. Agr. Tech. Bull.* 652, 1939.

Vennard, J. K.: "Elementary Fluid Mechanics," 4th ed., Wiley, New York, 1961.

PROBLEMS

10-1 What is the flow rate in a 48-in. circular corrugated-metal pipe on a slope of 0.003 if the depth of flow is 19 in.?

10-2 At what depth will 8.8 cfs flow in a 36-in.-diameter concrete pipe on a slope of 0.004?

10-3 What is the flow rate in a 15-in.-diameter concrete pipe on a slope of 0.004 if the depth is 3.0 in.?

10-4 Repeat Prob. 10-2 for the case of a 36-in.-diameter corrugated metal pipe.

10-5 It is necessary to discharge 20 cfs in a cast-iron pipe 36 in. in diameter. What is the minimum possible slope which can be employed if the pipe is to flow at 0.8 depth?

10-6 On what slope should one construct a 10 ft-wide rectangular channel ($n = 0.013$) so that critical flow will occur at a depth of 4.0 ft?

10-7 Is the flow of Prob. 10-2 subcritical or supercritical?

10-8 Water flows steadily at 16.0 cfs in a very long triangular flume which has side slopes 1 on 1. The bottom of the flume is on a slope of 0.0040. At a certain section A the depth of flow is 2.00 ft. Is the flow at this section subcritical or supercritical? At another section B, 200 ft downstream from A, the depth of flow is 2.50 ft. Approximately what is the value of Manning's n? Find d_n. Under what conditions can this flow occur?

10-9 Water flows at 300 cfs in a 10-ft wide open channel of rectangular cross section. The bottom slope is adverse, i.e., it rises 0.2 ft per 100 ft in the direction of flow. If the water depth decreases from 7.0 ft to 5.5 ft in a 500-ft length of channel determine Manning's n.

10-10 The flow in a long 12-ft wide rectangular channel which has a constant bottom slope is 1000 cfs. A computation using the Manning equation indicates that the normal depth of flow for this flow rate is 7.0 ft. At a certain section A the depth of flow in the channel is 3.0 ft. Will the depth of flow increase, decrease, or remain the same as one proceeds downstream from section A? Sketch a physical situation where this type of flow will occur.

10-11 A rectangular channel is 8 ft wide, has an 0.008 slope, discharge of 150 cfs, and $n = 0.014$. Find d_n and d_c. If the actual depth of flow is 5 ft, what type of profile (Fig. 10-5) exists? Where might this condition occur in engineering practice?

10-12 A rectangular channel is 3 m wide and ends in a free outfall. If the discharge is 10 cu m/sec, slope is 0.0025, and $n = 0.016$, find d_n, d_c, and the water-surface profile for a distance of 150 m upstream from the outfall.

10-13 A rectangular drainage channel is 15 ft wide and is to carry 500 cfs. The channel is lined with rubble masonry and has a bottom slope of 0.0015. It discharges into a stream which may reach a stage 10 ft above the channel bottom during floods. Calculate d_n, d_c, and the distance from the channel outlet to the point where normal depth would occur under this condition.

10-14 A trapezoidal channel with a bottom width of 10 ft and side slopes of 2 horizontal to 1 vertical has a horizontal curve with a radius of 100 ft and without superelevation. If the discharge is 800 cfs and the water surface at the inside of the curve is 5 ft above the channel bottom, find the water-surface elevation at the outside of the curve. Assume the flow is subcritical.

10-15 What would be the cross section of greatest hydraulic efficiency for a trapezoidal channel with side slopes of 1.5 horizontal to 1 vertical if the design discharge is 350 cfs and the channel slope is 0.0005? Use $n = 0.025$.

10-16 It is desired to measure a discharge which may vary from 0.5 to 2.5 cfs with a relative accuracy of at least 0.5 percent throughout the entire range. A stage recorder which is accurate to the nearest 0.001 ft is available. What is the maximum width of rectangular sharp-crested weir that will satisfy these conditions? Assume the weir has end contractions and $C_w = 3.5$. Neglect velocity of approach.

10-17 What is the maximum permissible vertex angle of a V-notch weir that will satisfy the conditions of Prob. 10-16?

10-18 With a head of 0.5 ft on a 60° V-notch weir, what error in the measured head will produce the same percentage error in the computed flow rate as an error of 1° in the vertex angle?

10-19 The stilling-well depth-measurement scale of a Parshall flume with a 4-ft throat gives an h_a reading that is 0.05 ft too large. Compute the percentage errors in flow rate when the observed h_a readings are 0.5 ft and 2.0 ft.

10-20   It is desired to measure the flow in a canal which may carry between 25 and 100 cfs. The flow is to be measured to an accuracy of 2 percent, and the available water-level recorder is accurate to 0.01 ft. The canal is on a very flat slope and the head loss in the measuring device should be as small as possible. What type and size of flow-measuring device would you recommend?

10-21   Analyze the water-surface profile in a rectangular channel with reinforced-concrete lining. The channel is 6 ft wide, the flow rate is 250 cfs, and the channel slope is 0.002. A sharp-crested rectangular weir is located at the downstream end of the channel with its crest 4 ft above the channel bottom.

10-22   Compute the water-surface profile in the channel of Illustrative Example 10-3 if the channel slope is 0.010.

10-23   Solve Prob. 10-21 if the channel slope is 0.025.

10-24   Analyze the water-surface profile in a long rectangular channel with reinforced-concrete lining ($n = 0.013$). The channel is 10 ft wide, the flow rate is 400 cfs, and there is an abrupt change in channel slope from 0.0150 to 0.0016.

10-25   An irrigation district plans to convey 500 cfs of water from a diversion dam to their distribution works by a canal which can be cut into a very tight clay. Leakage is not considered serious. Between the two ends of the canal there is a drop of 100 ft in a distance of 10,000 ft. Design a suitable open-channel system to carry the water.

10-26   Repeat Prob. 10-24 for a flow rate of 150 cfs.

10-27   Given an open channel with a parabolic cross section ($x = 1.0$ m, $y = 1.0$ m) on a slope of 0.02 with $n = 0.015$, find the normal depth and the critical depth for $Q = 2.0$ cu m/sec.

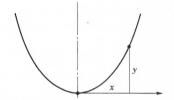

10-28   When a varied flow of 2.0 cu m/sec occurs in the channel of Prob. 10-27, find the distance between sections where the depths are 1.20 m and 1.50 m. Which section is farther downstream? What type of water-surface profile is this?

Chapter 11

Pressure conduits

A pressure conduit is a conduit which is flowing full. Such conduits are often less costly than canals or flumes because they can generally follow a shorter route. If water is scarce, pressure conduits may be used to avoid loss of water by seepage and evaporation which might occur in open channels. Pressure conduits are preferable for public water supplies because of the reduced opportunity for pollution. Since the hydraulic engineer deals almost exclusively with turbulent flow in pipes, the discussion of this chapter will be limited to this type of flow.

HYDRAULICS OF PRESSURE CONDUITS

Writing the energy equation between sections A and B of Fig. 11-1 gives

$$z_A + \frac{p_A}{\gamma} + \frac{V_A{}^2}{2g} + h_p = z_B + \frac{p_B}{\gamma} + \frac{V_B{}^2}{2g} + h_L \tag{11-1}$$

in which z is the vertical distance above an arbitrary horizontal datum, p/γ is the pressure head, V the average velocity of flow, h_p the energy imparted to the water by the pump in foot-pounds per pound of fluid, and h_L the total head loss between sections A and B in foot-pounds per pound of fluid. The symbol $+h_p$ would be replaced by $-h_t$ if a turbine were in the line instead of a pump.

11-1 Head loss by pipe friction The head loss by pipe friction may be found by the Darcy-Weisbach equation

$$h_L = f \frac{L}{D} \frac{V^2}{2g} \tag{11-2}$$

in which L and D are the length and diameter of the pipe, respectively, and f

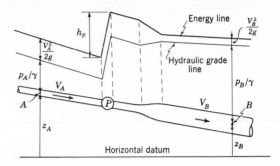

FIG. 11-1   Definition sketch for pipe flow.

is a friction factor. This factor is a function of the relative roughness of the pipe and the Reynolds number ($N_R = DV/v$) of the flow. If N_R is less than 2100, the flow is laminar; if over 3000, the flow is turbulent; between these values a transitional type of flow exists. The *relative roughness* e/D of a pipe depends on the absolute roughness e of the interior surface and the diameter D. Curves for determining the friction factor f and values of absolute roughness e based on extensive tests[1] of commercial pipe are shown in Fig. 11-2.

The Manning equation (Sec. 10-1) is also applicable to turbulent flow in pressure conduits and gives good results if the roughness coefficient n is accurately estimated.[2] The nomograph of Fig. 10-2 may be used for solving this equation. An empirical formula for pipe flow which is widely used is the Hazen-Williams equation

$$V = 1.318 C_H R^{0.63} S^{0.54} \tag{11-3}$$

in which C_H is a coefficient (Table 11-1) and S is the slope of the energy line. The Hazen-Williams formula is commonly solved by nomographs.[3]

ILLUSTRATIVE EXAMPLE 11-1   Find the head loss in 1000 ft of 6-ft-diameter smooth concrete pipe carrying 80 cfs of water at 50°F.

By the Darcy-Weisbach formula,

$$V = \frac{Q}{A} = \frac{80 \times 4}{\pi \times 6^2} = 2.83 \text{ ft/sec}$$

$$N_R = \frac{DV}{v} = \frac{6 \times 2.83}{1.41 \times 10^{-5}} = 1.21 \times 10^6$$

viscosity @ 50°

---

[1] L. F. Moody, Friction Factors for Pipe Flow, *Trans. ASME*, Vol. 66, pp. 671–684, 1944.

[2] It should be noted that n increases as R gets larger. See J. B. Franzini and P. S. Chisholm, Current Practice in Hydraulic Design of Conduits, *Water and Sewage Works*, Vol. 110, pp. 342–345, October, 1963.

[3] E. W. Steel, "Water Supply and Sewerage," 4th ed., p. 136, McGraw-Hill, New York, 1960.

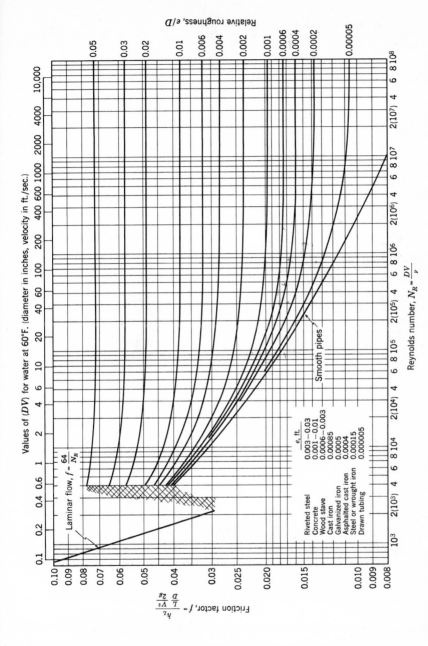

FIG. 11-2 Moody diagram for estimating f for pipes.

TABLE 11-1   Values of Hazen-Williams Coefficient C_H

Pipe material	C_H
Concrete (regardless of age)	130
Cast iron:	
New	130
5 yr old	120
20 yr old	100
Welded steel, new	120
Wood stave (regardless of age)	120
Vitrified clay	110
Riveted steel, new	110
Brick sewers	100
Asbestos-cement	140

Assume $e = 0.001$ ft; then $e/D = 0.001/6 = 0.00017$ and $f = 0.014$ from Fig. 11-2.

$$h_L = f\frac{L}{D}\frac{V^2}{2g} = 0.014\frac{1000}{6}\frac{2.83^2}{64.4} = 0.29 \text{ ft}$$

By the Chézy-Manning formula, using $n = 0.013$,

$$h_L = SL = \frac{V^2 n^2 L}{2.22 R^{4/3}} = \frac{2.83^2 \times 0.013^2 \times 1000}{2.22 \times 1.5^{4/3}} = 0.35 \text{ ft}$$

By the Hazen-Williams formula, using $C_H = 130$,

$$h_L = SL = \frac{V^{1.85}L}{(1.318 C_H)^{1.85} R^{1.17}} = \frac{2.83^{1.85} \times 1000}{(1.318 \times 130)^{1.85} \times 1.5^{1.17}} = 0.32 \text{ ft}$$

The main element of uncertainty in all methods is the selection of the roughness coefficient. Tests indicate that the absolute roughness of most metal pipes increases more or less linearly with the length of use. The carrying capacity of ordinary tar-dipped cast-iron pipe will decrease about 30 percent in 20 yr. The rate of change of the hydraulic characteristics of a pipe depends on the pipe material and the chemical nature of the fluid it conveys. It is customary in selecting pipes to use higher values of e and n and lower values of C_H than are indicated for new pipe, to allow for loss in capacity with age.

11-2   Minor losses in pipelines   Minor losses in pipelines are caused by abrupt changes in the flow geometry as a result of changes in pipe size, bends, valves, and fittings of all types. In long pipelines these minor losses can often be neglected without serious error, but they may be quite important in short pipes. Minor losses are generally greater where flow decelerates than where an increase in velocity occurs because of the eddies created by separation of the flow from the conduit boundary. Minor losses in turbulent flow vary approximately with the square of the velocity and are usually expressed as a function of velocity head (Table 11-2). It should be noted that the velocity head is lost at submerged discharge (a special case of sudden enlargement). Head loss at submerged discharge can be decreased by installing a diverging section of pipe to lower the discharge velocity. Draft tubes of reaction turbines make use of this principle.

11-3   Flow with negative pressure   Figure 11-3 shows the hydraulic conditions when water is pumped from a reservoir and discharged through a nozzle. Taking the datum at the surface of the reservoir and writing the energy equation between A and B,

$$h_p = h_L + h_{L_m} + z_B + \frac{V_B{}^2}{2g} \qquad (11\text{-}4)$$

This indicates that the head developed by the pump is used up in overcoming pipe friction h_L and minor losses h_{L_m}, in lifting the water a distance z_B, and in imparting the velocity head $V_B{}^2/2g$ to the water.

The pressure head at any section of a pipe is indicated by the vertical distance from the section to the hydraulic grade line. If the absolute pressure at any point in a system falls to the vapor pressure e_w, water vapor and dissolved gases will collect in high spots and obstruct the flow. The lowest pressure in the system of Fig. 11-3 occurs at point C on the suction side of the pump, and flow would cease when

$$\frac{p_C}{\gamma} = -\left(z_p + \frac{V_{AC}{}^2}{2g} + h_{L_{AC}}\right) = -\frac{p_{\text{atm}} - e_w}{\gamma} \qquad (11\text{-}5)$$

This condition limits the rate at which water may be pumped. To increase the discharge above this limiting rate, the pump may be placed at a lower

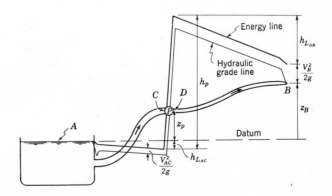

FIG. 11-3   Conditions of flow with negative pressure.

TABLE 11-2 Minor Losses in Pipe Lines

a. Enlargements

$$\left[\text{Values of } K_L \text{ in } h_{L_m} = K_L \frac{(V_1 - V_2)^2}{2g}\right]$$

θ^*	$\dfrac{D_2}{D_1} = 3$	$\dfrac{D_2}{D_1} = 1.5$
10	0.17	0.17
20	0.40	0.40
45	0.86	1.06
60	1.02	1.21
90	1.06	1.14
120	1.04	1.07
180	1.00	1.00

*The angle θ is the angle in degrees between the sides of the tapering section.

b. Abrupt Contractions

$$\left(\text{Values of } K_L \text{ in } h_{L_m} = K_L \frac{V_2^2}{2g}\right)$$

$\dfrac{D_2}{D_1}$	K_L
0	0.5
0.4	0.4
0.6	0.3
0.8	0.1
1.0	0

c. Pipe Entrance from Reservoir

$$\text{Bell-mouth } h_L = 0.04 \frac{V^2}{2g}$$

$$\text{Square-edge } h_L = 0.5 \frac{V^2}{2g}$$

d. Bends

$$\left(\text{Values of } K_L \text{ in } h_{L_m} = K_L \frac{V^2}{2g}, \text{ the head loss in excess of that in a straight pipe of equal length}\right)$$

$\dfrac{\text{Radius of bend}}{\text{Pipe diameter}}$	Deflection angle of bend		
	$90°$	$45°$	$22.5°$
1	0.50	0.37	0.25
2	0.30	0.22	0.15
4	0.25	0.19	0.12
6	0.15	0.11	0.08
8	0.15	0.11	0.08

e. Valves and Fittings†

$$\left(\text{Values of } K_L \text{ in } h_{L_m} = K_L \frac{V^2}{2g}\right)$$

Globe valve (wide open)	10
Swing check valve (wide open)	2.5
Gate valve (wide open)	0.2
Gate valve (half open)	5.6
Return bend	2.2
Standard tee	1.8
Standard 90° elbow	0.9

†From Flow of Fluids, *Tech. Paper* 410, Crane Co., 1965.

elevation or the suction pipe increased in diameter or decreased in length. At sea level under ordinary pressures and temperatures the maximum permissible negative pressure is theoretically equal to about 33 ft of water. The condition of Eq. (11-5) represents only a theoretical limit since cavitation (Sec. 9-2) or the release of dissolved gases will occur before the pressure drops to the vapor pressure.

ILLUSTRATIVE EXAMPLE 11-2   In Fig. 11-3 assume $z_p = 10$ ft, length of pipe from reservoir to pump $= 1000$ ft, from pump to nozzle $= 5000$ ft, $f = 0.02$, and diameter of pipes 3 ft and 2 ft, respectively. Neglecting minor losses, compute the maximum theoretical rate at which water may be pumped if the atmospheric pressure is 13.8 psi and the water temperature is 80°F. At this temperature, $e_w = 0.5$ psi (see Appendix).

$$\frac{p_c}{\gamma} = -\left(10 + \frac{V^2}{2g} + 0.02 \times \frac{1000}{3}\frac{V^2}{2g}\right) = \frac{-(13.8 - 0.5)}{62.4} \times 144$$

From this expression V is found to be 13.2 ft/sec, and hence the limiting discharge $Q = AV = (\pi/4) \times 3^2 \times 13.2 = 93.5$ cfs.

Subatmospheric pressures always occur in a pressure conduit which is above the hydraulic grade line. For design purposes the maximum negative pressure head should not exceed about two-thirds of that given by Eq. (11-5). Pipelines are frequently provided with vacuum pumps or air valves at their summits to remove gases which may accumulate.

11-4   Flow in branching pipes   In Fig. 11-4 flow will take place from reservoirs A and B to C. If the energy line at junction D were above the water-surface elevation in reservoir B, flow would occur into reservoir B. The flow distribution can be determined by writing the energy and continuity equations as follows:

$$z_A = z_D + \frac{p_D}{\gamma} + f_A \frac{L_A}{D_A}\frac{V_A^2}{2g} \tag{11-6a}$$

$$z_B = z_D + \frac{p_D}{\gamma} + f_B \frac{L_B}{D_B}\frac{V_B^2}{2g} \tag{11-6b}$$

$$z_C = z_D + \frac{p_D}{\gamma} - f_C \frac{L_C}{D_C}\frac{V_C^2}{2g} \tag{11-6c}$$

$$\frac{\pi D_A^2 V_A}{4} + \frac{\pi D_B^2 V_B}{4} = \frac{\pi D_C^2 V_C}{4} \tag{11-6d}$$

If the direction of flow in any pipe is reversed, the corresponding head-loss term is negative. If the pipe sizes, lengths, and controlling elevations are

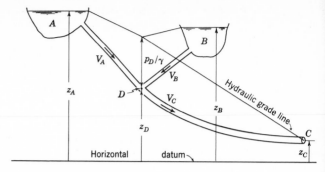

FIG. 11-4 · *Definition sketch for branching pipes.*

known, these equations can be solved by trial. A value of p_D/γ is assumed, and the velocities computed from the first three equations must satisfy Eq. (11-6d).

11-5   Flow in pipe systems   If two pipes are in parallel (Fig. 11-5) the head loss through each pipe must be the same. If the pipe characteristics are known, the distribution of flow in the two pipes can be calculated by equating the head losses and applying the continuity equation. For pipes in series the total head loss is equal to the summation of the individual head losses. If there are several pipes in parallel, or in series, or a combination of both, it is convenient to express the head loss in the pipe system by

$$h_L = KQ^x \tag{11-7}$$

where K depends on the arrangement, length, diameter, and roughness of the pipes as well as the fluid properties. On the basis of the Manning equation the exponent x in Eq. (11-7) would be 2, while for the Hazen-Williams equation $x = 1.85$. The Darcy-Weisbach formula [Eq. (11-2)] gives values of x varying from 1.75 for smooth pipes to 2.0 for rough pipes at high N_R.

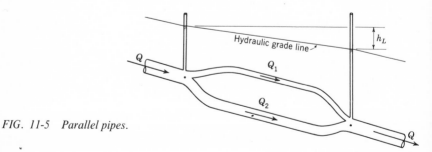

FIG. 11-5   *Parallel pipes.*

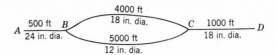

FIG. 11-6   *Sketch for Illustrative Example* 11-3.

ILLUSTRATIVE EXAMPLE 11-3   Using $n = 0.013$ and neglecting minor losses, express the head loss through the pipe system of Fig. 11-6 in the form of Eq. (11-7).

From Eq. (10-4),

$$h_L = S \times L = \frac{n^2 V^2 L}{2.21 R^{4/3}} = \frac{n^2 Q^2 L}{2.21 R^{4/3} A^2}$$

$$h_L \ A \ \text{to} \ B = \frac{(0.013)^2 Q^2 (500)}{2.21 (2/4)^{4/3} (\pi)^2} = 0.0098 \ Q^2$$

$$h_L \ C \ \text{to} \ D = \frac{(0.013)^2 Q^2 (1000)}{2.21 (1.5/4)^{4/3} (1.76)^2} = 0.088 \ Q^2$$

h_L upper pipe $= h_L$ lower pipe

$$Q_1^2 L_1 / R_1^{4/3} A_1^2 = Q_2^2 L_2 / R_2^{4/3} A_2^2$$

$$Q_1^2 (4000)/0.375^{4/3} 1.76^2 = Q_2^2 (5000)/0.25^{4/3} 0.785^2$$

$$Q_1 = 3.3 Q_2 ; Q = Q_1 + Q_2 = 4.3 Q_2$$

$$h_L \ B \ \text{to} \ C = (0.013)^2 Q_2^2 (5000)/2.21(0.25)^{4/3} 1.76^2$$

$$= 0.778 Q_2^2 = 0.181 Q^2$$

Finally,

$$h_L \ A \ \text{to} \ D = 0.28 Q^2$$

ILLUSTRATIVE EXAMPLE 11-4   Determine the exponential expression for head loss when water at 50°F flows in a 12-in.-diameter clean cast-iron pipe. Use the Darcy-Weisbach formula, and assume a velocity range from 3 to 12 ft/sec.

From Fig. 11-2, $e = 0.00085$ ft and $e/D = 0.00085$. For $V = 3$ ft/sec, $N_R = 214,000$, $f = 0.020$, and $Q = (\pi/4) \times 1 \times 3 = 2.36$ cfs. For $V = 12$ ft/sec, $N_R = 856,000, f \doteq 0.019$, and $Q = 9.45$ cfs. Hence

$$\text{At 3 ft/sec:}\quad h_L = 0.02 \frac{L}{D} \frac{3^2}{2g} = K \times 2.36^x$$

$$\text{At 12 ft/sec:}\quad h_L = 0.019 \frac{L}{D} \frac{12^2}{2g} = K \times 9.45^x$$

Simultaneous solution of these equations yields $x = 1.96$ and $K = 0.00052L$. Hence for the specified conditions of flow $h_L = 0.00052 L Q^{1.96}$.

11-6 Pipe networks In any pipe network two conditions must be satisfied: (1) *the algebraic sum of the pressure drops around any closed loop must be zero*, and (2) *the flow entering a junction must equal the flow leaving it.* The first condition states that there can be no discontinuity in pressure, i.e., the pressure drop through any route between two junctions must be the same. The second condition is a statement of the law of continuity.

Network problems are usually solved by methods of successive approximation since any analytical solution requires the use of many simultaneous equations, some of which are nonlinear. A procedure suggested by Hardy Cross[1] requires that the flow in each pipe be assumed so that the principle of continuity is satisfied at each junction. A correction to the assumed flow is computed successively for each pipe loop in the network until the correction is reduced to an acceptable magnitude. If Q_a is the assumed flow and Q the true flow in a pipe, the correction Δ is $Q - Q_a$ and

$$Q = Q_a + \Delta \tag{11-8}$$

Expressing head loss by Eq. (11-7), the condition that the head loss around any closed loop be zero gives

$$\Sigma K(Q_a + \Delta)^x = 0 \tag{11-9}$$

Expanding this summation,

$$\sum KQ_a^x + \sum xK\Delta Q_a^{x-1} + \frac{x-1}{2}\sum xK\Delta^2 Q_a^{x-2} + \cdots = 0 \tag{11-10}$$

If Δ is small, the third and all succeeding terms of the expansion may be neglected. Hence,

$$\Sigma KQ_a^x + \Delta\Sigma xKQ_a^{x-1} = 0 \tag{11-11}$$

where Δ has been removed from the summation since it is the same for all pipes of the loop. Solving for Δ gives

$$\Delta = -\frac{\Sigma KQ_a^x}{\Sigma|xKQ_a^{x-1}|} \tag{11-12}$$

In applying Eq. (11-12), the direction of flow as well as the rate of flow in the circuit must be assumed. The numerator of Eq. (11-12) is then the algebraic sum of the head loss in the circuit, with due regard for sign. The correction Δ must be applied in the same sense to each pipe in the loop. If

---

[1] Hardy Cross, Analysis of Flow in Networks of Conduits or Conductors, *Univ. Illinois Bull.* 286, November, 1936.

the counterclockwise direction is assumed positive, Δ is added algebraically to flows assumed in the counterclockwise direction and subtracted from flows in the clockwise direction. In Eq. (11-9), Δ was given the same sign in all pipes. Because of this, the denominator of Eq. (11-12) is taken as the absolute sum without regard to the signs of the individual items in the summation. Since the Hardy Cross method assumes a constant value of x, it is usually assumed as 1.85 or 2 and is not varied as the Darcy-Weisbach formula would require. Minor losses are generally neglected, but they can be introduced by substituting an equivalent length of pipe. The system must be divided into two or more loops, such that each pipe in the network is included in at least one loop. Values of Δ are computed for each loop, and the assumed flows are corrected. Repeated adjustments are made until the desired accuracy is attained.

ILLUSTRATIVE EXAMPLE 11-5   Determine the flow in each pipe of the network shown in Fig. 11-7, using $f = 0.02$ throughout.
    Taking $x = 2$,

$$h_L = f \frac{L}{D} \frac{V^2}{2g} = f \frac{L}{D} \frac{1}{2g} \left(\frac{4Q}{\pi D^2} \right)^2 = \frac{8fL}{\pi^2 g D^5} Q^2 = KQ^2$$

Hence $K = 0.81 fL/gD^5$, and the K value for each pipe is:

Diameter, in.	3	4	5	6	7	8
K	1030	368	160	80.4	22.4	11.5

The assumed flows are indicated on the figure in parentheses. For loop $AEDB$

$$\Delta_1 = -\frac{(1030 \times 0.5^2) + (11.5 \times 0.1^2) - (22.4 \times 0.2^2) - (368 \times 0.7^2)}{2[(1030 \times 0.5) + (11.5 \times 0.1) + (22.4 + 0.2) + (368 \times 0.7)]}$$

$$= -0.05 \text{ cfs}$$

and for loop BDC

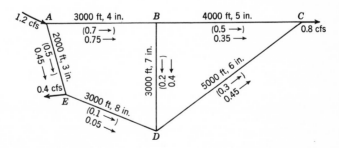

FIG. 11-7   Sketch for Illustrative Example 11-5.

$$\Delta_2 = -\frac{(22.4 \times 0.2^2) + (80.4 \times 0.3^2) - (160 \times 0.5^2)}{2[(22.4 \times 0.2) + (80.4 \times 0.3) + (160 \times 0.5)]} = +0.15 \text{ cfs}$$

The corrected flows appear on the figure below the first assumed flows. Recomputing Δ for each loop yields $\Delta_1 = +0.001$ cfs and $\Delta_2 = -0.001$ cfs.

An electrical analogy is sometimes used to solve complex pipe-network problems. Ohm's law is not applicable to turbulent flow in pipes since it makes the discharge (current) a function of the first power of the potential gradient. Special tubes, called *fluistors*, have been developed[1] which cause voltage to vary with the 1.85 power of the current. With these tubes in the circuit, the pressure and flow distribution can be determined by measuring voltage and current at the desired points.

Another way of solving a pipe-network problem is to use a digital computer. Programming takes time and care; but once set up, there is great flexibility. The effect of changing a pipe size, for example, can be readily determined by simply replacing the old data card for that pipe with a new one. The speed of the computer and the accuracy that can be obtained are the outstanding advantages of this approach.[2]

11-7   Power in fluid flow   When water flows in a pipe at Q cfs, the rate of flow can be expressed as γQ lb/sec. Multiplying γQ by head h (ft-lb/lb), one obtains γQh ft-lb/sec, a unit of power. Applying the proper conversion units,

$$\text{Horsepower} = \frac{\gamma Qh}{550} \tag{11-13}$$

where h is the significant head. In finding the power put into the flow by a pump, $h = h_p$. When finding the power loss from pipe friction, $h = h_L$, and to compute the horsepower of a jet, $h = \dfrac{V^2}{2g}$.

ILLUSTRATIVE EXAMPLE 11-6   If the spillway of Illustrative Example 9-2 is 100 ft long, determine the horsepower dissipated in the hydraulic jump formed at the toe, when the head on the spillway is 8 ft.

In this case the significant head to be used in Eq. (11-13) is $\Delta E = E_1 - E_2 = 31.0$ ft.

$$\text{Horsepower loss} = \frac{\gamma Q(\Delta E)}{550} = \frac{62.4(100 \times 88.3)(31.0)}{550} = 31,000$$

---

[1] M. S. McIlroy, Pipe Networks Studied by Nonlinear Resistors, *Trans. ASCE*, Vol. 118, pp. 1055–1067, 1953.

[2] Robert Epp and A. G. Fowler, Efficient Code for Steady-State Flows in Networks, *J. Hydraulics Div., ASCE*, Vol. 96, pp. 43–56, January, 1970.

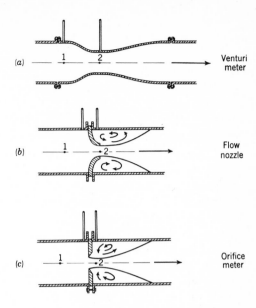

FIG. 11-8   Three types of constriction meters for pipe flow.

MEASUREMENT OF FLOW IN PRESSURE CONDUITS

Flow measurement in pressure conduits may be accomplished by various methods. These methods have their own characteristics which may make one preferable under given conditions.

11-8   Differential head meters   The flow of fluid through a constriction in a pressure conduit results in a lowering of pressure at the constriction. The drop in piezometric head between the undisturbed flow and the constriction is a function of the flow rate. The venturi meter, flow nozzle, and orifice meter (Fig. 11-8) are *constriction meters* which make use of this principle. The difference in piezometric head may be measured with a differential manometer or pressure gages. In order that such an installation may function properly, a straight length of pipe at least 10 diameters long should precede the meter. Straightening vanes may also be installed in the conduit just upstream from the meter to suppress disturbances in the flow. The head loss through a venturi meter is considerably less than for the other two types of meters.

A venturi meter is a machined casting of considerable size and is relatively costly as compared with a nozzle or orifice meter, which may be readily inserted between the flanges of a pipeline. Applying the Bernoulli equation between sections 1 and 2 of Fig. 11-8, neglecting head loss, expressing V in terms of Q, and solving for Q gives

$$Q_{\text{theoretical}} = \frac{A_2}{\sqrt{1 - (A_2/A_1)^2}} \sqrt{2g\left(z_1 + \frac{p_1}{\gamma} - z_2 - \frac{p_2}{\gamma}\right)} \qquad (11\text{-}14)$$

This equation may be modified to

$$Q = C_d A_2 \sqrt{2g\left[\Delta\left(z + \frac{p}{\gamma}\right)\right]} \qquad (11\text{-}15)$$

The discharge coefficient C_d corrects for the head loss between sections 1 and 2 and accounts for the geometric characteristics of the meter. The location of the pressure taps and the Reynolds number of the flow affect the value of the coefficient. Standard specifications[1] for the various types of constriction meters and their coefficients are available. However, for high accuracy, a meter should be calibrated in place.

Another type of differential head meter is the *bend meter*, which utilizes the pressure difference between the inside and outside of a pipe bend. The meter is simple and inexpensive. An elbow already in the line may be used without causing added head loss. For best results a bend meter should be calibrated in place. The meter equation will be

$$Q = C_d A \sqrt{2gh} \qquad (11\text{-}16)$$

where h is the difference in piezometric head between the outside and the inside of the bend at the mid-section and A is the cross-sectional area of the pipe. Lansford[2] suggests that for a standard 90°-flanged elbow the coefficient C_d is given within 10 percent by $\sqrt{r/2D}$, where D is the pipe diameter, and r, the center-line radius of the bend, is less than $1.5D$.

Pitot-static tubes and pitometers (Fig. 11-9) may also be classed as differential head meters. These devices indicate the velocity head at a point in the pipe cross section. If measurements are made at intervals across the pipe, the velocity profile can be established and the flow rate computed. By use of a relation between flow rate and center-line velocity, a single central measurement may provide an approximate discharge figure. The general equation for pitot-static tubes and pitometers is

$$V = C_I \sqrt{\frac{2g(p_s - p')}{\gamma}} \qquad (11\text{-}17)$$

where p_s and p' are as indicated in Fig. 11-9. Pitometers and pitot-static

---

[1] "Fluid Meters: Their Theory and Application," 5th ed., American Society of Mechanical Engineers, 1959.

[2] W. M. Lansford, The Use of an Elbow in a Pipe Line for Determining the Rate of Flow in a Pipe, *Univ. Illinois Eng. Exp. Sta. Bull.* 289, December, 1936.

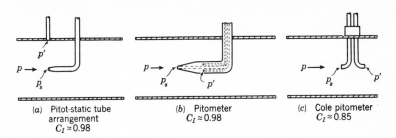

(a)   Pitot-static tube
       arrangement
       $C_I \approx 0.98$

(b)   Pitometer
       $C_I \approx 0.98$

(c)   Cole pitometer
       $C_I \approx 0.85$

FIG. 11-9   Various types of pitot-tube flow-measuring devices.

tubes are not often used for continuous flow measurement, but rather for calibrating other types of meters in place and for intermittent measurements.

Differential head meters have no moving parts and hold their initial accuracy as long as they are kept clean. The head differences are difficult to measure at low flows. Several manufacturers make recording and indicating devices which permit direct reading of the flow rate and total flow through differential meters.

11-9   Mechanical meters   Mechanical meters are ordinarily used to measure total volume of flow. Most such meters do not have timing devices permitting direct reading of flow rate. Mechanical meters are of limited accuracy, and their accuracy decreases with use because of wear of the moving parts. They are not suited for measurement of very low flow rates because the liquid may pass the meter without moving the mechanical elements. Because of high head loss they are not often used for flow rates above 10 cfs. Mechanical meters are widely used in water-distribution systems, where their low cost and small size are substantial advantages over other recording meters. Two types of mechanical meters in common use are *displacement meters* and *inferential meters*. A simple displacement meter consists of a piston which moves back and forth with the passing fluid. A counter registers the number of piston strokes, and the flow volume is the product of the number of strokes and the displacement per stroke. The nutating disk meter, widely used in municipal water systems, consists of a disk which oscillates in a measuring chamber. For each oscillation a known volume of water moves through the meter. A stem at right angles to the disk trips a lever connected to a counter. New disk meters are usually accurate within 1 percent, but they underregister considerably after some use.

Inferential meters consist of propellers whose speed of rotation is a function of the velocity of flow. The number of revolutions is determined by a counting mechanism and is related to discharge by calibration. This type of meter is often provided with guide vanes which may be adjusted to change the calibration. If the surface of the blades becomes worn or coated with deposits, the calibration may change.

11-10   Chemical tracers   The Allen[1] *salt-velocity method* uses the increased conductivity of salt water as a means of timing the travel of a salt solution through a length of pipe. A concentrated sodium chloride solution is suddenly injected into the conduit at an injection station. At two downstream stations, electrodes are projected into the flow and connected to a recording ammeter and direct-current power source. The difference in the times of passage of the center of area of the current-time graph at each station is the time of travel between the stations. From this time, the velocity and discharge can be computed for the known distance. The *salt-dilution method* (Sec. 2-9) is also used for flow measurement in pipes.

FORCES ACTING ON PIPES

Pipes must be designed to withstand stresses created by internal and external pressures, changes in momentum of the flowing water, external loads, and temperature changes, and to satisfy the hydraulic requirements of the project.

11-11   Internal pressures   The internal pressure within a conduit is caused by static pressure and water hammer. Internal pressure causes circumferential tension in the pipe walls which is given approximately by

$$\sigma = \frac{pr}{t} \tag{11-18}$$

where σ is the tensile stress, p the pressure (static plus water hammer), r the internal radius of the pipe, and t the wall thickness.

11-12   Water hammer   When a liquid flowing in a pipeline is abruptly stopped by the closing of a valve, dynamic energy is converted to elastic energy and a series of positive and negative pressure waves travel back and forth in the pipe until they are damped out by friction. This phenomenon is known as *water hammer*. At the instant the valve of Fig. 11-10 is closed, the element of water x_1 just upstream from the valve will be compressed by the water flowing against it. This results in a pressure rise which causes a portion of the pipe surrounding the element to stretch. In the next instant the forward motion of element x_2 is stopped, and it, too, is compressed by the remaining water in the pipe, which still possesses forward motion. The process is repeated on successive elements until in a relatively short time the pressure wave has traveled back to the reservoir, and all the water in the pipe is at rest. A pressure in excess of hydrostatic

---

[1] C. M. Allen and E. A. Taylor, The Salt Velocity Method of Water Measurement, *Trans. ASME*, Vol. 45, pp. 285–341, 1923.

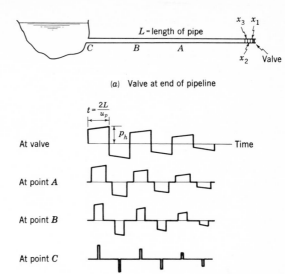

(a)   Valve at end of pipeline

(b)   Water hammer pressures at selected points for the
case of abrupt valve closure

FIG. 11-10   The nature of water hammer.

cannot be maintained at the junction of pipe and reservoir, and the pressure at C drops to normal as some of the water in the pipe flows back into the reservoir.

As the reduction in pressure travels back down the pipe toward the valve, the pipe contracts and the water expands until normal pressures exist throughout the pipe. The inertia of the water flowing into the reservoir results in the discharge of more water than the excess originally stored in the pipe. This causes negative pressures, and a rarefaction wave travels back up the pipe from the valve to the reservoir. Since a pressure less than hydrostatic cannot be maintained at C, a wave of normal pressure again travels back down the pipe as water flows from the reservoir into the pipe. This results in a return to the conditions which initiated the water hammer, and the process is repeated until damped out by friction. The pressure variation at several points along the pipe as a result of instantaneous valve closure is shown in Fig. 11-10b.

The velocity u of a pressure wave in any medium is the same as the velocity of sound in that medium and is given by[1]

$$u = \sqrt{\frac{E}{\rho}} \tag{11-19}$$

---

[1] For a derivation of this see J. K. Vennard, "Elementary Fluid Mechanics," 4th ed., App. III, p. 547, Wiley, New York, 1961.

where E is the modulus of elasticity of the medium and ρ is the density. For water under ordinary conditions u is about 4700 ft/sec. The velocity of a pressure wave created by water hammer is less than that given by Eq. (11-19) because of the elasticity of the pipe. If longitudinal extension of the pipe is prevented while circumferential stretching takes place freely, the velocity of a pressure wave u_p is given by[1]

$$ u_p = \sqrt{\frac{E}{\rho}} \sqrt{\frac{1}{1 + (ED/E_p t)}} \tag{11-20} $$

where E_p is the modulus of elasticity of the pipe walls, D is the pipe diameter, and t is the wall thickness.

If the valve of Fig. 11-11 is closed instantaneously, a pressure wave travels up the pipe with the velocity u_p. In a short interval of time dt an element of water of length $u_p\,dt$ is brought to rest. Applying Newton's second law and neglecting friction,

$$ F\,dt = M\,dV \tag{11-21} $$

$$ -A\,dp\,dt = \rho A u_p\,dt\,dV \tag{11-22} $$

$$ -dp = \rho u_p\,dV \tag{11-23} $$

Since velocity is reduced to zero, $dV = -V$ and dp equals the pressure p_h caused by water hammer. Hence,

$$ p_h = \rho u_p V \tag{11-24} $$

The total pressure at the valve immediately after closure is $p_h + p$, where p is the static pressure in the pipe.

If the length of the pipe is L, the wave travels from valve to reservoir and back in time $t = 2L/u_p$. This is the time that a positive pressure will be maintained at the valve (Fig. 11-10). If the valve is closed gradually, a series of small pressure waves is transmitted up the pipe. These waves are reflected at the reservoir and return down the pipe as waves of normal

---

[1] J. S. McNown, Surges and Water Hammer, chap. VII in Hunter Rouse (ed.), "Engineering Hydraulics," Wiley, New York, 1950.

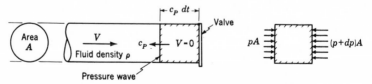

FIG. 11-11   Definition sketch for analysis of water hammer in pipes.

pressure. If the valve is completely closed before the reflected wave returns from the reservoir, the pressure increase is given approximately by Eq. (11-24). If the closure time t_c is greater than t, negative pressure waves will be superimposed on the positive waves and the full pressure will not be realized. The water-hammer pressure p_h' developed by gradual closure of the valve when $t_c > t$ is given approximately by

$$p_h' \approx \frac{t}{t_c} p_h = \frac{2Lp_h}{u_p t_c} = \frac{2LV\rho}{t_c} \qquad (11\text{-}25)$$

More complete analysis of water hammer may be found in the references cited at the end of this chapter.

ILLUSTRATIVE EXAMPLE 11-7   Water flows at 5 ft/sec from a reservoir into a 36-in. steel pipe which is 7500 ft long and has a wall thickness of 1 in. Find the water-hammer pressure developed by closure of a valve at the end of the line if the closure time is (a) 1 sec, (b) 8 sec.

$$u_p = 4700 \times \sqrt{\frac{1}{1 + (300{,}000 \times 36)/(30 \times 10^6 \times 1)}} = 4020 \text{ ft/sec}$$

$$t = \frac{2L}{u_p} = 2 \times \frac{7500}{4020} = 3.73 \text{ sec}$$

$$\text{If } t_c < 3.73 \text{ sec}, p_h = \frac{62.4 \times 4020 \times 5}{32.2 \times 144} = 270 \text{ psi.}$$

$$\text{If } t_c = 8 \text{ sec}, p_h' \approx \frac{3.73}{8} \times 270 = 126 \text{ psi.}$$

Water-hammer pressures can be greatly reduced by use of slow-closing valves, automatic relief valves, air chambers, and surge tanks. If the velocity of flow is increased suddenly by the opening of a valve or starting of a pump, a situation opposite to water hammer develops. When the valve is opened, an expansion of the water upstream from the valve occurs, and a wave of rarefaction travels up the pipe. While this problem is similar to that of the positive water hammer, friction cannot be neglected because of its effect in retarding the development of flow in the pipe.

An interesting penstock failure occurred at the Oigawa[1] power station in Japan in June 1950 where the abrupt closure of a butterfly valve created water-hammer pressures that split a 25-ft length of pipe just upstream of the valve. The subsequent excessive discharge from the broken penstock formed

---

[1] C. C. Bonin, Water Hammer Damage to Oigawa Power Station, *Trans. ASME*, Vol. 82, ser. A, pp. 111–119, 1960.

a vacuum upstream and about 175 ft of pipe collapsed through buckling action.

11-13   Forces at bends and changes in cross section   A change in the direction or magnitude of flow velocity is accompanied by a change in the momentum of the fluid. The force required to produce this change in momentum comes from the pressure variation within the fluid and from forces transmitted to the fluid from the pipe walls. Figure 11-12b shows a free-body diagram of the forces acting on the water contained in a horizontal pipe bend. Applying the impulse-momentum principle gives

$$p_1 A_1 - F_x - p_2 A_2 \cos \theta = \rho Q(V_2 \cos \theta - V_1) \qquad (11\text{-}26)$$

$$F_y - p_2 A_2 \sin \theta = \rho Q V_2 \sin \theta \qquad (11\text{-}27)$$

where p_1 and p_2 and V_1 and V_2 represent the pressure and average velocity in the pipe at sections 1 and 2, respectively. For a pipe bend of uniform section, $V_1 = V_2$, and $p_1 \approx p_2$.

A similar analysis can be applied to the forces acting on the water in the region of a pipe contraction (Fig. 11-13). The forces F_x and F_y of Eqs. (11-26) and (11-27) are those transmitted from the pipe to the water. An

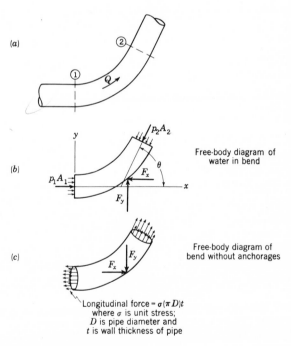

(a)

(b)   Free-body diagram of water in bend

(c)   Free-body diagram of bend without anchorages

Longitudinal force = $\sigma(\pi D)t$
where σ is unit stress;
D is pipe diameter and
t is wall thickness of pipe

FIG. 11-12   Forces at a horizontal pipe bend.

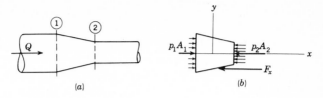

FIG. 11-13   Forces at a contraction in a pipeline.

equal and opposite force must be developed by stresses in the pipe wall. These stresses may be eliminated or reduced by providing an anchorage at the bend, contraction, or enlargement.

11-14   Temperature stresses   Longitudinal stresses of considerable magnitude may develop in pipes exposed to large changes in temperature. The change in length δ of a pipe of length L when subjected to a temperature change ΔT is

$$\delta = \alpha L \, \Delta T \tag{11-28}$$

where α is the coefficient of thermal expansion of the pipe material. If this change in length is prevented, longitudinal stresses will develop. From mechanics of materials it is known that in the elastic range

$$\sigma = E\epsilon = E\frac{\delta}{L} \tag{11-29}$$

where ϵ is the unit strain (elongation per unit length), E is the modulus of elasticity, and σ is the resulting unit stress. Combining Eqs. (11-28) and (11-29) gives

$$\sigma = E\alpha \, \Delta T \tag{11-30}$$

This indicates the longitudinal stress that would result when a pipe with fixed ends is subjected to a temperature change. Expansion joints are usually provided to reduce temperature stress.

ILLUSTRATIVE EXAMPLE 11-8   Find the longitudinal stress in a steel pipe caused by a temperature increase of 50°F. Assume that longitudinal expansion is prevented. For steel $E = 30 \times 10^6$ psi, and $\alpha = 6.5 \times 10^{-6}$ ft/ft/°F.

$$\sigma = E\alpha \, \Delta T = 30 \times 10^6 \times 6.5 \times 10^{-6} \times 50 = 9750 \text{ psi} \qquad \text{(compression)}$$

11-15   Flexural stresses   An unsupported pipe acts as a beam with loads resulting from the weight of the pipe, weight of water in the pipe, and any superimposed loads. The stresses resulting from beam action may be determined by the usual methods of analysis applied to beams. A pipe is a fairly efficient beam section, and stresses resulting from beam action alone are usually negligible except for long spans or when there are large superimposed loads. A rigorous analysis of the combined stresses resulting from internal pressures, external loads, temperature changes, and beam action involves application of the principles of elasticity.

11-16 External loads on buried pipes Pipes are often placed in an excavated trench which is backfilled, or they are laid on the ground surface and covered with earth. In either case a vertical load is imposed on the pipe. If a load is superimposed on the fill, a portion of it will be transferred to the buried pipe. The magnitude of the load thus produced depends on the rigidity of the pipe, the bedding, and the character of the fill material. Our understanding of the effect of external loads on buried pipes is based mainly on experiments made at Iowa State College[1] under the direction of Dean Marston.

Rigid pipes (concrete, cast iron, vitrified clay) cannot deform materially without cracking. On the other hand, *flexible* steel pipe can deform considerably without structural damage. If the pipe is placed in a narrow trench, not wider than about 2 or 3 pipe diameters, settlement of the fill is resisted by the pipe and by shearing stresses between the backfill and the wall of the trench (assuming no cohesion). For rigid pipes in narrow trenches, the load w in pounds per foot of pipe has been found to be

$$w = c\gamma B^2 \tag{11-31}$$

where B is the trench width at the top of the pipe, γ is the specific weight of the fill material, and c is a coefficient characteristic of the fill material and the ratio of cover depth to width of trench (Table 11-3). Tables that give loads on buried pipe are available in the literature.[2]

The deformation of flexible pipe under load (Fig. 11-14a) develops lateral pressures which help to resist the load. The deformation also permits some bridging of the fill material above the pipe. As a result, the load transmitted to a flexible pipe is less than that for a rigid conduit. The empirical formula for the load on a buried flexible pipe in a narrow trench is

$$w = c\gamma BD \tag{11-32}$$

where D is the outside diameter of the pipe.

If a pipe is placed on undisturbed ground and covered with fill (as a highway culvert), the fill adjacent to the pipe is deeper than that over the

---

[1] M. G. Spangler, Underground Conduits—An Appraisal of Modern Research, *Trans. ASCE*, Vol. 113, pp. 316–374, 1948.

[2] "Clay Pipe Engineering Manual," National Clay Pipe Institute, Los Angeles, 1968.

TABLE 11-3 Values of the Coefficient c for Eqs. (11-31) and (11-32)

Fill material Specific weight, lb/cu ft	Sand and gravel 100	Saturated topsoil 100	Clay 120	Saturated clay 130
Cover depth Trench width	Values of c			
1.0	0.84	0.86	0.88	0.90
2.0	1.45	1.50	1.55	1.62
3.0	1.90	2.00	2.10	2.20
4.0	2.22	2.33	2.49	2.65
5.0	2.45	2.60	2.80	3.03
6.0	2.60	2.78	3.04	3.33
7.0	2.75	2.95	3.23	3.57
8.0	2.80	3.03	3.37	3.76
9.0	2.88	3.11	3.48	3.92
10.0	2.92	3.17	3.56	4.04
12.0	2.97	3.24	3.68	4.22
14.0	3.00	3.28	3.75	4.34

pipe and can, therefore, settle a greater distance (Fig. 11-14b). Under these conditions, generally referred to as *embankment* or *broad-fill* conditions, a portion of the weight of the adjacent prisms of fill is transferred to the central prism by shear, and the load on the pipe is greater than for trench conditions. The equations for the load on a buried pipe under embankment conditions is

$$w = c_p \gamma D^2 \tag{11-33}$$

Values of c_p depend on the type of pipe and the character of the foundation and backfill. Typical values for c_p are given in Table 11-4.

If the conduit can deform or the ground under the conduit settles more than the adjacent ground, the load is reduced since less is transferred to the central prism from the adjacent fill. Piles or other structures which resist the settlement of the conduit while allowing the adjacent prisms to settle should be avoided.

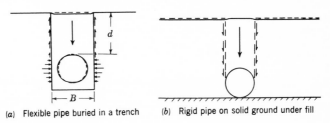

(a) Flexible pipe buried in a trench   (b) Rigid pipe on solid ground under fill

FIG. 11-14 Loads on a buried pipe.

TABLE 11-4   Values of the Coefficient c_p in Eq. (11-33)

Cover depth / Pipe diam	Rigid pipe, unyielding base, noncohesive backfill	Flexible pipe, average conditions
1.0	1.2	1.1
2.0	2.8	2.6
3.0	4.7	4.0
4.0	6.7	5.4
6.0	11.0	8.2
8.0	16.0	11.0

As the width of trench increases, the load on the pipe increases according to Eqs. (11-31) and (11-32) until it equals that given by Eq. (11-33). Hence if the trench width is more than about twice the diameter of the pipe, the load should be taken as the lesser of the two values computed by Eq. (11-31) or (11-32) and by Eq. (11-33).

The portion of a superimposed load which is transmitted to a buried conduit may be found approximately by Boussinesq's equation for the distribution of stress in an elastic solid. Assuming the fill surface to be horizontal, this equation is

$$p = \frac{3d^3 P}{2\pi Z^5} \tag{11-34}$$

where p is the unit pressure at any point in the fill at a depth d below the surface and a slant height Z from the load P. The total load on a unit length of conduit may be found by integrating Eq. (11-34) over the projected area of the pipe. This integration is normally accomplished by subdividing the projected area of the pipe into small squares, computing the load on each subdivision, and adding these values to obtain a total. Impact from moving loads on unsurfaced roads may nearly double the computed loads. Paving greatly reduces the effect of impact. Note that the effect of superimposed loads decreases rapidly as the depth of cover increases.

ILLUSTRATIVE EXAMPLE 11-9   A 3-ft-diameter steel pipe is buried in a trench 4 ft wide. The backfill is clay (specific weight 120 lb/cu ft), and the top of the pipe is 6 ft below the surface of the fill. The pipe passes at right angles under a one-lane road which carries a vehicle whose loading (including impact) consists of two concentrated 1800-lb loads located 5 ft apart transverse to the roadway. Find the maximum vertical force exerted on a unit length of this pipe.

Taking $c = 1.2$, the load caused by the backfill is

$$w = c\gamma BD = 1.2 \times 120 \times 4 \times 3 = 1730 \text{ lb/ft}$$

The slant height Z from one of the wheel loads to a point on the pipe midway between the loads is $Z = \sqrt{6^2 + 2.5^2} = 6.5$ ft. The pressure on the pipe as a result of a single wheel load is

$$p = \frac{3d^3 P}{2\pi Z^5} = \frac{3 \times 6^3 \times 1800}{2\pi \times 6.5^5} = 16.0 \text{ psf}$$

The total force on the pipe per foot of length is $1730 + 2(16.0 \times 3) = 1826$ lb/ft. A more accurate estimate of the effect of the superimposed load could have been made by analyzing smaller unit areas, but this hardly seems justified in this case.

ILLUSTRATIVE EXAMPLE 11-10 An 8-ft-diameter rigid concrete pipe rests on unyielding ground and is covered with sand (specific weight 100 lb/cu ft) to a depth of 6 ft. Directly above the pipe is a concentrated load of 10,000 lb. Find the vertical force on a 1-ft length of the pipe.

Taking $c_p = 0.9$ (Table 11-4), the pressure caused by the fill is

$$w = 0.9 \times 100 \times 8^2 = 5760 \text{ lb/ft}$$

Subdivide the projected area of a 1-ft length of pipe into eight 1-ft squares. The pressure on the squares on one side of the center line is the following:

Distance from center line to mid-point of square, ft	0.5	1.5	2.5	3.5
Slant height to mid-point of square, ft	6.02	6.18	6.40	6.95
Computed pressure by Eq. (11-34), psf	130.2	114.2	96.2	63.5

Total pressure on 1-ft length of pipe is

$$5760 + 2(130 + 114 + 96 + 64) = 6568 \text{ lb}$$

11-17 Strength of pipe Because of the complex nature of the combined stress in pipes, the stresses are rarely analyzed in detail except for large and important structures. Table 11-5 presents ASTM standards for the crushing strength of various types of pipe in three-edge bearing (Fig. 11-15b). Strengths from sand-bearing tests (Fig. 11-15a) will be about 50 percent more than those for three-edge bearing. The effective strength of pipe is greatly influenced by bedding conditions. Figure 11-16 shows several types of bedding and indicates the strength ratios, i.e., the factors by which the three-edge bearing strength is multiplied to find the effective strength in place. It is customary to use two-thirds of the effective strength as the design strength. In general, all buried pipe should be placed on a bed which has been rounded to fit the pipe, and the backfill should be carefully placed and thoroughly tamped around and above the top of the pipe. This

TABLE 11-5   Crushing Strength of Clay and Concrete Pipe by the Three-edge Bearing Method*
(All Strengths in Pounds per Linear Foot)

Internal diameter, in.	Clay Standard	Clay Extra strength	Plain concrete Standard	Plain concrete Extra strength	Class I	Class II	Class III	Class IV	Class V
4	1200	2000	1000	2000					
6	1200	2000	1100	2000					
8	1400	2200	1300	2000					
10	1600	2400	1400	2000					
12	1800	2600	1500	2250		1500	2000	3000	3750
15	2000	2900	1750	2750		1875	2500	3750	4690
18	2200	3300	2000	3300		2250	3000	4500	5620
21	2400	3850	2200	3850		2625	3500	5250	6560
24	2600	4400	2400	4000		3000	4000	6000	7500
27	2800	4700				3375	4500	6750	8440
30	3300	5000				3750	5000	7500	9380
33	3600	5500				4125	5500	8250	10,220
36	4000	6000				4500	6000	9000	11,250
39						4825	6500	9750	
42						5250	7000	10,500	13,120
48						6000	8000	12,000	15,000
54						6750	9000	13,500	16,880
60					6000	7500	10,000	15,000	18,750
66					6600	8250	11,000	16,500	20,620
72					7200	9000	12,000	18,000	22,500
—					—	—	—	—	—
108					10,800	13,500	18,000	27,000	33,750

The "Reinforced concrete ultimate strength" heading spans Class I through Class V.

*Data from 1968 Book of ASTM Standards as follows: clay, Specifications C13 and C200; plain concrete, Specification C14; reinforced concrete, Specification C76, American Society for Testing and Materials, Philadelphia, Pa.

is especially true of flexible pipes since the lateral pressure of the backfill adds appreciably to the strength of the pipe. Angular bedding material provides better support than rounded particles; optimum support is provided by $\frac{1}{4}$ to $\frac{3}{4}$-inch crushed stone.[1]

MATERIALS FOR PRESSURE CONDUITS

The principal pipe materials are steel, cast iron, concrete, wood, asbestos-cement, and vitrified clay. Relative economy plays a large part in the

[1] J. S. Griffiths and C. Keeney, Load-Bearing Characteristics of Bedding Materials for Sewer Pipe, J. Water Pollution Control Federation, Vol. 39, pp. 1–11, April, 1967.

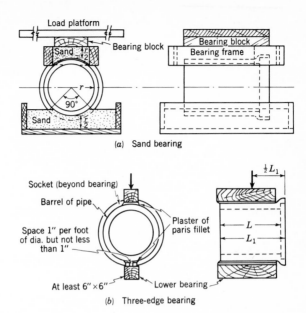

FIG. 11-15   *Methods for determining the crushing strength of pipe. (ASTM Specification C-14)*

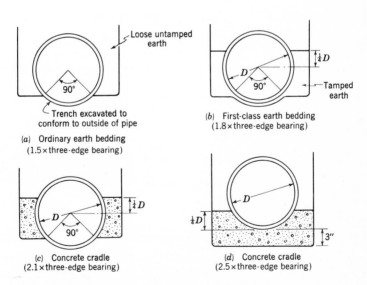

FIG. 11-16   *Some methods of laying sewer pipe and the relative bearing values developed as compared to the three-edge bearing tests.*

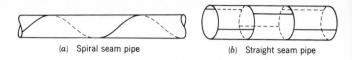

(a) Spiral seam pipe   (b) Straight seam pipe

FIG. 11-17   Two methods of fabricating steel pipe.

selection of pipe material, but availability of skilled labor for construction and accessibility of the site may be influencing factors.

11-18   Steel pipe   Steel pipe has been used in all sizes up to more than 20 ft in diameter. Steel pipe in sizes of 0.5 to 12 in. in diameter is often a continuous tube formed by drawing over a mandrel. In sizes under 42 in., steel pipes are sometimes made of long, narrow steel plates which are bent to shape and welded (or riveted) along a spiral joint (Fig. 11-17a). This type of pipe has considerable flexural strength and can be used for spanning small ravines. Larger sizes are usually built on the job by welding or riveting steel plate (Fig. 11-17b). Welded pipe is much smoother than riveted pipe and with modern welding techniques is superior in strength. Consequently very little new riveted pipe is being installed.

Circumferential corrugations considerably strengthen the pipe wall and permit large-diameter pipes with very thin walls. Such metal pipe is commonly galvanized and manufactured in diameters from 8 through 84 in. A 36-in.-diameter metal corrugated pipe weighs only 51 lb/ft compared to over 500 lb/ft for a rigid pipe of the same size. The lower weight greatly reduces the cost of transporting the pipe from the factory to the field site. Steel bands or stiffening rings are sometimes shrunk on large steel pipes to aid in resisting bursting pressures. The wall thickness required to withstand internal pressures may be computed from Eq. (11-18). The working stress for steel is usually taken as 16,000 psi. Buried steel pipes are not usually provided with expansion joints since they are not subject to large temperature changes. Pipes exposed to the atmosphere may, however, require expansion joints to minimize temperature stresses. A simple expansion joint is shown in Fig. 11-18.

In the range of sizes encountered in engineering practice, steel-pipe sizes vary by 2-in. increments from 4- to 24-in. diameter and by 6-in. increments from 30- to 72-in. diameter. The internal diameter of steel pipe depends on the wall thickness. The life of any pipe material depends very much on the conditions to which it is exposed (Sec. 11-20), but properly protected steel pipe should have a life of at least 40 yr under ordinary conditions. Protection is commonly provided internally by a centrifugally spun coating of cement mortar.

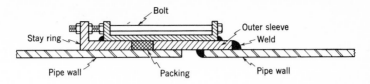

FIG. 11-18   Typical expansion joint for steel pipe.

11-19   Cast-iron pipe   Cast-iron pipe[1] is widely used for city water-distribution systems because of its high resistance to corrosion and consequent long life. Under normal conditions cast-iron pipe can be expected to last 100 yr. The usual length of a pipe section is 12 ft, but lengths up to 20 ft can be obtained. Cast-iron pipe is made in several thickness classes for various pressures up to a maximum of 350 psi. Cast-iron pipes are usually dipped in a bituminous compound for protection against corrosion and to improve their hydraulic qualities; larger sizes may be provided with a lining of cement mortar.

A common joint for cast-iron pipe is the bell and spigot (Fig. 11-19). A few strands of jute are wrapped around the spigot before it is inserted into the bell, and then more jute is packed into the joint. Finally, the space between the bell and spigot is filled with molten lead, which is tightly calked into the joint after cooling. Patented compounds of sulfur and other materials and neat cement mortar are also used for joints. These materials are cheaper than lead, but the joints are usually less flexible. Flanged pipe is

---

[1]Cast-iron pipe is commercially available in diameters 2, 3, and 4 in., by 2-in. increments to 20 in., 24 in., and by 6-in. increments to 60 in.

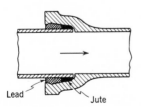

FIG. 11-19   Bell-and-spigot joint for cast-iron pipe.

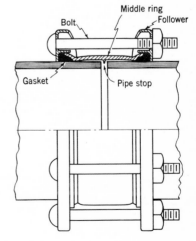

FIG. 11-20   Dresser coupling.

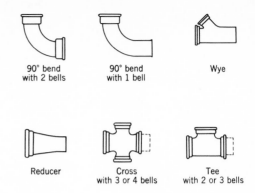

FIG. 11-21   *Some standard bell-and-spigot fittings.*

used for pumping stations, filter plants, and other locations where it may be necessary to disjoint the pipe. Flanged couplings must be fitted perfectly and provided with a gasket if they are to be watertight. Because of the skilled labor required for lead joints, numerous mechanical couplings are finding extensive use and have largely replaced bell-and-spigot joints. These couplings are bolted together and designed to avoid the careful fit required of ordinary flanged couplings and to permit flexibility in pipe placement. One of the most common is the Dresser coupling (Fig. 11-20). Some of the standard fittings required with cast-iron pipe are shown in Fig. 11-21.

11-20   Corrosion of metal pipes   Metal pipes are subject to chemical corrosion. In its simplest form, corrosion occurs when iron enters solution as positive ions and combines with the negative ions of water to form ferrous hydroxide. If the water contains oxygen, the ferrous hydroxide is oxidized to ferric hydroxide, an insoluble, red-brown precipitate. The initial rust coating which forms on the pipe tends to protect it from further corrosion, but the coating is not impermeable, and some corrosion usually continues. Water with a large amount of dissolved carbon dioxide is an active corrosive agent. Corrosion of iron pipe results in the formation of tubercles of ferric hydroxide on the inside of the pipe. This deposit (known as *tuberculation*) decreases the pipe area and increases pipe roughness, thus greatly reducing the hydraulic carrying capacity. Cast-iron pipes of small diameter have had their capacities reduced as much as 50 percent in 5 yr by tuberculation. Such pipes can be reconditioned with a cement lining centrifugally cast in place by special machines.

Corrosion of metal pipes may result from electrolytic action (electrolysis). Electrolysis is often caused by the galvanic action[1] resulting when dissimilar

[1] Rolf Eliassen and James C. Lamb, III, Mechanism of Internal Corrosion of Water Pipe, *J. Am. Water Works Assoc.*, Vol. 45, pp. 1281–1294, 1953.

metals are immersed in water. The rate of electrolysis depends on the dissimilarity of the two metals as indicated by their relative position in the electrochemical series. A metal which is high in the series is dissolved and deposited on the other metal. This type of corrosion may occur in water systems between pipes and fittings of different metals or between the pipe metal and impurities in the pipe metal. Pipes laid in soil that has a high electrical conductivity are particularly vulnerable to electrolysis. *Stray-current* electrolysis may occur where electric currents leave a pipe. This was a serious problem when water mains and electric traction lines were close together. The electrolysis resulting from alternating current is negligible compared with that from direct current because of the reversal of polarity in the alternating-current field.

Corrosion of metal pipes may be reduced or eliminated by protective coatings of paint, galvanizing, bituminous compounds, or cement linings. Red lead paint or zinc pigments offer some protection and are used on the exterior of exposed metal pipes. The rate of corrosion of zinc is about one-fifth that of steel, and galvanizing by dipping the pipe in molten zinc is an effective corrosion control except for highly acid waters. Galvanized pipe is widely used for small service lines in distribution systems but is too expensive for large pipes.

Large pipes are usually protected by a bituminous coating or cement lining. Numerous commercial bituminous compounds are available for both hot and cold application. Two coats are sometimes desirable to increase the thickness of the coating, and the outside of the pipe is often wrapped with asbestos felt saturated with asphalt. It is important that the covering be complete without any bare spots where corrosion can start. Mortar for pipe linings is commonly 1 part cement to 2 parts sand. It may be applied in the factory or field by use of special machines. Cement linings are somewhat more vulnerable to cracking than are bituminous linings.

Electrolytic corrosion is often controlled by *cathodic protection*, the physical act of reversing the electrochemical force. This is often accomplished by making the pipe the negative pole of a direct-current circuit with anodes, usually blocks of magnesium or zinc, buried in the ground near the pipe. Another method employs anodes energized by a direct-current power supply. The anode is connected to the positive terminal of the power source and the pipe to the negative terminal. By making the pipe negative with respect to its surroundings, the current flow is to the pipe, and the migration of metallic ions from the pipe is retarded. Corrosion control is a specialized field and experts should be consulted whenever problems are encountered.

11-21   Concrete pipe   Precast-concrete pipe is available in sizes up to 72 in. diameter, and sizes up to 180 in. have been made on special order. Precast pipes are reinforced except in sizes under 24 in. diameter. The

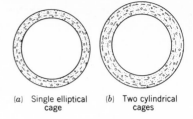

FIG. 11-22   Types of steel reinforcement for concrete pipe.

(a) Single elliptical cage

(b) Two cylindrical cages

reinforcement may take the form of spirally wound wire or elliptical hoops (Fig. 11-22a). In large pipes the reinforcement usually consists of two cylindrical cages (Fig. 11-22b). Precast-concrete pipe is usually made by rotating the form rapidly about the pipe axis. The centrifugal force presses the mortar tightly against the forms and results in a high-density watertight concrete. For low heads, concrete pipe is usually joined with a mortar-calked bell-and-spigot joint, but for high pressures the *lock joint* (Fig. 11-23) or some other special joint is required. For heads above 100 ft a welded steel cylinder is often cast in the pipe for watertightness.

Because of the better control in its manufacture, precast pipe is usually of higher quality and need not be so thick as cast-in-place pipe of the same size. Because of the need to move plant and forms over long distances, cast-in-place pipe is relatively expensive and is normally used only for pipe sizes not available in precast form or where transportation difficulties make use of precast pipe impossible. For gravity flow, No-joint concrete pipe has been developed in California. This pipe has been constructed in sizes 24 to 72 in. A special pipe-laying machine with a slip form is used. Rates of production vary from 40 to 120 ft/hr. Though this pipe is not reinforced, the experience record to date has been good. Concrete pipe should last at least 35 to 50 yr under average conditions. Alkaline water may cause rapid deterioration of thin concrete sections. Concrete pipes carrying wastewater may be subject to sulfide corrosion (Sec. 19-4) and may be short-lived unless proper precautions are taken.

11-22   Vitrified-clay pipe   Clay pipe is not often used as pressure

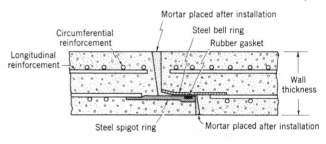

FIG. 11-23   Lock joint for concrete pressure pipe.

pipe but is widely used in sewerage and drainage for flow at partial depth. The main advantage of clay pipe is that it is virtually corrosion-free, has a long life, and its smooth surface provides high hydraulic efficiency. Use of clay pipe under pressure is usually prevented by its low strength in tension and the difficulty of securing watertight joints. The most common joint for clay pipe is the bell-and-spigot flexible compression joint in which precision mated surfaces are in tight contact with one another. Rubber-sleeve couplings held in place with corrosion-resistant steel bands are occasionally used with plain-end clay pipe, but more often the joints of this type of pipe are left open to permit passage of water either into or out of the pipe.

Clay pipe is most commonly made in 3-ft lengths but 2-, 2.5-, and 4-ft lengths can be obtained. Inside diameters vary by 2-in. increments from 4 to 12 in. and by 3-in. increments above 12 in. Clay pipe in diameters greater than 36 in. is rarely used. Because of the dimension changes while the pipe is in the kiln, liberal tolerances in all dimensions are necessary.

11-23 Asbestos-cement pipe Asbestos-cement pipe is made from asbestos, silica, and cement converted under pressure to a dense, homogeneous material possessing considerable strength. The asbestos fiber is thoroughly mixed with the cement and serves as reinforcement. This type of pipe is available in diameters of from 4 to 36 in. in 13-ft lengths. The pipe is made in various grades, the strongest being intended for internal pressures up to 200 psi. Asbestos-cement pipe is assembled by means of a special coupling which consists of a pipe sleeve and two rubber rings which are compressed between the pipe and the interior of the sleeve (Fig. 11-24).

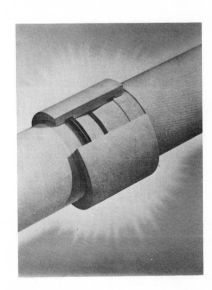

FIG. 11-24   Ring-tite coupling for asbestos-cement pipe. (Johns-Mansville Co.)

The joint is as resistant to corrosion as the pipe itself and is flexible enough to permit as much as 12° deflection in laying pipe around curves.

Asbestos-cement pipe is light in weight and can be assembled without skilled labor. It can be joined to cast-iron pipe with lead or sulfur-base compounds. It is easily cut and can be tapped and threaded for service connections. The hydraulic efficiency of asbestos-cement pipe is high. However, the rubber-joint seals may deteriorate if exposed to gasoline or other petroleum products. The pipe is easily damaged by excavating tools and does not have much strength in bending.

11-24 Miscellaneous types of pipe Other materials used for pipe include copper, wrought iron, plastics, asphaltic fiber, wood, and brick. Copper and wrought iron are used for small-diameter pressure pipe. Copper is quite expensive but may be advantageous in situations where corrosion is likely to occur. Plastic pipe[1] is corrosion free and light in weight. Its low strength, however, precludes it from being used in large sizes. Moreover, it is embrittled by temperatures exceeding 140°F. In sizes up to about 6 in., plastic pipe is finding some use in low-pressure service for street mains, house plumbing, sprinkler systems, and house connections to sewers. A large variety of polymers are used; hence there is a wide variation in the characteristics of various plastic pipes. Since 1965, several types of lightweight, corrosion free, fiber-glass-reinforced pipe have been developed. These have a composite structure of polyester resin mortar reinforced with continuous fiber-glass elements. This pipe has been manufactured in diameters from 8 to 48 in. and in standard lengths of 10 and 20 ft. It has been used for gravity-flow and pressure conduits with pressures as high as 150 psi. About the only use for asphaltic-fiber pipe is as house connections to sewers. Asphaltic-fiber pipe is inexpensive but generally does not hold up well.

APPURTENANCES FOR PRESSURE CONDUITS

11-25 Gates and valves A large number of different types of valves are required for the proper functioning of a pipeline. *Gate valves* are used to regulate the flow in the pipe. These valves (Fig. 11-25) are similar to the gate valves used in dams but not so large. Both manually and mechanically operated gate valves are used. On large pipelines, valves of smaller diameter than the pipe are used for economy. The saving in valve cost must be balanced against the increased head loss through the valve, including contraction and reexpansion.

Check valves (Fig. 11-26) permit flow in one direction only. They may be

---

[1] Robert J. Sweitzer, Developments in Plastics and Plastic Pipe, *J. Am. Water Works Assoc.*, Vol. 52, pp. 1251–1262, 1960.

FIG. 11-25   Gate valve. Outside-screw type, cast-iron gate valve. (Iowa Valve Co.)

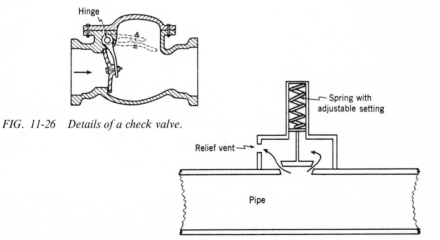

FIG. 11-26   Details of a check valve.

FIG. 11-27   Pressure-relief valve in the open position.

installed on the discharge side of a pump to prevent backflow when the pump is stopped. Check valves are also required at interconnections between a polluted water system and a potable water system to prevent entry of pollution into the pure water. Some state health boards require two check valves and a gate valve at such interconnections for safety. If water discharges from a pipe into a river or tidal estuary where water levels may vary, a form of check valve called a tide gate (Fig. 18-7) may be used to prevent backflow into the pipe when the river or estuary is at a high level. The simplest check valve is a flap which closes under its own weight when flow in the permissible direction ceases. *Drain* or *blow-off valves* are necessary at the low points of a pipeline to permit the pipe to be drained for inspection and repair.

Water-hammer pressures in a pipe can be minimized by use of *pressure-relief valves* (Fig. 11-27), adjusted to open automatically at a predetermined pressure. This type of valve is used on small pipelines where the escape of a relatively small amount of water will alleviate water-hammer pressures. Either a spring or a counterweight may be used to set the opening pressure for relief valves.

Air-inlet valves open automatically when the pressure in a pipe drops below some predetermined value and allow air to enter the pipe. This provides protection against the possible collapse of the pipe when internal pressure is well below the atmospheric pressure.

Figure 11-28 shows water discharging from a reservoir through a pipeline with gate valves at A and B. If valve B is suddenly closed, water-hammer pressures can be relieved by a pressure-relief valve upstream from B. Negative pressure may occur when, with steady flow in the pipe, valve A is closed. Water in the pipe downstream from A will continue to flow, creating a vacuum in the pipe. Thus an air-inlet valve downstream from A is necessary for protection against negative pressures. Relief valves must be properly located to have free access to air at all times and must be protected against jamming by snow, ice, or dust.

Occasionally it is necessary to interconnect a high- and a low-pressure water system. A *pressure-regulating valve* (Fig. 11-29) at the junction permits flow from the high-pressure system to the low-pressure system only when

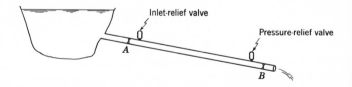

FIG. 11-28   *Air-relief valves installed on a pipeline.*

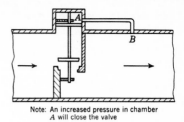

FIG. 11-29   Schematic diagram of a pressure-
             regulating valve.

Note: An increased pressure in chamber
A will close the valve

pressure on the low-pressure side is not excessive. When the pressure at B is
too high, the pressure in chamber A closes the valve.

11-26   Surge tanks   Surge tanks are installed on large pipelines to
relieve excess pressure caused by water hammer and to provide a supply of
water to reduce negative pressure if a valve is suddenly opened. A simple
surge tank is a vertical standpipe connected to a pipeline (Fig. 11-30). The
valve might represent turbine gates which may open or close rapidly with
changes in load on the generators. With steady flow in the pipe, the water
level y_1 in the surge tank is below the static level ($y = 0$). When the valve
is suddenly closed, water rises in the surge tank. The water surface in the
tank then fluctuates up and down until damped out by fluid friction.

The energy equation for unsteady flow, neglecting fluid friction and
velocity head in the surge tank and loss at pipe and surge-tank entrances,
can be written as

$$y + f\frac{L}{D}\frac{V^2}{2g} + \frac{L}{g}\frac{dV}{dy}\frac{dy}{dt} = 0 \tag{11-35}$$

and the continuity equation as

$$AV = A_s\frac{dy}{dt} \tag{11-36}$$

where y is the water level in the surge tank measured from the static level
(positive upward), L, f, and D are characteristics of the pipe between the

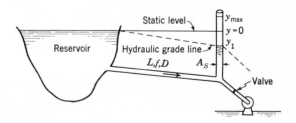

FIG. 11-30   Definition sketch for a surge-tank analysis.

tank and the reservoir, and A_s is the cross-sectional area of the surge tank. Combining these two equations, integrating, and solving for V yields

$$V^2 = \frac{2gAD^2}{LA_s f^2}\left(1 - \frac{fA_s}{AD}y\right) - Ce^{-(fA_s/AD)y} \tag{11-37}$$

which expresses the relation between velocity in the pipe and water-surface level in the tank over the interval from valve closure to the top of the first surge. Equation (11-37) may be used to estimate the maximum height of surge y_{max} by finding the constant of integration C for steady-state conditions at the instant of closure ($y = y_1$) and then solving for y_{max} when $V = 0$. Since the derivation neglected tank friction and velocity head and entrance losses and assumed instantaneous valve closure with all water-hammer effects dissipated in the tank and not the pipe, the estimate of y_{max} will be too large. However, the results provide a conservative estimate for preliminary design of simple surge tanks.

Surge tanks are usually open at the top and of sufficient height so that they will not overflow. In some cases overflow is permitted if the water can be disposed of without damage. The simple surge tank is often adequate for low heads, but under high heads some modifications to improve the damping action and minimize cost are desirable. A restricted entry (Fig. 11-31a) increases the damping action and reduces the initial surge. If the tank is closed at the top, the air cushion absorbs a portion of the water hammer and the tank need not be so high as a simple surge tank. Differential surge tanks (Fig. 11-31c and d) have a vertical riser about the same diameter as the pipe. Flow of water into the main tank is limited by the capacity of openings in the riser or an opening around the base of the riser. Water levels fluctuate more rapidly in the riser than in the tank, with the result that the fluctuations are out of phase and the oscillations in the riser are damped out more quickly than in a simple surge tank. The diameter of the outer tank of a differential

FIG. 11-31   Types of surge tanks.

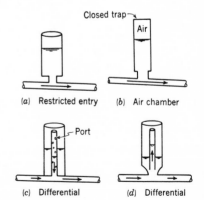

(a) Restricted entry   (b) Air chamber

(c) Differential   (d) Differential

surge tank need be only about 70 percent as great as that of a simple surge tank to achieve the same effect.

The bottom of the surge tank must be far enough below reservoir level so that the tank contains water at all times to prevent air from entering the pipe. Surge tanks reduce the pressures only in the pipe from the tank to the reservoir, and hence the tank should be as close to the turbine as possible. A surge tank is not economical on an extremely high-head line[1] because of the great height of tank required. In such cases, a decrease in delivery of water to the turbine is best accomplished by use of deflection nozzles or bypasses, which do not require an immediate decrease in flow through the pipe itself. Surge tanks are often built partially or wholly underground. An underground surge tank on the Appalachia project of the Tennessee Valley Authority (TVA) is 233 ft high with a riser 16 ft in diameter and a tank 66 ft in diameter. Exposed surge tanks are usually built of steel or reinforced concrete. Surge chambers for tunnels may be excavated in the rock above the tunnel if geologic conditions are favorable. In cold climates a surge tank must be protected from freezing. Electric heating units have been used successfully for this purpose.

INVERTED SIPHONS

11-27   The term *inverted siphon* is applied to a pressure pipe carrying the flow of a canal or sewer across a depresssion. Actually no siphon action is involved, and the terminology *sag pipe* or *depressed sewer* would be both more descriptive and more accurate. Flow in a sag pipe is under pressure and follows the principles of flow in pressure conduits. Assuming that the elevation of the water surface at entrance and exit is fixed, the design of a sag pipe for clear water involves nothing more than selecting a pipe size which will carry the required maximum flow with a head loss no greater than the difference in entrance and exit elevations. A suitable transition structure between the open channel and the sag pipe at entrance and exit is also required.

If the water contains suspended solids, the minimum velocity in the sag pipe should be great enough to prevent deposition of these solids at the bottom of the sag. This usually requires a minimum velocity of about 3 ft/sec. If the flow rate in the sag pipe is to be reasonably steady, no difficulty will be encountered in maintaining adequate velocity. If the flow is to be quite variable, it may become difficult to select a single pipe which will assure a satisfactory velocity at low flow rates and a satisfactory head loss at high

[1] Charles Jaeger, Economics of Large Modern Surge Tanks, *Water Power*, Vol. 10, No. 5, pp. 172–177, May, 1958.

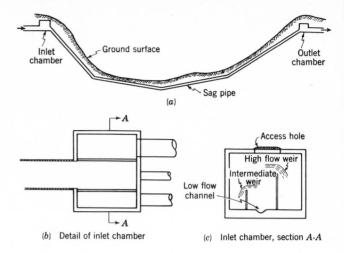

(a)

(b) Detail of inlet chamber

(c) Inlet chamber, section A-A

FIG. 11-32   A multiple-pipe inverted siphon, or sag pipe.

flow rates. In this case, two or more pipes will be used, the smallest designed to carry the minimum flow at adequate velocity and the larger pipe or pipes designed to carry additional increments of flow. In the case of multiple pipes, flow from the main channel would be directed into the smallest pipe until the flow becomes so large that it can spill over side weirs into the next larger pipe. The exit must be so designed that water from the small pipe cannot back up and flow into a larger pipe. Figure 11-32 illustrates a typical layout for a multiple-pipe installation.

BIBLIOGRAPHY

Camp, T. R., and J. C. Lawler: Water Distribution, sec. 37 in C. V. Davis and K. E. Sorenson (eds.), "Handbook of Applied Hydraulics," 3d ed., McGraw-Hill, New York, 1969.

"Cast Iron Pipe Handbook," Cast Iron Pipe Research Association, 2d ed., Chicago, 1952.

"Clay Pipe Engineering Manual," National Clay Pipe Institute, Los Angeles, 1968.

"Concrete Pipe Handbook," American Concrete Pipe Association, Arlington, Va., 1967.

Daugherty, R. L., and J. B. Franzini: "Fluid Mechanics with Engineering Applications," 6th ed., McGraw-Hill, New York, 1965.

Hinds, Julian: Canals, Flumes, Covered Conduits, Tunnels, and Pipe Lines, sec. 10 in C. V. Davis (ed.), "Handbook of Applied Hydraulics," 2d ed., McGraw-Hill, New York, 1952.

Parmakian, John: "Waterhammer Analysis," Prentice-Hall, Englewood Cliffs, N.J., 1955.

Rich, G. R.: Water Hammer, sec. 27, and Surge Tanks, sec. 28, in C. V. Davis and K. E. Sorenson (eds.), "Handbook of Applied Hydraulics," 3d ed., McGraw-Hill, New York, 1969.

Streeter, V. L., and E. B. Wylie: "Hydraulic Transients," McGraw-Hill, New York, 1967.

Troskolanski, Adam T.: "Hydrometry—Theory and Practice of Hydraulic Measurements," Pergamon Press, New York, 1960.

Uhlig, H. H.: "Corrosion and Corrosion Control," Wiley, New York, 1963.
Vennard, J. K.: "Elementary Fluid Mechanics," 4th ed., Wiley, New York, 1961.

PROBLEMS

11-1 Find the head loss in 100 ft of 3-ft-diameter welded-steel pipe carrying 15 cfs of water at 60°F by each of the three equations given in the text.

11-2 When first installed between two reservoirs a 4-in.-diameter metal pipe of length 6000 ft conveyed 0.20 cfs of water. If after 15 yr a chemical deposit had reduced the effective diameter of the pipe to 3.0 in., what then would be the flow rate? Assume f remains constant. Assume no change in reservoir levels. What would be the flow rate if in addition to the diameter change, f had doubled in value?

11-3 In using the Darcy-Weisbach equation for flow in a pressure conduit, what percentage error is introduced in Q when f is misjudged by 20 percent?

11-4 Find the head loss in a pipeline consisting of 200 ft of 4-in. steel pipe, a 90° bend on 24-in. radius, 4-in. gate valve (fully open), 100 ft of 4-in. steel pipe, expansion to 6 in. with a 20° taper, 300 ft of 6-in. steel pipe, abrupt contraction to 3-in. diameter, and 50 ft of 3-in. steel pipe. The discharge rate is 1.5 cfs.

11-5 What is the maximum theoretical flow rate for a pump situated 15 ft above a reservoir if the suction pipe is 100 ft of 8-in. steel pipe ($e = 0.00015$ ft), the discharge pipe is 500 ft of 6-in. cast-iron pipe, and discharge occurs 20 ft above the pump center line? Find also the maximum desirable flow rate. Assume elevation 5000 ft, water temperature 60°F.

11-6 Three pipes, each of length 200 ft with $f = 0.03$, have the following diameters: AB, 2 ft; BC, 1 ft; and AC, 3 ft. A flow of 15.0 cfs enters junction A and splits up with 6 cfs going through pipe AB and 9 cfs through AC. Find the rate of flow in pipes BD, CE, and BC. Neglect head loss at junctions.

11-7 In a reservoir system such as that of Fig. 11-4, $z_a = 200$ ft, $z_b = 230$ ft, $z_c = 120$ ft, $z_d = 140$ ft, $BD = 3000$ ft of 4-in. cast-iron pipe, $AD = 2000$ ft of 1-in. steel pipe, $DC = 500$ ft of 6-in. cast-iron pipe. Using $f = 0.025$ and neglecting minor losses, determine the flow in each pipe.

11-8 Three reservoirs A, B, and C whose water surface elevations are 50, 40, and 22 ft respectively are interconnected by a pipe system with a common junction at elevation 25 ft. The pipes are as follows: from A to junction, $L = 800$ ft, $d = 3$ in.; from B to junction, $L = 500$ ft, $d = 10$ in.; from C to junction, $L = 1000$ ft, $d = 4$ in. Assume $f = 0.02$ for all pipes and neglect minor losses and velocity heads. Determine whether the flow is into or out of reservoir B.

11-9 A pipe system connects two reservoirs whose difference in elevation is 15 m. The pipe system consists of 300 m of 60 cm concrete pipe, branching into 600 m of 30 and 45 cm pipe in parallel, which join again to a single 60-cm line 1500 m long. What would be the flow in each pipe? Assume $f = 0.030$ for all pipes.

11-10 What single pipe 1000 ft long would be the equivalent of a network $ABCD$ if $AB = 500$ ft of 8-in. pipe, $BC = 500$ ft of 24-in. pipe, $CD = 1000$ ft of 36-in. pipe, and $DA = 800$ ft of 12-in. pipe? Flow enters the system at A and leaves at C. Assume f is the same for all pipes.

11-11 Refer to Fig. 11-6. Suppose $p_A/\gamma = 6.5$ ft, $p_D/\gamma = 20$ ft, $z_A = z_D$. A pump in the 4000-ft pipe (flow from left to right) develops 30 ft of head. Find the flow rate in each pipe. Assume $n = 0.013$ for all pipes.

11-12 A pipe network similar in plan to Fig. 11-7 consists of pipes as follows: AB, 5000 ft, 12-in.; BC, 3000 ft, 6-in.; CD, 8000 ft, 24-in.; DE, 7000 ft, 8-in.; EA, 4000 ft, 10-in.; BD, 7000 ft, 12-in. A flow of 4000 gpm enters the system at D, while outflow at the junctions is as follows:

A, 500 gpm; *B*, 300 gpm; *C*, 1000 gpm; *E*, 2200 gpm. Find the flow in each pipe and the pressure at each junction if the head at *D* is 400 ft. Take $f = 0.023$.

11-13 A pipe network consists of the following pipes: *AB*, 3000 ft, 8-in.; *BC*, 3000 ft, 12-in.; *CD*, 10,000 ft, 36-in.; *DE*, 8000 ft, 24-in.; *EF*, 5000 ft, 6-in.; *FA*, 4000 ft, 8-in.; *BE*, 10,000 ft, 6-in.; *BF*, 8000 ft, 12-in. Inflow at *D* = 6 cfs. Outflows are *A*, 2 cfs; *B*, 1 cfs; *E*, 1.5 cfs; *C*, 0.5 cfs; *F*, 1 cfs. Assume the Manning *n* = 0.015, and find the flow in each pipe and the pressure at each junction if the pressure at *D* is 120 psi.

11-14 Compute the horsepower the pump of Illustrative Example 11-2 must put into the water when the flow rate is 72 cfs. Assume $z_B = 18$ ft, diameter of nozzle = 14 in. Under these conditions what is the pressure in the pipe on the suction side of the pump?

11-15 A steel pipeline ($e = 0.00015$ ft) 2 ft in diameter and 2 mi long discharges freely at its lower end under a head of 200 ft. What water-hammer pressure would develop if a valve at the outlet were closed in 5 sec? 60 sec? Wall thickness = 0.2 in. For both cases of closure compute the stress that would develop in the walls of the pipe. If the working stress of steel is taken as 16,000 psi, what would be the minimum time of safe closure?

11-16 A horizontal steel pipe 4 ft in diameter and 6500 ft long takes off from a reservoir 200 ft below the reservoir water surface through a square-edged entrance and extends to a valve that is to be closed in 20 sec. During closure the flow rate is reduced from 275 to 0 cfs. Express the required thickness of the pipe in terms of *x*, the distance from the reservoir. Use 16,000 psi as the working stress of steel. Assume $u_p = 3000$ ft/sec.

11-17 What force would be exerted by the water in a pipe at a point where a 120-cm pipe changes to a 100-cm pipe if the discharge is 10 cu m/sec and the pressure head just upstream from the transition is 15 m?

11-18 What resisting force would be required of an anchor at a 60° bend (deflection) in a 24-in. pipe carrying 40 cfs of water if the pressure is 20 psi?

11-19 A 1000-ft length of 24-in. steel pipe is laid at a time when the temperature is 85°F. If the minimum temperature expected in the line is 32°F, what force can be expected in each bolt of a flanged connection at one end? Assume both ends of the pipe are firmly anchored. A standard 24-in. pipe flange has sixteen $\frac{5}{8}$-in. bolts. The pipe is of no. 10 gage steel (0.1345 in. thick). What is the stress in the pipe?

11-20 What is the maximum bending stress in a 12-in. steel pipe with an unsupported span of 20 ft? The pipe has a wall thickness of 0.1793 in. and a weight of 23.21 lb/ft. Assume the pipe to be full of water and that there are free end conditions.

11-21 Find the external loads on a 2-ft-diameter rigid pipe from 12 ft of sand and gravel backfill when the pipe is placed in trenches of various widths. Plot load per foot versus trench width, letting trench widths range from 3 ft to infinity. Assume the base of the trench is un-yielding. At what minimum trench width does *embankment* condition occur?

11-22 Repeat Prob. 11-21 for depths of fill of 6 and 16 ft.

11-23 A 4-ft-diameter flexible pipe is used as a culvert in a highway fill. The pipe is covered with 8 ft of backfill ($\gamma = 110$ lb/cu ft). A four-wheeled vehicle is to pass over the fill. Each wheel load can be considered to be a concentrated load of 6000 lb. In plan, the four loads form a rectangle 6 by 12 ft. Find the maximum load that would be exerted on a 1-ft length of this pipe. Assume average conditions, and neglect any effect of a pavement placed on top of the fill. What effect would a pavement have? Repeat for case where depth of cover is 4 ft.

11-24 What type of bedding would you recommend for a 24-in.-diameter standard-strength clay pipe when in a 3-ft trench under 12 ft of cover? When in a 4-ft trench under 12 ft of cover? When in a 3-ft trench under 18 ft of cover? In all cases assume backfill is saturated topsoil.

11-25 What class of 4-ft-diameter reinforced-concrete pipe is required in a trench of width 5 ft if ordinary earth bedding is to be used? The backfill is clay ($\gamma = 120$ lb/cu ft) to a depth of 20 ft.

11-26 A 48-in. steel pipe 3000 ft long supplies water to a small power plant. What height would be required for a simple surge tank 6 ft in diameter situated 50 ft upstream from the valve at a point where the center line of the pipe is 120 ft below the water surface in the reservoir if the tank is to protect against instantaneous closure of the wicket gates at the plant? The gates are 150 ft below reservoir level, and the discharge is 250 cfs. Take $f = 0.014$. There is a bell-mouthed entrance to the pipe from the reservoir. The surge tank is not to overflow.

11-27 Repeat Prob. 11-26 for the case where the surge tank is to have a diameter of 10 ft.

11-28 Using the data of Prob. 11-26 find the diameter of surge tank that will produce a surge requiring a tank height of 180 ft.

Chapter 12

Hydraulic machinery

The two types of hydraulic machinery of interest to the water-resources engineer are pumps and turbines. A pump converts mechanical energy into hydraulic energy, while a turbine serves the opposite purpose. Though most machines are single purpose, the reversible pump-turbine[1] operates as either a pump or turbine depending on the direction of rotation. There are many types of pumps and turbines. Each type has its own characteristics, and for a given set of operating conditions there is a type and size of hydraulic machine best suited to the job. The design of hydraulic machinery is a highly specialized field, and the emphasis of this chapter will be on the selection of a machine for a given use.

TURBINES

There are two basic types of turbines. In the *impulse turbine*, a free jet of water impinges on the revolving element of the machine, which is exposed to atmospheric pressure. In a *reaction turbine*, flow takes place under pressure in a closed chamber. Although the energy delivered to an impulse turbine is all kinetic, while the reaction turbine utilizes pressure energy as well as kinetic energy, the action of both turbines depends on a change in the momentum of the water so that a dynamic force is exerted on the rotating element, or *runner*.

12-1 Description of impulse turbines The impulse turbine (Fig. 12-1) is sometimes called a tangential waterwheel or a Pelton wheel after the man who developed the basic design now in use. The wheel has a series of

---

[1] Thaddeus Zowski, Reversible Pump-Turbines, sec. 26 in C. V. Davis and K. E. Sorensen (eds.), "Handbook of Applied Hydraulics," 3d ed., pp. 78–90, McGraw-Hill, New York, 1969.

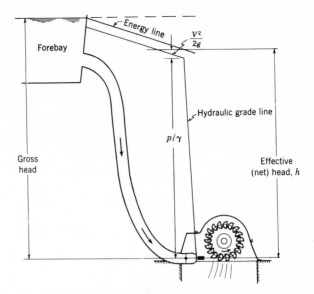

FIG. 12-1 Definition sketch for impulse-turbine installation.

split buckets located around its periphery (Fig. 12-2). The buckets may be individually cast and bolted to the central spider, or the entire runner may be cast as a single unit. When the jet strikes the dividing ridge of the bucket, it is split into two parts, which discharge at both sides of the bucket. Only one jet is used on small turbines, but two or more jets impinging at different points around the wheel are often used on large units. The jets are usually produced by a *needle nozzle* similar to a needle valve (Sec. 9-23). The speed of the wheel is kept constant under varying load by changing the discharge of the jet.

Since water hammer might occur in the supply pipe if the nozzle were suddenly closed, some nozzles are provided with a bypass valve which opens whenever the needle valve is closed quickly. The same effect can be obtained with a jet deflector, whose position can be adjusted to deflect the jet away from the wheel when the load drops. A governor is required to actuate the nozzle and bypass or deflector units. A simple governing mechanism is shown in Fig. 12-3. If the speed of the rotor increases, *A* rises and permits oil to enter chamber *B*, thus closing the nozzle slightly. If speed drops, the operation is reversed. The actual governing mechanism is much more complicated since a compensating arrangement is required to prevent over-travel of the servomotor piston to provide very sensitive regulation.

The generator rotor is usually mounted on a horizontal shaft between two bearings with the runner installed on the projecting end of the shaft.

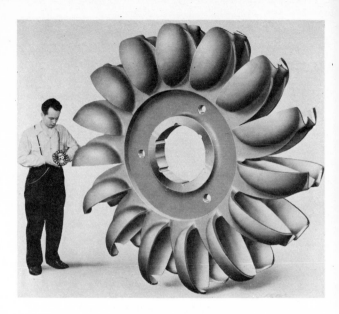

FIG. 12-2   Impulse wheels. (Baldwin, Lima, Hamilton Co.)

This is known as a *single-overhung* installation. Often runners are installed on both sides of the generator (*double-overhung* construction) to equalize the bearing loads. Impulse turbines are provided with housings to prevent splashing, but the air within the housing is substantially at atmospheric pressure. Some modern wheels are mounted on a vertical axis below the generator and are driven by jets from several nozzles spaced uniformly around the periphery of the wheel. This arrangement simplifies shaft and bearing design as well as details of the jets and jet deflectors, though piping for the multiple nozzles becomes complicated.

For good efficiency the width of the bucket should be three to four times the jet diameter and the wheel diameter 15 to 20 times the jet diameter.

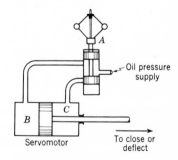

FIG. 12-3   Schematic diagram of a simple governing mechanism.

The diameters of impulse turbines range up to about 15 ft. Theoretically maximum efficiency would result if a bucket completely reversed the relative velocity of the jet. This is not possible because the water must be deflected to one side to avoid interfering with the following bucket, and the bucket angle β is usually about 165° (Fig. 12-4b). The notch in the bucket prevents the jet from striking the back of the bucket as it moves into position.

12-2   Action of the impulse turbine   Figure 12-4a shows a portion of an impulse turbine rotating with a velocity u at the center line of the buckets. A plan view of one of the buckets is shown in Fig. 12-4b. Position A represents the instant of entry of water into the bucket, while position B indicates the moment the water leaves the bucket. The true path of the water is shown, and its velocity changes from V_1 at entry to V_2 at exit. A vector diagram of the velocities at entry is shown in Fig. 12-4c. The vector v represents the velocity of the water relative to the bucket. Assuming friction to be negligible, the magnitude of the water velocity relative to the bucket remains constant, but its direction at discharge must be tangential to the bucket (Fig. 12-4d). Applying the impulse-momentum principle, the force exerted by the water on the bucket in the direction of motion is

$$F = \rho Q(V_1 - V_2 \cos \alpha) \qquad (12\text{-}1)$$

where Q is the discharge of the nozzle. In terms of relative velocities

$$F = \rho Q(v - v \cos \beta) = \rho Q(V_1 - u)(1 - \cos \beta) \qquad (12\text{-}2)$$

since the relative velocity $v = V_1 - u$.

The power transmitted to the buckets from the water is the product of the force and the velocity of the body on which the force is acting. Hence,

$$P = Fu = \rho Q(V_1 - u)(1 - \cos \beta)u \qquad (12\text{-}3)$$

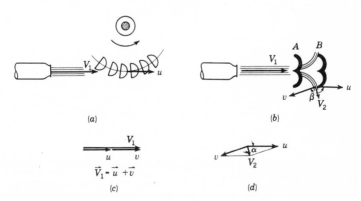

(a)   (b)

$V_1 - \vec{u} + \vec{v}$

(c)   (d)

FIG. 12-4   *Hydraulic relationships for an impulse turbine.*

There is no power developed when $u = 0$ or when $u = V_1$. For a given turbine and jet the maximum power occurs at an intermediate u which can be found by differentiating Eq. (12-3) and equating to zero. Thus

$$\frac{dP}{du} = \rho Q(1 - \cos \beta)(V_1 - u - u) = 0 \tag{12-4}$$

from which $u = V_1/2$. Thus the greatest hydraulic efficiency (neglecting fluid friction) occurs when the peripheral speed of the wheel is half of the jet velocity. Tests of impulse turbines show that, because of energy losses, the best operating conditions occur when u/V_1 is between 0.43 and 0.48.

Since the nozzle is considered an integral part of a turbine, the effective head h acting on the turbine is the sum of the pressure head p/γ and the velocity head at the inlet to the nozzle as shown in Fig. 12-1. The brake horsepower delivered by a turbine to the generator can be expressed as

$$\text{Brake horsepower} = \frac{\gamma Q h e}{550} \tag{12-5}$$

where e is the overall efficiency of the unit. Energy losses in an impulse turbine include loss through the nozzle, hydraulic-friction and eddy losses in the bucket, kinetic energy of the water leaving the buckets, bearing friction, and air resistance. The peak efficiency of well-designed impulse wheels is on the order of 85 to 90 percent.

12-3 Reaction turbines The flow through a reaction turbine may be radially inward, axial, or mixed (partially radial and partially axial). The two types of reaction turbines in general use are the *Francis* turbine and the *propeller* turbine. In the usual Francis turbine water enters the scroll case (Fig. 12-5) and moves in to the runner through a series of guide vanes (Fig. 12-6) with contracting passages which convert pressure head to velocity head. These vanes are adjustable so that the quantity and direction of flow can be controlled. The vanes, known as *wicket gates*, are operated by moving a shifting ring to which each gate is attached. If the load on the turbine drops rapidly, the governor actuates the mechanism which closes the gates. A relief valve or a surge tank is generally necessary to prevent serious water-hammer pressures.

Flow through the usual Francis runner (Fig. 12-7) is at first inward in the radial direction, gradually changing to axial. Francis turbines are usually mounted on a vertical axis, although horizontal axes are sometimes used. The scroll case of a Francis turbine is designed to decrease the cross-sectional area in proportion to the decreasing flow rate passing a given section and thus to maintain constant velocity. For heads under about 40 ft, Francis turbines are often used with an open flume setting (Fig. 12-8).

FIG. 12-5   *Scroll case for a reaction turbine being bored. (Baldwin, Lima, Hamilton Co.)*

FIG. 12-6   *Wheel-case assembly for a reaction turbine, showing wicket gates and gate-control mechanism. (Allis Chalmers Mfg. Co.)*

FRANCIS
TURBINE
(runner)

FIG. 12-7   *Cast-steel reaction turbine runner for the Cheoah Plant of the Carolina Aluminum Co. Designed for 31,000 hp at a head of 188 ft. (Allis Chalmers Mfg. Co.)*

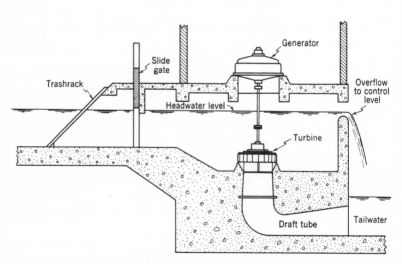

FIG. 12-8   *Typical open flume setting for a reaction turbine at low heads.*

After leaving the runner, the water enters a *draft tube* which has a gradually increasing cross-sectional area to reduce the discharge velocity. The simplest and most efficient draft tube is the vertical type (Fig. 12-9a), which is usually of steel plate. The elbow draft tube is often formed in the concrete of the powerhouse structure (Fig. 12-9b) and is used when space does not permit the vertical type. The draft tube reduces the discharge velocity and hence the loss of head at exit. To prevent separation of flow from the walls of the draft tube, the vertex angle of the cone of divergence should be less than 10°. To prevent cavitation, the vertical distance z_1 (Fig. 12-10) from the tailwater level to the draft-tube inlet should be limited (Sec. 12-6) so that at no point within the turbine will the absolute pressure drop to the vapor pressure of the water.

The propeller turbine (Fig. 12-11), an axial-flow machine with its runner confined in a closed conduit, is commonly set on a vertical axis, though it may be set on a horizontal or slightly inclined axis. The usual runner has four to eight blades mounted on a hub, with very little clearance between the blades and the conduit wall. The blades have free outer ends like a marine propeller. A *Kaplan* turbine is a propeller turbine with movable blades whose pitch can be adjusted to fit existing operating conditions. The adjustment is accomplished by a mechanism in the runner hub which is actuated hydrauli-

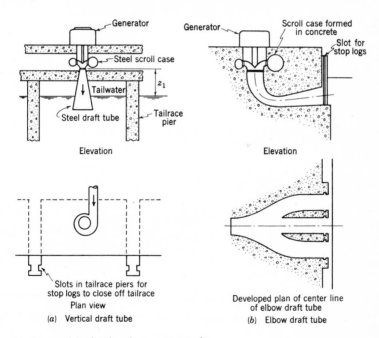

FIG. 12-9   *Two types of draft tubes for reaction turbines.*

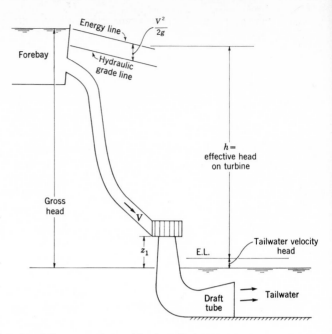

FIG. 12-10   *Definition sketch for a reaction-turbine installation.*

cally by the governor in synchronization with guide-vane adjustments. With an axial-flow turbine, the generator may be set outside the water passageway as in Fig. 12-11; or it may be placed in a streamlined watertight steel housing mounted in the center of the passageway.

12-4   Action of the reaction turbine   A general understanding of the action of a Francis turbine can be obtained by considering the flow through a radial-flow runner (Fig. 12-12). The velocity of the water as it enters the blade is essentially tangent to the exit end of the guide vane. The velocity at the entrance edge of the runner blade is $u_1 = \omega r_1$, where ω is the rotative speed of the runner in radians per second. For best operating conditions the water leaving the guide vanes should pass smoothly into the runner. To accomplish this, the rotative speed of the runner must be such that the velocity v_1 of the water relative to the blade is tangential to the blade. From the vector diagram the component of V_1 tangential to the runner at entry is

$$V_{t_1} = \omega r_1 + V_{r_1} \cot \beta_1 \tag{12-6}$$

where V_{r_1} is the radial component of V_1. At exit,

$$V_{t_2} = \omega r_2 + V_{r_2} \cot \beta_2 \tag{12-7}$$

Assuming that the force exerted on all the blades is the same, the torque T exerted on the runner by the water is

$$T = \rho Q(V_{t_1} r_1 - V_{t_2} r_2) \tag{12-8}$$

where Q is the total discharge through the turbine. Hence, the power P transmitted to the turbine from the water is

$$P = T\omega = \rho Q \omega (V_{t_1} r_1 - V_{t_2} r_2) \tag{12-9}$$

The radial components V_{r_1} and V_{r_2} necessary to solve Eqs. (12-6) and (12-7) are found from the continuity equation

$$Q = 2\pi r_1 Z V_{r_1} = 2\pi r_2 Z V_{r_2} \tag{12-10}$$

FIG. 12-11   Cross section of a 13,500-hp propeller turbine installation. (Allis Chalmers Mfg. Co.)

where Z represents the height of the turbine blades.

A similar analysis may be used for a mixed-flow runner in which the direction of flow changes from radial to axial. The analysis must, however, be modified to account for the fact that the inflow and outflow velocity vectors do not lie in a plane perpendicular to the axis of the runner. An analysis of a propeller turbine is rather complex and will not be presented here.

The effective head h acting on a reaction turbine is defined in Fig. 12-10. The overall efficiency e of a reaction turbine is sometimes expressed as the product of the hydraulic efficiency e_h, the mechanical efficiency e_m, and the volumetric efficiency e_v. Thus

$$e = e_h e_m e_v \qquad (12\text{-}11)$$

Hydraulic efficiency accounts for hydraulic-friction and eddy losses through the scroll case, guide vanes, runner, and draft tube and the kinetic energy at draft-tube exit. Mechanical efficiency accounts for bearing friction and *disk friction*, the drag on the runner in the clearance spaces, while volumetric efficiency accounts for water which bypasses the runner through the clearance spaces without doing any work. Usually less than 2 percent of the water

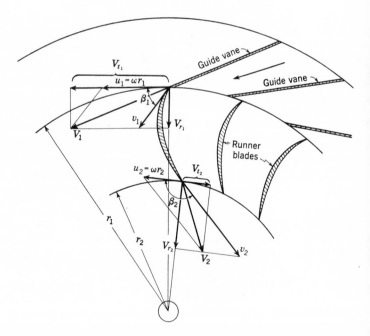

FIG. 12-12   Vector diagrams for a reaction turbine.

entering a reaction turbine is ineffective. Under best operating conditions, reaction turbines have shown efficiencies between 90 and 95 percent.

12-5 Turbine laws and specific speed If the peripheral speed of a turbine runner is expressed as $u_1 = \phi\sqrt{2gh}$, then

$$\phi = \frac{u_1}{\sqrt{2gh}} = \frac{\omega r_1}{\sqrt{2gh}} = \frac{(2\pi N/60)(D/2)}{\sqrt{2gh}} = \frac{DN}{153.3\sqrt{h}} \qquad (12\text{-}12)$$

where D is the nominal diameter of the runner in feet, N the rotative speed in revolutions per minute, and h the total effective head in feet. For any turbine there is a value of ϕ which gives the highest efficiency. Designating this as ϕ_e, its magnitude for the various turbine types is about

> Impulse wheel,[1] 0.43 to 0.48
> Francis turbine, 0.6 to 0.9
> Propeller turbine, 1.4 to 2.0

If it is desired to operate a particular turbine at peak efficiency, that is, $\phi = \phi_e$, Eq. (12-12) shows that the rotative speed N of the turbine should vary with $\sqrt{h}$. Since most turbines are connected to electric generators, it is necessary that N be constant.[2] In order to operate a turbine at constant speed, it is apparent that ϕ cannot be maintained constant if there is a variation in head, and thus efficiency will drop if the head is either more or less than the design head for the given load. An increase in load creates an increase in torque against which a turbine acts. This causes the turbine to slow down, but opening the needle valve or wicket gates to increase Q will maintain the rotative speed constant. Decreases in load will be accompanied by a corresponding reduction in the opening of the gates to maintain constant speed. Under normal operating conditions, the wicket gates of a reaction turbine are about three-fourths open to permit an increase in gate opening should there be an increase in the power demand. A change in gate opening changes the velocity vectors at entrance and exit from the runner. Water will no longer enter the blade tangentially and the resulting turbulence will increase the head loss in the runner. These factors cause a drop in the efficiency of a Francis turbine when operating at part load (Fig. 12-13). A fixed-blade propeller turbine operates even more poorly at part load than a Francis turbine. A Kaplan turbine, however, maintains good efficiency under part load since the pitch of the blades can be adjusted to the changing

---

[1] Note that ϕ_e is approximately equal to the ratio u/V_1 of Sec. 12-2.

[2] In the United States, 60-cycle current is most common, and under such conditions the rotative speed of the turbine in revolutions per minute is given by $N = 7200/n$, where n is the number of poles in the generator and must be an even integer. Most generators have from 12 to 96 poles.

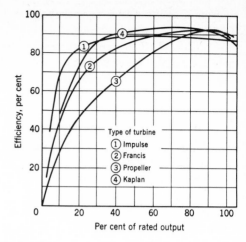

FIG. 12-13   Efficiency vs. load for typical turbines.

flow conditions. Variation in flow results in some change in the energy loss of an impulse turbine and a corresponding change in efficiency, but the effect is not large until the load falls far below the rated output.

If a comparison is made between two turbines of identical geometric shape but of different size, it can be concluded that, since $A \propto DZ \propto D^2$ and $V_r \propto V \propto (2gh)^{1/2}$,

$$Q \propto D^2 h^{1/2} \tag{12-13}$$

Since power $P = \gamma Qhe$, it follows that

$$P \propto D^2 h^{3/2} = KD^2 h^{3/2} \tag{12-14}$$

where K is a constant of proportionality. Equating the D's of Eqs. (12-12) and (12-14),

$$\frac{153.3\phi h^{1/2}}{N} = \sqrt{\frac{P}{Kh^{3/2}}} \tag{12-15}$$

or

$$153.3\phi K^{1/2} = \frac{NP^{1/2}}{h^{5/4}} \tag{12-16}$$

Assuming the turbine to operate at its most efficient speed ($N = N_e$) and letting $153.3\phi_e K^{1/2} = n_s$ gives

$$n_s = \frac{N_e P^{1/2}}{h^{5/4}} \tag{12-17}$$

where n_s is the *specific speed* of the turbine; N_e is expressed in rpm, P in horsepower, and h in feet.

Two turbines of identical geometric shape but of different size will have the same specific speed. Hence, n_s serves to classify a turbine as to type. Specific speeds in the range of best efficiency for the various types of turbines are shown in Fig. 12-14. An impulse wheel is a low-speed turbine, i.e., it has a lower rotative speed than a Francis turbine when both are developing the same power under the same head at optimum operating conditions. Comparing an impulse turbine ($n_s = 6$, $\phi = 0.45$) and a Francis turbine ($n_s = 24$, $\phi = 0.75$) for the same output and head, the rotative speed of the Francis turbine will be about four times that of the impulse wheel. The relative size of the two turbines is determined from Eq. (12-12) as

$$\frac{\phi_I}{\phi_F} = \frac{D_I N_I}{D_F N_F} = \frac{6D_I}{24 D_F} = \frac{0.45}{0.75} \quad \text{from which} \quad D_I = 2.4 D_F$$

Thus the rotative speeds vary in the ratio of $4:1$, while the respective diameters vary in the ratio of $1:2.4$. This is not surprising since water is in contact with only a small portion of an impulse wheel at any instant, while the passages of a Francis turbine are filled with water at all times. Hence, more power is developed by a reaction wheel of a given diameter than by an impulse wheel of the same size and operating under the same head. The trend of hydraulic-turbine design is toward higher specific speed, leading to smaller and higher speed machines for the same design requirements and output.

Two turbines of identical geometric shape but of differing size will have practically the same efficiency if operated so that their ϕ values are the same,

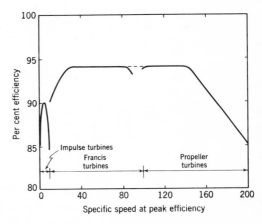

FIG. 12-14   Maximum turbine efficiency as a function of specific speed. (Moody)

since the velocity diagrams will be proportional. Actually the efficiencies will not be precisely the same. The efficiencies of impulse turbines change very little with size, but reaction turbines have somewhat higher efficiencies in larger sizes because of decreased frictional effects and improved volumetric efficiency.

12-6 Cavitation in turbines   Cavitation[1] is undesirable because it results in *pitting* (Fig. 12-15), mechanical vibration, and loss of efficiency. Impulse-wheel buckets may suffer some damage from cavitation, but reaction-turbine runners are usually more seriously affected. In reaction turbines, cavitation is apt to occur on the back sides of the runner blades near their trailing edges. Cavitation may be avoided by designing, installing,

---

[1] R. E. B. Sharp, Cavitation of Hydraulic-turbine Runners, *Trans. ASME*, Vol. 62, pp. 567–575, 1940.

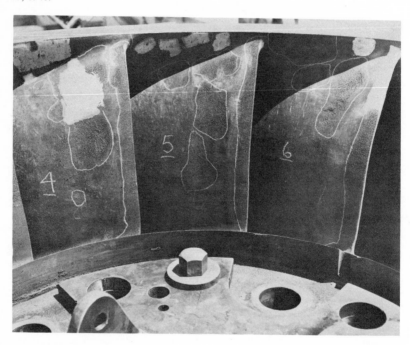

FIG. 12-15   *Cavitation pitting of Francis wheel and scroll case at Mammoth Pool Powerhouse after $2\frac{1}{2}$ yr of operation. Conditions of service are relatively severe. Turbine rating: 88,000 hp at effective head of 950 ft; operating speed 360 rpm. The bright shiny spots are stainless-steel welds that withstood cavitation pitting for over a year. (Courtesy of Southern California Edison Co.)*

TABLE 12-1   Relation between Cavitation Parameter σ_c and
Turbine Specific Speed n_s

n_s	40	80	120	160	200
σ_c	0.10	0.40	0.55	0.80	1.50

and operating the turbine in such a manner that at no point will the local
absolute pressure drop to the vapor pressure of the water. The most critical
factor in the installation of reaction turbines is the vertical distance from the
runner to the tailwater (*draft head*).[1] In comparing the cavitation characteristics of hydraulic machines it is convenient to define a cavitation parameter as

$$\sigma = \frac{(p_{atm}/\gamma) - (e_w/\gamma) - z_1}{h} \tag{12-18}$$

where z_1 and h are as defined in Fig. 12-10. The term $p_{atm}/\gamma - e_w/\gamma$ represents the height to which water will rise in a water barometer. At sea level
with 70°F water temperature $(p_{atm}/\gamma) - (e_w/\gamma) = 33.1$ ft. The minimum
value of σ at which cavitation occurs is denoted by σ_c. Its value can be
determined experimentally for a given machine or model by noting the
operating conditions under which cavitation first occurs as evidenced by the
presence of noise, vibration, and drop in efficiency.

It follows from Eq. (12-18) that the maximum permissible elevation
above tailwater for the setting of a turbine is given by

$$z_1 = p_{atm}/\gamma - e_w/\gamma - \sigma_c h \tag{12-19}$$

Typical values of σ_c versus n_s for reaction turbines are presented in Table 12-1.

Equation (12-17) shows that, for a given head and power, the rotative
speed of a turbine increases with its specific speed. Since water velocity
increases with rotative speed, the higher the specific speed of the turbine,
the lower the maximum head under which it can operate without cavitation
hazard. Figure 12-16 shows recommended limits of safe specific speed for
various heads based on experience with existing power plants. Propeller-
turbine runners are sometimes set below tailwater level to reduce the possibility of cavitation, in which case the draft head is negative. The use of
stainless steel and aluminum bronzes for turbine runners[2] will increase
resistance to damage from pitting.

---

[1] A positive draft head indicates that the runner is above tailwater.

[2] W. J. Rheingans, Recent Developments in Francis Turbines, *Mech. Eng.*, *Vol.* 74, pp. 189–
196, 1952.

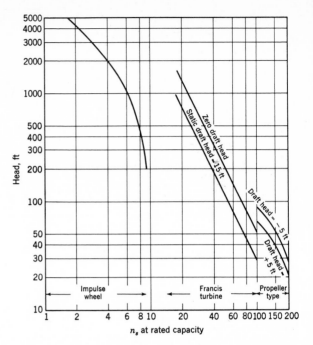

FIG. 12-16   *Recommended limits of specific speed for turbines under various effective heads at sea level with water temperature at 80° F. (After Moody)*

ILLUSTRATIVE EXAMPLE 12-1   Determine the maximum permissible elevation above tailwater for the setting of a Francis turbine ($n_s = 80$, $\sigma_c = 0.40$) to operate under a net head of 55 ft at an elevation of 5000 ft with water temperature at 60°F.

At 5000-ft elevation,

$$\frac{p_{atm}}{\gamma} = \frac{12.2 \times 144}{62.4} = 28.2 \text{ ft}$$

At 60°F,

$$\frac{e_w}{\gamma} = \frac{0.26 \times 144}{62.4} = 0.5 \text{ ft}$$

$$z_1 = 28.2 - 0.5 - 0.40(55) = 5.7 \text{ ft}$$

This result compares closely with the recommended limits of Fig. 12-16.

12-7   Selection of turbines   Inspection of Eq. (12-17) indicates that at high heads for a given speed and power output, a low-specific-speed machine such as an impulse turbine is required. On the other hand, propeller turbines with high n_s are indicated for low heads. An impulse turbine may, however, be suitable for a low-head installation if the flow rate (or power) is small, but often under such conditions the required size of the impulse wheel is prohibitive. Impulse wheels have been used for heads as low as 50 ft if the capacity is small, but they are commonly employed for heads greater than 1000 ft. One of the highest head turbines is at the Reisseck Power Plant[1] in Austria where the effective head is 5800 ft. The limiting head for Francis turbines is about 1000 ft because of possible cavitation and the difficulty of building casings to withstand such high pressures. By choosing a high speed of operation and hence a high-specific-speed turbine, the runner size and first cost are reduced. However, there is some loss of efficiency at high specific speeds. Nevertheless, the modern trend is toward the selection of high-specific-speed machines.

ILLUSTRATIVE EXAMPLE 12-2   A turbine is to be selected for an installation where the net head is 600 ft and the permissible flow is 200 cfs. What type of turbine should be chosen?

Assuming 90 percent efficiency, the available power is

$$\frac{\gamma Qhe}{550} = \frac{62.4 \times 200 \times 600 \times 0.9}{550} = 12,200 \text{ hp}$$

Assuming an operating speed of 100 rpm,

$$n_s = \frac{100 \sqrt{12,200}}{600^{5/4}} = 3.7$$

This indicates that an impulse wheel should be selected. The required wheel diameter is found from Eq. (12-12), assuming $\phi = 0.45$,

$$D = \frac{153.3 \sqrt{600} \times 0.45}{100} = 16.8 \text{ ft}$$

A wheel diameter of 16.8 ft would be quite large, and a smaller size could be selected by increasing the rotative speed. If $N = 150$, $n_s = (3.7 \times 150)/100 = 5.6$ and $D = (16.8 \times 100)/150 = 11.2$ ft. Other combinations of n_s and D could be used with other speeds; however, in accordance with Fig. 12-14, n_s should be less than about 7.2 to ensure high efficiency. Another possible solution is two identical turbines with $n_s = 4$ operating at 150 rpm. Generally it is desirable to have more than one turbine at an installation so that the plant can continue operation while a unit is out of service for repairs.

---

[1] G. A. Bovet, Modern Trends in Hydraulic-turbine Design in Europe, *Trans. ASME*, Vol. 75, pp. 975–993, 1953.

PUMPS

There are many different types of pumps, but the two types which the hydraulic engineer will encounter most frequently are *roto-dynamic* and *displacement* pumps. A roto-dynamic pump has a rotating element which imparts energy to the water in an action which is the reverse of that of a reaction turbine. Displacement pumps include the *reciprocating* type, in which a piston draws water into a cylinder on one stroke and forces it out on the next, and the *rotary* type, in which two cams or gears mesh together and rotate in opposite directions to force water continuously past them. In addition, there are *jet pumps*, *air-lift pumps*, and *hydraulic rams*, which may prove useful under special conditions.

12-8 Description of roto-dynamic pumps   The rotating element of a roto-dynamic pump is called the *impeller*. The impeller may be shaped to force water outward in a direction at right angles to its axis (*radial flow*), to give the water an axial as well as radial velocity (*mixed flow*), or to force the water in the axial direction alone (*axial flow*) (Fig. 12-17). Radial-flow and mixed-flow machines are commonly referred to as *centrifugal pumps* while axial flow machines are called *axial-flow pumps*. Radial-flow impellers are sometimes arranged so that water can enter the *eye* of the impeller from both sides. Such an impeller is called a *double-suction impeller*. It has the same effect as two single-suction impellers placed back to back and results in doubling the capacity without increasing impeller diameter. Radial- and mixed-flow impellers may be either open or closed. The *open impeller* consists of a hub to which the vanes are attached, while the *closed impeller* has plates (or *shrouds*) on each side of the vanes. The open impeller does not have as high an efficiency as the closed impeller, but it is less likely to become clogged and hence is adapted to handling liquids containing solids.

The casing of radial-flow pumps may be either the *volute* type or the *turbine* type (Fig. 12-18). A volute casing is designed to produce an equal velocity of flow around the circumference of the impeller and to reduce the velocity of the water as it enters the discharge pipe. In the turbine-type pump, the impeller is surrounded by stationary guide vanes which reduce the velocity of the water and convert velocity head to pressure head. The casing surrounding the guide vanes is usually circular and concentric with the impeller.

A *single-stage* pump has only one impeller. A *multistage* pump has two or more impellers arranged in such a way that the discharge from one impeller enters the eye of the next impeller (Fig. 12-19). A *deep-well turbine pump* has several impellers on a vertical shaft suspended from the prime mover at the surface. This type of pump is installed in a well casing of limited size, and the entire assembly must be of small diameter. A *submersible* pump is one in which the pump and electric motor are placed below the water surface

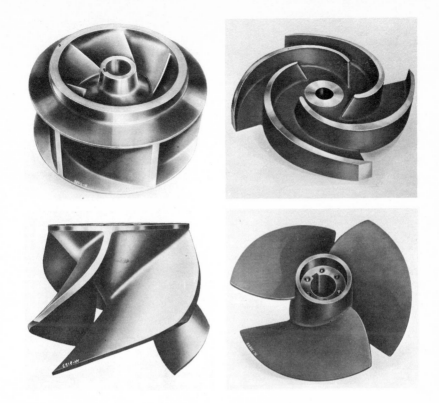

FIG. 12-17   *Types of pump impellers. Top left: closed radial. Top right: open radial. Bottom left: mixed flow. Bottom right: propeller. (Worthington Pump Co.)*

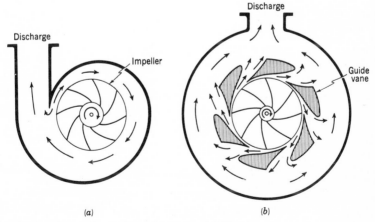

FIG. 12-18   *Types of centrifugal pumps. (a) Volute. (b) Turbine.*

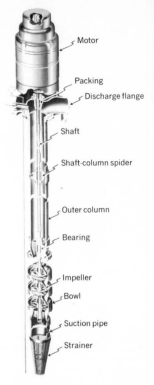

Motor

Packing

Discharge flange

Shaft

Shaft-column spider

Outer column

Bearing

*FIG. 12-19   A three-stage, deep-well turbine pump.
(Johnston Pump Co.)*

Impeller

Bowl

Suction pipe

Strainer

of a well. Delivery of water to the surface is through a riser pipe on which the assembly is suspended. One advantage of the submersible pump is the elimination of the long drive shaft from the ground surface to the pump. This reduces bearing friction and provides an unobstructed pipe for delivery of water to the surface.

A centrifugal pump should be *primed* before it is started. Priming consists in filling the casing with water so that air trapped in the pump does not hinder its operation and reduce its efficiency. A pump located below the source of supply will not ordinarily require priming, although some air may be trapped in the casing of pumps mounted on horizontal shafts. The simplest method of priming is to fill the pump with water from an outside source while permitting the displaced air to escape through an exhaust valve. Some pumps are provided with vacuum pumps to remove the air from the casing. All centrifugal pumps are designed to pass a little water through the seal between the shaft and the casing to lubricate and cool the packing and prevent it from burning out.

Proper arrangement of the suction and discharge piping is necessary if a centrifugal pump is to operate at best efficiency. For economy the diameter of the pump casing at suction and discharge is often smaller than that of the

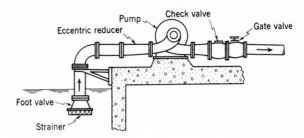

FIG. 12-20   *Typical centrifugal-pump installation.*

pipe to which it is attached. If there is a horizontal reducer between the suction and the pump, an eccentric reducer (Fig. 12-20) should be used to prevent air accumulation. A *foot valve* (check valve) may be installed in the suction pipe to prevent water from leaving the pump when it is stopped. The discharge pipe is usually provided with a check valve and a gate valve. The check valve prevents backflow through the pump if there is a power failure. Suction pipes taking water from a pump or reservoir are usually provided with a screen to prevent entrance of debris which might clog the pump.

Axial-flow pumps (Fig. 12-21) usually have only two to four blades and, hence, large, unobstructed passages which permit handling of water containing debris without clogging. The blades of some large axial-flow pumps are adjustable to permit setting the pitch for best efficiency under existing conditions.

12-9   Action of a centrifugal pump   Figure 12-22 shows the impeller of a centrifugal pump which is rotating counterclockwise at the rate of ω radians/sec. Water enters the impeller with a velocity V_1 and leaves with velocity V_2. The water velocity relative to the vanes is v_1 at entrance and

FIG. 12-21   *Typical axial-flow pump installation.*

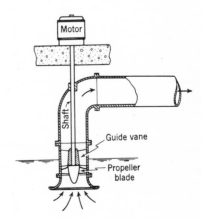

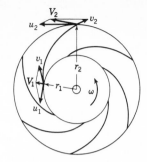

FIG. 12-22   *Vector diagrams for a radial-flow impeller.*

v_2 at exit. It can be seen that the situation is the inverse of that in a reaction turbine (Sec. 12-4). Hence, the torque T exerted by the impeller on the water is given by Eq. (12-8) with the signs changed, i.e.,

$$T = \rho Q(V_{t_2} r_2 - V_{t_1} r_1) \tag{12-20}$$

The water horsepower transferred from the impeller to the water is

$$\text{Water horsepower} = \frac{\omega T}{550} = \frac{\gamma Q h}{550} \tag{12-21}$$

where Q is the flow rate through the impeller in cubic feet per second and h is the net head developed by the pump. From Fig. 12-25, h is

$$h = (z_d - z_s) + \left(\frac{p_d}{\gamma} - \frac{p_s}{\gamma}\right) + \left(\frac{V_d^2}{2g} - \frac{V_s^2}{2g}\right) \tag{12-22}$$

where the subscripts s and d refer to suction and discharge sides of the pump, respectively. The shaft horsepower that must be supplied to the pump shaft is

$$\text{Shaft horsepower} = \frac{\text{water horsepower}}{e} \tag{12-23}$$

where e is the overall efficiency of the pump. Overall efficiencies slightly higher than 90 percent have been achieved by some centrifugal pumps operated at rated head and discharge (i.e., optimum conditions for a given speed). Under usual operating conditions centrifugal pumps have efficiencies between 50 and 85 percent. In general the larger pumps have higher peak efficiencies as in the case of turbines. A pump having a rated capacity of 100 gpm may have a maximum efficiency of only 60 percent, while the maximum efficiency of a 1000-gpm pump is more likely to be near 80 percent. The losses in a centrifugal pump are similar to those in a reaction turbine. Hydraulic losses include fluid friction in the water passages and turbulence caused by the sudden change in the water velocity as it leaves the impeller. Volumetric loss includes flow which escapes past the sealing rings, while mechanical losses include bearing friction.

Every centrifugal and axial-flow pump has operating characteristics which depend on its design and speed of operation. A typical characteristic curve (Fig. 12-23) indicates the relation between head, discharge, and efficiency at a particular speed of operation. A series of such curves for different rotative speeds gives a complete picture of the operating characteristics of the pump. The head at zero discharge is called the *shutoff head.* As discharge is increased, the head produced by the pump may rise or fall slightly depending on the pump design, but eventually the head developed by any centrifugal or axial-flow pump will drop as discharge increases. For a given speed of operation there is a particular discharge (or head) where efficiency is a maximum; this discharge is the *normal discharge,* or *rated capacity,* of the pump at that speed. If the quantity of water to be delivered by the pump varies, the flow may be regulated by throttling with a valve on the discharge line. This results in reduced efficiency since maximum efficiency at a given speed occurs at some fixed value of Q. The rate of flow may be changed by varying the pump speed with belts and pulleys or with a variable-speed motor. The preferred way to regulate flow, however, is by providing several pumps in parallel so that a variable number may be operated at capacity, depending on flow requirements. With this arrangement it is possible to operate all pumps near peak efficiency.

Most pump manufacturers run laboratory tests to check the quality of their product. For large installations, model pumps[1] are often built and

---

[1] Robert L. Daugherty, Centrifugal Pumps for Colorado River Aqueduct, *Mech. Eng.,* Vol. 60, pp. 295–299, 1938.

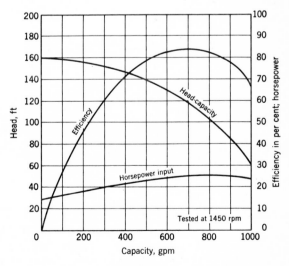

FIG. 12-23   *Characteristic curves for a typical centrifugal pump.*

tested before the prototype is constructed. Custom-made pumps are designed to meet specifications established by the purchaser, and an acceptance test is run on such pumps soon after installation to see if they meet the specifications.

12-10 Specific speed of centrifugal and axial-flow pumps The equations developed for turbines in Sec. 12-5 are equally applicable to pumps. However, Eq. (12-17) for the specific speed of a turbine in terms of power developed is of little value when dealing with pumps since it is the discharge capacity of a pump which is of interest. Noting that $P \propto Qh$ and neglecting constant factors, Eq. (12-17) may be written as

$$n'_s = \frac{NQ^{1/2}}{h^{3/4}} \tag{12-24}$$

where n'_s is the specific speed of the pump. For pumps, Q is usually expressed in gallons per minute,[1] while N is the rotative speed in revolutions per minute when delivering the flow Q against the head h. Specific speeds of pumps are determined by the operating characteristics at the point of maximum efficiency. For double-suction pumps, the discharge to be used in Eq. (12-24) is one-half of the total discharge of the unit. For multistage pumps, h is the head per stage.

Figure 12-24 shows several typical impellers in section and their corre-

---

[1] The specific speed of large pumps is occasionally computed using discharge in cubic feet per second, which changes the numerical values.

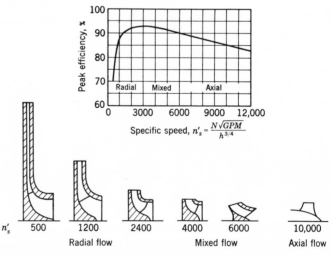

FIG. 12-24   *Shape and maximum efficiency of impellers as a function of specific speed.*

sponding specific speeds. Radial-flow impellers generally have specific speeds between 500 and 3500, mixed-flow between 3500 and 7500, and axial-flow from 7500 to 12,000. As in the case of turbine runners, two pump impellers having the same geometric shape have the same specific speed although their sizes may differ. The curve of Fig. 12-24 indicates the variation of peak efficiency of centrifugal and axial-flow pumps with specific speed. It should be noted that pumps with specific speed below about 800 tend to be inefficient.

Changing the speed of an impeller will change all velocity vectors at impeller entrance and exit in direct proportion to the change in speed. Since the cross-sectional area of the impeller passages is fixed, the discharge of a centrifugal pump will vary directly with impeller speed. Since Q is proportional to $h^{1/2}$, the head developed by a centrifugal pump varies as the square of the speed and hence the power required varies as the cube of the speed since $P \propto Qh$. Thus the operating characteristics of a centrifugal pump are related to speed as follows:

$$Q \propto N \qquad h \propto N^2 \qquad P \propto N^3 \tag{12-25}$$

Since pump efficiency varies with speed, these relations are not quite correct. However, if the speed is varied by a factor of not more than 2, the resulting error is quite small. Operating speeds commonly used with centrifugal pumps vary from 30 to 3500 rpm.

Equation (12-24) indicates that pumping against high heads requires a low-specific-speed pump. For very high heads and low discharges the required specific speed may fall below the values for normal design and result in a pump with a low efficiency. To overcome this problem, the head may be distributed among a number of pumps in series, or a multistage unit may be used. The head per stage is usually limited to about 200 ft, although some pumps in use develop more than 400 ft of head per stage. A multistage pump is generally less expensive than a series of individual pumps, but this may be offset by the very high pressures developed in the multistage pump. When several pumps are spaced more or less uniformly along a pipeline, excessive pressures in the system can be avoided.

The total discharge of several pumps in parallel is the sum of the individual pump discharges. When pumps are installed in parallel, it is important that they have the same operating characteristics. Their shutoff heads should be about the same, or one or more of the pumps may be relatively ineffective. For example, suppose pump A with a shutoff head of 120 ft is placed in parallel with pump B, whose characteristics are shown in Fig. 12-23. If the head to be developed exceeds 120 ft, pump A would pump no water. If head were less than 120 ft, pump A would be effective, but the efficiency of pump B would drop since its rated capacity of 700 gpm is at a head of

120 ft. Since shutoff head depends on speed of operation, it is possible that these two pumps can be made to operate more efficiently together by changing the speed of one or both pumps.

ILLUSTRATIVE EXAMPLE 12-3   Determine the specific speed of the pump whose operating characteristics are shown in Fig. 12-23. If this pump were operated at 1200 rpm, what head and discharge would be developed at rated capacity, and what power would be required?

At its rated capacity of 700 gpm this pump develops 120 ft of head when operating at 1450 rpm. Thus

$$n'_s = \frac{1450 \times 700^{1/2}}{120^{3/4}} = 1060$$

If the efficiency remains constant with the speed change, at 1200 rpm

$$Q = {}^{1200}\!/_{1450} \times 700 = 580 \text{ gpm}$$
$$h = ({}^{1200}\!/_{1450})^2 \times 120 = 82.2 \text{ ft}$$
$$P = ({}^{1200}\!/_{1450})^3 \times 26 = 14.8 \text{ hp}$$

12-11   Cavitation in centrifugal and axial-flow pumps   Like turbines, centrifugal and axial-flow pumps are subject to cavitation. It is most likely to occur near the point of discharge from radial- and mixed-flow impellers and near the blade tips of axial-flow impellers. Cavitation places a limit on the head against which a pump can operate satisfactorily. The limiting head depends on the specific speed of the pump and the *suction lift* h_z (the elevation difference between the energy line at suction and the eye of the impeller as shown in Fig. 12-25). The recommended limiting heads for single-stage single-suction centrifugal pumps are indicated in Fig. 12-26 on the basis of experience with operating pumps. The axial-flow pump is more vulnerable to cavitation than other pumps because of its high speed and is therefore restricted to fairly low heads.

For pumps, the cavitation parameter is defined as

$$\sigma = \frac{(p_{atm}/\gamma) - (e_w/\gamma) - h_z}{h} = \frac{NPSH}{h}$$
$$= \frac{(p_s/\gamma) + (V_s^2/2g) - (e_w/\gamma)}{h} \quad (12\text{-}26)$$

where the p's are expressed as absolute pressures. The numerator, *NPSH*, *net positive suction head*, is the difference between the total absolute head at the suction side of the pump and the vapor pressure head. For any given pump there is a certain minimum value of *NPSH* below which cavitation

h_z = suction lift

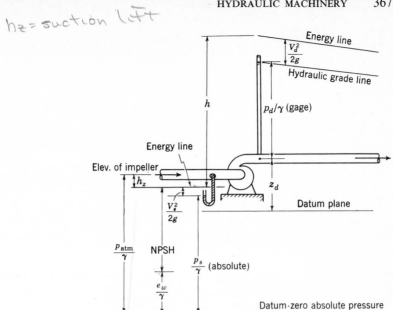

FIG. 12-25   Definition sketch for head developed by a pump.

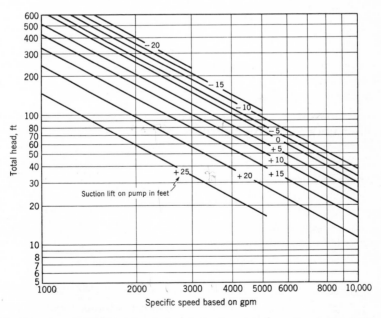

FIG. 12-26   Recommended limiting heads for single-stage, single-suction pumps as a function of specific speed and suction lift. At sea level with water temperature of 80°F. (Moody)

will occur. This can be expressed in terms of the critical value of the cavitation parameter as

$$NPSH = \sigma_c h \qquad (12\text{-}27)$$

where h is the head delivered by the pump. Values of σ_c depend on the design of the pump but typically range from 0.05 for $n_s' = 1000$ to about 1.0 for $n_s' = 8000$.

ILLUSTRATIVE EXAMPLE 12-4   A mixed-flow centrifugal pump ($n_s' = 4000$, $\sigma_c = 0.30$) is to develop 50 ft of head. Find the maximum permissible suction lift on the pump if the pump is at sea level and water at 80°F.

$$NPSH = \sigma_c h = 0.3(50) = 15 \text{ ft}$$

$$h_z = -NPSH + \frac{p_{\text{atm}}}{\gamma} - \frac{e_w}{\gamma} = -15 + (33.9 - 1.2) = 17.7 \text{ ft of lift}$$

This result compares favorably with the value of permissible suction lift given in Fig. 12-26.

If the pump had been at elevation 5000 ft and the water at 100°F, then

$$h_z = -15 + (28.2 - 2.2) = 11 \text{ ft of suction lift}$$

12-12   Reciprocating pumps   The simplest type of reciprocating pump is the hand-operated well pump (Fig. 12-27). This pump must be primed before it will operate and is limited to suction lifts of about 20 ft. On the downstroke the piston valve opens, permitting water to enter the space above the piston. On the upstroke the piston valve closes, and the water above the piston is lifted to the spout. At the same time the check valve opens, and water is drawn into the space below the piston. By placing the piston cylinder in the well, perhaps even below the waterline, pumps of this type can develop a high lift.

The pump of Fig. 12-27 is a *single-acting* pump in which discharge occurs only on alternate piston strokes. *Double-acting* pumps discharge with each stroke and provide a more uniform flow. Since a definite volume of water is discharged with each stroke, the delivery rate of a reciprocating pump depends only on the speed and is independent of head. The efficiency of a reciprocating pump depends upon the slippage, or amount of water which leaks between the piston and the cylinder walls during a stroke. If the valves and packing are in good condition, slippage should be less than 10 percent of total output.

A *deep-well reciprocating pump* has three parts (Fig. 12-27)—power head, cylinder, and drop pipe and rods. The power head is usually at ground level directly over the well. The power head may be a steam cylinder or a crank-shaft connected to a prime mover by gears or belt drive. The cylinder, or

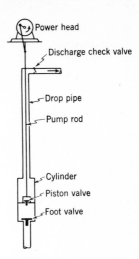

Power head

Discharge check valve

Drop pipe

Pump rod

Cylinder

Piston valve

Foot valve

FIG. 12-27   Deep-well reciprocating pump.

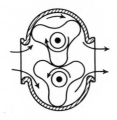

FIG. 12-28   Rotary pump of the three-lobe type.

pumping element, is placed in the well at the lowest expected water level. The drop pipe transmits the water from the cylinder to the surface, while the rods transmit the oscillatory motion from the power head to the plunger. A single-rod pump has one plunger, which may be either single- or double-acting. The double-rod pump has two pump rods (a solid rod inside of a hollow rod) each of which connects to its own plunger. The two plungers work one above the other in the same cylinder. Deep-well reciprocating pumps are widely used for oil wells.

A reciprocating pump must be primed before it will operate, and a supply of water for priming is generally provided in the installation. Before starting a reciprocating pump, the discharge valve should be opened to prevent high pressures which may damage the discharge pipe or cylinder. Many reciprocating pumps are provided with an air chamber on the discharge side of the pump. The air in the chamber compresses and expands on each stroke to cause a more uniform flow in the discharge pipe.

Many large pumping stations equipped with reciprocating pumps were built during the nineteenth century. Enormous pumping engines—some as much as three stories high—were employed. The development of the electric motor and centrifugal pump provided a compact pumping unit which has virtually eliminated reciprocating pumps with their pulsating flow from water-supply service. Centrifugal pumps are better suited to pumping water containing solids than are reciprocating pumps. Generally reciprocating pumps are advantageous only where high-head pumping is required or where their potentially higher efficiency overcomes their high initial and maintenance costs.

12-13   Rotary pumps   The rotary pump (Fig. 12-28) is a displacement pump consisting of two cams or gears which mesh together and rotate in opposite directions. The rotating elements fit the casing closely, and the water trapped between them and the casing is forced through the pump as they rotate. A definite amount of water depending on the size and shape of the gears or cams is passed with each revolution. Water containing grit is very injurious to a rotary pump as the wear will destroy the seal between the cams and casing. Rotary pumps are best adapted to low pressures with discharges less than 500 gpm, although rotary pumps have operated at pressures up to 1000 psi and discharges over 30,000 gpm. Rotary pumps are self-priming and are often used to prime large centrifugal or reciprocating pumps. The flow from a rotary pump is nearly free from pulsations. Since they have no valves, rotary pumps are simpler to construct and easier to maintain than reciprocating pumps.

The overall efficiency of rotary pumps is determined by the slip, or leakage, between rotor and casing. The slip is determined by vacuum at suction, discharge pressure, and condition of the pump. An increase of the vacuum at suction will reduce the efficiency of a rotary pump because gas entrained in the water will expand and occupy more of the pump displacement. An increase in discharge pressure tends to force water back through the clearances to the suction side of the pump and results in lowered efficiency. Rotary pumps are often used for fire-protection systems for buildings and for small domestic water systems.

12-14   Air-lift pumps   An air-lift pump (Fig. 12-29) uses compressed air to deliver water from wells. The air is forced into the well through a small air pipe and discharged into the *eduction pipe* at the bottom of the well. The resulting mixture of air and water in the eduction pipe is lighter than the water outside the pipe and hence is forced upward by hydrostatic pressure. This type of pump has been used for lifts as great as 500 ft, but its efficiency is usually only between 25 and 50 percent. The air-lift pump will operate best if the ratio h_p/h_s varies from about 2 when h_p is 500 ft to about 0.5 when $h_p = 50$ ft. To satisfy this condition, a well may have to be deep-

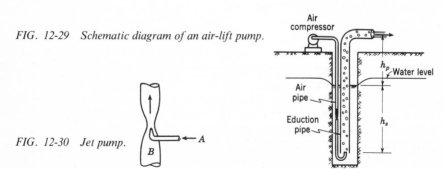

FIG. 12-29   *Schematic diagram of an air-lift pump.*

FIG. 12-30   *Jet pump.*

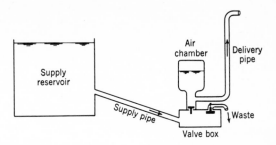

FIG. 12-31   Schematic layout for a hydraulic ram.

ened, thus increasing the cost. In spite of low efficiency an air lift can deliver large amounts of water from small-diameter wells. The air lift is not harmed by sandy water and is particularly suitable for use in crooked or damaged wells. A separator is often placed at the discharge end of the eduction pipe to remove the air from the water. The reclaimed air is usually cooler than atmospheric air and can be recompressed more cheaply. An air-lift pump is not adapted to raising water much above ground level, and if this is necessary, a second pump may be required.

12-15   Jet pumps   The *jet pump*, or *ejector*, is shown in Fig. 12-30. Steam, compressed air, or water forced through pipe *A* discharges from a nozzle in the throat of pipe *B* at high velocity. The resulting jet creates a suction which draws the water up pipe *B*. The efficiency of jet pumps is rarely over 25 percent, but they are compact, light in weight, and can handle muddy water. Jet pumps are sometimes used in construction work for dewatering trenches and are widely used for pumping from small wells.

12-16   The hydraulic ram   The *hydraulic ram* (Fig. 12-31) lifts water by utilizing the impulse developed when a moving mass of water is suddenly stopped. A relatively large amount of water must be available at moderate head to lift a small volume to a higher head. A supply pipe conveys water from the source to the valve box of the hydraulic ram. The valve box contains two automatic valves, a waste valve opening downward and a delivery valve opening upward. Above the delivery valve there is an air chamber to the base of which the delivery pipe is attached. A steady flow of water occurs through the waste valve with the delivery valve closed. If the waste valve is suddenly closed, water-hammer pressures will develop, forcing the delivery valve open and permitting some water to pass up the delivery pipe. When the wave of negative pressure returns from the reservoir, the delivery valve closes automatically and the waste valve opens. A gradually accelerating flow through the waste valve occurs until the net force exerted upward on the valve exceeds the weight of the valve and it closes automatically to begin a new cycle. The air chamber serves the same purpose as it does on a reciprocating pump, i.e., it reduces the fluctuations in flow through the delivery pipe.

The hydraulic ram is wasteful of water, but it may be advantageously employed in situations where no other outside source of power is available. The ratio of wasted water to pumped water for a well-designed hydraulic ram will vary from 6:1 to 2:1 depending upon the supply head, the lift, and other factors. In cases where a portion of the water is required at an elevation below the ram, there may be no waste at all. One objection to the hydraulic ram is that it produces a considerable amount of noise. Some rams are designed to lift water from a source other than the one which provides water for the supply pipe. The two waters can be kept separate so that pure water may be lifted with energy obtained from polluted water without any contamination of the pure water.

12-17 Selection of pumps The operating characteristics of the various types of pumps have been discussed in the preceding sections. For ordinary pumping the roto-dynamic pump is most commonly used as it provides satisfactory and economic service. At very low flow rates the rotary pump may be just as satisfactory and less costly if the water to be pumped is free of grit. The discharge of a centrifugal pump varies with the head, and a variable-speed drive is necessary if constant discharge is to be maintained under varying head. A reciprocating pump overcomes this difficulty since its discharge depends only on the speed of the pump. Reciprocating pumps are high in first cost and difficult to maintain in efficient operating condition. They are best adapted for use under very high heads. Centrifugal pumps are well suited to pumping wastewater and water containing solids, but displacement pumps are not generally used for such duty. The hydraulic ram, although wasteful of water, finds occasional use where water is plentiful and outside power unavailable.

The type of well pump selected depends on the depth of the well and the desired flow. A hand-operated reciprocating pump is satisfactory for shallow wells if only a small and intermittent discharge is required. For deeper wells, a deep-well turbine pump or a deep-well reciprocating pump may be used. The air-lift or jet pump is suitable for wells where the discharge is not large. Jet pumps are also used for dewatering excavations, although self-priming centrifugal pumps are more commonly used for this purpose.

BIBLIOGRAPHY

Addison, Herbert: "Centrifugal and Other Rotodynamic Pumps," 3d ed., Chapman & Hall, London, 1966.

Brown, J. Guthrie (ed.): "Hydro-electric Engineering Practice," 2d ed., Vol. II, Part A, Water Turbines, Blackie, Glasgow, 1970.

Daugherty, R. L., and J. B. Franzini: "Fluid Mechanics with Engineering Applications," 6th ed., McGraw-Hill, New York, 1965.

Hicks, T. G.: "Pump Selection and Application," McGraw-Hill, New York, 1957.

Karassik, Igor, and Roy Carter: "Centrifugal Pumps: Selection, Operation, and Maintenance," F. W. Dodge, New York, 1960.

Kovalev, N. N.: "Hydroturbines Design and Construction," translated from Russian, Office of Technical Service, U.S. Department of Commerce, Washington, D.C., 1965.

Lazarkiewicz, S., and A. T. Troskolanski: "Impeller Pumps," Pergamon Press, New York, 1965.

Leliavsky, Serge: "Irrigation and Hydraulic Design," Vol. III, chap. 13, pp. 617–752, Chapman & Hall, London, 1960.

Mechanical Design of Hydro Plants, *Tech. Rept.* 24, Vol. 3, U.S. Tennessee Valley Authority, 1960.

Moody, Lewis F., and T. Zowski, Hydraulic Machinery, sec. 26 in C. V. Davis and K. E. Sorensen (eds.), "Handbook of Applied Hydraulics," 3d ed., McGraw-Hill, New York, 1969.

Nechleba, Miroslav: "Hydraulic Turbines, Their Design and Equipment," Artia, Prague, 1957.

Norrie, D. H.: An Introduction to Incompressible Flow Machines, Arnold, London, 1963.

"Standards for Centrifugal Pumps," 11th ed., Hydraulic Institute, New York, 1965.

Stepanoff, A. J.: "Centrifugal and Axial Flow Pumps," 2d ed., Wiley, New York, 1957.

PROBLEMS

12-1 An impulse wheel with a single nozzle of 12-in. diameter receives water through a 36-in.-diameter riveted-steel pipe ($e = 0.01$ ft) which is 1000 ft long. The water level in the reservoir is 650 ft above the nozzle of the turbine. If the overall turbine efficiency is 85 percent and the loss coefficient of the nozzle is 0.04, what maximum power output can be expected? How much additional power is possible if a 48-in. riveted-steel pipe is used?

12-2 A reaction turbine is supplied with water through a 150-cm pipe ($e = 1.0$ mm) which is 40 m long. The water surface in the reservoir is 20 m above the draft-tube inlet which is 4.5 m above the water level in the tailrace. If the turbine efficiency is 92 percent and the discharge is 12 cu m/sec, what is the power output of the turbine in kilowatts?

12-3 The turbine of Prob. 12-2 has a vertical draft tube whose diameter changes from 6 to 10 ft in a length of 18 ft. Compute the absolute pressure head at the inlet to the draft tube. Assume standard sea-level atmospheric pressure. Might one expect cavitation in the turbine under these conditions? Specific speed of the turbine is 60.

12-4 At rated capacity, the flow through a small radial reaction turbine, Fig. 12-12, is 4.0 cfs and its efficiency is 90 percent. The head on the machine is 29.6 ft. Its dimensions are $r_1 = 0.8$ ft, $r_2 = 0.2$ ft, $\beta_1 = 60°$, $\beta_2 = 140°$, and $Z = 0.4$ ft. Determine the specific speed of this turbine. Also compute ϕ and compare it with typical values given in the text.

12-5 A turbine is to be installed at a point where the net available head is 175 ft, and the available flow will average 1000 cfs. What type of turbine would you recommend? Specify the operating speed and number of generator poles for 60-cycle electricity if a turbine with the highest tolerable specific speed to safeguard against cavitation is selected. Assume static draft head of 10 ft and 90 percent turbine efficiency. Approximately what size of turbine runner is required?

12-6 For the conditions of Prob. 12-5 select a set of two identical turbines to be operated in parallel. Specify the speed and size of the units.

12-7 What is the least number of identical double-overhung impulse turbines that can be used at a powerhouse where the net available head is 1200 ft and $Q = 1650$ cfs? Assume turbine efficiency is 90 percent and speed of operation is 277 rpm. Specify the size and specific speed of the units.

12-8 Repeat Prob. 12-7 but operate at 450 rpm instead of 277 rpm.

12-9 Determine the specific speed of a roto-dynamic pump which is rated at 3000 gpm under a head of 80 ft at 1760 rpm. What would be the head and capacity of this pump if operated at 1200 rpm? For each rotative speed note the maximum tolerable suction lift as recommended in Fig. 12-26.

12-10 A pump is required to deliver 1600 gpm against a head of 350 ft at a rotative speed of 1150 rpm. Would a two-stage pump probably be more efficient than a single-stage?

12-11 A centrifugal pump with a 12-in.-diameter impeller is rated at 600 gpm against a head of 80 ft when rotating at 1750 rpm. What would be the rating of a pump of identical geometric shape with a 6-in.-diameter impeller? Assume pump efficiencies and rotative speeds are identical.

12-12 Select the specific speed of the pump or pumps required to lift 25 cfs of water 300 ft through 10,000 ft of 2-ft-diameter pipe ($f = 0.020$). The pump rotative speed is to be 2250 rpm. Consider the following cases: single pump, two pumps in series, three pumps in series, two pumps in parallel, three pumps in parallel.

12-13 Select the number and type (n_s) of centrifugal pumps for a pumping station that must deliver 4600 cfs against a head of 150 ft at 180 rpm. Assume $NPSH$ on pumps is $+15$ ft. Use identical single-stage pumps in parallel.

12-14 A pump is required to lift 400 gpm against a head of 600 ft. If the minimum desirable specific speed from the efficiency standpoint is 1000 and the motor speed is to be 2000 rpm, how many stages would you recommend? Also what is the maximum permissible static lift?

12-15 A very long pipe connects two reservoirs whose water levels are the same. Identical pumps A and B are connected to the pipe in parallel. The operating characteristics of each of these pumps are as shown in Fig. 12-23. The pumps are operated at 1450 rpm. When only one pump is operating, the flow rate was found to be 910 gpm. How much flow will there be when both pumps operate? Assume the flow occurs at high Reynolds number so that friction factor f remains constant. Find also the rates of delivery if the pump speed is reduced to 1200 rpm.

12-16 A pump is installed to deliver water from a reservoir of surface elevation zero to another of elevation 300 ft. The 12-in.-diameter suction pipe ($f = 0.020$) is 100 ft long and the 10-in.-diameter discharge pipe ($f = 0.026$) is 5000 ft long. The pump characteristic at 1200 rpm is defined by $E_p = 375 - 24Q^2$ where E_p, the pump head, is in feet and Q is in cubic feet per second. Compute the rate at which this pump will deliver water under these conditions assuming the setting is low enough to avoid cavitation.

12-17 Repeat Prob. 12-16 determining the flow rate if two such pumps were installed in series. Repeat for two pumps in parallel.

12-18 The pump of Fig. 12-23 is placed in a 6-in.-diameter pipe ($f = 0.020$) 850 ft long that is used to lift water from one pond to another. The difference in water-surface elevations between the ponds fluctuates from 50 to 140 ft. Plot a curve showing delivery rate versus water-surface elevation difference.

12-19 A double-acting piston pump with a 12-in. stroke and an 8-in.-diameter cylinder is operated at 90 revolutions per minute. If the discharge is 424 gpm, what is the percent slip?

12-20 A centrifugal pump driven by an electric motor lifts water through a total height of 150 ft from reservoir to discharge. The pump efficiency is 78 percent and the motor efficiency is 89 percent. The lift is through 1000 ft of 4-in.-diameter pipe and the pumping rate is 300 gpm. If $f = 0.025$ and power costs 6 mills per kilowatt-hour, what is the cost of power for pumping a million gallons of water? An acre-foot? A thousand cubic meters?

Chapter 13

Engineering economy
in water-resources planning

EUGENE L. GRANT*

Water-resources engineering design involves many choices among physically feasible alternatives. Generally speaking, each choice among alternatives should be made on economic grounds. Each alternative that is given serious consideration should be expressed in money units before the choice is made. In fact, unless alternatives can be expressed in money units, the items involved in such choices are incommensurable. For instance, the money unit is the only measuring unit that can be applied to such diverse items as steel pipe, kilowatthours of electricity, hours of skilled and unskilled labor, and reduction of the hazard of flood damages.

Writers on engineering economy often quote the classic questions that Gen. John J. Carty applied to engineering proposals when he was chief engineer of the New York Telephone Company. These questions were *Why do this at all? Why do it now? Why do it this way?* These may be viewed as different aspects of the general question, *Will it pay?*

As these questions suggest, most engineering proposals involve major sets of alternatives, with many subordinate sets of alternatives involved in each major alternative. For example, consider the question of whether to use hydro or thermal power to add needed capacity to an electric-power system. For the hydro alternative, a number of different possible sites may

*Emeritus Professor of Economics of Engineering, Stanford University.

be available. At a particular site, there may be different possible designs for a diversion dam, different types and sizes of conduit to carry water from the point of diversion to the power plant, different possibilities for number, type, and size of turbines at the power plant, and so on. Each subalternative is likely to have its own subalternatives. At every stage in the analysis, it is necessary to consider relative economy in order to make a rational choice among the possible alternatives and thus to establish the most favorable overall design.

13-1   Social importance of economy in water-resources engineering decisions   The following quotation from the report of the President's Water Resources Policy Commission[1] emphasizes the importance of sound economic decisions on major water projects:

> Once they are completed, major water control structures can be altered only with difficulty, or not at all. There are only a relatively few suitable dam sites, and once they are appropriated, the possibilities for economic multiple-purpose development are very limited. Once an irrigation project is developed, it cannot be moved because unfavorable soil or climate factors are discovered. There is a sobering finality in the construction of a river basin development; and it behooves us to be sure we are right before we go ahead.

Although the public importance of sound economic analysis in major river-basin projects is generally recognized, engineers and others are not always aware of the social costs arising from unsound decisions in what seem to them to be minor matters. Certain "minor" problems occur so frequently that their aggregate importance is large. For example, the decision on the capacity to be provided in a highway-drainage crossing. Although the extra costs resulting from an uneconomical decision for a single culvert or bridge may be relatively small, a design procedure that is not economically sound which is applied to all the drainage crossings on a state highway system may increase total highway user costs by a substantial amount.

13-2   Steps in an engineering economy study   It is helpful to think of an economy study as involving the following steps:

1. Each alternative that seems promising should be identified and clearly defined in physical terms.
2. In so far as practicable, the physical estimates for each alternative should be translated into money estimates. Generally speaking, money estimates should be made of those receipts and disbursements which will be influenced by the choice among the alternatives. Estimates should be made of the dates as well as the magnitudes of the receipts and disburse-

---

[1] "A Water Policy for the American People. The Report of the President's Water Resources Policy Commission, 1950," Vol. 1, p. 18, 1950.

ments. This requires estimates of the lives and salvage values, if any, of the structures and other assets required for each alternative. It also calls for a decision regarding the length of the study period—the period of time for which the economy study is to be made.

3. Usually the money estimates need to be placed on a comparable basis by appropriate conversions that make use of the mathematics of compound interest. These conversions should use as an interest rate the minimum attractive rate of return that is appropriate in the particular circumstances (Sec. 13-4).

4. A choice (or recommendation for a choice) among the alternatives must be made. This choice is properly influenced both by the comparison in terms of money units and by other matters that it has not been practicable to reduce to money terms (so-called "irreducibles" or "intangibles").

Within the space limitations of this chapter, it is impossible to cover the subject matter of a textbook on engineering economy. It is assumed that the reader is familiar with the common manipulations of compound-interest mathematics required for simple economy studies. This chapter reviews certain general principles for economy studies and illustrates these principles with numerical examples from the field of hydraulic engineering. The chapter also introduces certain topics of particular importance in the economic analysis of water projects. Economic matters relative to particular types of water projects (e.g., irrigation, flood mitigation) are discussed in the later chapters dealing with such projects.

13-3 A simple example of an annual cost comparison Two alternative plans are considered for a section of an aqueduct. Plan A uses a tunnel; plan B uses a section of lined canal and a section of steel flume. In plan A, the estimated first cost of the tunnel is $450,000, its estimated annual maintenance cost is $4000, and its estimated life is 100 yr. Estimated first costs and lives for the elements of plan B are canal (not including lining), $120,000, 100 yr; canal lining, $50,000, 20 yr; flume, $90,000, 50 yr. The annual maintenance cost is $10,500. The interest rate to be used in the economy study is 6 percent per annum. The study period is 100 yr. All salvage values are assumed to be negligible. There are no estimated revenue differences between the two plans, and no other cost differences are anticipated. (For instance, there is no expected difference in water loss between the two alternatives.)

The comparison of equivalent annual costs for the two plans is as follows:

Plan A

Capital recovery cost for tunnel = $450,000 × 0.06018 = $27,081
Annual maintenance cost =   4,000
Total annual cost = $31,081

Plan B

Capital recovery cost for canal = $120,000 × 0.06018 = $ 7,222
Capital recovery cost for canal lining = $ 50,000 × 0.08718 =   4,359
Capital recovery cost for flume = $ 90,000 × 0.06344 =   5,710
Annual maintenance cost =   10,500
Total annual cost = $27,791

In the preceding tabulation, the first cost figures for tunnel, canal, lining, and flume are multiplied by compound-interest factors that depend on their respective estimated lives and on the 6 percent interest rate. The appropriate factor to convert an investment into an equivalent annual cost is designated as the *capital-recovery factor* and may be computed from the expression $i(1 + i)^N/[(1 + i)^N - 1]$ where i represents the interest rate per annum (expressed as a decimal fraction) and N represents the years of estimated life. Table 13-1 contains capital recovery factors for a number of values of i and N.

When any present sum of money is multiplied by the capital-recovery factor for N yr and interest rate i, the product is an annual figure sufficient to repay exactly the present sum in N yr with interest rate i. For example, the tabulated costs in plan B show $4359 as the capital-recovery cost of $50,000, including 6 percent interest on each year's unpaid balance.

The comparison of the $31,081 annual cost of plan A with the $27,791

TABLE 13-1   Capital Recovery Factors

Years	Interest rate, percent						
	2	3	4	5	6	8	10
5	0.21216	0.21835	0.22463	0.23097	0.23740	0.25046	0.26380
10	0.11133	0.11723	0.12329	0.12950	0.13587	0.14903	0.16275
15	0.07783	0.08377	0.08994	0.09634	0.10296	0.11683	0.13147
20	0.06116	0.06722	0.07358	0.08024	0.08718	0.10185	0.11746
25	0.05122	0.05743	0.06401	0.07095	0.07823	0.09368	0.11017
30	0.04465	0.05102	0.05783	0.06505	0.07265	0.08883	0.10608
35	0.04000	0.04654	0.05358	0.06107	0.06897	0.08580	0.10369
40	0.03656	0.04326	0.05052	0.05828	0.06646	0.08386	0.10226
45	0.03391	0.04079	0.04826	0.05626	0.06470	0.08259	0.10139
50	0.03182	0.03887	0.04655	0.05478	0.06344	0.08174	0.10086
60	0.02877	0.03613	0.04420	0.05283	0.06188	0.08080	0.10033
70	0.02667	0.03434	0.04275	0.05170	0.06103	0.08037	0.10013
80	0.02516	0.03311	0.04181	0.05103	0.06057	0.08017	0.10005
90	0.02405	0.03226	0.04121	0.05063	0.06032	0.08008	0.10002
100	0.02320	0.03165	0.04081	0.05038	0.06018	0.08004	0.10001

annual cost of plan B is representative of the innumerable annual cost comparisons that may be made in engineering economy studies. It is common for alternative designs to require different investments in fixed assets. In this example, the total investment figures are $450,000 and $260,000, respectively. The extra investment of $190,000 produces certain advantages that have a money value, such as lower maintenance costs and longer lives for the assets. In this instance, the annual cost comparison tells us that these advantages are insufficient to justify the extra investment. Plan B therefore should be selected unless plan A has some other advantages—so-called irreducibles or intangibles—that were not reflected in the estimates of costs and lives and that are believed to be sufficient to justify extra costs of $3290 a year.

It should be emphasized that annual cost calculations of the type here illustrated are valid regardless of the scheme of financing to be employed. Even though it is convenient to explain the capital-recovery factor in terms of a loan transaction, the computed totals for plans A and B are the equivalent annual costs whether the entire first cost of the proposed assets is to be borrowed, whether it is to be entirely equity capital, or whether some combination of borrowing and equity capital is contemplated. The preceding statement is subject to the qualification that the interest rate used in the economy study needs to be the rate that is most appropriate for the particular circumstances.

13-4 Selecting an interest rate for an economy study The conclusions from a comparison of annual costs obviously are greatly influenced by the interest rate used in computing the costs. If plans A and B had been compared using an interest rate of 2 percent (rather than 6 percent), plan A would have appeared to be more economical by a margin of about $4760 a year. If 8 percent had been used, the margin of superiority of plan B would have been nearly $7500 a year. The two plans have equal annual costs at an interest rate of slightly over 4 percent.

The choice of the interest rate to be used in economy studies to guide the design of an engineering project will therefore have considerable influence on the design selected. With a very low rate, such as 2 percent, many proposed investments (such as the tunnel of plan A) will appear to be economical even though the same investments would seem unduly costly with an interest rate of 6 or 8 percent. In effect, the decision to use a particular rate, such as the 6 percent that was used in comparing plans A and B, is a decision that the rate selected is the minimum attractive rate of return, all things considered. It is desirable that the foregoing viewpoint be recognized when an interest rate is selected.

In private enterprise, the interest rate used in economy studies should ordinarily be not less than a figure that reflects the overall cost of capital to the enterprise—equity capital as well as borrowed capital. In privately owned public utility companies, such as electric utilities and water utilities, a good measure of this figure is the "fair return" permitted by the regulatory commission that controls utility rates. Although fair-return rates properly vary from company to company and from time to time, they have most often been in the range from 6 to 8 percent. Perhaps 7 percent may be thought of as a typical figure for privately owned utilities.

In competitive private enterprise, the cost of capital often is considerably higher than this 7 percent figure. Moreover, the appropriate minimum attractive rate of return to use in economy studies for a particular enterprise may properly be influenced by a limitation in the funds that can be secured for investments in new fixed assets. Often there are proposals for many more such investments than it is possible to finance. In such instances, the use of a relatively high interest rate in economy studies (10 percent, for example) has the effect of eliminating the least productive of the proposed investment opportunities and of conserving the limited funds to use in the most productive places.

In contrast to the foregoing practice in private enterprise, the common practice in economy studies for public works in the United States has been to use an interest rate equal to the bare cost of borrowed money for the public body in question. Such rates naturally differ from time to time and from one public body to another. Rates now used in studies for federal river-basin projects are typically between 4 and 5 percent.[1] Rates used in studies for municipal projects have sometimes been even lower than these federal rates. Because interest received by holders of bonds of local political subdivisions is exempt from federal income taxation in the United States, cities, counties, states, and local improvement districts have often been able to borrow at lower rates than those paid by the federal government. In effect, such local borrowing has had the benefit of a hidden subsidy. Thus some cities have sold bonds at interest rates as low as 3 percent—occasionally even less.

13-5 Estimated lives of hydraulic structures A bulletin of the Internal Revenue Service gives estimated average lives for many thousands of different types of industrial assets.[2] The lives (in years) given for certain elements of hydraulic projects are listed in Table 13-2.

---

[1] These rates have differed among the various federal agencies as well as from time to time. By action of the Water Resources Council the interest rate used in estimating project reimbursements is based on the average yield on federal government bonds having approximately the same maturity period as the economically useful life of the project in question. Where this computed average rate is not a multiple of $\frac{1}{8}$ percent, the next lower multiple of $\frac{1}{8}$ percent shall be used.

[2] Income Tax Depreciation and Obsolescence, Estimated Useful Lives and Depreciation Rates, *U.S. Bur. Internal Revenue Bull.* F, 1942. See also Depreciation Guidelines and Rules, *Pub.* 456, rev. ed., Internal Revenue Service, August, 1964.

TABLE 13-2   Lives (in Years) for Elements of Hydraulic Projects

Barges	12	Pipes:	
Booms, log	15	Cast-iron	
Canals and ditches	75	2–4 in.	50
Coagulating basins	50	4–6 in.	65
Construction equipment	5	8–10 in.	75
Dams:		12 in. and over	100
Crib	25	Concrete	20
Earthen, concrete, or masonry	150	Steel	
Loose rock	60	Under 4 in.	30
Steel	40	Over 4 in.	40
Filters	50	Transite, 6 in.	50
Flumes:		Transmission lines	30
Concrete or masonry	75	Tugs	12
Steel	50	Wood-stave	
Wood	25	14 in. and larger	33
Fossil fuel power plants	28	3–12 in.	20
Generators:		Pumps	18–25
Above 3000 kva	28	Reservoirs	75
1000–3000 kva	25	Standpipes	50
50 hp–1000 kva	17–25	Tanks:	
Below 50 hp	14–17	Concrete	50
Hydrants	50	Steel	40
Marine construction equipment	12	Wood	20
Meters, water	30	Tunnels	100
Nuclear power plants	20	Turbines, hydraulic	35
Penstocks	50	Wells	40–50

Such country-wide estimates of average lives may be helpful even though they are not necessarily the most appropriate figures to use in any given instance. Moreover, a conservative life estimate to use in an economy study usually should be shorter than the full expected service life. Many factors combine to cause economic lives to be considerably shorter than full service lives. Thus a competitive private industry might use a 10-yr payoff period for assets having expected lives of 20 yr or more. Costs of public river-basin projects often have been computed on the basis of a 50-yr life for all project elements expected to last 50 yr or more.

In this connection, it should be noted that, for long-lived assets, a large difference in estimated life has less effect on annual cost than a moderate difference in interest rate. For example, assume a given life estimate is increased from 35 to 100 yr and at the same time the interest rate used is increased from 4 to 5 percent; the increase of annual cost due to the higher interest rate is greater than the reduction of cost due to the estimate of longer life.

13-6   Taxes and other investment charges   Capital recovery cost is sometimes referred to as interest plus depreciation or as interest plus amortization. Regardless of the way it is described, this cost is proportional to investment. Some types of economy studies are simplified by combining this cost with certain other costs proportional to investment.

For assets subject to fire or casualty insurance, the cost of this insurance is an investment charge. For assets subject to general property taxes based on assessed valuation, this tax cost is an investment charge. To determine the tax rate to use in an economy study, the stated tax rate applied to assessed valuation should be multiplied by the expected ratio of assessed valuation to first cost.

For persons and corporations subject to income taxes, such taxes also require consideration in economy studies. Any adequate treatment of this complex subject would require much more space than is available here. One general statement that is accurate as far as it goes is that in economy studies for regulated public utilities, income taxes may properly be treated as investment charges. To estimate the ratio of income taxes to first cost, it is necessary to estimate the prospective income tax rate, the fair-return rate allowed by the regulatory body, the average interest rate paid on the utility's long-term debt, the utility's ratio of borrowed capital to total capital, and the estimated life of the asset in question. For a case in which the fair return is 6 percent, the income tax rate 52 percent, the interest on debt 3 percent, the life 50 yr, and with half the utility's capital borrowed, the income tax will be about 3.5 percent of first cost if salvage value is nil.

In competitive private industry, the treatment of income taxes in economy studies is even more complex.[1] A somewhat oversimplified statement is that, in order to secure a return of i after income taxes, the return before taxes must be i divided by one minus the tax rate. For example, with a tax rate of 52 percent, the return before income taxes must be 14.6 percent in order to secure 7 percent after income taxes.

Mention has already been made of the difference in interest rates used in economy studies for private industry and for public works. Differences in the economic analysis of private and public works are accentuated by the existence of general property taxes and income taxes. For example, consider total-investment charges on a dam having an estimated 50-yr life. As part of a public works project using an interest rate of 4 percent, this dam would be subject to investment charges of 4.66 percent of first cost. If it were built for a privately owned public utility having a cost of capital (fair return) of 6 percent, a general property tax of 1.5 percent, and an income tax amounting to 2.54 percent of first cost, its investment charges would be 10.38 percent.

---

[1] For a good discussion of income taxes in economy studies see Eugene L. Grant and W. Grant Ireson, "Principles of Engineering Economy," 5th ed., Ronald, New York, 1967.

13-7   The relationship between expected frequencies of extreme events and the economic design of hydraulic structures   Many problems of hydraulic design are related to infrequent extreme events such as extreme values of streamflow. At the time designs are made, the dates of the extreme events are completely unpredictable. Extreme values can be predicted only in the probability sense, i.e., in the sense of relative frequency in the long run. One of the objectives of hydrologic studies is, of course, to predict such long-run frequencies.

The required type of economic analysis may be illustrated by a brief consideration of the problem of determining the most economic capacity of storm sewers. The greater the capacity of a storm sewer, the larger its first cost; consequently, the larger its annual investment charges. On the other hand, the greater the capacity, the less frequently will the storm sewer be overflowed. Presumably each overflow will involve some damage to property with resulting costs to the property owners. The economic objective in the design should be to minimize the sum of the annual costs of the sewer (investment charges, maintenance costs, etc.) and the average annual costs of damages by overflow. The more serious the adverse consequences of an overflow, the greater the justifiable investment to reduce the frequency of overflow. For example, in a particular city it might be concluded that the most economical capacity in a residential district would be a capacity that would result in overflow on the average 1 yr out of 3; in the commercial district it might be economical to have large enough capacity to limit overflow to 1 yr in 20.

For many types of hydraulic structures, it turns out to be economical to design against extreme events so infrequent that they may not occur at all during the life of the structure. For instance, a structure with an estimated life of 50 yr might be designed to withstand a flood to occur only once in 500 yr. In this type of situation, the costs associated with the occurrence of the extreme event cannot be expressed as average annual damages expected during the life of the structure due to occurrences of the event. In place of average annual damages, it is necessary to estimate the annual cost of the risk of damage. This annual cost is the product of the estimated probability that the event will be equaled or exceeded in any one year and the estimated cost of the adverse consequences if the event occurs. Table 13-3 illustrates an economy study based on this point of view. This table is adapted from an example cited by Allen Hazen.[1]

The table records an economy study for a large electric utility that acquired a small hydroelectric plant in connection with the purchase of a small utility property. It seemed to the engineers of the large company

---

[1] Allen Hazen, "Flood Flows," Wiley, New York, 1930.

TABLE 13-3   Annual   Costs   for   Different   Proposed   Spillway   Capacities   (After Hazen)

Spillway capacity, cfs	Probability of greater flood in any one year	Cost of enlarging spillway capacity to provide for this flood	Annual investment charges at 10.38%	Annual "cost of flood risk"	Sum of annual costs
1700	0.05	No cost	$ 0	$20,000	$20,000
2000	0.02	$ 30,000	3,114	8,000	11,114
2300	0.01	46,000	4,775	4,000	8,775
2700	0.005	62,000	6,436	2,000	8,436
3000	0.002	81,000	8,408	800	9,208
3300	0.001	104,000	10,795	400	11,195
3600	0.0005	130,000	13,494	200	13,694

that the spillway capacity of 1700 cfs provided by the dam at the plant was inadequate; it was believed that a flood exceeding this capacity would overtop and wash out a section of earth dam whose replacement cost was $400,000.

In order to determine what, if any, increase in spillway capacity was justified, the engineers estimated the costs of increasing the spillway capacity by various amounts. They also estimated the long-run frequencies of floods of various magnitudes, expressed as probabilities of occurrence in any one year. Annual investment charges on the spillway enlargement were assumed as 10.38 percent (capital recovery based on a 50-yr life with interest at 6 percent, property taxes at 1.5 percent, income taxes at 2.54 percent). The annual "cost of the risk" was calculated as the product of the expected damage of $400,000 if the spillway capacity were exceeded and the estimated probability that the capacity would be exceeded in any one year. The total annual costs are lowest for an investment of $62,000. This protects against a flood of 2700 cfs, expected—on the average—once in 200 yr.

In the majority of flood problems, damage is not constant but is an increasing function of flood magnitude. With a 1700-cfs spillway capacity, a 2000-cfs flow might cause $300,000 of damage while a 3000-cfs flow might do $460,000 damage. In this case, the area under the damage vs. probability curve must be determined.[1]

ILLUSTRATIVE EXAMPLE 13-1   Flow-probability and flow-damage data for the project of Table 13-3 are given below. Find the average annual flood damage.

---

[1] Joseph B. Franzini, Flood Control—Average Annual Benefits, *Consulting Eng.*, Vol. XVI, pp. 107–109, May, 1961.

Peak flow, cfs	Probability of flow being equaled or exceeded in any year	Expected damage, dollars
1700	0.05	0
2000	0.02	200,000
2300	0.01	320,000
2700	0.005	400,000
3000	0.002	460,000
3300	0.001	500,000
3600	0.0005	540,000

Compute the cost of flood risk as follows:

Range of peaks, cfs	Average damage, dollars	Probability of flow in interval	Annual damage, dollars
1700–2000	100,000	(0.05–0.02)	3000
2001–2300	260,000	(0.02–0.01)	2600
2301–2700	360,000	(0.01–0.005)	1800
2701–3000	430,000	(0.005–0.002)	1290
3001–3300	480,000	(0.002–0.001)	480
3301–3600	520,000	(0.001–0.0005)	260
		Average annual damage	9430

The probability of occurrence of a flood peak between 2701 and 3000 cfs is 0.003. Since such a flood would cause about $430,000 damage, the average risk in any year is 0.003 (430,000) = $1290. The total for all intervals, $9430, is the average annual flood damage. If a larger spillway were assumed, new estimates of damage would be required.

13-8   Contrast between economy studies for private enterprise and for public works   From whose viewpoint should an economy study be made? Generally speaking, in any study for a business enterprise in competitive industry, the viewpoint will be that of the owners of the enterprise. For example, consider a choice among different designs for hydraulic structures for a private corporation. To make such a choice, estimates should be made of the influence of different designs on the prospective receipts and disbursements by the corporation. Receipts and disbursements by others may usually be viewed as irrelevant. This viewpoint, which reflects the owner's interests, is essential if the corporation is to survive in competition with others. Competition itself serves the public interest through its stimulus to technological progress, cost reduction, and improvements in the standard of living.

In contrast, an engineering economy study for a governmental unit should properly be made from the viewpoint of all the persons affected. For example, consider a proposal that a municipal water department undertake water softening. In judging the merits of this proposal, it is necessary to estimate the benefits to the general public from the delivery of softer water. It is not sufficient to evaluate this proposal solely from the point of view of the costs and revenues of the municipal water department.

In certain types of federal public works projects in the United States, a formal evaluation of the public benefits is required by law. For example, the Flood Control Act of 1936 included the following clause:

It is hereby recognized that destructive floods upon the rivers of the United States, upsetting orderly processes and causing loss of life and property, including the erosion of lands, and impairing and obstructing navigation, highways, railroads, and other channels of commerce between the states, constitute a menace to national welfare; that it is the sense of Congress that flood control on navigable waters or their tributaries is a proper activity of the Federal Government in cooperation with States, their political subdivisions and localities thereof; that investigations and improvements of rivers and other waterways, including watersheds thereof, for flood-control purposes are in the interest of the general welfare; that the Federal Government should improve or participate in the improvement of navigable waters or their tributaries, including watersheds thereof, for flood-control purposes if the benefits to whomsoever they may accrue are in excess of the estimated costs, and if the lives and social security of people are otherwise adversely affected.[1]

From the viewpoint of economic analysis, the key phrase in the foregoing is *benefits to whomsoever they may accrue*. As applied to a proposed flood-mitigation project in any river basin, this implies that a good deal of investigation is required regarding the damages caused by floods of various magnitudes. To secure a money figure for benefits that can be compared with costs, the average annual reduction of damages from each flood-mitigation proposal must be expressed in money terms.

The estimation of benefits of proposed water projects has often been a highly controversial matter. There are obvious difficulties in getting the necessary facts to permit estimation of the way in which a project will affect all the public. Moreover, it is hard to trace the chain of effects throughout the economic system.[2]

¶ [1] "United States Code," p. 2964, 1940.

[2] A good brief statement of the assumptions and methods of estimating benefits and costs of federal flood irrigation, navigation, hydro power, and multipurpose projects is given in the following pamphlet: "Policies, Standards and Procedures in the Formulation, Evaluation, and Review of Plans for Use and Development of Water and Related Land Resources," 87th Cong. 2d Sess., Senate Doc. 97, May, 1962.

13-9   Example of an economic analysis of benefits and costs of alternate public works projects   The average annual damage from floods in a certain river basin is estimated to be $400,000. Estimates are made for several alternate proposals for flood-mitigation works. Certain channel improvements would increase the capacity of the stream to carry flood discharge. There are two possible sites, A and B, for a dam and storage reservoir. Because dam site A is located in the reservoir area for B, one or the other of these sites may be used, but not both. Either site may be used alone or may be combined with channel improvement. It is also possible to use channel improvement alone. Table 13-4 shows the estimated first cost of each project, the estimated annual damages due to floods with each project, the annual investment charges, using an interest rate of 3 percent, and the estimated annual disbursements for operation and maintenance. The life of the channel improvements is estimated as 25 yr; the life of the dam and reservoir is estimated as 100 yr. The final column of the table gives the sum of annual damages and annual project costs. This sum is a minimum for project III, the development at site B alone.

The more conventional way to analyze such public works proposals is by means of the benefit-cost ratio. Table 13-5 illustrates such an analysis. The benefits (to the flood sufferers) are the estimated annual reductions in flood damages. The costs (to the government and therefore to the taxpayers) are the annual investment charges plus annual disbursements for operation and maintenance.

The five benefit-cost ratios of Table 13-5 do not in themselves provide enough information to make an economic choice among the five projects. To use the benefit-cost ratio as a sound basis for project formulation, additional calculations are necessary. The additional benefits added by each separable increment of costs should be computed, and the ratios of the increments of benefits to the corresponding increments of costs should be

TABLE 13-4   Economic Analysis of Proposals for Flood Mitigation

Project	Investment	Average annual flood damages	Annual investment charges	Annual operation and maintenance	Sum of annual damages and project costs
No flood mitigation at all	$ 0	$400,000	$ 0	$ 0	$400,000
I. Channel improvement alone	500,000	250,000	28,720	100,000	378,720
II. Development at site A alone	3,000,000	190,000	94,950	60,000	344,950
III. Development at site B alone	4,000,000	125,000	126,600	80,000	331,600
IV. Site A with channel improvement	3,500,000	100,000	123,670	160,000	383,670
V. Site B with channel improvement	4,500,000	60,000	155,320	180,000	395,320

TABLE 13-5   Benefit-cost Analysis of Flood-mitigation Proposals

Project	Annual benefits	Annual costs	Benefit-cost ratio	Benefit minus costs
I	$150,000	$128,720	1.17	$21,280
II	210,000	154,950	1.36	55,050
III	275,000	206,600	1.33	68,400
IV	300,000	-283,670	1.06	16,330
V	340,000	335,320	1.01	4,680

determined. Such an analysis will lead to selection of project III, just as in the analysis shown in Table 13-4. The preceding statement assumes that extra costs are justifiable whenever the resulting benefits exceed the extra costs but are not justified if the resulting benefits are less than the extra costs. In other words, the most economical design is the one that gives the greatest excess of benefits over costs.

Thus project II adds benefits of $60,000 over project I, whereas costs are increased by only $26,230; the ratio of extra benefits to extra costs is 2.29. Similarly, the extra $51,650 of costs of project III over project II are justified by increased benefits of $65,000; the incremental benefit-cost ratio is 1.26. But projects IV and V are clearly uneconomical as compared with project III because the added benefits are considerably less than the extra costs required to produce the benefits.

Figure 13-1 shows the benefits and costs of a hypothetical problem graphically. From point A to B, benefits exceed costs, i.e., the benefit-cost ratio exceeds one. The curve of benefits minus costs (i.e., net benefits) shows a maximum at C. Beyond this point, each dollar of costs returns less than a dollar of benefits, i.e., Δ benefits/Δ costs < 1. The maximum ratio of benefits to costs occurs at D, and this should be the limit of project size if it is desired to obtain a maximum rate of return on the investment. The increment from D to C is, however, economic since the rate of return on the increment exceeds the minimum attractive rate of return.

As suggested in the previous paragraph, a project may be evaluated in terms of *rate of return*. Using the estimated cost and benefit stream over the period of the project life, one may determine the rate of return on investment represented by the excess of benefits over costs. Projects may be ranked in merit on rate of return, and the decision to proceed can be based on a minimum acceptable rate of return as well as on a benefit-cost ratio.

13-10   Capital budgeting   There is almost always a limit on the funds available for water-resources projects during a period of time. The limit may be set by limitations on tax revenue or bonded indebtedness. Water

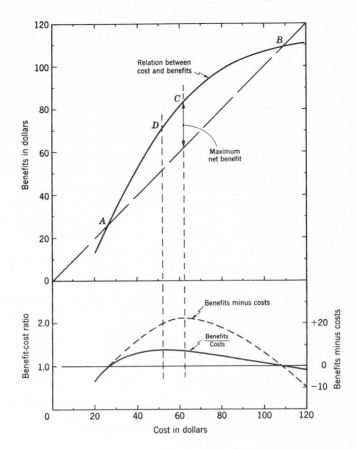

FIG. 13-1   *Relation between costs and benefits for a hypothetical project.*

projects are always competing with schools, roads, public health, defense, social security and other activities for the same funds. No government could possibly afford to build all the feasible water projects within its jurisdiction in any single year.

Project analysis should, therefore, do more than indicate that the project is above some marginal limit which makes it "economically feasible." The analysis should serve to rank all possible projects in order of priority so that those projects offering the highest return can be built first. Marglin shows[1] that in the presence of a budgetary limitation, only those projects with

[1] Stephen A. Marglin, Economic Factors Affecting System Design, chap. 4 in Maass and others, "Design of Water Resources Systems," Harvard, Cambridge, Mass., 1962.

benefit-cost ratios in excess of $1 + \lambda$ should be constructed. The factor λ is set such that the summation of costs of all projects with benefit-cost ratios greater than $1 + \lambda$ is equal to the available funds.

The budgetary constraint may take two forms: (1) a limitation on the capital investment in any year or (2) a limitation on the net expenditure (investment + operation and maintenance − income to the funding agency) in any year. The first form will encourage planners to minimize capital expenditure at the cost of increased operation, maintenance, and replacement (OMR) costs. If OMR costs are small, the decision between the two forms of constraint is academic. If OMR costs are large, the second form of constraint requires that OMR costs return benefits equal to $1 + \lambda$, and planners must consider the effect of all future OMR costs as well as capital investment. In general, the constraint will be applied as a means of rationing present funds and the limitation on construction costs only is most appropriate.

The principle of ranking projects on the basis of economic analysis is an important one if an agency is to obtain maximum return on its investment. Governments find such ranking difficult to reconcile with political pressures which encourage uniform areal distribution of projects rather than maximization of return. Intangible factors such as the need to stimulate the economy within a region also indicate a departure from the strict capital-budgeting rules.

BIBLIOGRAPHY

Bullinger, C. E.: "Engineering Economic Analysis," 3d ed., McGraw-Hill, New York, 1958.
Eckstein, O.: "Water Resources Development: The Economics of Project Evaluation," Harvard, Cambridge, Mass., 1958.
Grant, E. L., and P. T. Norton, Jr.: "Depreciation," Ronald, New York, 1949.
Grant, E. L., and W. G. Ireson: "Principles of Engineering Economy," 5th ed., Ronald, New York, 1970.
Hirshleifer, J., J. C. DeHaven, and J. W. Milliman: "Water Supply—Economics, Technology, and Policy," The University of Chicago Press, Chicago, 1960.
James, L. D., and R. R. Lee: "Economics of Water Resources Planning," McGraw-Hill, New York, 1971.
Krutilla, J. V., and O. Eckstein: "Multiple Purpose River Development," Johns Hopkins, Baltimore, 1958.
Maass, Arthur, and others: "Design of Water Resource Systems," Harvard, Cambridge, Mass., 1962.
McKean, R. N.: "Efficiency in Government through Systems Analysis, with Emphasis on Water Resource Development," Wiley, New York, 1958.
Woods, B. M., and E. P. De Garmo: "Introduction to Engineering Economy," 2d ed., Macmillan, New York, 1953.

PROBLEMS

13-1 Section 13-4 states that if plans A and B of the preceding section should be compared, with an interest rate of 1.5 percent, plan A would appear to be more economical by nearly $2700 a year, whereas at 8 percent plan B would be superior by nearly $20,000 a year. Make the necessary calculations to verify these figures.

13-2 Show the necessary calculations to verify the statement in the final sentence of Sec. 13-5.

13-3 An engineer examining Table 13-5 states, "I would select plan II because it gives the highest benefit-cost ratio of all of the plans." Comment on this statement.

13-4 Another engineer examining Table 13-5 states, "I would select plan V, as it gives the highest possible benefits and it is an economically sound project because its benefits are greater than its costs." Comment on this statement.

13-5 Compare the five flood-mitigation projects of Tables 13-4 and 13-5, using an interest rate of 4 percent.

13-6 Compare the five flood-mitigation projects of Tables 13-4 and 13-5, using an interest rate of 6 percent.

13-7 Compare the conclusions of the economic analysis of the five flood-mitigation projects with the interest rate of 3 percent used in Sec. 13-9 and with the 4 percent and 6 percent rates used in your solution to Probs. 13-5 and 13-6. What generalizations are suggested by this comparison with reference to the influence of the interest rate on such studies?

13-8 Projects X, Y, and Z have costs of 100, 180, and 140, respectively. The benefit-cost ratios are 1.5, 2.2, and 2.5. Which of these projects would you select? Why?

13-9 Intermittent flooding has occurred along a reach of river which has a capacity of 5000 cfs. Study shows that the average annual cost of increasing the capacity by lining and straightening could be approximated by $C = 100Q - (400 + 4Q^2)$ where C is in thousands of dollars and Q is the increased capacity in thousands of cfs. The average annual benefits in thousands of dollars would be $B = -1050 + 230Q - 0.8Q^3$. These expressions apply in the range from 5000 to 10,000 cfs. What is the optimum capacity? Making reasonable assumptions estimate the first cost of this project. State all assumptions clearly.

13-10 A pump is to raise 700 gpm against a static head of 75 ft through 4000 ft of pipe. Determine the most economical pipe size if the power cost is 1.8 cents per kilowatthour; the pump will operate 18 hr/day, and the cost of pipe per foot is $1.80 for 4-in., $2.45 for 6-in., $3.15 for 8-in., $3.70 for 10-in., $4.15 for 12-in., and $4.55 for 14-in. Assume a minimum attractive return of 8 percent and a life of 20 yr. Take $f = 0.03$ and pump efficiency as 80 percent.

13-11 An impulse turbine is to be installed at a point where 2 mi of pipeline will be required. The available head is 1200 ft, and the turbine capacity is 50 cfs. The turbine will operate 300 days/yr at full load and 60 days/yr at half load and will be out of operation 5 days/yr. Turbine efficiency is 85 percent at full load and 80 percent at half load. If the steel pipe will cost 17 cents per pound and the power produced by the turbine is worth 0.7 cent per kilowatthour, what is the most economical pipe size? Assume a 30-yr life for the installation and a 6 percent minimum attractive return. Neglect water hammer and assume pipe of constant thickness.

13-12 A water department is considering the purchase of a motor to operate a pump, requiring 80 hp. Two motors are to be considered. One is a gasoline motor costing $2500 and developing 95 hp. This motor is expected to have a salvage value at the end of 10 yr of $500. Its annual costs for fuel and attendance are estimated at $4500. A 100-hp electric motor will cost $3500 installed, and it will have a salvage value of $500 at the end of 20 yr. Annual power and attendance costs are estimated at $4000. Which motor would be the best selection on the basis of an interest rate of 6 percent? Are there any irreducibles that would tend to alter this decision?

13-13 A canal is to carry 500 cfs for 300 days during the year. The canal will have the most efficient cross section, with side slopes of 2 horizontal to 1 vertical. The canal can be excavated at a price of $0.40 per cubic yard. A concrete lining, if provided, will cost $1.20 per square yard. Annual maintenance cost of the unlined canal (cleaning and killing weeds) is estimated at $1000 per mile. Maintenance charges of $5000 every 12 yr are estimated for repair of the lined canal. If the interest rate is 5 percent, the life of either alternative 75 yr, the estimated seepage loss from the unlined canal 1 cu ft/sq ft/day, and the price of water $4.50 per acre-foot, would you recommend lining the canal? Land slope = 0.0025. Freeboard = 1 ft.

13-14 A small irrigation project is considering the design of the head gates for its main canal. Manually operated gates will cost $7800. These gates will require daily adjustment on each of the 280 days of the irrigation season. It is estimated that it will take a man about 2 hr to travel to the intake, adjust the gates, and return. His pay rate will be $2.25 per hour. Transportation costs for the 12-mi trip (one way) are estimated at $0.10 per mile. The alternative is motor-operated gates, which can be controlled electrically from headquarters. Installation costs are estimated at $25,000, and annual costs for maintenance and power are estimated at $450. If the minimum attractive return is 8 percent, what project life would be necessary for the two alternatives to break even? What life would be necessary if the minimum attractive return were 3 percent?

13-15 A 12-in. well extends 150 ft below the ground surface. The undisturbed water level is 60 ft below the ground surface, and the transmissibility of the aquifer is 2000 gpd/ft. The value of water is $3 per acre-foot, power costs 2 cents per kilowatthour, and the pump and motor efficiency is 73 percent. Using Eq. (4-12) and assuming $r_2 = 500$ ft, determine the break-even point for pumping costs in terms of well discharge. Ignore entrance losses into the well.

13-16 Ten water-resource projects are under consideration. The estimated annual costs, annual benefits, and B/C ratios for the projects are as follows:

Project	Average annual costs	Average annual benefits	B/C
P	$ 40,000	$ 50,000	1.20
Q	60,000	60,000	1.00
R	25,000	37,000	1.48
S	80,000	94,000	1.17
T	100,000	118,000	1.18
U	72,000	90,000	1.25
V	50,000	66,000	1.32
W	60,000	50,000	0.83
X	70,000	100,000	1.43
Y	80,000	120,000	1.50

Which projects should be built if budgetary limitations restrict the annual costs to: (a) $75,000; (b) $200,000; (c) $300,000?

13-17 Alternate bridges are under consideration for a particular river crossing in a remote location on a major highway. There is no development near the bridge, so that if it were to fail, the only loss would be that of the bridge plus any economic loss caused by inability to transport goods because of the washout. In case of washout, assume zero salvage value. All alternatives have a useful structural life of 30 yr, and the cost of their maintenance is negligible. Interest rate is 5 percent. The alternatives, their estimated costs, and the flood at which each will fail are as follows:

Alternate	First cost	Washout flood, yr
A	$150,000	100
B	120,000	60
C	105,000	40

From an economic viewpoint which alternate ought one to select? Consider the three cases in which the estimated economic loss from washout due to inability to transport goods are (a) zero, (b) $200,000, and (c) $500,000, respectively.

Chapter 14

Irrigation

Irrigation is the application of water to soil to supplement deficient rainfall to provide moisture for plant growth. The first use of irrigation by primitive man is lost in the shadows of time, but it must have marked an important step forward in the march of civilization. Only about one-third of the earth's surface receives enough precipitation in a normal year to mature food crops, and much of this area is unsuited for agriculture.

The first irrigation on the American continent antedates the coming of the white man. Some ditches constructed by the Hohokam Indians in the Salt River Valley of Arizona prior to A.D. 1400 are still in use. Spanish colonists under Don Juan de Onate constructed irrigation works in New Mexico in 1598. The reclamation of the western United States dates from 1847, when Mormon pioneers built their first irrigation dam across City Creek at Salt Lake City, Utah. From this project, which irrigated 5 acres of land, irrigation in the United States[1] had been extended to over 45 million acres by 1965. Irrigation accounts for 83 percent of the water consumptively used in the United States.

14-1 Land classification The first step in planning an irrigation project is to establish the capability of the land to produce crops which provide adequate returns on the investment in irrigation works. *Arable land* is land which, when properly prepared for agriculture, will have a sufficient yield to justify its development. *Irrigable land* is arable land for which a water supply is available. The U.S. Bureau of Reclamation general specifications for land classification appear in Table 14-1. To be suitable for irrigation farming, the soil must have a reasonably high water-holding capacity (Sec. 14-6) and must be readily penetrable by water. The infiltra-

---

[1] "The Nation's Water Resources," U.S. Water Resources Council, 1968.

TABLE 14-1   U.S. Bureau of Reclamation General Land Classification Specifications

	Class 1, *arable*	Class 2, *arable*	Class 3, *arable*
Texture	Sandy loam to friable clay loam	Loamy sand to very permeable clay	Loamy sand to permeable clay
Depth to sand or gravel	36 in. plus of free-working fine sandy loam or heavier, or 42 in. of sandy loam	24 in. plus of free-working fine sandy loam or heavier, or 30–36 in. sandy loam	18 in. plus of free-working fine sandy loam or heavier, or 24–30 in. of lighter soil
Depth to impermeable shale or raw soil	60 in. plus or 54 in. with 6 in. of gravel over impervious material, or sandy loam throughout	48 in. plus or 42 in. with 6 in. of gravel over impervious material, or loamy sand throughout	42 in. plus or 36 in. with 6 in. of gravel over impervious material, or loamy sand throughout
Depth to penetrable lime zone	18 in. with 60 in. penetrable	14 in. with 48 in. penetrable	10 in. with 36 in. penetrable
Alkalinity at equilibrium*	Exchangeable sodium generally less than 15% for all land classes, but may be higher or lower depending on the type of clay minerals		
Salinity at equilibrium*	Electrical conductivity of saturation extract less than 4 millimhos per cm	Electrical conductivity of saturation extract less than 8 millimhos per cm	Electrical conductivity of saturation extract less than 12 millimhos per cm
Slopes	Smooth slopes up to 4% with large areas in same plane	Smooth slopes up to 8% in large areas in the same plane, or rougher slopes less than 4% in general gradient	Smooth slopes up to 12% in large areas in the same plane, or rougher slopes less than 8% in general gradient
Surface	Requires little leveling and no heavy grading	Moderate grading required, but in amounts found feasible in comparable areas	Heavy and expensive grading required in spots, but in amounts found feasible in comparable irrigated areas
Cover (rocks and vegetation)	Insufficient to affect productivity, or clearing cost small	Sufficient to reduce productivity and interfere with farming; clearing possible at moderate cost	Requires expensive but feasible clearing
Drainage†	No drainage requirement expected	Some drainage expected, but at reasonable cost	Considerable drainage required. Considered expensive but feasible

See page 396 for footnotes.

TABLE 14-1   (continued)

Class 4, *limited arable*

Includes irrigable lands which are adaptable to a narrow range of crops

---

Class 5, *nonarable*

Includes lands which require additional studies to determine their irrigability and lands re-
classified as temporarily nonproductive pending construction of corrective works and reclamation through application of these works

---

Class 6, *nonarable*

Includes lands which do not meet the minimum requirements and small areas of arable land lying
within larger bodies of nonarable land

---

*Equilibrium conditions based on projected use of a specific irrigation water supply.

†Drainage study to be made by drainage engineers. Drainage requirements indicated refer to
expenditures by the farmer. Additional drainage facilities, as required, will be provided as part
of project costs.

tion rate should be low enough, however, to avoid excessive loss of water by
percolation below the root zone. The soil must be deep enough to allow
root development and permit drainage. It must be free of black alkali, a
sodium-saturated condition, and free of salts not susceptible to removal by
leaching. Finally the soil must have an adequate supply of plant nutrients
and be free of toxic elements (Sec. 14-9).

Land slopes should be such that excessive erosion will not occur. Steep
slopes are also conducive to water losses by surface runoff unless the soil is
quite permeable. Land on moderate slopes but with an irregular surface
may be leveled if the soil is sufficiently thick. Where the soil is thin, the
leveling operation may remove productive soil and leave areas of relatively
barren soil at the surface. Impermeable substrata may lead to a perched
water table, which if close to the surface may require expensive drainage
facilities (Chap. 18). Removal of excess water from the root zone is essential
to avoid accumulation of salts and permit the aeration required by most
plants. Lands located in depressions or valley floors may present drainage
problems because of the lack of natural drainage outlets.

The land should be so located that irrigation is possible without excessive
pumping or transmission costs. The general layout and size of the area
should be conducive to division into field units which permit effective
farming practices. The land should be adaptable to more than one crop
since changing economic or technological factors may force changes in
cropping practice. Climate is an important factor in land evaluation. A
year-round, frost-free period permits double or triple cropping and corre-

spondingly greater return per acre. A short growing season limits the return and the types of crops which can be grown.

14-2   Crop water requirements   Having established the suitability of an area for irrigation, the next step is the determination of water requirements. The total water requirement consists of the water needed by the crop plus the losses associated with the delivery and application of the water. The best source of information on overall water requirements is often the experience of good irrigators operating under conditions similar to those of the project area. Such information must be selected with care since it is common practice to use excessive amounts of water if an abundant supply is available. Dissimilarity in soil, climate, or underlying geology can also greatly change the total water requirement.

If no direct determination of total water requirement is possible, an estimate may be made by first estimating *consumptive use* (Sec. 2-16) and adjusting this value for other losses. Consumptive use may be determined experimentally by planting a crop in a lysimeter or tank of soil and keeping an accounting of water added and soil-moisture changes. The consumptive use is equal to the water added plus or minus any change in soil moisture. A similar determination may be made in a field plot if the groundwater table is far enough below the surface so that it supplies no water for the plants. In field-plot experiments, deep percolation of applied water must be avoided, usually by applying only small amounts at each irrigation.

The overall consumptive use for large areas may be estimated by calculating the hydrologic balance for the area [Eq. (2-7)]. Table 14-2 gives some values of consumptive use for selected crops and native plants. The consumptive use depends on the crop, soil fertility, available moisture, climate, and irrigation methods and varies considerably from one study to another.

In the absence of data which can be transferred to the project area, consumptive use may be estimated on the basis of pan evaporation or meteorological data. Seventy percent of pan evaporation approximates the potential evapotranspiration. The Jensen-Haise[1] formula is

$$E_t = (0.014T - 0.37)R_s \tag{14-1}$$

where E_t is the potential evapotranspiration in inches per day and R_s is the solar radiation expressed in inches per day based on 1487 cal/sq cm $= 1$ in. T is the mean air temperature in degrees Fahrenheit.

Many tests indicate the existence of an optimum consumptive use which produces a maximum crop yield. Typical curves from some of these experi-

---

[1] M. E. Jensen and H. R. Haise, Estimating Evapotranspiration from Solar Radiation, *J. Irrigation and Drainage Div., ASCE*, Vol. 89, pp. 15–41, December, 1963.

TABLE 14-2   Consumptive Use by Various Crops and Native Plants

Crop	Region	Consumptive use, ft/yr	Method	Authority
Alfalfa	Bonners Ferry, Idaho	2.8	Tank	Criddle-Marr
Alfalfa	Los Angeles, Calif.	3.1	Field	Blaney
Beets	Scottsbluff, Nebr.	2.0	Field	Bowen
Citrus	Los Angeles, Calif.	1.9	Field	Blaney
Cotton	Shafter, Calif.	2.5	Field	Beckett-Dunshee
Cotton	State College, N. Mex.	2.4	Tank	Israelson
Grass weeds	San Bernardino, Calif.	1.8	Tank	Blaney-Taylor
Greasewood	Escalante Valley, Utah	2.1	Tank	White
Mixed	San Luis Valley, Colo.	1.6	Eq. (14-2)	Blaney-Rohwer
Mixed	Carlsbad, N. Mex.	2.4	Eq. (14-2)	Blaney-Morin
Mixed	Uncompahgre, Colo.	.2.3	Eq. (14-2)	Lowry-Johnson
Potatoes	San Luis Valley, Colo.	1.3	Tank	Blaney-Israelson
Rushes	Ft. Collins, Colo.	4.4	Tank	Parshall
Tamarisk	Safford, Ariz.	5.1	Tank	Turner-Halpenny
Tules	King Island, Calif.	7.5	Tank	Stout
Peaches	Ontario, Calif.	2.5	Field	Blaney
Walnuts	Santa Ana, Calif.	2.1	Field	Beckett
Wheat	San Luis Valley, Colo.	1.2	Tank	Blaney-Israelson
Wheat	Bonner's Ferry, Idaho	1.5	Tank	Criddle-Marr
Wild hay	Gray's Lake, Idaho	2.6	Tank	Criddle-Marr

ments are shown in Fig. 14-1 as relations between water applied and resulting yield. The shape of such curves varies with climate, crop, soil, and other factors. It appears that maximum dry matter production occurs when water use equals potential evapotranspiration.[1] The cost of water and other fixed charges on the farm enter the determination of the most economic use of water. The point of maximum yield is not necessarily the best goal.

14-3 Crop-irrigation requirement The *crop-irrigation requirement* is that portion of the consumptive use which must be supplied by irrigation. It is the consumptive use less the effective precipitation. Winter precipitation is effective only to the extent that it remains in the soil until the growing season. Average moisture retention per foot of depth for various soil types is given in Table 14-3. Effective winter precipitation is the actual precipitation or the available moisture storage, whichever is less. Only storage in the root zone, which usually extends to a depth of 3 or 4 ft, should be considered.

Precipitation during the growing season is effective only when it remains

---

[1] R. J. Hanks, H. R. Gardner, and R. L. Florian, Plant growth evapotranspiration relations for several crops in the Central Great Plains, *Agronomy Journal*, Vol. 61, pp. 30–34, 1969.

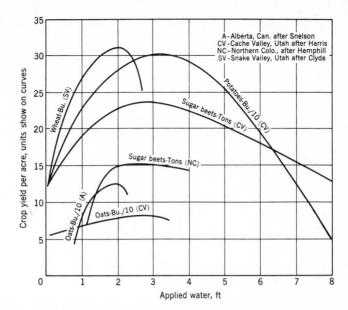

FIG. 14-1   Some examples of the variation of crop yield with applied water.

in the soil and is available to plants. Rainfall-runoff relations (Sec. 3-6) might be used to estimate soil-moisture accretion storm by storm, but this would be a tedious job if done for a long record. More in accord with the accuracy of estimates of water requirements is the assumption of a linear variation from 100 percent effectiveness for the first inch of rain in a month to zero effectiveness for all rain over 6 in. in a month. Such an approach assumes a maximum effective precipitation of 3.5 in./month. This is obviously only an approximation. The effective growing-season precipitation is the sum of the monthly values of effective precipitation. The average annual effective precipitation for the period of record is subtracted from the esti-

TABLE 14-3   *Moisture Storage in Various Soils**

Soil type	Storage, in./ft
Clay	2.5
Silt loam	2.0
Sandy loam	1.5
Fine sand	1.0
Coarse sand	0.5

*See Appendix for additional information on soils.

mated annual consumptive use to determine the annual crop-irrigation requirement. Modern computer simulation procedures (Chap. 3) offer a much more realistic approach and an opportunity to consider net consumptive requirements in terms of probability.

It is usually necessary to determine monthly increments of the crop-irrigation requirement in order to design a distribution system capable of delivering the water required in the period of highest demand. Here again the best guide is the experience of irrigation projects operating under similar conditions to those of the project under design. Typical monthly percentages of demand for some projects are shown in Table 14-4. The estimated annual demand may be distributed on the basis of these data.

14-4   Farm delivery requirement   It is virtually impossible to operate any irrigation project without waste or loss of water. Losses at the farm during irrigation include deep seepage and surface runoff. In light sandy soils without an impermeable subsoil, a considerable amount of water may percolate downward beyond the root zone and so become useless for crops. Percolation loss may be minimized by applying small amounts of water at each irrigation so that the storage capacity of the soil reservoir is not exceeded. Sprinkler irrigation may result in percolation losses as low as 5 percent of applied water, while flooding over an extended period of time may result in a loss of three-fourths of the water. The usual range of percolation loss is from 15 to 50 percent of applied water.

When irrigation water is applied at a rate in excess of the infiltration capacity of the soil, it may flow across the field and be wasted as surface

TABLE 14-4   *Monthly Consumptive-use Percentages*

Month	Boise, Idaho*	West Stanislaus district, Calif.	Mesilla Valley, N. Mex.*
January	1.4	0	1.1
February	1.8	1.2	2.5
March	3.2	5.5	5.4
April	6.8	13.4	8.6
May	11.8	13.4	9.3
June	16.3	16.0	13.2
July	17.3	20.8	18.3
August	16.8	18.0	17.6
September	11.4	9.3	11.8
October	5.9	2.0	6.4
November	4.1	0.3	3.6
December	3.2	0.1	2.2

*U.S. Bur. Reclamation Manual, Vol. IV, Water Studies, Chap. 4.1, 1951.

runoff at the downslope side. Steep slopes or soils of low permeability favor high rates of surface runoff. Surface runoff should not exceed about 5 percent of the applied water with proper irrigation methods.

The amount of water q_f in acre-feet per acre per year that must be delivered to the farm is

$$q_f = \frac{U_c - P_{\text{eff}}}{1 - L_f} \tag{14-2}$$

where U_c is the consumptive use and P_{eff} the effective precipitation, both expressed in feet per year, and L_f the farm loss expressed as a decimal. Additional water may be necessary for leaching salts from the soil or to prevent salt accumulation if the irrigation water is highly mineralized (Sec. 14-9).

The ratio of irrigation water consumed ($U_c - P_{\text{eff}}$) to q_f is called the farm efficiency. Average efficiencies[1] are usually between 40 and 60 percent although with careful choice of irrigation method, application rate, and irrigation frequency to fit the soil conditions, efficiencies above 80 percent are possible under favorable conditions.[2] As water shortages become more severe, efficient irrigation practices will become more important if agriculture is to compete with municipal and industrial uses for water.

14-5   Diversion requirement   In addition to farm losses, some water will be lost in transit to the farm (*conveyance loss*). This loss consists of evaporation from the canal, transpiration by vegetation along the canal bank, seepage from the canal,[3] and operational waste. Evaporation and transpiration losses are ordinarily small and are usually neglected. Evaporation loss may be estimated as discussed in Sec. 2-14, and transpiration may be treated as consumptive use (Sec. 14-2). Operational waste includes water discharged through wasteways because of refusal by users to take the total flow, leakage past gates, and losses from overflow or breakage of canal banks. The magnitude of operational waste depends on the care that is exercised in the operation of the system but should be less than 5 percent. The largest factor in conveyance loss is seepage (Sec. 10-13). The diversion requirement may be taken as the sum of the farm delivery and the estimated conveyance loss in acre-feet. As an alternative, conveyance losses may be estimated as a fraction of the diversion, and the diversion requirement q_d in acre-feet per year is then

---

[1] M. E. Jensen, Evaluating Irrigation Efficiency, *J. Irrigation and Drainage Div., ASCE,* Vol. 93, pp. 83–98, March, 1967.

[2] J. Keller, Effect of Irrigation Method on Water Conservation, *J. Irrigation and Drainage Div., ASCE,* Vol. 91, pp. 61–72, June, 1965.

[3] A. Bandini, Economical Problems of Irrigation Canals: Seepage Losses, *J. Irrigation and Drainage Div., ASCE,* Vol. 92, pp. 35–57, December, 1966.

$$q_d = \frac{q_f A}{1 - L_c} \qquad (14\text{-}3)$$

where L_c is the conveyance loss in decimals and A is the gross area irrigated in acres. With open ditches, conveyance loss will usually range between 25 and 40 percent of the diversion. Conveyance losses may be virtually eliminated by using a pipe system, and economy would result if the added cost were offset by the value of the water saved.

14-6 Soil-water relationships   Water applied during irrigation enters the soil, and plants in turn extract water from the soil for their growth. The soil is actually a reservoir in which water is stored for use by plants between irrigations. The storage and movement of this soil water are important factors in irrigation planning.

Water may be present in the zone of aeration in three different conditions. It may be moving in the large soil pores under the influence of gravity; it may be under the action of capillarity in small pore spaces; or it may be retained about individual soil particles by molecular attraction (hygroscopic water). The capillary water can be removed from soil only by applying a force sufficient to overcome the capillary forces, while hygroscopic moisture can be removed only by heating. This division of soil moisture into three classes might lead one to expect a sharp discontinuity in moisture retention as each class is depleted. If the moisture content of a soil sample is measured after equilibrium has been reached under various negative pressures, a curve of moisture content vs. negative pressure applied[1] is a smooth curve (Fig. 14-2). This reflects the wide range of pore sizes in the soil and indicates a gradual transition in the magnitude of the moisture-holding forces in soil. Such a curve can also be established by measuring the equilibrium moisture in the soil at various levels above a free water surface (Fig. 14-3). The height above the free water surface is a measure of the capillary tension existing in the soil.

Curves such as Fig. 14-2 show that a greater effort is necessary to remove each successive increment of soil moisture as the amount of residual moisture is depleted. Certain critical points on this curve have been defined arbitrarily to aid in describing soil-water characteristics. The *field capacity* is the moisture content of the soil after free drainage has removed most of the gravity water. Since drainage of water will continue for some time after a soil is saturated, field capacity must be defined in terms of a specified drainage period, usually 2 to 5 days. A more precise definition of field capacity is desirable, and a promising approach seems to be a relation in terms of a

---

[1] R. K. Schofield, The pF of Water in the Soil, *Trans. Third Intern. Congr. Soil Sci.*, Vol. 2, p. 37–48, 1935.

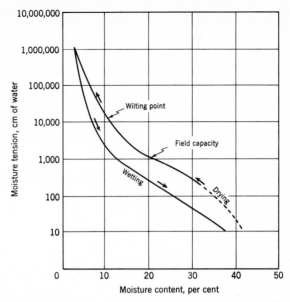

FIG. 14-2   Relation between moisture content and capillary tension in Greenville (Utah) loam. (After Schofield)

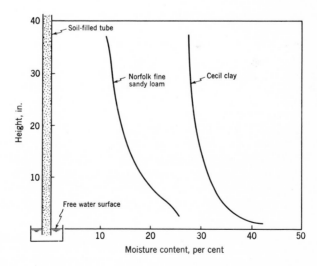

FIG. 14-3   Variation of soil-moisture content above a free water surface. (After Buckingham)

definite tension. Colman[1] showed that, for 120 California soils, the field capacity is approximately equal to the moisture retained at a tension of $\frac{1}{3}$ atm. Veihmeyer and Hendrickson[2] have shown that the *moisture equivalent*, or moisture retained after centrifuging a sample $\frac{3}{8}$ in. deep for 30 min at a speed equivalent to a force of $1000g$, is nearly equal to the field capacity in fine-grained soils.

The other end of the moisture scale from the agricultural viewpoint is the *permanent wilting point*. This represents the moisture content at which plants can no longer extract sufficient water from the soil for growth. The wilting point has been determined for many years by testing soil samples with young plants. Recent studies indicate that the wilting point is closely indicated by the moisture retained against a tension of 15 atm. Table 14-5 gives average values of field capacity and wilting point for the major soil types. The difference between the moisture content at these levels is the *available water*, the soil moisture which is useful to plants. It must be emphasized that there is a wide range in the moisture characteristics of soils within each of the major classes. Theoretically, the most efficient irrigation[3] is achieved if, when the moisture content of the soil approaches the wilting point, irrigation water is supplied in an amount sufficient to raise the soil moisture to the field capacity within the root zone.

14-7 Movement of soil moisture The movement of soil moisture is in many ways analogous to the transfer of heat. The general equation of moisture movement (rate of flow per unit of cross-sectional area) may be stated as

$$q = Q/A = -k\frac{\partial H}{\partial s} \tag{14-4}$$

where H is the total head, k the hydraulic conductivity of the soil, and s distance along the line of flow. In the general case, H is the algebraic sum of the capillary, gravitational, and vapor-pressure heads. *Capillary head* is the suction required to pull water from the soil. The *gravity head* is the height above some datum. The *vapor-pressure head* is a function of the vapor-pressure gradients in the soil which result from temperature differences within the soil mass. The conductivity k is the quantity of water which will move through a unit cross section in 1 sec under a unit gradient. As would be expected, k depends on the size of the soil pores, but it also de-

[1] E. A. Colman, A Laboratory Procedure for Determining the Field Capacity of Soils, *Soil Sci.*, Vol. 63, p. 277, 1947.

[2] F. J. Veihmeyer and A. H. Hendrickson, The Moisture-equivalent as a Measure of the Field Capacity of Soils, *Soil Sci.*, Vol. 32, pp. 181–193, 1931.

[3] W. O. Pruitt and M. C. Jensen, Determining When to Irrigate, *Agr. Eng.*, Vol. 36, pp. 389–393, June, 1955.

TABLE 14-5   Typical Moisture Values for Various Soil Types

Soil	Field capacity	Wilting point	Available water	Specific weight, lb/cu ft (dry)
	percent of dry weight of soil			
Sand	5	2	3	95
Sandy loam	12	5	7	90
Loam	19	10	9	85
Silt loam	22	13	9	80
Clay loam	24	15	9	80
Clay	36	20	16	75
Peat	140	75	65	20

creases as the moisture content of the soil decreases. Because of the difficulty of determining H and k as a function of moisture content, Eq. (14-4) may not always be useful. The essence of the equation is that moisture tends to move from regions of high moisture content to regions of low moisture content. The mathematical theory[1] of soil moisture movement (a form of unsteady unsaturated flow) has been developed largely since 1950. It is based on a partial differential equation of the diffusion type that is nonlinear and generally solved by numerical methods[2] with a digital computer.

When rain or irrigation water is applied to a soil surface, both gravity and capillary potential tend to cause its downward movement by infiltration. If the water table is close to the surface and sufficient water is supplied, the moisture may reach the water table and add to the groundwater. If the water table is deep or the applied water insufficient, the moisture may never reach the groundwater as it may be removed by evapotranspiration before it reaches the water table. The presence of a relatively impermeable subsoil checks the downward movement of the water. If the supply of water is adequate, a perched water table, or zone of saturation, may form above the impermeable layer. This may create a drainage problem since most plants do not prosper with their roots in a saturated zone. The proper quantity and rate of application of irrigation water are thus controlled by the moisture-holding and transmission capacities of the soil. It should be noted that these soil characteristics may change as a result of chemical action with saline

[1] E. E. Miller and A. Klute, The Dynamics of Soil Water, chap. 13 in R. M. Hagan, H. R. Haise, and T. W. Edminster (eds.), "Irrigation of Agricultural Lands," American Society of Agronomy, Madison, Wis., 1967.

[2] J. Rubin, Numerical Method for Analyzing Hysteresis-affected, Post Infiltration Redistribution of Soil Moisture, *Proc. Soil Sci. Soc. of Amer.*, Vol. 31, pp. 13–20, 1967.

water or fertilizer or as a result of compaction by movement of heavy farm equipment.

14-8   Measurement of infiltration and soil moisture Infiltration rates into irrigated soils are commonly measured by the *tube method.* A metal tube 8 to 12 in. in diameter is forced into the soil to a depth of 18 to 24 in., with a few inches projecting above the soil surface. Water is applied in this tube, and the rate of disappearance is measured. Since air must escape from the soil as water enters, the simple tube just described offers favorable infiltration conditions, for displaced air may rise outside the tube. To minimize this effect, a buffer strip may be provided with a concentric ring of larger diameter than the tube (Fig. 14-4). A single determination of infiltration rate may be misleading because of local variations in soil characteristics, and several measurements are necessary to establish average characteristics.

The oldest method of measuring soil moisture is to remove a sample of soil and determine its loss in weight when oven-dried. This loss expressed as a percentage of the weight of the dry soil is the moisture content of the soil. If the specific weight of the soil is known, the moisture content may be expressed as inches per foot of depth. In-place measurements of soil moisture can be made by electrical-resistance methods.[1] A resistance element consists of two electrodes embedded in a porous dielectric such as plaster of paris, nylon, or fiber glass. When the element is buried in the soil, it maintains a moisture equilibrium with the surrounding soil, and the resistance between the electrodes varies with the moisture content of the material composing

---

[1] E. A. Colman and T. M. Hendrix, The Fiberglass Electrical Soil-moisture Instrument, *Soil Sci.,* Vol. 67, pp. 425–438, 1949.

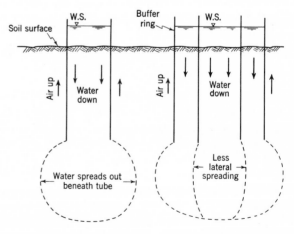

FIG. 14-4   The tube method of measuring infiltration.

the block. Measurements of this resistance provide an indication of the soil moisture. A calibration curve for an element may be developed by taking soil samples near the installation for laboratory analysis, or by embedding a similar element in a small container of the soil and taking simultaneous readings of resistance and weight of the container as the soil dries. If the chemical content of the water changes a new calibration curve is needed. Elements of this type have been used by farmers to indicate the need for irrigation. The resistance readings of an element are about the same at field capacity (or wilting point) in most soils. Thus, without specific calibration, it is possible to know approximately when the moisture is at one of these values.

The most recent development for the measurement of soil moisture is the neutron-scattering device.[1] A fast neutron source is lowered into a prepared aluminum access tube in the soil. Fast neutrons lose much more energy when they collide with atoms of low atomic weight than when they collide with heavier atoms. The hydrogen in water is usually the only atom of low atomic weight in the soil. The fast neutrons which are slowed by collision with the hydrogen of the soil water are detected with a slow neutron counter which is part of the device lowered into the access hole. The higher the count the higher the hydrogen concentration and, hence, the higher the moisture content. Once a soil has been calibrated, time and depth variations in moisture content can easily be determined. The neutron-scattering device samples moisture in a sphere of soil surrounding the neutron source, and its indications are somewhat poorer near the surface where the sampled volume is distorted.

14-9   Irrigation water quality   Not all water is suitable for irrigation use. Unsatisfactory water may contain (1) chemicals toxic to plants or to persons using the plants as food, (2) chemicals which react with the soil to produce unsatisfactory moisture characteristics, and (3) bacteria injurious to persons or animals eating plants irrigated with the water.

Actually it is the concentration of a compound in the soil solution which determines the hazard, and soil solutions are 2 to 100 times as concentrated as the irrigation water. Hence, criteria based on the salinity of the irrigation water can only be approximate. At the beginning of irrigation with undesirable water no harm may be evident, but with the passage of time the salt concentration in the soil may increase as the soil solution is concentrated by evaporation. Free drainage of soil allows the downward movement of salts and helps to prevent serious accumulations. Artificial drainage of soil (Chap. 18) may be necessary for this reason if natural drainage is inadequate.

---

[1] W. Gardner and D. Kirkham, Determination of Soil Moisture by Neutron Scattering, *Soil Sci.*, Vol. 73, pp. 391–401, 1952.

High salt concentrations may sometimes be avoided by mixing the salty water with better-quality water from another source so that the final concentration is within safe limits. Precipitation during the nongrowing season will help to leach salts from the soil. It may, however, become necessary to apply an excess of irrigation water so that deep percolation will prevent undesirable salt accumulation in the soil. If the salinity of the irrigation water is C and the quantity applied is Q, the total salt applied to the field is CQ. The salinity C_s of the soil solution after the consumptive use U_c is taken from the soil is $QC/(Q + P_{eff} - U_c)$. Hence the theoretical quantity of water Q of salinity C required to maintain the soil solution at concentration C_s is

$$Q = \frac{C_s(U_c - P_{eff})}{C_s - C}$$

(14-5)

Because of variations in soil characteristics and salt concentrations, actual applications for leaching are usually higher than indicated by Eq. (14-5).

A large number of elements may be toxic to plants or animals. Traces of boron are essential to plant growth, but concentrations[1] above 0.5 mg/l are considered deleterious to citrus, nuts, and deciduous fruits. Some truck crops, cereals, and cotton are moderately tolerant to boron, while alfalfa, beets, asparagus, and dates are quite tolerant. Even for the most tolerant crops a concentration of boron exceeding 4 mg/l is considered unsafe. Boron is present in many soaps and thus may become a critical factor in the use of wastewater for irrigation. Selenium, even in low concentration, is toxic to livestock and must be avoided.

Salts of calcium, magnesium, sodium, and potassium may also prove injurious in irrigation water. In excessive quantities these salts reduce the osmotic activity of plants, preventing the absorption of nutrients from the soil. In addition they may have indirect chemical effects on the metabolism of the plant and may reduce soil permeability, preventing adequate drainage or aeration. The effect of salts on the osmotic activity of plants depends largely on the total amount of salts in the soil solution. The critical concentration in the irrigation water depends upon many factors. However, amounts in excess of 700 mg/l are harmful to some plants, and more than 2000 mg/l of dissolved salts is injurious to almost all crops.

Most normal soils of arid regions have calcium and magnesium as the principal cations, with sodium representing generally less than 5 percent of the exchangeable cations. If the sodium percentage in the soil is increased to 10 percent or more, the aggregation of soil grains breaks down and the soil

---

[1] Concentrations are commonly expressed in milligrams of salt per liter of water. This is numerically equal to the old term, parts per million by weight.

becomes less permeable, crusts when dry, and its pH increases toward that of alkaline soils. Since calcium and magnesium will replace sodium more readily than vice versa, irrigation water with a low *sodium-adsorption ratio* (SAR) is desirable. The SAR is defined as

$$SAR = \frac{Na^+}{\sqrt{(Ca^{++} + Mg^{++})/2}} \tag{14-6}$$

where the concentration of the ions is expressed in *equivalents per million* (epm).[1] The SAR indicates the relative activity of the sodium ions in exchange reactions with the soil. An irrigation water with a high SAR will cause the soil to tighten up.

The U.S. Department of Agriculture[2] has classified irrigation waters into four groups (Fig. 14-5) with respect to sodium hazard depending on the SAR value and the specific conductance (Sec. 15-12). By adding gypsum, $CaSO_4$, to the water or directly to the soil, the SAR value can be reduced. Observations of water quality in streams or groundwater may not be sufficient to judge their suitability for irrigation. Evaporation from reservoirs increases salt concentration in surface water and leached salts from irrigation may progressively raise the concentration of salts in groundwater. These changes must be considered in the planning phase.

Bacterial contamination of water is normally not serious from the irrigation viewpoint unless severely contaminated water is used on crops which are eaten uncooked. Raw wastewater is used for irrigation in many countries, but in the United States its use is frowned upon except for nursery stock, cotton, and other crops processed after harvesting.[3] Most states have regulations governing the use of wastewater for irrigation.

14-10 Irrigation methods   There are four basic methods of applying irrigation water to fields—flooding, furrow irrigation, sprinkling, and subirrigation. Numerous subclasses exist within each of these basic methods. *Wild flooding* consists in turning the water onto natural slopes without much control or prior preparation. It is usually wasteful of water, and unless the land is naturally smooth, the resulting irrigation will be quite uneven. Wild flooding is used mainly for pastures and fields of native hay on steep slopes where abundant water is available and crop values do not warrant more expensive preparations. *Controlled flooding* may be accomplished from *field ditches* or by use of *borders, checks,* or *basins.* Flooding from field ditches is

---

[1] Equivalents per million are determined by dividing the concentration of the salt in milligrams per liter (ppm) by the combining weight. See Appendix.

[2] L. V. Wilcox, Classification and Use of Irrigation Waters, *U.S. Dept. Agr. Circ.* 969, 1955.

[3] P. W. Eastman, Municipal Wastewater Reuse for Irrigation, *J. Irrigation and Drainage Div., ASCE,* Vol. 93, pp. 25–31, September, 1967.

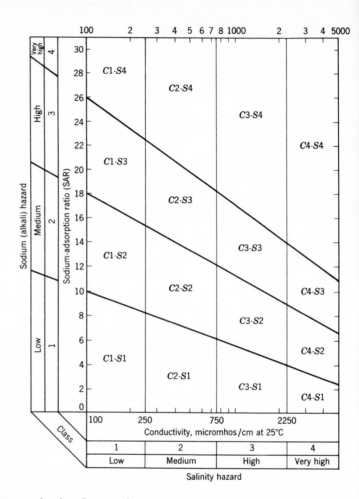

FIG. 14-5   *Diagram for classification of irrigation waters.*

often adaptable to lands with topography too irregular for other flooding methods. It is relatively inexpensive because it requires a minimum of preparation. Water is brought to the field in permanent ditches and distributed across the field in smaller ditches spaced to conform to the topography, soil, and rate of flow (Fig. 14-6). Under ideal conditions, the ditch spacing and flow rate should be such that the water will just infiltrate in the time it is flowing across the field. If the flow is too rapid, some of the water will not have time to infiltrate and surface waste will occur at the lower edge of the field. If flow is too slow, excessive percolation will occur near the ditch, and too little water will reach the lower end of the field.[1]

The border method of flooding requires that the land be divided into strips 30 to 60 ft wide and 300 to more than 1000 ft long. The strips are separated by low levees, or *borders*. Water is turned into each strip through a head gate along one of the narrow sides and flows downhill the length of the strip. Preparation of land for border strip irrigation is more expensive than for ordinary flooding, but this may be offset by a decrease in water waste because of the improved control. *Check flooding* is accomplished by turning water into relatively level plots, or checks, surrounded by levees. If the land is initially level, the plots may be rectangular but with some initial slope the checks will usually follow the contours. Check flooding is useful in very permeable soils where excessive percolation might occur near a supply ditch. It is also advantageous in heavy soils where infiltration would be inadequate in the time required for the flow to cross the field. In check flooding the check is filled with water at a fairly high rate and allowed

---

[1] L. S. Willardson and A. A. Bishop, Analysis of Surface Irrigation Application Efficiency, *J. Irrigation and Drainage Div.*, *ASCE*, Vol. 93, pp. 21–36, June, 1967; P. H. Berg, Methods of Applying Irrigation Water, *J. Irrigation and Drainage Div.*, *ASCE*, Vol. 86, pp. 71–82, September, 1960.

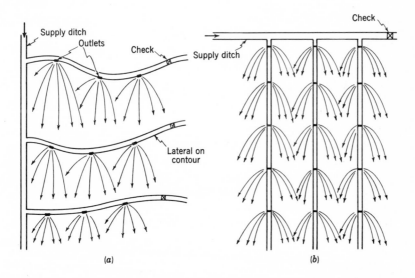

FIG. 14-6   Methods of flooding from field ditches: (a) from contour laterals, (b) from downslope ditches.

to stand until the water infiltrates. The *basin-flooding* method is check flooding adapted to orchards. Basins are constructed around one or more trees depending on topography, and the flow is turned into the basin to stand until it infiltrates. Portable pipes or large hoses are often used in place of ditches for conveying water to the basins.

Furrow irrigation is widely used for row crops, and small furrows, called *corrugations*, have been used for forage crops such as alfalfa. The furrow is a narrow ditch between rows of plants. An important advantage of the furrow method is that only 0.2 to 0.5 as much surface area is wetted during irrigation as compared with flooding, and evaporation losses are correspondingly reduced. Furrow irrigation is adapted to lands of irregular topography. Customarily the furrows are run normal to the contours, although this should be avoided on steep slopes where soil erosion may be severe. Spacing of furrows is determined by the proper spacing of the plants. Furrows vary from 3 to 12 in. deep and may be as much as 1500 ft long. Excessively long furrows may result in too much percolation near the upper end and too little water at the downslope end. Water may be diverted by an opening in the bank of the supply ditch, but many farmers now use small siphons made out of 4-ft lengths of plastic or aluminum tubing about 2 in. in diameter. These siphons are easily primed by immersion in the ditch and provide a uniform flow to the furrow without the necessity of damaging the ditch bank.

The development of lightweight pipe with quick couplers resulted in a rapid increase in *sprinkler irrigation*[1] after World War II. Sprinkler irrigation offers a means of irrigating areas which are so irregular that they prevent use of any surface-irrigation methods. By using a low supply rate, deep percolation or surface runoff and erosion can be minimized. Offsetting these advantages is the relatively high cost of the sprinkling equipment and the permanent installations necessary to supply water to the sprinkler lines. Very low delivery rates may also result in fairly high evaporation from the spray and the wetted vegetation. In recent years, high labor costs for surface irrigation have increased the attractiveness of sprinkler irrigation. During the period from 1958 to 1967 the total irrigated acreage in the United States changed from 36 million to 45 million acres, an increase of 26 percent, while sprinkler use rose from 3.7 million to 7.6 million acres, a 130 percent increase.[2] Sprinkling may be accomplished with fixed perforated pipe, rotating sprinkler heads, or fixed sprinkler heads. It is impossible to get completely uniform distribution of water around a sprinkler head, and

---

[1] Verne H. Scott, Sprinkler Irrigation, *Calif. Agr. Exp. Sta. Extension Serv. Circ.* 456, University of California, 1956.

[2] M. N. Langley, Trends to Sprinkler Irrigation, *Reclamation Era*, Vol. 55, pp. 3–4, August, 1969.

spacing of the heads must be planned to overlap spray areas so that distribution is essentially uniform.

Trickle irrigation in which water (and nutrients) are continuously applied at the base of each plant through small plastic tubes virtually eliminates all evaporation from soil, deep percolation, and runoff, and leads to very high efficiency in the use of water. Because of low water use, soil salinity problems are reduced. The technique of trickle irrigation was developed in Israel.

In a few areas soil conditions are favorable to *subirrigation*.[1] The required conditions are a permeable soil in the root zone, underlain by an impermeable horizon or a high water table. Water is delivered to the field in ditches spaced 50 to 100 ft apart and is allowed to seep into the ground to maintain the water table at a height such that water from the capillary fringe is available to the crops. Low flow rates are necessary in the supply ditches; and free drainage of water must be permitted, either naturally or with drainage works, to prevent waterlogging of the fields. The irrigation water should be of good quality to avoid excessive soil salinity. Subirrigation results in a minimum of evaporation loss and surface waste and requires little field preparation and labor. Subirrigation has been employed in the Egin Bench project, Idaho; in Cache Valley, Utah; in San Luis Valley, Colorado; and in the delta of the Sacramento and San Joaquin Rivers in California.

14-11 Supplemental irrigation Even in humid regions where rainfall is usually adequate for crop growth, drought periods of several days and longer do occur. Such droughts may have a serious effect on crop production if they occur when seeds require moisture for germination or during other critical periods in plant growth. Many farmers in humid regions of the country have installed equipment for supplying irrigation water during droughts. Sprinkler irrigation is most common because it can be introduced without prior preparation of the land. Row crops can often be irrigated by the furrow method with perforated pipe in lieu of supply ditches. In some instances the return in increased crops in a single year has repaid the investment. In general, however, supplemental irrigation must be viewed as a long-term investment in insurance against serious drought. Sprinkler equipment has in some instances been used as a means of frost protection. If a crop is wet when freezing temperatures occur, the water must be frozen before the plant temperature can be lowered below the freezing point. Because of the high heat of fusion of water, this offers considerable protection against air temperatures as low as 25°F.

14-12 Irrigation structures Many structures are necessary for the effective operation of the complicated systems of canals and ditches in an

[1] R. L. Fox, J. T. Phelan, and W. D. Criddle, Design of Subirrigation Systems, *Agr. Eng.*, Vol. 37, pp. 103–107, 1956.

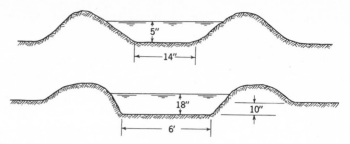

FIG. 14-7 Typical cross sections for farm ditches.

irrigation project. The main structural features of canals are described in Chap. 10. The principal structures encountered on the farm are the ditches and pipes for conveyance; checks, dams, and gates to regulate the flow; and dividers to separate a stream into prescribed portions.

The elevation of farm ditches must be sufficient to permit gravity flow to the field. Hence, they are rarely fully excavated but are made in broad, shallow dikes (Fig. 14-7). Gates for regulating flows in the ditch system may be of wood, steel, or concrete. Steel or wooden gates are used for temporary installations where the gate is replaced each time the ditch is rebuilt. Permanent gates (Fig. 14-8) are often of concrete with wooden or metal flashboards. In order to raise the level of the water in a ditch to make diversion into a field possible, a check may be required. These may be wooden or steel dams, but often a timber laid across the channel and supporting a piece of heavy canvas with its lower end weighted down with earth is sufficient.

Division boxes (Fig. 14-9) may be used to distribute flow to several channels. If an accurate division of flow is required, a symmetrical Y divider or a division weir (Fig. 14-10) may be used. For accurate flow division, the divider must be installed in a long, straight channel in order that the velocity distribution across the channel may be reasonably uniform. A proportional division of flow from an underground pipe may be accomplished by bringing the water into a stilling basin with overflow weirs discharging into separate channels (Fig. 14-11).

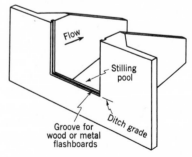

FIG. 14-8 Permanent type of concrete gate or check structure.

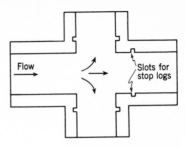

FIG. 14-9   Plan view of a simple three-way
outlet box.

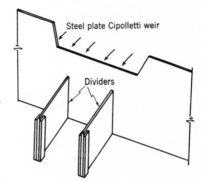

FIG. 14-10   Division weir. Dividers may be of
steel or timber protected by sheet
metal.

To avoid loss of water and the annual cost of ditch construction, many
farmers are turning to the use of pipe distribution systems.[1] Concrete pipe
is widely used for permanent underground installation with riser pipes at
intervals to bring the water to the surface. The risers may discharge through
top boxes, designed to control erosion at the outlet, into small field ditches
which convey the water to the furrows or basins. Special valves called
alfalfa, or *orchard*, *valves* can be attached to the top of the riser pipes to
control the flow. Portable hydrants (Fig. 14-12) can be installed on the top
of the alfalfa valves to direct flow in a definite direction or to permit connection of small portable pipes. If water is pumped into the pipes, a *pump
stand* (Fig. 14-13) constructed of pipe sections is usually necessary as a
surge tank. Valves may be provided in the pump stand to permit control of
flow into two or more outlets. Vertical pipes carried about 2 ft above the
hydraulic grade lines are recommended at junctions and changes in direction
for the same purposes.

Ordinary unreinforced-concrete pipe which under ASTM specification
C14 is tested at a maximum pressure of 15 psi is commonly used in pipe

---

[1] Arthur F. Pillsbury, Concrete Pipe for Irrigation, *Calif. Agr. Exp. Sta. Extension Serv.
Circ.* 418, University of California.

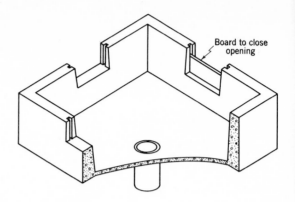

FIG. 14-11   *Concrete proportional division box.*

irrigation systems. Excess discharge pressure may cause erosion around the riser and often results in wasted energy in pumping. When pipelines are installed for irrigating hillsides (Fig. 14-14), the size should be such that the hydraulic grade line is not much above ground level. Gates are installed at intervals of elevation no greater than the safe working pressure of the pipe with risers which will overflow before an excess pressure can be exerted on the line. To irrigate between *A* and *B* (Fig. 14-14), valve *B* is closed, and the riser at *A* controls the pressure. To irrigate between *B* and *C*, valve *C* is closed, and pressure control is at *B*.

Portable steel and aluminum pipe are also employed for irrigation. In addition to use with sprinkler systems, pipes with orifices at intervals along their length are used to distribute water to furrows. Small sheet-metal slide

FIG. 14-12   *Portable irrigation hydrant.*
            (*Snow Manufacturing Co.*)

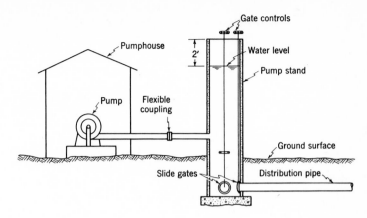

FIG. 14-13   Typical irrigation pump house and pump stand.

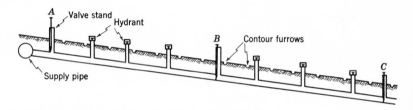

FIG. 14-14   Pipe-distribution system for irrigating hillsides.

gates close off orifices not in use. Portable pipe is often used to carry water from a well to a field. In addition to conservation of water, use of pipes increases the area available for cultivation and permits a flexibility of operation which is difficult to obtain with ditch irrigation.

Considerable labor is involved in operating an irrigation system, and some of the inefficiencies of irrigation result from inadequate attention to such operation. Consequently much attention is now being given to automation of irrigation using float activated systems which can operate unattended.[1]

14-13   Legal aspects of irrigation   A single farm may sometimes be irrigated by diversion of flow from a small stream at a cost which an individual farmer can bear. Projects requiring expensive storage or conveyance works are usually beyond the capability of an individual to develop and require some sort of group effort. The first large irrigation projects in the United States were privately developed under two different types of organiza-

---

[1] H. R. Haise, E. G. Kruse, and L. Erie, Automating Surface Irrigation, *Agr. Eng.*, Vol. 50, pp. 212–216, April, 1969.

tion. The earliest projects were *irrigation cooperatives*, or *mutual irrigation companies*, in which farmers banded together to construct and maintain the necessary works. These companies were often unincorporated, and division of water was on the basis of labor contributed to the construction. Later, incorporated mutual companies were organized in which the participating farmers were shareholders and water was sold at cost in quantities based on established water rights.

During the irrigation boom of the late nineteenth century, many *commercial irrigation companies* were organized. These companies, financed by stock sale to nonresident as well as resident investors, constructed irrigation projects and obtained water rights so that they might sell water. These speculative operations ran into many difficulties. Farmers sometimes let their land lie idle until the water company went bankrupt for lack of water sales. They then purchased the bankrupt company at a fraction of its value and organized a new company, either commercial or mutual. Other speculators sometimes purchased land and held it idle to force a water company out of business. The most effective commercial operation was the development company, which owned both water and land and developed the project for the potential profit from the increased land values. Many commercial irrigation enterprises are still in operation, but many more have been converted to mutual companies.

A quasi-public *irrigation district* is now the most common type of organization. An irrigation district is an organization formed in accordance with state laws[1] on the basis of a majority vote of the voters of the district. Funds to finance district operations are raised by water tolls, bond issues, assessments on the irrigable land within the district, and assessments on lands benefiting indirectly from the district operations as, for example, the business centers of the area. Districts are usually governed by directors elected by the district voters, and their management of district affairs is subject to regulations prescribed by state law.

Early federal laws made it possible to settle on public lands, but since title to the land remained with the government until the land was reclaimed, it was impossible to use the land as security for loans to finance the irrigation works needed. The Carey Act (1894) attempted to correct this by allotting each Western state 1 million acres of public land for development. State laws varied but, in general, provided that individuals or corporations might, by contract with the state, construct irrigation works and sell water rights to the settlers on the land. On the whole, the Carey Act was a disappointment because the cost of reclamation was too great for most potential developers.

The Reclamation Act (1902) provided a revolving fund for federal

---

[1] The first such law was passed in California in 1887.

development of irrigation in the 16 Western states.[1] The fund was initially to consist of all money received from the sale of public land in these states, and was to be used for projects at the discretion of the Secretary of Interior, the cost to be repaid by the settlers over a period of 10 yr, thus making the money again available for new construction. Subsequent acts have increased the statutory repayment period to 40 yr with a 10-yr development period before repayment must begin. Longer repayment periods are possible with special congressional approval. Project costs chargeable to navigation or flood mitigation need not be repaid by water users, and revenue from power sales is credited against the irrigation costs. Since 1915 congressional approval for expenditures from the fund has been necessary. The act of 1920 provided that 52.5 percent of royalties from public-land oil leases be included in the Reclamation Fund. The 1926 act limited the land for which a single owner could obtain water from federal projects to 160 acres, to encourage development of small farmsteads instead of large, corporation-owned ranches. Special acts govern major projects such as the Boulder Canyon and Columbia Basin projects.

14-14   Some economic aspects of irrigation   The large water requirement inherent in irrigating large areas is augmented by wasteful irrigation practices, conveyance losses, and leaching requirements. That part of the applied water which eventually returns to the stream after irrigation is called *return flow*. Because of evaporative loss and additional salts leached from the soil, salt concentrations in streams of irrigated regions tend to increase downstream. The actual consumptive use of irrigation water is that part returned to the atmosphere by evaporation and transpiration, i.e., diverted water less return flow and accretion to deep groundwater. However, the water which is not consumptively used has been degraded by increased salt content and has lost economic value.

Many federal reclamation projects sell water to the irrigator at prices well below cost. This subsidy encourages irrigated agriculture and may help the economic development of the region. Such subsidies also lead to irrigation of marginal crops on marginal land and encourage wasteful use of water. Subsidized irrigation may limit the supply of water for industry, which yields much higher returns per unit of water than agriculture. The relation of the expected crops from irrigated agriculture to the existing surpluses must also be considered in evaluating a proposed irrigation project.

14-15   Planning the irrigation project   No two irrigation projects are identical, and no absolute outline of procedure for project design is feasible. The list which follows summarizes in general terms the steps which are required for most projects:

---

[1] Texas was included as the seventeenth state at a later date.

1. Land classification (Sec. 14-1)
2. Estimate of irrigation water requirement (Secs. 14-2 to 14-5).
3. Determination of sources of available water (Chaps. 2 to 4).
4. Analysis of chemical quality of available water (Sec. 14-9).
5. Design of storage reservoir to assure necessary water (Chap. 7).
6. Design of dam and spillway for storage reservoir or diversion works (Chaps. 8 and 9).
7. Design of distribution works (Chaps. 10 and 11).
8. Economic analysis of the project to determine whether the estimated cost is returnable from the potential benefits, and financial analysis to establish a repayment plan (Chap. 13).
9. Establishment of legal title to water (Chap. 6).
10. Establishment of the organization which will operate the project (Sec. 14-13). In some cases this is a necessary first step, since the operating organization may also design the project.

BIBLIOGRAPHY

Childs, E. C.: "An Introduction to the Physical Basis of Soil Water Phenomena," Wiley, New York, 1969.
Clyde, George D.: Irrigation in the United States, *Trans. ASCE*, Vol. CT, pp. 311–342, 1953.
Golze, Alfred R.: "Reclamation in the United States," McGraw-Hill, New York, 1952.
Hagan, R. M., H. R. Haise, and T. W. Edminster: "Irrigation of Agricultural Lands," American Society of Agronomy, Madison, Wis., 1967.
Houk, Ivan E.: "Irrigation Engineering," Wiley, New York, 1956.
Huffman, Roy E.: "Irrigation Development and Public Power Policy," Ronald, New York, 1953.
"Irrigation Water Requirements," Vol. IV of "Water Studies," chap. 4.1, U.S. Bureau of Reclamation, Denver.
Israelson, O. W., and V. E. Hansen: "Irrigation Practices and Principles," 3d ed., Wiley, New York, 1962.
Leliavsky, Serge: "Irrigation and Hydraulic Design," 3 vols., Chapman & Hall, London, 1960.
Quality of Surface Waters for Irrigation, Western United States, *U.S. Geol. Surv. Water Supply Paper* 1485, 1956.
"Reclamation Project Data," U.S. Bureau of Reclamation, 1961.
Roe, Harry B.: "Moisture Requirements in Agriculture," McGraw-Hill, New York, 1950.
Ruttan, V. O., "The Economic Demand for Irrigated Acreage," Johns Hopkins, Baltimore, 1965.
White, Gilbert F.: "The Future of Arid Lands," American Association for the Advancement of Science, Washington, D.C., 1956.
Zimmerman, J. D.: "Irrigation," Wiley, New York, 1966.

PROBLEMS

14-1 Using Table 14-1 as a guide, evaluate the suitability of some assigned area for irrigation. What irrigation methods would you consider most feasible for this area? Why? If sufficient data are available, analyze the economic feasibility of irrigation for this area.

14-2 A canal to carry 250 cfs is to be located in land having a slope of 22 ft per 1000 ft. The canal is to be trapezoidal with side slopes of $1\frac{1}{2}$ horizontal to 1 vertical, bottom width of 8 ft, depth of 5 ft, and $n = 0.020$. Excavation costs 35 cents per cubic yard, and the minimum allowable depth of cut above a drop is 2 ft. Drop structures 3 ft high cost $1200 each, and higher structures add an additional $150 per foot. What is the most economic height of drop under these conditions? Assume freeboard of 1 ft, i.e., use a water depth of 4 ft.

14-3 Assuming an annual consumptive use for cotton of 2.4 ft/yr, monthly distribution corresponding to that for the Mesilla Valley (Table 14-4), a farm efficiency of 48 percent, and conveyance loss of 30 percent, compute the annual and maximum monthly water requirements for a 160-acre farm. Assume a growing season from April 1 to September 1 and monthly precipitation as follows:

Month	Jan.	Feb.	Mar.	Apr.	May	June	July	Aug.	Sept.	Oct.	Nov.	Dec.
Precipitation, in.	0.3	0.4	0.3	0.2	0.3	0.5	1.8	1.7	1.3	0.7	0.6	0.5

14-4 A project is utilizing irrigation water with a salinity of 500 mg/l. The consumptive use is 2.7 ft/yr, and the effective rainfall is 7.5 in. If it is desired to maintain the salinity of the soil solution at 2000 mg/l, what is the irrigation-water requirement?

14-5 Samples of two waters yielded the following chemical analyses:

Chemical	Colorado River below Hoover Dam	Rio Grande at El Paso, Tex.
Sodium	114 mg/l	410 mg/l
Calcium	102 mg/l	106 mg/l
Magnesium	32 mg/l	30 mg/l
Specific conductance	1230 micromhos	2860 micromhos

Assuming these samples represent average conditions, compute the SAR for each water and classify them according to the USDA classification of Fig. 14-5. How much gypsum (calcium sulfate) would be required to reduce the SAR of an acre-foot of Rio Grande water to 5?

14-6 A pipe-irrigation system similar to that of Fig. 14-14 is to be installed on a hillside which has a 10 percent slope. The total length of run is to be 850 ft, and the pipe is to have a minimum cover of 18 in. The invert of the pipe at the entrance must be 4 ft below the ground surface to meet the main supply pipe. The hydraulic gradient in the supply pipe at the junction will be 2 ft above ground level. The total area to be irrigated is 10 acres, average depth of irrigation 0.5 ft, and a complete irrigation should take not more than 24 hr. Maximum hydrant spacing is to be 50 ft. Draw a sketch of the system, showing pipe sizes, hydrant locations, and gate stands. Draw the energy grade line and the hydraulic grade line when all hydrants between the pipe entrance and the first gate stand are discharging equally. Assume the head loss at each hydrant is equal to the loss in sudden expansion, that is, $K_L = 1.0$ (Table 11-2a). This is an approximation which is very conservative. For more detail see Rouse (editor), "Engineering Hydraulics," pages 436 to 437.[1] Use $n = 0.013$ for concrete pipe.

14-7 Compute the water-hammer pressure if a gate in one of the stands of the pipeline of Prob. 14-6 is shut off instantaneously while the full capacity of the pipe is being discharged.

14-8 If the price of wheat is $1.48 per bushel, irrigation water $7.50 per acre-foot, and

[1] Wiley, New York, 1950.

the fixed cost of operating the farm $10 per acre, what is the economic production of wheat on the basis of Fig. 14-1? What is the economic production for a water cost of $10 per acre-foot?

14-9 A farmer is using plastic siphons with 2 in. internal diameter and 4 ft long to divert from a supply ditch to a field for furrow irrigation. If the furrows are triangular with side slopes of 1:1 and longitudinal slopes of 0.001 and the water level in the supply ditch is 18 in. above the bottom of the furrow, what is the discharge of a single siphon? Assume uniform flow in the furrow and $n = 0.030$.

14-10 A rectangular, contracted, sharp-crested weir is used for measurement of flow in an irrigation ditch. A farmer wishes to divide the flow at the weir into two portions, one of which is to be twice the other portion. If the weir crest is 2.0 m long and the head is 0.6 m, where should he locate a thin partition to accomplish this division? What error would result when the head was 0.3 m?

14-11 What are the essential features of your state laws regarding the formation of irrigation districts?

14-12 A 6-in.-diameter steel irrigation pipe with capped end has 1-in. orifices at 5-ft intervals. Assuming equal discharge from all orifices and neglecting head loss at the orifices, express the head loss in a 100-ft length of pipe as a function of the velocity at entrance. Assume $f = 0.020$ all along the pipe. Plot the energy and hydraulic grade lines.

14-13 Given a sandy loam with a dry weight of 88 lb/cu ft and wilting point and field capacity of 7 and 15 percent respectively of the dry weight. If the root-zone depth is 4.0 ft, compute the inches of moisture in the root zone at (a) the wilting point and (b) the field capacity.

14-14 The root zone of a certain soil has a field capacity of 7 in. and a wilting point of 4.0 in. The consumptive use of crops for July is 0.22 in./day. Assuming no rainfall, how often ought a farmer to irrigate? How much water should be applied at each irrigation if there is to be no deep percolation?

14-15 Precipitation and consumptive-use data for citrus in a certain area are:

Month	U_c, in.	Precipitation, in.		
		1957	1958	1959
January	0.3	0.0	1.1	2.0
February	0.8	0.6	2.0	1.2
March	1.6	1.2	0.0	2.1
April	2.5	1.8	1.3	1.1
May	2.8	0.0	1.0	0.1
June	3.9	3.1	2.2	0.8
July	5.4	1.6	0.0	2.2
August	5.0	0.3	0.8	0.0
September	3.4	0.0	0.0	0.0
October	1.9	2.0	3.4	1.7
November	1.1	1.3	1.6	3.8
December	0.6	0.0	1.8	2.0

The growing season is April 1 to September 30, moisture storage capacity of the soil is 6.0 in., and the available soil moisture on April 1, 1957 was 4.2 in. Assume that it is desirable to maintain available soil moisture above 4.0 in. in April and May, above 3.0 in. in June and July, and above 2.0 in. in August and September. Find the irrigation requirement on a monthly basis.

14-16 A farmer in the Cache Valley of Utah raises sugar beets on an 80-acre plot and can market them at $4 per ton. His fixed operating costs are $4000 per year, and the cost of water is $3.50 per acre-foot. Using Fig. 14-1, determine the total amount of water the farmer should use annually for optimum return. What amount of water should be used if the water cost is $20 per acre-foot, and all other costs remain the same?

14-17 A stream with a flow rate of 500 cfs has a dissolved salt concentration of 300 mg/l. How many pounds of salt pass the station each hour? How many kilograms per day?

14-18 During the irrigation season 50 cfs are diverted from a stream which has a flow of 300 cfs above the point of diversion. The salt content of the diverted water averages 200 mg/l. Several miles downstream from the irrigated area the streamflow is 280 cfs with a salt content of 232 mg/l. How many pounds of salt are leached from the soil each day? State all necessary assumptions.

Chapter 15

Municipal and industrial
water supply

The water requirement of a modern city is so great that a community system capable of supplying a sufficient quantity of potable water is a necessity. The first step in the design of a waterworks system is a determination of the quantity of water that will be required, with provision for the estimated requirements of the future. Next a reliable source of water must be located, and finally a distribution system must be provided. Usually water at its source is not fit for drinking, and water-purification facilities are ordinarily included as an integral part of the system.

WATER REQUIREMENTS

Water use varies from city to city, depending on the population, climatic conditions, industrialization, and other factors. In a given city, use varies from season to season and from hour to hour. The planning of a water supply system requires that the probable water use and its variations be estimated as accurately as possible.

    15-1 Uses of water In 1965 an average of about 270 billion gallons of water was withdrawn daily from surface and underground sources in the United States.[1] Municipal and industrial use (excluding thermal power) accounts for 56 percent of this total (Table 15-1). Much of the withdrawn water is returned to surface streams in the form of wastewater or irrigation returns. Actual consumption of water by evapotranspiration or by being incorporated in an industrial product is much less than the withdrawal.

---

[1] "The Nation's Water Resources," U.S. Water Resources Council, Washington, D.C., 1968.

TABLE 15-1   *Approximate Classification of Water Use in the United States in 1965**

	Average daily flows in billions of gallons		
Users	Withdrawal	Return	Consumption
Self-supplied industry	131	126.5	4.5
Agriculture	113	46.7	66.3
Municipalities	20	14.1	6.9
Total	270	186.3	77.7

*"The Nation's Water Resources," p. 4–1, U.S. Water Resources Council, Washington, D.C., 1968.

Municipal and industrial water consumption is only about 15 percent of the total consumptive use.

Municipal uses of water may be broken down into various classes. *Domestic use* is that water which is used in private residences, apartment houses, etc., for drinking, bathing, lawn sprinkling, and sanitary purposes. Domestic use in the United States averages about 75 gallons per capita per day (gpcd). Extensive lawns and gardens will result in greatly increased consumption during dry periods. *Commercial and industrial use* is that water used by commercial establishments and industries and averages about 65 gpcd. In small residential communities the commercial and industrial use may be as low as 10 gpcd, but in industrial cities it may run as high as 100 gpcd. *Public use* is water required in parks, civic buildings, schools, hospitals, churches, street washing, etc. In most cities about 20 gpcd is used in this manner. Water which leaks from the system, meter slippage, unauthorized connections, and all other unaccounted-for water is classified as *loss and waste*. It is often estimated at about 20 gpcd, but proper construction and careful maintenance will reduce this figure. Over a 16-yr period Chicago located and stopped leaks totaling over 285 million gallons per day (mgd). Eighty-five per cent of this total was attributed to leaks at pipe joints and in house service pipes.[1]

15-2 Factors affecting water use The average daily water use in cities in the United States varies between 40 and 500 gal per person. Among the many factors which influence withdrawals are:

Climate More water is used in warm, dry climates than in humid

---

[1] W. W. Deberard, Chicago Water System Serves Growing Population with Decreased Pumpage, *Civil Eng. (N. Y.)*, Vol. 20, pp. 17–21, November, 1950.

climates for bathing, lawn watering, air conditioning, etc. In extremely cold climates water may be wasted at faucets to prevent freezing of pipes.

Characteristics of population   Water use is influenced by the economic status of the customers. The per capita use of water in slum areas will be much less than that in high-class residential districts. In unsewered districts consumption may be as low as 10 gpcd.

Industry and commerce   Manufacturing plants often require large amounts of water. The actual amount depends on the extent of the manufacturing and the types of industry. Table 15-2 lists the use in some industries. Some industries develop their own water supply and place little or no demand on a municipal system. Zoning of the city affects the location of industries and may help in estimating future industrial demands. Commercial districts include office buildings, warehouses, and stores. The per capita demand in such areas is not high, averaging about 10 gpd per full-time employee.

About 80 percent of industrial water is for cooling and need not be of high quality. Water used for process purposes or for boiler feed must be of good quality. In some cases industrial water must have a lower content of dissolved salts than can be permitted in drinking water. The location of industry is often much influenced by the availability of water supply. However, when other factors dictate the plant location, water requirements may be reduced far below the industry average. Extensive recirculation of water has reduced water requirements at the Kaiser steel mill in Fontana, California, to 1400 gal/ton of steel—about 4 percent of the average requirement indicated in Table 15-2.

Water rates and metering   If water costs are high, people may be more conservative in water use and industries will often develop their own

TABLE 15-2   *Use of Water for Selected Industrial Products*

Product	Unit of production	Typical water use, gal per unit
Beer	Barrel	470
Canned apricots	Case of No. 2 cans	80
Canned lima beans	Case of No. 2 cans	250
Coke	Ton	3,600
Oil refining	Barrel	770
Paper	Ton	39,000
Leather (tanned)	Ton	16,000
Rayon hosiery	Ton	18,000
Woolens	Ton	140,000
Steel	Ton	35,000
Electricity-steam	Kw-hr	80

supply to obtain cheaper water. Metered customers are more likely to repair leaks and use water with discretion. The installation of meters in some communities has reduced water use by as much as 40 percent.

Size of city   Per capita use tends to be higher in large cities than in small towns. The difference results from greater industrial use, more parks, greater commercial use, and perhaps more loss and waste in the larger cities.

15-3 Estimates of water use   The first step in the design of a water-works system is an estimate of the requirement for water. Ordinarily an average of about 160 gpcd is assumed, but this figure may be altered considerably by local conditions. Previous records in the city under study or data from similar cities in the area are the best guide in selecting a value of per capita use for design. There has been a steady upward trend in water use in the United States as a result of such appliances as air conditioners, automatic washing machines, dishwashers, home garbage-disposal units, and the like. It is entirely possible to reduce per capita use without sacrificing amenities by use of redesigned plumbing fixtures and water-using appliances, and by more efficient lawn watering and garden use. As additional water supplies for cities become more difficult to obtain, water conservation will become more common.

After deciding on an average per capita requirement, an estimate of the future population of the city must be made to determine average total use. The economic aspects of the problem determine how far into the future the population estimate should be projected. The basic question to be answered is: "Is it cheaper in the long run to design and build the system to meet the demand expected at some future date, or to build now for a short time and plan to make additions as future needs develop?" The answer is often a compromise. Some portions of the project may be more economically built to ultimate size immediately, while other portions are left for future expansion. For example, if wells are to be used for the water source, the design period may be quite short since other wells can be drilled when needed. On the other hand, if a dam must be built, the design period may be quite long.

Most discussions on population estimating suggest five "methods" for extrapolating the past population curve of a city (Fig. 15-1) on the assumption of (1) uniform growth rate, (2) uniform percentage growth, (3) decreasing growth rate, (4) graphical extension, and (5) comparison with growth of other cities.[1] Clearly no simple extension of the past can accurately predict the future.

The 1938 median forecast by the U.S. Census Bureau predicted a United States population of 154 million in 1980. This point was reached in 1951.

---

[1] Paul M. Reid, Problems and Techniques of Population Forecasting, *J. Am. Water Works Assoc.*, Vol. 50, pp. 655–660, 1958.

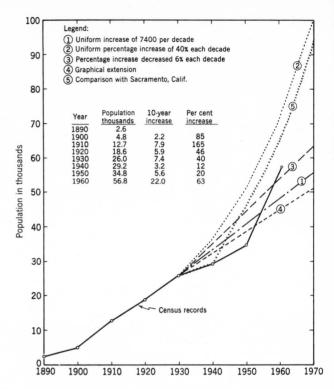

Legend:
① Uniform increase of 7400 per decade
② Uniform percentage increase of 40% each decade
③ Percentage increase decreased 6% each decade
④ Graphical extension
⑤ Comparison with Sacramento, Calif.

Year	Population thousands	10-year increase	Per cent increase
1890	2.6		
1900	4.8	2.2	85
1910	12.7	7.9	165
1920	18.6	5.9	46
1930	26.0	7.4	40
1940	29.2	3.2	12
1950	34.8	5.6	20
1960	56.8	22.0	63

Census records

FIG. 15-1 *Extension of the population curve for Bakersfield, California, by various methods. Point of departure 1930.*

As late as 1950, the official median estimate of the probable peak population of the United States was 165 million but by 1968 the actual figure was 200 million. This sudden change in population trend resulted mainly from earlier marriages and more children per family. War, changes in industrial techniques, new scientific discoveries, new attitudes on family size, and many other factors can cause abrupt changes in population trends of nations and cities, and no known methods of population forecasting can anticipate these changes.

An engineer required to make a population estimate for a city should consider among other things any physical limits on the city's growth. Surrounding incorporated areas or geographical barriers may limit growth to that which can occur through use of undeveloped land within the city limits or multifamily apartments in lieu of single-family residences. Small towns close to large cities may often look forward to a rapid influx of commuters unless transportation difficulties interfere. Facilities or resources favorable

TABLE 15-3   Typical Population Densities

	Persons per acre
Residential areas:	
Single-family dwellings	5–30
Multiple-family dwellings	30–100
Apartments	100–1000
Commercial areas	15–30
Industrial areas	5–15

to industrial development presage continued growth. The preference of local citizens as reflected in zoning ordinances suggests possible growth trends. Areas dedicated to single-family residences will reach their limiting population sooner than areas permitting large apartment developments. The engineer must assemble all pertinent facts together with opinions of local residents who have watched the area's growth. Comparison with similar cities may be helpful, but the development of a single industry or a new transportation facility may change the similarity almost overnight.

The ultimate population of a geographic area may be predicted on the basis of the prospective development as indicated by the master plan for the community. Such estimates, of course, are disrupted by any changes in the zoning pattern. Typical population densities for various types of development are shown in Table 15-3.

Because of the uncertainty in any population forecast, a range of estimates[1] should be prepared and preliminary plans developed to cope with the range of possibilities. The best plan is the one which offers the flexibility to adjust to future developments while at the same time satisfactorily dealing with current problems.[2]

15-4   Fluctuations in use   The use of water in a community varies almost continually. In midwinter the average daily use is usually about 20 percent lower than the annual daily average, while in summer it may be 20 to 30 percent above the daily average. Seasonal industries such as canneries may cause wide variation in water demand during the year. For most communities the maximum daily use will be about 180 percent of the average daily use throughout the year.

Within any day there is much less use at night than during daytime hours (Fig. 15-2). In the early morning the hourly use is at a minimum of 25 to 40 percent of the average hourly use for the day. Near noon the

[1] E. G. Altouney, The Role of Uncertainties in the Economic Evaluation of Water Resources Projects, *Rept.* EEP-7, Department of Civil Engineering, Stanford University, August, 1963.
[2] R. F. Scarato, Time-Capacity Expansion of Urban Water Systems, *Water Resources Research*, Vol. 5, pp. 929–936, October, 1969.

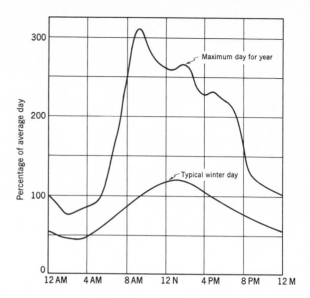

FIG. 15-2   *Hourly variation in water consumption at Palo Alto, California.*

demand usually hits a peak of about 150 to 200 percent of the average hourly demand for the day. In residential communities in summer there may be midmorning and late-afternoon peaks created by lawn watering. The peak hourly use for a city whose average annual use is 160 gpcd may be about

$$160 \times 1.80 \times 2.00 \approx 576 \text{ gpcd}$$

A final use of water which must be considered is for *fire fighting*. The annual volume of water used for fire fighting is small, but during fires the rate of use may be quite high. The National Board of Fire Underwriters recommends the allowance for fire flow shown in Table 15-4. Amounts given by this table represent those required in the business districts of a community. Required flows in residential areas vary from 500 to 3000 gpm depending on population density. If this flow cannot be maintained for the required time, the fire insurance rate for the community may be adjusted upward.[1] The flow required for fire fighting is added to the average use for the maximum day to determine what is presumed to be the maximum rate of use during a fire.

---

[1] The National Board of Fire Underwriters has a standard rating schedule with which all aspects of the water-supply system, fire-fighting equipment, and fire-alarm system are graded as a basis for fixing fire insurance rates.

TABLE 15-4   *Fire-flow Capacity Recommended by National Board of Fire Underwriters**

Population	Fire flow, gpm	Duration, hr
1,000	1,000	4
2,000	1,500	6
3,000	1,750	7
4,000	2,000	8
5,000	2,250	9
6,000	2,500	10
10,000	3,000	10
20,000	4,350	10
40,000	6,000	10
60,000	7,000	10
80,000	8,000	10
100,000	9,000	10
150,000	11,000	10
200,000	12,000	10

*"Standard Schedule for Grading Cities and Towns of the United States with Reference to Their Fire Defenses and Physical Conditions," National Board of Fire Underwriters, New York, 1956.

WATER-DISTRIBUTION SYSTEMS

Extensive distribution systems are needed to deliver water to the individual consumer in the required quantity and under a satisfactory pressure. This distribution system is often the major investment of a municipal waterworks.

15-5   Types of water-distribution systems   If topographic conditions are ideal, *gravity distribution* is used. This requires a reservoir at a sufficient elevation above the city so that water can reach any part of the distribution system with adequate pressure without pumping. If pumping is necessary, water may be pumped directly into closed distribution lines, or it may be pumped into distribution reservoirs which serve to equalize pumping rates throughout the day and provide for peak use.

A treelike distribution system with many dead ends is unsatisfactory because water may become stagnant at the extremities of the system. Moreover, if repairs are necessary a large district must be cut off from water. Finally with a locally heavy demand, or during a fire, head loss may be excessive unless the pipes are quite large. These difficulties are minimized with a gridiron, or *belt-line*, layout (Fig. 15-3).

A *single-main system* is one in which a single main serves both sides of a

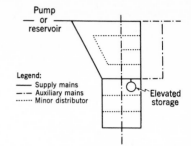

FIG. 15-3   A typical water-distribution system.

street. In a *double-main system* there is a main on each side of the street. One pipe supplies fire hydrants and domestic service on its side of the street, while the other (and smaller) pipe serves only domestic needs on the other side. The chief advantage of the two-main system is that repairs can be made without interfering with traffic and without damage to the pavement.

15-6   Pressure requirements in water-distribution systems   In designing water-distribution systems pressure requirements for ordinary use and for fire fighting must be considered. In residential districts fire pressures of 60 psi at the hydrant are recommended. In commercial districts a minimum pressure of 75 psi is tolerable, but higher pressures must be provided in districts with tall buildings. The American Water Works Association recommends a normal static pressure of 60 to 75 psi throughout a system. Many cities use fire-department motor pumpers to develop the necessary fire pressure so that normal operating pressure can be less than that quoted above. The maintenance of high pressure in mains means increased pumping costs and usually also increased leakage. Some large cities have installed dual systems in business districts, a low-pressure system for ordinary use and a high-pressure system (150 to 300 psi) for fire fighting only. Other cities use standby pumps to raise the pressure in the entire system whenever a fire occurs.

Faucet pressures of 5 psi are satisfactory for most domestic needs. Assuming a maximum pressure loss of 5 psi in the meter, about 20 psi in the house service pipe and plumbing, and with the main about 5 ft below ground level, a total pressure of about 35 psi in the main is adequate for residential districts with one- and two-story houses. Allowing about 5 psi for additional stories, a pressure of 75 psi should be satisfactory for buildings up to 10 stories in height. Many cities require owners of tall buildings to install booster pumps in order to avoid the need for very high pressures in the mains. Unnecessarily high pressures should be avoided since leakage loss in the mains and from leaky plumbing fixtures will be increased.

15-7   Distribution reservoirs   Distribution reservoirs are used to provide storage to meet fluctuations in use, to provide fire storage, and to stabilize pressures in the distribution system. The reservoir should be located as close to the center of use as possible. The water level in the reservoir must be high enough to permit gravity flow at satisfactory pressures to the system which it serves. In large cities several distribution reservoirs may be located at strategic points throughout the city. Water is usually pumped into a distribution reservoir when the demand is low and withdrawn by gravity flow during periods of high demand. The required capacity of a distribution reservoir is established by the use characteristics of the district which it serves (Chap. 7).

Elevated storage may be advantageously employed for pressure stabilization. Figure 15-4a indicates the hydraulic grade line at time of high use in a system with an elevated tank which is poorly located. Pressure will be quite low at the far end of the system. An improvement in pressure conditions would result if the elevated tank were situated in the high-consumption district as indicated in Fig. 15-4b. The resulting hydraulic grade lines indicate that adequate pressures are maintained throughout the system at all times.

Various types of distribution reservoirs are built to meet the topographic and structural conditions encountered. If hills of adequate elevation exist in or near the town, a surface reservoir, either below ground level or of the cut-and-fill type, is usually the best selection. Small reservoirs may be simple excavations lined with gunite, asphalt, an asphalt membrane, or butyl rubber.[1] Larger reservoirs require a concrete lining, with side walls designed as retaining walls to resist external soil loads with reservoir empty.

Most surface reservoirs are covered to prevent contamination by animals, birds, and human beings, and, by shutting out sunlight, the growth of algae.

---

[1] Robert T. Chuck, Largest Butyl Rubber Lined Reservoir, *Civil Eng.*, Vol. 40, No. 5, pp. 44–47, May, 1970.

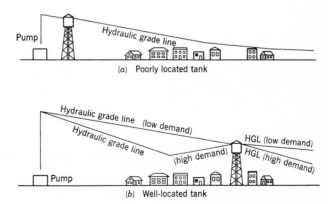

(a) Poorly located tank

(b) Well-located tank

FIG. 15-4   *Location of elevated storage tanks.*

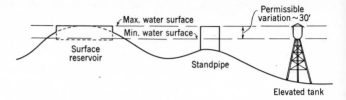

FIG. 15-5   Types of elevated storage for water distribution.

Reservoir roofs may be of wood, concrete, or steel. Precast slabs of light-weight concrete are widely used. In at least one instance a reservoir roof has formed part of a city street. An open distribution reservoir should have a high fence around it to keep out trespassers.

If the topography does not permit sufficient head from a surface reservoir, a standpipe or elevated tank may be used to gain the necessary height (Fig. 15-5). Steel, reinforced concrete, and timber are used for the construction of standpipes. A steel standpipe (Fig. 15-6a) is made of steel plates which are joined together by welding or riveting. Prestressed construction is extensively used for concrete standpipes to minimize cracking. Prestressing is usually accomplished by wrapping the tank with a continuous wire, which is then covered with mortar. Since large variations in pressure are undesirable in a distribution system, fluctuation of the water level in a stand-pipe is usually limited to 30 ft or less. Generally standpipes over 50 ft high are not economical since the lower portion of the standpipe serves only to support the upper useful portion. The economic limit of height for stand-pipes is reached when the supporting structure for an elevated tank becomes less costly than the lower ineffective portion of the standpipe. There are many different types of elevated tanks.[1] Figure 15-6b shows an elevated

---

[1] James O. Jackson, Advancements in Steel Storage Facilities, *J. Am. Water Works Assoc.*, Vol. 52, pp. 365–372, 1960.

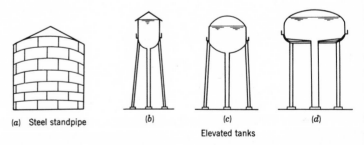

(a)   Steel standpipe   (b)   (c)   (d)

Elevated tanks

FIG. 15-6   (a) Standpipe, (b) hemispherical-bottom tank, (c) hemiellipsoidal tank, and (d) radial-cone tank.

tank which has a hemispherical bottom, cylindrical shell, and conical roof. This type of tank is available in standard sizes ranging from 5000 to 200,000 gal capacity. For a 30-ft head range the maximum capacity of such a tank is about 60,000 gal. The tank of Fig. 15-6c has a hemiellipsoid top and bottom with a cylindrical shell in between. This type of tank is built in sizes ranging from 25,000 to 1,500,000 gal and for a 30-ft head range has a limiting capacity of about 200,000 gal. Large-capacity tanks such as that of Fig. 15-6d have radial girders supported by tower columns. The bottoms of such tanks may consist of a series of radial cones or may be the combination of a sphere segment and a torus. The largest radial-cone-bottom tanks have capacities up to 3 million gallons within a head range of 30 ft.

The selection of type, number, and location of distribution reservoirs is an economic problem in which annual cost of reservoirs, pipe, and pumping should be minimized. Surface reservoirs can be larger, and hence fewer might provide adequate storage. However, if suitable surface sites are not favorably located in the distribution area, longer supply mains may be necessary. In addition larger mains may be required to keep pressures at proper levels.

15-8 Design of the pipe system   The design of a pipe system requires a detailed map of the city, showing contours (or all controlling elevations) and the location of present and future streets and lots. After studying the topography and selecting the location of distribution reservoirs, the city may be divided into districts, each to be served by a separate distribution system. The probable maximum use (allowing for fire and future growth) for each subarea of the city must be estimated. A skeleton system of supply mains leading from a distribution reservoir is assumed (Fig. 15-3). These supply mains must be large enough to deliver the expected requirements with adequate pressure. The Hardy Cross method (Sec. 11-6) is used to determine the discharge and head loss in each pipe of the network. The effect of the flow in auxiliary mains is neglected at first but may be accounted for later. Flow in the supply-main network is analyzed for fire use located in several different areas of the district to check the adequacy of the system under various patterns of withdrawal. In selecting supply mains, possible future capacity requirements should be considered. It may prove much wiser to anticipate future requirements than to replace the main with a larger one at some future date.

After the supply-main network is selected, distribution mains are added to the system. The hydraulic computations can be only approximate since all factors affecting the flow cannot possibly be accounted for. Minor distribution mains which serve fire hydrants should be at least 6 in. in diameter in residential areas and 8 or 10 in. in diameter in high-value districts. Street mains serving only domestic needs are normally 2- or 4-in. pipe.

The valve layout in a distribution system is an important part of the design. Air-relief valves should be provided at summits and drain valves at low points. Numerous other types of valves are required, but by far the most common are gate valves, which should be not more than $\frac{1}{4}$ mi apart. A closer spacing is preferable in order to minimize the area cut off from water during repairs. Gate valves over 12 in. in size are usually placed in manholes to permit inspection and are provided with concrete supports to prevent settling. Small valves are accessible from the street through cast-iron valve boxes in which a special wrench can be inserted. Valves can become inoperable because of corrosion or sediment accumulation and should be inspected at least once a year. Standardized locations for gate valves (such as the northeast corner of an intersection) permit them to be found readily in emergencies. *Altitude valves* to prevent the overflow of elevated tanks are usually designed to operate automatically.[1] Pressure-regulating valves (Fig. 11-29) may be used to divide the distribution system into various pressure zones.

Fire hydrants should be not more than 500 ft apart to avoid excessive head loss in fire hose. They are usually much closer in high-value districts. Hydrants are preferably placed at intersections so that they can be used in all directions from the corner. There are several types of fire hydrants and many designs within each type. *Flush hydrants* are located in pits below the ground surface. They are not recommended for regions of heavy snowfall. *Wall hydrants* project from the wall of a building and are used extensively in commercial districts. *Post hydrants*, which extend about 3 ft above the ground near the curb line, are most easily located. The hydrant is usually placed on a concrete block to eliminate settling and braced to resist the lateral forces of the flowing water. Hydrants are provided with one or more $2\frac{1}{2}$-in. hose outlets and a 4-in. pumper connection if fire-department pumpers are to be used. In cold climates the valve is located below ground level so that the barrel contains no water except when in use. A drain valve opens automatically when the hydrant valve is closed, to permit escape of water after use and avoid damage by freezing. This type of hydrant can be designed in such a way that the valve is not likely to open if the hydrant is broken off by a car. In warm climates the hydrant barrel may contain water at all times, and an individual valve is provided for each outlet.

15-9 Construction and maintenance of water-distribution systems
The basic requirements of pipes for water-distribution systems are adequate strength and maximum corrosion resistance. Cast-iron, cement-lined steel, plastic, and asbestos-cement compete in the small sizes, while steel and reinforced concrete are competitive in the larger sizes. In cold climates pipes

---

[1] William F. End, Jr., Automation and Automatic Valves, *J. Am. Water Works Assoc.*, Vol. 49, pp. 301–308, 1957.

should be far enough below ground to prevent freezing in winter.[1] For even the coldest parts of the United States a depth of 5 ft is generally more than adequate. In warm climates the pipes need be buried only sufficiently to avoid damage from traffic loads. Service connections to cast-iron or asbestos-cement pipe are made by tapping the distribution main with a special tapping machine which provides a threaded hole $\frac{1}{2}$ to 2 in. in diameter. A corporation cock is then installed with a flexible gooseneck pipe leading to the service pipe. The gooseneck prevents damage if there is unequal settlement between the main and the service pipe. Service pipes leading from the main to the consumer are usually copper tubing or galvanized steel. For single-family dwellings $\frac{3}{4}$- to $1\frac{1}{4}$-in. pipe is common for service pipes, but larger sizes may be needed for apartment houses or business establishments.

When a new pipe is first filled, all hydrants and valves are opened so that air can escape freely. Filling is done slowly and may require several days for large systems. Excessive pressures can develop if the air is not properly driven out of the system. When a steady, uninterrupted stream issues from a hydrant, it is closed. The procedure is continued until all valves and hydrants are closed and the system is full of water.

Leakage from distribution systems will vary with the care exercised in construction and the age and condition of the system. Construction contracts usually specify an allowable leakage in the range from 50 to 250 gpd per inch of pipe diameter per mile of pipe. The test is made by closing off a length of pipe between valves and all service connections to the pipe. Water is introduced through a special inlet, and normal working pressure is maintained for at least 12 hr while leakage is measured. In an operating system the total loss is estimated from the difference between measured input to the system and metered deliveries to the customers. There are several possible methods of locating a specific leak. Patented leak detectors use audiophones to pick up the sound of escaping water or the disturbance in an electrical field caused by saturated ground near the leak. Similar devices may be used to locate the pipe itself if the exact location is unknown. If pressure gages are installed along a given length of pipe from which there are no take offs, a change in slope of the hydraulic gradient will indicate a leak. In some instances the escaping water itself or unusually lush vegetation may show the location of a leak. If the leak is due to a faulty joint, it may be necessary only to repack and recalk the joint. If the pipe itself is cracked, the entire length of pipe may have to be replaced. The location of all pipes, valves, and appurtenances should be entered on maps. This information is essential in case repairs are required at a future date.

While pipe is being handled and placed, there are many opportunities

[1] W. L. Shannon, Prediction of Frost Penetration, *J. New Engl. Water Works Assoc.*, Vol. 59, pp. 356–363, 1945.

for pollution. Hence, it is necessary to disinfect a new system or an existing system after repairs or additions. Disinfection is usually accomplished by introducing chlorine, calcium hypochlorite, or chlorinated lime in amounts sufficient to give an immediate chlorine residue of 50 mg/l (Sec. 15-19). The chemical is introduced slowly and permitted to remain in the system for at least 12 and preferably 24 hr before it is flushed out. The flushing may be accomplished by opening several fire hydrants.

The hydraulic efficiency of pipes will diminish with time because of tuberculation, incrustation, and sediment deposits. Flushing will dislodge some of the foreign matter, but to clean a pipe effectively a scraper must be run through it. The scraper may be forced through by water pressure or pulled through with a cable. Cleaning, even though costly, may pay off with increased hydraulic efficiency and increased pressures throughout the system. The effects of cleaning may last only a short time, and in many cases the pipes are lined with cement mortar[1] after cleaning to obtain more permanent results.

15-10   Pumping required for water-supply systems   Some municipalities can rely completely on gravity to bring water from its source to the customer. Most communities, however, must do some pumping. Pumps are required to deliver water from wells and in most communities the topography is such that water must be lifted to the distribution reservoirs and elevated tanks. In many municipal distribution systems booster pumps must be installed on the mains to increase the available pressure.

Centrifugal pumps are better adapted than displacement pumps for use in water-distribution systems. Since pump efficiency varies with the load, two or more pumps are often installed in parallel so that the number of pumps in operation can be varied depending on the flow through the pumping station. Each pump should have a check valve in its discharge to prevent backflow when the pump is stopped.

Electric motors are the best source of power for pumping. They are compact and well-adapted to automatic control or remote operation. By scheduling pumping operations at off-peak hours the power cost may be held to a minimum. Gasoline or diesel engines or motor generators are frequently used as standby power for plants using electricity as a power source. With adequate high-level storage, standby power needs are minimized.

15-11   Cross connections   A cross connection occurs when the drinking-water supply is connected to some source of pollution. For example,

---

[1] Small Water Mains Cleaned and Lined in Place, *Eng. News-Record*, Vol. 139, pp. 457–459, Oct. 2, 1947. Machine Lining Applied to Large Pipe, *Eng. News-Record*, Vol. 144, May 4, 1950.

if a community has a dual water-distribution system, one for fire fighting and the other for domestic consumption, the two may be interconnected so that domestic water may be used to supplement the other system in case of fire. Such an arrangement is dangerous, for contaminated water from the fire-fighting supply may get into the drinking-water system even though the two systems are normally separated by closed valves. The preferred method for interconnecting dual systems is the air break (Fig. 15-7), although double check valves are sometimes used.

Cross connections may occur in private residences, apartment houses, and commercial buildings, especially with old-style plumbing fixtures. If the water inlet of a plumbing fixture is below the overflow drain or rim, a reduced pressure in the water system may cause back siphonage. This situation is illustrated in Fig. 15-8, where a reduction in system pressure (which might be caused by opening valve A) can result in flow back into the supply pipe from the laundry tub and flushometer. Cross connections of this type can be eliminated by prohibiting the use of underrim water inlets and requiring the use of vacuum breakers (air-inlet valves) in the water lines leading to flushometer toilets. Other sources of cross connections around a household include bathtubs, fishponds, swimming pools with underrim inlets, and lawn sprinklers that become submerged when used.

In Santa Ana, California, during a period of heavy rain in 1924, waste-water backed up into the drain from a pump and got into the water-supply system. This contamination resulted in 10,000 cases of diarrhea and 369 of typhoid. Drains from a drinking-water system to a sewer should always be designed so that wastewater cannot back up into the water supply. Sewers should not be placed above or close to a water pipe because of the danger of infiltration into the water pipe. For the same reason drinking water should never be conveyed in clay or concrete pipes which are not under pressure.

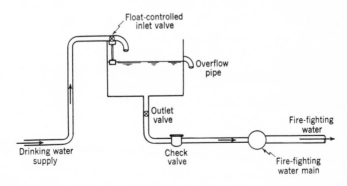

FIG. 15-7   Air break in a dual water system.

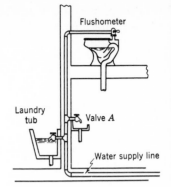

Flushometer

Laundry tub

Valve *A*

Water supply line

FIG. 15-8   Possible household cross connection.

WATER PURIFICATION[1]

Safe drinking water is essential to the health and welfare of a community, and many water systems must have some means of water purification. To be safe for drinking, water must be free of disease-producing bacteria (pathogens). In addition, the water should not possess undesirable tastes, odors, color, turbidity, or chemicals.

The principal waterborne diseases are typhoid fever, dysentery, gastroenteritis and infectious hepatitis. Between 1900 and 1950, deaths from typhoid fever in the United States dropped from 36 to 0.2 per hundred thousand of population. A substantial portion of this decrease resulted from improvement in purity of water supplies. Typhoid fever death rates in small communities are generally higher than in large cities because of the poorer supervision and control over the water supply of small towns.

15-12   Physical characteristics of water   If water contains a considerable amount of suspended clay or organic material, it will be turbid (have a muddy appearance). The turbidity depends on the fineness of the particles and their concentration. It is determined in the laboratory with a turbidimeter, which measures the interference to the passage of light rays through the water sample. The standard of comparison is the Jackson turbidimeter[2] in which the depth of water required to cause the image of the flame of a standard candle to disappear is a measure of the turbidity.

Water sometimes contains considerable color resulting from certain types of organic matter leached from soil or decaying vegetation. Color is deter-

---

[1] This section on water purification has been revised with the assistance of Prof. Perry L. McCarty of Stanford University.

[2] "Standard Methods for Examination of Water and Wastewater," 12th ed., American Public Health Association, New York, 1965.

mined by a visual comparison with Nessler tubes, glass tubes containing solutions of different standard color intensity. Tastes and odors in water are caused by decomposed organic material and volatile chemicals. Tastes and odors are measured by diluting the sample until the taste or odor is barely detectable by human observation. Water for drinking should be practically free of color and should not possess any undesirable tastes or odors.

The specific conductance of a water sample is determined by measuring its electrical resistance between two electrodes and comparing this with the resistance of a standard solution of potassium chloride at 25°C. Specific conductance gives a good indication of the dissolved-salts content of a water. For most waters the dissolved-salts[1] content in mg/l $\approx$ 0.65 × specific conductance in micromhos at 25°C. The exact value of the coefficient depends on the type of salts in the water.

15-13   Chemical characteristics of water   In the chemical analysis of a water, the total solids are determined by evaporating a sample and weighing the dry residue. The suspended solids are found by filtering a sample of the water. The difference between total solids and suspended solids represents the dissolved solids. The pH indicates the acidic or basic nature of a water and is defined as the logarithm of the reciprocal of the hydrogen-ion concentration in moles per liter. Pure water is balanced with respect to H^+ and OH^- ions, and contains 10^{-7} moles per liter of each. Thus, the pH of neutral water is 7. Waters with pH less than 7 are acidic and those with pH larger than 7 are basic. The pH of a water may be found with a potentiometer which measures the electrical potential exerted by the H^+ ions or with color-indicator dyes such as methyl orange or phenolphthalein.

The alkalinity of a water is a measure of its capacity to neutralize acids. Alkalinity in natural waters is related to the concentration of bicarbonates, carbonates, and hydroxides present. Total alkalinity is expressed in terms of equivalent calcium carbonate in milligrams per liter. Acidity of water is expressed in terms of the amount of calcium carbonate required to neutralize the water. Standard methods of testing for alkalinity, acidity, and the content of various chemicals in water may be found elsewhere in the literature.[2]

The compounds dissolved in water may be beneficial or detrimental.[3] If the chloride concentration is greater than 250 mg/l, the water will have

---

[1] The concentration of chemicals in water is expressed in terms of milligrams per liter (mg/l). Formerly the term "parts per million" (ppm) was widely used but this was often misinterpreted.

[2] "Standard Methods for Examination of Water and Wastewater," 12th ed., American Public Health Association, New York, 1965.

[3] Public Health Service Drinking Water Standards, 1962, *U.S. Pub. Health Serv. Pub.* 956, 1963.

an undesirable taste. A high sulfate concentration has a laxative effect on the human body. Hydrogen sulfide (H_2S) imparts an undesirable taste and odor to water. Lead and barium salts are toxic, and only very low concentrations of these salts can be tolerated by the human system. A fluoride concentration of about 1.0 mg/l results in fewer cavities in the teeth of children, but a concentration greater than 1.5 mg/l causes the enamel to be mottled. The evidence seems to indicate that the beneficial effects of fluoride are gained mainly during the early years of childhood. Many communities now add fluorides to their water to bring the concentration up to 0.8 or 1.0 mg/l.

Hard waters contain calcium and magnesium ions which react with soap to form scum. *Temporary hardness* is caused by bicarbonates of calcium and magnesium. If water with temporary hardness is boiled, carbon dioxide is driven off and an insoluble carbonate is precipitated. This reaction is the source of boiler scale. *Permanent hardness* is caused by the sulfates, chlorides, and nitrates of calcium and magnesium and is not removed by boiling. Hardness is expressed in milligrams per liter or grains per gallon[1] of equivalent calcium carbonate. For boilers and laundries hardness should be less than 50 mg/l for satisfactory operation. Underground waters quite frequently have a hardness in excess of 300 mg/l. Surface waters are usually softer since they do not have as much opportunity for contact with minerals.

Iron and manganese in concentrations greater than 0.3 mg/l and 0.05 mg/l, respectively, are undesirable as they cause discoloration of clothing which is laundered in the water. Moreover, incrustations will be formed in water mains by the deposit of ferric hydroxide and manganese oxide.

ILLUSTRATIVE EXAMPLE 15-1   The chemical analysis of a water indicates the presence of cations in the following concentrations:

Na^+ 20 mg/l
Mg^{++} 60 mg/l
Ca^{++} 45 mg/l

Compute the hardness (as equivalent $CaCO_3$).
Combining weights (from Appendix) are

Ca 20, Mg 12, $CaCO_3$ 50

Hence,

$$\text{Hardness} = 45 \times \frac{50}{20} + 60 \times \frac{50}{12} = 362.5 \text{ mg/l}$$

---

[1] There are 7000 grains in a pound, and 1 gal of water weighs 8.33 lb. Hence, 1 grain/gal is equivalent to 17.1 ppm which in turn equals 17.1 mg/l.

15-14   Bacterial and microscopical characteristics of water   Bacteria vary in size from about 1 to 4 microns[1] and cannot be seen with the naked eye. Disease-causing bacteria are called *pathogenic bacteria. Nonpathogenic bacteria* are harmless. *Aerobic bacteria* require oxygen for survival, while *anaerobic bacteria* thrive in the absence of free oxygen. *Facultative bacteria* are those which can live either with or without free oxygen. *Escherichia coli* (*colon bacilli*, or *coliforms*) are among the bacteria which inhabit the intestines of warm-blooded animals. They are harmless bacteria which are excreted with feces, but their presence indicates the possibility that pathogenic bacteria are also present. A water which has been contaminated with sewage will contain coliform organisms.

The presence of coliform organisms in water is detected by mixing a sample with lactose broth and incubating for 48 hr at 37°C (body temperature). If carbon dioxide gas is evolved, the sample may contain coliforms. This is known as the *presumptive test*. Since harmless organisms of the coliform group live longer in water than pathogens, it is presumed that a water is safe if it is negative to the coliform test. Water that fails the presumptive test does not necessarily contain coliforms; and it is subjected to a further test, the *confirmed test*, that removes extraneous bacteria that might have affected the presumptive test. If a water fails in the confirmed test, it probably contains coliform organisms. This can be further confirmed by conducting the *completed test*.

In testing for coliforms, the standard sample consists of five representative 10-ml or 100-ml quantities. The number of samples per month specified by the U.S. Public Health Service depends on the population served by the water system. In general there should be no more than one coliform organism in 100 ml of water for the water to be considered safe for drinking. Samples are usually obtained at several different points in the distribution system. Bacteria reproduce by fission (cell division), hence a water sample to be examined for bacterial content must be tested soon after it is collected.

Another method of testing for coliforms is to filter the water sample through a sterile membrane on which bacteria will be retained. The membrane is then put in contact with nutrients that will permit only the growth of coliform colonies. After a 20-hr incubation period the colonies can be counted. This procedure is simple, and it can be performed in the field; it does not take as much time as the presumptive and confirmed tests.

In addition to bacteria, water may contain other types of undesirable microscopic organisms. *Algae* are one-celled plants which impart taste and odor to water. They can be seen growing in clear streams and ponds and grow only in the presence of sunlight. Excessive algae growth may be con-

---

[1] A micron (one-millionth of a meter) is approximately 0.00004 in.

trolled by application of copper sulfate or chlorine. *Fungi* are plants which grow without sunlight and at times will infest water mains, producing unpleasant tastes and odors and even causing clogging. One type of *protozoa* (one-celled animals), *Endamoeba histolytica*, is the cause of amoebic dysentery. *Crenothrix* is a genus of bacteria which possesses characteristics of the larger organisms. *Crenothrix* thrives in waters containing iron in solution, and may result in clogging of water mains.

15-15   Methods of water purification   Various methods are used to make water safe and attractive to the consumer. The method selected depends mostly on the character of the raw water. Much of the suspended material can be removed in a sedimentation basin, where the larger particles will settle under the action of gravity. This process is called *plain sedimentation.* The effectiveness of sedimentation may be increased by mixing chemicals with the water to form a flocculent precipitate which carries the suspended particles down as it settles. This process is called *chemical coagulation.*

Some of the finer particles in the water may still be in suspension after sedimentation. A large portion of these can be removed by *filtration.* Filtration is almost always preceded by sedimentation in order that the filters do not become too rapidly clogged. Turbid water should be clarified by chemical coagulation and filtration. Since pathogenic bacteria may pass through both the sedimentation basin and the filter, *disinfection* is essential. Disinfection (the killing of harmful bacteria) is usually achieved by chlorination.

Water which has been treated by sedimentation, filtration, and disinfection will be safe but not necessarily attractive. Unpleasant tastes and odors in water may have to be removed by *aeration*, addition of activated carbon, or other means. Hardness in water may be removed by a process known as softening. Whether or not a water should be softened depends on its hardness and intended use.

Water may be subjected to other treatments, such as *fluoridation*, the addition of soluble fluorides for the control of dental caries; *liming* to control acidity and reduce corrosive action; and *recarbonation*, the addition of carbon dioxide to prevent the deposition of calcium carbonate scale.

15-16   Plain sedimentation   The rate of settling of a particle in water depends on the viscosity and density of the water as well as the size, shape, and specific gravity of the particle. Warm water is less viscous, and a particle will settle more rapidly than in cold water. Suspended inorganic particles found in water have a specific gravity ranging from 2.65 for discrete sand particles to about 1.03 for flocculated-mud particles. The specific gravity of suspended organic matter ranges from 1.0 to about 1.4. Chemical flocs have a similar range of specific gravity depending on the amount of entrained water in the floc.

Figure 15-9 shows the settling velocities of discrete spherical particles in quiescent water at 68°F. Settling velocities in a sedimentation basin will be considerably less than shown in Fig. 15-9 because of the nonsphericity of the particles, the upward displacement of the fluid created by the settling of other particles, and convection currents.

Water purification by sedimentation aims to provide conditions such that the suspended material in water can settle out. Storage reservoirs serve as rudimentary sedimentation basins, but because of density currents, disturbances caused by wind, and other factors they cannot always be relied upon to give proper clarification. Sedimentation basins constructed for the

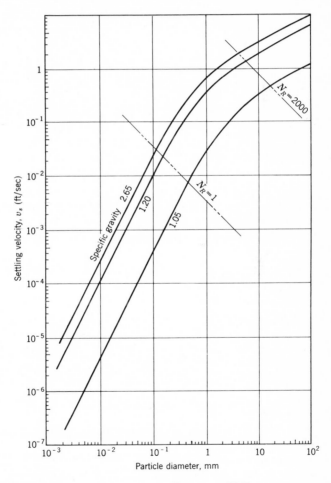

FIG. 15-9   Settling velocities of spherical particles in still water at 68°F.

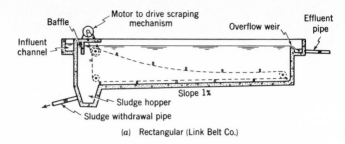

(a)   Rectangular (Link Belt Co.)

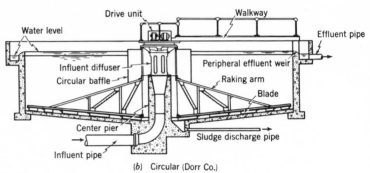

(b)   Circular (Dorr Co.)

*FIG. 15-10   Two types of sedimentation basins: (a) rectangular (Link-Belt Co.) and
(b) circular (Dorr-Oliver Co.).*

specific purpose of removing suspended material from the water are generally
of reinforced concrete and may be rectangular or circular in plan (Fig. 15-10).
The *retention period* (or *detention time*) is the average time required for the
water to flow through the basin. In order to get a fairly high percentage of
removal of the suspended material, it is desirable that the basin be properly
designed.

It is assumed that the sediment is uniformly distributed as the water
enters the basin (Fig. 15-11) at a uniform velocity, V. The path of a discrete
particle is given by the vector sum of its settling velocity v_s and V. Let it be
assumed that all particles whose paths of travel are above line AB will pass
through the basin. From geometric considerations it can be seen that

$$\frac{V}{v_s} = \frac{L}{H} \qquad (15\text{-}1)$$

But $V = Q/BH$ where B is the width of the basin.
Hence,

$$\frac{Q}{BHv_s} = \frac{L}{H} \qquad (15\text{-}2)$$

and

$$v_s = \frac{Q}{BL} \tag{15-3}$$

Consideration of the assumed criterion for removal indicates that all particles with $v_s > Q/BL$ will be removed. The quantity Q/BL, the discharge per plan area, is known as the *overflow rate*. Typical overflow rates for sedimentation basins are around 500 to 1000 gpd/sq ft.

The percentage of particles that will be removed depends on the gradation of their sizes and their settling velocities. According to the theory presented, all particles entering at level h with $v_s \geq (h/H)(Q/BL)$ will be removed. Hence, if the particle distribution at entry is uniform, h/H of all particles with $v_s \geq (h/H)(Q/BL)$ will be removed in the basin. Thus, with a given Q, increasing the plan area of the basin will reduce the overflow rate and increase the efficiency. Theoretically depth has no effect on the efficiency of removal.

The flow should be uniformly distributed through the cross section of the basin. If currents permit a substantial portion of the water to pass directly through the basin without being retained for the intended time, the flow is said to be "short-circuited." Properly located baffles near the entrance to the basin will distribute the flow uniformly and reduce or eliminate dead space in the basin (Fig. 15-12). Sedimentation basins are usually uncovered, but in severe climates they may be covered to prevent trouble from ice or wind. A well-designed sedimentation basin will remove 50 to 80 percent of the suspended solids contained in the water.

Many sedimentation basins have been designed to operate intermittently, i.e., they are shut down when necessary, and the sediment which has accumu-

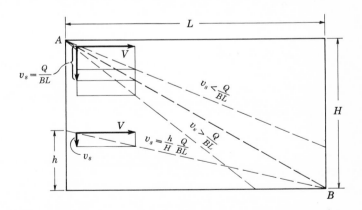

FIG. 15-11   Schematic diagram of rectangular sedimentation basin.

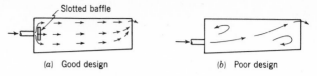

(a) Good design   (b) Poor design

FIG. 15-12   Flow in a rectangular sedimentation basin.

lated in the basin is flushed by hose. Basins which operate continuously are
more common. They are provided with some type of automatic cleaning
mechanism, usually a slow-moving scraper which pushes the sediment into a
hopper from which it can be drawn off from time to time, either by gravity
or pumping, and discharged to a suitable disposal area.

Detention periods ranging from 1 to 10 hr have been widely used. Sedimentation basins are usually 10 to 15 ft deep though some are as shallow as
6 ft and others as deep as 20 ft. With proper design, shallow basins will give
good performance.[1] For rectangular basins a width of 30 ft is common.

ILLUSTRATIVE EXAMPLE 15-2   One million gallons of water per day passes through
a sedimentation basin which is 20 ft wide, 50 ft long, and 10 ft deep. (a) Find the detention
time for this basin? (b) What is the average velocity of flow through the basin? (c) If
the suspended-solids content of the water averages 40 ppm, what weight of dry solids
will be deposited every 24 hr, assuming 75 percent removal in the basin? (d) Compute
the overflow rate.

(a) Detention time $= \dfrac{\text{volume of tank}}{\text{flow per unit time}}$

$$= \frac{20 \times 50 \times 10}{1 \times 10^6} \times 7.48 \times 24 = 1.8 \text{ hr}$$

(b) Velocity $= \dfrac{Q}{A} = \dfrac{1 \times 10^6}{20 \times 70} \, \dfrac{1}{7.48 \times 3600 \times 24} = 0.0077 \text{ ft/sec}$

(c) $\dfrac{40}{10^6} \times 10^6 \times 8.33 \times 0.75 = 250 \text{ lb/day}$

(d) Overflow rate $= \dfrac{Q}{BL} = \dfrac{10^6}{50 \times 20} = 1000 \text{ gpd/sq ft}$

15-17   Sedimentation with coagulation   If the suspended solids of
water are fine or colloidal in size, chemicals are often used to effect more
complete removal of the suspended matter. The coagulants react with the

---

turbidity particles to form a flocculent precipitate. The most common coagulant is alum $[Al_2(SO_4)_3 \cdot 14H_2O]$ which reacts with the alkalinity in the water to form an aluminum-hydroxide floc. If the water does not contain the required alkalinity, it may be necessary to introduce lime (CaO) or soda ash (Na_2CO_3) in addition to the alum to get proper flocculation. Activated silica, sometimes added to the water, provides nuclei for floc formation. In the process, the finely divided colloidal suspensions are converted to settleable solids by agglomeration. The individual floc particles collide with one another, and flocs of increasing size are formed. In quiescent water the flocs grow slowly. Flocculation is enhanced by gentle agitation of the water to increase the collision rate.[1] Violent agitation is detrimental as flocs will be disintegrated by the high shear forces that are produced.

The usual dosage of alum will be 10 to 40 mg/l (approximately 80 to 300 lb per million gallons). The amount of auxiliary chemical used depends on the character of the water. Ferrous sulfate ($FeSO_4$) and ferric chloride ($FeCl_3$) are also used as coagulants. They form iron hydroxide precipitates. The ferrous salt requires the use of lime as an auxiliary chemical, or it may be converted to the ferric form by addition of chlorine. The amount of chemical used is found through trial by placing samples of the raw water into a series of jars and adding different amounts of chemical to each. After a few seconds of vigorous mixing and several minutes of slow mixing the character of the flocs and their settleability are observed and the optimum dosage selected. Since the quality of the water may change, it is desirable to conduct the test frequently. At some plants this may be done several times a day; at others where the water is less variable, only once every few days.

Chemicals can be fed into water by a solution-feed machine or a dry-feed machine depending upon the nature of the chemicals used. The chemicals must be widely dispersed throughout the water by proper mixing. This is most commonly achieved by mechanical means with rapidly rotating paddles in mixing chambers having detention times of 30 to 60 sec. The hydraulic jump is also an effective mixer, but it may not be feasible over a wide range of flow rates.

The flash mix must be followed by a 20- to 30-min period of gentle agitation to permit flocculation. Flocculators consisting of large slowly rotating paddles in relatively deep basins are best suited for high-capacity plants. A baffled chamber (Fig. 15-13) through which the velocity of flow is about 1 ft/sec will also provide a suitable environment for flocculation.

If carefully controlled, sedimentation with coagulation will remove about

---

[1] Thomas R. Camp, Flocculation and Flocculation Basins, *Trans. ASCE*, Vol. 120, pp. 1–16, 1955.

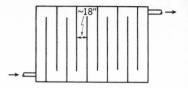

FIG. 15-13   Plan of a baffled mixing chamber.

90 percent of the suspended solids. Various manufacturers have come out with packaged units, generally circular in plan, that will provide clarification of water through sedimentation with coagulation. These units are particularly suited for small installations. Flexibility in the operation of sedimentation basins may be had by using several units in parallel. With such an arrangement one or more may be operated depending on the flow and shut down for repairs or cleaning if necessary.

15-18   Filtration   The usual filter consists of a layer of sand or crushed coal supported on a bed of gravel. The rapid sand filter is shown in Fig. 15-14. When water passes through the filter, suspended particles and flocculant material come in contact with the sand grains and adhere to them. This reduces the size of the water passages, and a straining action results. Bacteria are effectively removed by filtration. In time, more and more material is trapped in the sand bed, the pores clog up, and the hydraulic head loss through the bed becomes excessive. The filter is then backwashed to remove the trapped material. During backwashing the sand bed expands about 50 percent, and the material that had been filtered out of the water is dislodged by shearing action and carried off in the wash-water troughs. Water jets directed at the surface during backwash are often employed to free the

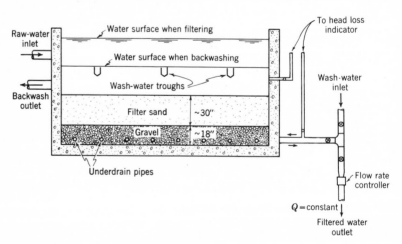

FIG. 15-14   Cross section of a rapid sand filter.

filtered material from the sand grains. The rate of wash-water rise should not exceed the settling velocity of the smallest particle to be retained in the filter and is usually 12 to 36 in./min; the period of backwash is generally 3 to 5 min. The amount of water required for washing a rapid sand filter varies from 1 to 5 percent of the total amount filtered. The filter must be backwashed every 1 to 3 days. The wash water is usually wasted to a sewer or stream, although sometimes it is discharged into a reservoir for later reclamation.

Various types of filter underdrains are employed in rapid sand filters. Perforated pipes are widely used for this purpose. A typical installation might consist of a 16-in.-diameter manifold pipe running lengthwise along the center of the filter bottom. Taking off from the manifold in both directions at right angles to it would be 4-in. laterals with $\frac{1}{2}$-in. holes every 6 in. The holes are placed near the bottom of the pipe to prevent clogging. Filter blocks, metal gratings, and special patented filter bottoms are also used. A properly designed filter bottom will give a uniform distribution of wash water. This will prevent mud balls from forming in the filter and lead to more effective filtration. A vigorous surface wash prior to backwashing is also used sometimes to prevent mud-ball formation. In order to achieve uniform distribution of wash water, the filter units should not be too large. A 4-mgd twin unit having an area of about 2000 sq ft is typical of the larger filter units.

From experience it has been found that a 24- to 30-in. layer of sand with uniform grain size (0.35- to 0.45-mm diameter) gives best results. The sand is frequently supported by a 12- to 18-in. layer of graded gravel. The usual rate of water application is about 2 gpd/sq ft. The discharge line from the filter is provided with a rate controller, a throttling device to maintain a uniform rate of flow through the filter. Too high a rate of flow, particularly immediately after backwashing, would permit some of the suspended material to be washed through the sand filter.

The filter should be provided with a head-loss indicator. Immediately after backwashing the head loss through a sand filter operating at the usual rate will be about 2 ft. This will gradually increase. When the head loss approaches 8 ft, the filter should be backwashed, otherwise difficulty with air binding may occur. Air binding is caused by excessive negative pressures in the sand bed. These permit gases to be released from the water and to lodge in the bed and underdrainage system. This reduces the capacity of the filter.

At a filter plant it is good practice to provide a regulating reservoir or *clear well* with a capacity of about 30 to 40 percent of the daily output so that one or more filters can be taken off line for several hours for maintenance. For small installations, such as industrial plants and swimming pools, pressure filters are employed. These are closed tanks containing a filter bed through which the water passes under pressure. Diatomaceous-earth filters are another type in which a layer of diatomaceous earth is built up on a supporting medium. Such filters, though not suitable for highly turbid water, can be made portable and hence are adaptable for emergency use. The rapid sand filter will remove about 90 percent of the turbidity and bacteria from water. Data on some representative rapid-sand-filter plants in the United States are presented in Table 15-5.

15-19 Disinfection More than 50 percent of the pathogens in water will die within 2 days, and 90 percent will be dead at the end of the week. Hence, reservoir storage is reasonably effective in controlling bacteria. A few pathogens, however, may live for 2 yr or longer; hence, disinfection is desirable. Chlorine is an ideal disinfectant. It forms hypochlorite when added to water and this has an immediate and disastrous effect on most forms of microscopic life. Liquid chlorine is obtained in pressure containers and is applied to the water through a chlorinator. Small chlorinators feed the gas directly into the water, while large-capacity chlorinators usually dissolve the gas in water and feed the solution. The chlorinator should be maintained at a temperature of at least 70°F to prevent condensation of chlorine gas in the feed lines. Both automatic and manual regulators for application of chlorine are available.

Some plants find it economical and safer to use calcium hypochlorite (bleaching powder) as a disinfectant. This chemical reacts with water to liberate hypochlorites as does chlorine. The amount of chlorine required depends on the amount of reducing inorganic and organic matter in the water. In general, most waters are satisfactorily disinfected if the free chlorine residual (hypochlorites) is about 0.2 mg/l 10 min after the chlorine is applied. A larger chlorine residual may cause tastes, while a smaller one cannot be relied upon. Chlorine is most effective if the pH of the water is low. When a water supply contains phenols, the addition of chlorine to the water will result in disagreeable tastes because of the formation of chlorophenol compounds. These tastes may be eliminated by adding ammonia to the water just before chlorination or by other means. The chlorine and ammonia combine to form chloramines, which are relatively stable disinfectants, but not as effective as hypochlorites. They do not react rapidly but continue their action for a long time. Hence, their disinfecting qualities may extend for a considerable distance into the distribution system.

Chlorination is practiced in a variety of ways depending on the quality of the raw water and other conditions. *Postchlorination*, the application of chlorine after treatment, is the customary method. *Prechlorination*, the application of chlorine before treatment, however, improves coagulation, reduces the load on the filters, and prevents the growth of algae. Prechlorination and postchlorination are often used together. Application of chlorine so as to leave an excessively large residual is known as *superchlorination* and is often employed to remove tastes and odors. Super-

TABLE 15-5 Data on Some of the Larger Rapid-sand-filter Plants in the United States*

| | | Chemical feed and mix | | | | | Sedimentation | | | Filtration | | | | | | |
City	Capacity, mgd	Type feed	Alum used, grains/gal	Lime used, grains/gal	Mixing time, min	Type of mix	Time, hr	Flow V, ft/min	Basin depth, ft	Filter rate, gpm/sqft	Unit size, mgd	Sand depth, in.	Sand size, mm	Gravel depth, in.	Wash rate, in./min	Wash water, per cent
Detroit, Mich.	320	Dry	0.74	0	3	Baffles	2.0	4.2	16	2.55	4	30	0.45	17	26–30	2.0
Milwaukee, Wis.	200	Dry	0.5	0.11	64	Mechanical	4.0	2.32	27	2.00	6.25	27	0.51	24	24	2.2
St. Louis, Mo.	160	Solution	1.2	4.7	45	Baffles	36.0	1.5	16–23	2.00	4	30	0.4	12	24	1.67
Toledo, Ohio	80	Dry	1.12	0.5	40	Mechanical	2.8	1.5	15	1.50	2	22	0.42	18	20	2.43
Louisville, Ky.	120	Dry	0.7†	0.31	...	None	2.0	5.7	17.5	2.00	6 and 3	26–30	0.4–0.5	14–24	22–36	2.0
Denver, Colo.	64	Dry	0.1 –1.0 0.05–0.5‡	0	20	Baffles	1.0	12.0	13.5	2.48	4.5	48†	0.62	15	25	1.4
Atlanta, Ga.	54	Dry	0.59	0.26	23	Baffles	9.3	0.7	14–24	2.00	3 and 5	24–27	0.4–0.5	18	30	1.3
Dallas, Tex.	48	Dry	1.0	5.2	12	Baffles	8.0	1.1	18	1.90	2	30	0.4–0.45	18	20	1.5
New Orleans, La.	40	Solution	0.75§	5.0	60	Baffles	18.0	0.55	13.75	1.95	4	30	0.33	9	24	0.3
Albany, N.Y.	32	Dry	1.5	0	20	Baffles	2.25	1.5	10–18	2.00	4	30	0.33	18	24	2.5
Richmond, Va.	30	Dry	2.0	0	10	Combined	10.0	6.0	10	1.90	3	26	0.43	16	24	1.0

*Data from Design and Operation Data on Large Rapid Sand Filtration Plants in the U.S. and Canada, *J. Am. Water Works Assoc.*, Vol. 48, pp. 819–853, July, 1956.

†Coal used as a filter medium.

‡$Na_2Al_2O_4$.

§$FeSO_4$.

chlorination must be followed by dechlorination, usually accomplished by treatment with sulfur dioxide or by passing the water through a filter of granulated activated carbon.

To test for residual chlorine, 5 ml of orthotolidine solution is added to 100 ml of the sample. If a yellow color results, the sample contains a chlorine residual. The deeper the yellow, the greater the residual. A quantitative determination of the chlorine residual can be made by comparison with standard colors. Chlorine frequently forms natural chloramines with nitrogen compounds in the water. These give a positive reaction in the orthotolidine test just as a free chlorine residual will. Addition of more chlorine will break down these chloramines and form a free chlorine residual at point B (Fig. 15-15). This point is known as the *break point*, and application of chlorine in amounts slightly above the break point (known as *break-point chlorination*) is sometimes desirable to control tastes and odors and to ensure more rigorous disinfection.

15-20   Removal of tastes and odors   Tastes and odors in water are caused by (1) dissolved gases such as hydrogen sulfide, (2) living organic material such as algae, (3) decaying organic material, (4) industrial wastes, and (5) chlorine either as a residual or in combination with phenol or decomposing organic matter.

Aeration, usually accomplished by spraying water from special nozzles or by permitting it to trickle over cascades, breaks the water into droplets and permits the escape of dissolved gases. Activated carbon can be used effectively to remove tastes and odors. It is made by heating paper-mill waste or sawdust in a closed retort and oxidizing it by means of air or steam to remove the hydrocarbons. The resulting product is very porous and possesses extremely good adsorptive properties. Activated carbon is sometimes used as a filter, though more often it is applied directly to the raw water in the mixing basin before coagulation. A dosage between 5 and 30 lb per million gallons is usually adequate. Chlorine, ozone, and potassium permanganate have also been used effectively to oxidize taste-producing material.

15-21   Water softening   The removal of hardness from water is not

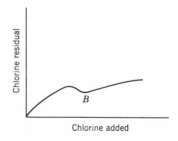

FIG. 15-15   *Variation of chlorine residual with chlorine supplied for a typical water.*

essential to make the water safe for use. The advantage lies chiefly in the reduction of soap consumption and a lowered cost in maintaining plumbing fixtures.[1] Whether or not the hardness of a water supply should be reduced depends on the relation between the cost of treatment and the saving and satisfaction to the customers. The two basic methods used for the removal of hardness are the *lime-soda process* and the *ion exchange process.*

In the lime-soda process, lime [$Ca(OH)_2$] and soda ash (Na_2CO_3) are added to the water. These react with the calcium and magnesium salts to form insoluble precipitates, calcium carbonate ($CaCO_3$) and magnesium hydroxide [$Mg(OH)_2$], which can be removed from the water by sedimentation. Typical chemical reactions are

$$Ca(HCO_3)_2 + Ca(OH)_2 \rightarrow 2CaCO_3 + 2H_2O$$

$$Mg(HCO_3)_2 + 2Ca(OH)_2 \rightarrow 2CaCO_3 + Mg(OH)_2 + H_2O$$

$$MgSO_4 + Ca(OH)_2 \rightarrow Mg(OH)_2 + CaSO_4$$

$$CaSO_4 + Na_2CO_3 \rightarrow CaCO_3 + Na_2SO_4$$

The lime serves to reduce the carbonate hardness and substitutes calcium salts for magnesium salts, while the soda acts on the noncarbonate hardness of the calcium salts. The sodium salts which are formed are soluble but are not objectionable in the amounts ordinarily resulting from the softening process. Most of the $CaCO_3$ and $Mg(OH)_2$ precipitate which is formed will settle out in a sedimentation basin, but some will remain as finely divided particles which may be deposited on a filter or in the pipes of the distributing system. To prevent this, the water should be *recarbonated* by passing carbon dioxide (CO_2) gas through it as it leaves the sedimentation tank. In this process the insoluble carbonates combine with CO_2 to reform soluble bicarbonates. Even though the water regains some of its hardness by this process, recarbonation is advisable.

The amount of lime and soda required for softening depends on the chemical quality of the water and the amount of hardness removal desired. Many hard waters contain low concentrations of sulfates, chlorides, and nitrates; hence lime is often the only chemical used. The apparatus used in the lime-soda process is similar to that for chemical coagulation. After sedimentation and recarbonation the water is usually passed through a sand filter. Alum is often added with the lime and soda to combine chemical coagulation and water softening into a single process. Special units have been designed which combine mixing, flocculation, and clarification into a

---

[1] Louis R. Howson, Economics of Water Softening, *J. Am. Water Works Assoc.*, Vol. 43, pp. 253–259, 1951; L. M. DeBoer and T. E. Larson, Water Hardness and Domestic Use of Detergents, *J. Am. Water Works Assoc.*, Vol. 53, pp. 809–822, 1961.

single structure. The usual arrangement of these units includes an inner chamber for mixing with an outer chamber for clarification. These units are relatively inexpensive and give good results if properly operated. The lime-soda process has three distinct disadvantages—a large quantity of sludge is formed, careful operation is essential if good results are to be obtained, and incrustation of pipes will result if the water is not properly recarbonated.

An *ion-exchange unit* resembles a sand filter in which the filtering medium is an ion-exchange resin rather than sand. Resins may be either natural (zeolites) or synthetic. As the hard water passes through the ion-exchange bed, there is an exchange of cations; the calcium and magnesium in the water are exchanged for the sodium in the resin. The sodium salts formed do not cause hardness. When a significant portion of the sodium in the resin has been replaced by calcium and magnesium, it is regenerated with a sodium chloride solution. The water passing through the bed during the regeneration cycle must be wasted, for it contains a high concentration of chlorides. The exchange reactions are as follows:

Softening:

$$\left. \begin{matrix} \text{Ca} \\ \text{Mg} \end{matrix} \right\} \left\{ \begin{matrix} (\text{HCO}_3)_2 \\ \text{SO}_4 \\ \text{Cl}_2 \end{matrix} \right. + \text{Na}_2 R \rightarrow \text{Na}_2 \left\{ \begin{matrix} (\text{HCO}_3)_2 \\ \text{SO}_4 \\ \text{Cl}_2 \end{matrix} \right. + \left. \begin{matrix} \text{Ca} \\ \text{Mg} \end{matrix} \right\} R$$

Regeneration:

$$\left. \begin{matrix} \text{Ca} \\ \text{Mg} \end{matrix} \right\} R + 2\text{NaCl} \rightarrow \text{Na}_2 R + \left. \begin{matrix} \text{Ca} \\ \text{Mg} \end{matrix} \right\} \text{Cl}_2$$

Ion-exchange units may be either gravity or pressure filters. The usual rate of raw-water passage through the bed is about 6 gal/sq ft/min. The ion-exchange process results in water of zero hardness. Since there is usually no need for such soft water, only a portion of the water passing through the treatment plant is softened and mixed with unsoftened water to obtain the desired water quality.

15-22 Desalting   In recent years there has been a rapidly increasing interest in the development of processes that will convert saline water to fresh water. In 1970, over 200 different plants throughout the world were producing a total of over 50 mgd of fresh water from saline sources. Though the cost of desalting has been reduced considerably it is still relatively high. However, in some areas desalting is competitive with other means of obtaining fresh water. Interest in desalting has been stimulated by the realization that freshwater supplies are quite limited in many areas. The U.S. Office of Saline Water has sponsored a continuous program of research on desalting since 1952. Several pilot plants are in operation, each using a different process.

Ocean water has a dissolved-salts content of about 35,000 mg/l. The largest desalting plant is at Kuwait on the Persian Gulf where over 5 mgd of fresh water are produced by flash distillation. Many groundwater supplies are brackish (dissolved salts 1000 to 3000 mg/l) and too salty for consumption. Some of the many processes for removing salts from water are discussed in the following paragraphs.

Distillation   Distillation of seawater has been practiced for many years. Recent research has been aimed at development of evaporators that would have minimum difficulty with scale formation. Various types of vapor-compression and multiple-effect flash systems show promise; however, as of 1970, the minimum cost of desalting by distillation was about $300 per acre-foot in plants of a few million gallons capacity. Estimates are that costs can be reduced to about $100 per acre-foot with large dual-purpose plants, which generate high-pressure steam for electric power and low-pressure steam for evaporator operation. Solar stills have been used for limited production in areas having abundant sunshine throughout the year. However, large-scale production by solar stills does not appear economically feasible within the near future.

Freezing   In the freezing process the temperature of the seawater is gradually lowered until ice crystals are formed. These are free of salt and can be separated from the brine. As of 1970 a cost of about $200 per acre-foot had been achieved by the freezing process.

Demineralization   Salts can be removed from water through use of ion exchangers similar to those used for softening. In demineralization two different resins are used, one for the removal of cations and the other for the removal of anions. The process is prohibitively expensive for use on seawater; however, it is well adapted for use on waters with salt content of less than 1000 mg/l, where costs of $100 to $200 per acre-foot have been estimated.[1]

Electrodialysis   By this method ions are removed by an electrochemical process wherein they diffuse under the action of an electric potential through membranes that are selectively permeable to different types of ions. The city of Coalinga,[2] California, formerly imported drinking water by railroad tank car at a cost of somewhat over $2000 per acre-foot. It now treats

---

[1] W. A. Homer, New Concepts for Desalting Brackish Water, *J. Am. Water Works Assoc.*, Vol. 60, pp. 869–881, 1968.

[2] E. S. Cary, H. J. Ongerth, and R. O. Phelps, Domestic Water Supply Demineralization at Coalinga, *J. Am. Water Works Assoc.*, Vol. 52, No. 5, May, 1960.

brackish well water (2000 mg/l) by electrodialysis at a cost of less than $400 per acre-foot. In the process there is about 1 gal of water wasted for each gallon produced. Cost of salt removal by electrodialysis is proportional to the amount of salt in the water. Because of extremely high cost, electrodialysis is unsuitable for use with seawater.

Reverse Osmosis This process also makes use of membranes, but ones which are selectively permeable to water, rather than to salts. By applying pressures as high as 1500 psi, fresh water is pushed through the membranes while leaving the salts behind. Flow rates through the membranes depend upon the initial salinity and have varied from about 20 gpd/sq ft for sea water to 30 gpd/sq ft for brackish waters. This process is still under development but has stimulated considerable interest because the energy cost is potentially very small. No change of state of the water is necessary.

Figure 15-16 shows approximately the cost of desalting as a function of salt content as of 1970. Table 15-6 gives a list of energy requirements for different desalting processes based on technology in 1964 and that expected by 1980. In using such figures, the costs of different types of energy must also be considered. The values indicate the technological improvements which are anticipated.

In the United States and Canada there are over 1000 water utilities having raw water with total dissolved salts in the 1000 to 3000 mg/l range.[1] It is likely that the first widespread application of desalting will be on brackish water. Ocean water, when converted to fresh water, generally has to be pumped considerable distances to the place of use, thus further adding to the cost of such water.

15-23   Selection of water-treatment method   The type of treatment

[1] Progress in Saline Water Conversion—Task Group Report, *J. Am. Water Works Assoc.*, Vol. 53, pp. 1091–1105, 1961.

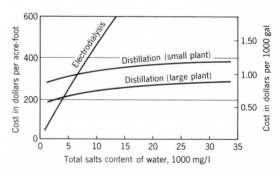

FIG. 15-16   Approximate cost of desalting as of 1970.

*TABLE 15-6   Energy Requirements for Six Desalting Processes**

Processes	Energy required (per 1000 gal of product water)			
	1964 technology		Estimate for 1980 technology†	
	Btu × 10^{-3}	kw-hr	Btu × 10^{-3}	kw-hr
Processes using heat				
Multistage flash distillation	1020	300	610	180
Long-tube vertical distillation (LTV)	1020	300	610	180
Processes using electricity‡				
Electrodialysis (brackish water only)	250	25	150	15
Vapor compression distillation	610	60	360	35
Freezing	610	60	360	35
Reverse osmosis	510	50	310	30

*W. S. Gillam and W. H. McCoy, Desalination Research and Water Resources, in K. S. Spiegler (ed.), *Principles of Desalination*, Academic Press, New York, 1966.
†The estimated 1980 energy requirements are for high-efficiency processes and are not applicable to processes using low-cost energy.
‡The energy values given for the "electrical" processes are the thermal energies for the appropriate electrical power generation at 33% plant efficiency.

required depends on the physical, chemical, and biological characteristics of the water. Water from deep wells, for example, is usually free of pathogenic bacteria, and no purification is necessary. Most well waters are hard; and, softening, together with removal of iron and manganese, may be desirable. Well waters are generally quite clear so there is no need for turbidity removal. If there is possibility of pollution, chlorination is advisable.

The relatively high turbidity of river water usually requires chemical coagulation and filtration. The turbidity of river water varies considerably throughout the year. It may be high during floods but low at other times. Some plants treating river water may have facilities for adding coagulants but use them only during floods. Storage in reservoirs will reduce the need for sedimentation. Many cities have also taken steps to reduce erosion in the watersheds tributary to their water source in order to minimize the requirements for clarification. Most surface waters are subject to contamination, and disinfection is usually essential.

Flow diagrams for several typical water-treatment plants are shown in Fig. 15-17. The plant of the Metropolitan Water District of Southern California (Fig. 15-17c) gives very complete treatment to hard water obtained from the Colorado River. The cost of water treatment depends on the

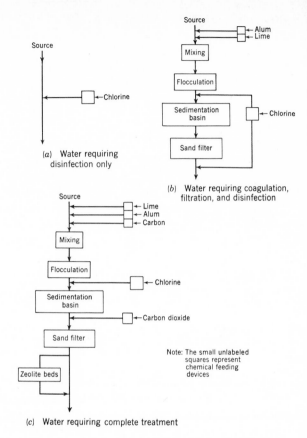

FIG. 15-17   Flow diagrams for some typical water-treatment plants.

type of treatment and the capacity of the treatment plant. Recent studies[1] indicate that the cost of complete treatment is about as given in Table 15-7.

15-24 Reservoir sanitation To minimize pollution of their water supply, some cities have successfully fenced off the entire tributary watershed and completely restricted human entry. This extreme is possible only on small watersheds and only if the land can be obtained before development makes its purchase unduly expensive. A much more common approach is to close the reservoir area itself to human use. Bathing, boating, fishing, picnicking, and other similar activities are banned. This approach may prevent direct pollution of the reservoir itself but does not control possible

[1] Louis Koenig, The Cost of Water Treatment By Coagulation, Sedimentation, and Rapid Sand Filtration, *J. Am. Water Works Assoc.,* Vol. 59, pp. 290–336, 1967.

TABLE 15-7   Approximate Cost (1967) of Water Purification (Complete Treatment)

Treatment plant capacity, mgd	Total cost of treatment per million gallons	Total cost of treatment per acre-ft
1	$170–240	$57–80
10	85–120	27–40
100	40–60	13–20

pollution of the water before it enters the reservoir, although natural purification may be a considerable aid. If the reservoir is large compared with the inflows, i.e., if water is stored for long periods of time, the reservoir itself acts as a settling basin in which much of the suspended material is eliminated.

Bacterial pollution is not the sole problem involved in reservoir sanitation. Organisms called *plankton*, which include algae, protozoa, rotifers, and crustaceans, are found in all natural waters and are a major source of tastes and odors in surface water supplies.[1] In addition, they cause clogging of intake screens and rapid sand filters. When these organisms die and settle to the bottom of reservoirs, they decompose and cause other water-quality problems such as anaerobic conditions, hydrogen sulfide production, and the solubilization of iron and manganese to unacceptable levels. The main method of control of such organisms is by treatment with copper sulfate before the plankton reach undesirable levels. If the copper content of the water is about 0.3 mg/l, most of the plankton will be killed. A dosage of copper sulfate of about 8 lb per million gallons (about 2.4 lb/acre-ft) is equivalent to 0.3 mg/l of copper. The chemical may be sprayed in crystal form from a boat, dropped by helicopter, or dispersed by dragging bags full of the crystals through the water.

Algae and other plankton are living organisms and require nutrients. Algae being plants must also have sunlight. Shallow reservoirs permit sunlight to reach a greater portion of the water than deep reservoirs. Shallow reservoirs or shallow areas of a large reservoir permit growth of aquatic plants whose decay provides nutrients for algae and other plankton. Fertilizer or manure in farmlands adjacent to a reservoir may add nutrients such as nitrogen and phosphorus to the water. Treated or untreated domestic or industrial wastes discharged to streams flowing into the reservoir are also major sources of nutrients. All these factors encourage the growth of plankton in a reservoir. Shallow, marshy areas should be avoided in reservoirs,

---

[1] Nutrient-Associated Problems in Water Quality and Treatment—Task Group Report, *J. Am. Water Works Assoc.*, Vol. 58, p. 1337, 1966.

not only because they encourage growth of plankton, but because they may also provide a breeding ground for mosquitoes.

The immediate effect of solar heating is found in the surface water of a reservoir. Any temperature changes at depths below a few feet result from circulation of the water within the reservoir. This circulation may be induced by temperature differences, density differences caused by salts or suspended sediments, wind acting on the reservoir surface, withdrawal of water through the sluiceways or intakes, and inflow from tributary streams. The changes in the temperature profile in a reservoir are the result of complex forces and are not easily predicted.

A typical annual temperature cycle might be visualized as beginning in the summer months, when the surface water of the reservoir is being heated by solar radiation and warm air. In general the surface temperature cycle will follow the cycle of air temperature with a lag depending on the size of the reservoir. As the surface water is warmed, its density decreases and no thermal circulation can be expected. Unless disturbed by other effects, a stratification develops with a slow decrease in temperature for some distance below the surface. At an intermediate depth a zone of rapid temperature decrease, called the *thermocline*, may occur, and beneath this the water temperature will be fairly constant with depth. Water in this lower zone of constant temperature is stagnant and the decomposition of plankton which settle into this zone can cause a depletion of the dissolved oxygen and the creation of undesirable anaerobic conditions.

As the air temperature falls with the approach of winter, the surface water of the reservoir is cooled and sinks toward the bottom, and warmer water from below rises to take its place. Eventually, if the reservoir is not too deep, a condition of almost uniform temperature with depth will develop. The reservoir is no longer stable so that wind action at the surface can cause the surface water to circulate to the bottom of the reservoir and the water from the bottom to rise to the surface in the fall *overturn*. In water-supply reservoirs this may bring water having a musty odor and taste to the intakes. Continued cooling at the surface results in cooling of the entire body by thermal convection until a temperature of 39.2°F is reached. At this point, water is at maximum density, and any further cooling results in decreasing the density of the surface water, which remains at the surface and if cooled sufficiently, may freeze.

If water temperature in the lower portion of the reservoir drops below 39.2°F during the winter, a spring overturn may occur when surface water, heated by rising temperature, reaches maximum density and sinks toward the bottom to be replaced by the colder but less dense water below. After the spring overturn, continued warming results in warming of the surface water and the development of the type of temperature profile described at the beginning of this cycle.

The temperature characteristics of a lake are important in their effect on water quality resulting from the overturning of the lake waters. The location of intakes for water-supply systems must therefore be made with possible thermal stratification and its effect on water quality in mind.

15-25 Waterworks systems Publicly owned waterworks in the United States outnumber privately owned ones about 3 to 1. In terms of population served the ratio is approximately 6:1. Most private waterworks are small because of the difficulty of financing large projects with private funds. Privately owned works raise funds by the sale of stocks or bonds. If private works are to serve an incorporated city, a franchise or contract between the city and water company must be obtained. Waterworks enjoy a monopoly and are subject to governmental regulation through public utility commissions, water-pollution control boards, and other bodies. The National Board of Fire Underwriters exerts great influence since it regulates fire insurance rates as a function of the quantity and location of water available for fire control.

The design of a complete waterworks system is the exception rather than the rule. Usually the work consists of an extension or improvement of existing facilities rather than design of a completely new project. In the early stages of a proposed project the engineer is required to prepare a report describing the project and presenting a cost estimate. In the case of a public system this report is presented to the governing body of the community for consideration. If the proposed project is acceptable, an election may be called to authorize a bond issue, the most common method of financing public waterworks systems. The bonds are secured by the taxing power of the municipality, by the physical properties of the utility, or by the revenue to be derived from the sale of water. Direct taxation, special assessments, and federal loans are also used as means of financing public water systems.

In planning a project the engineer should obtain legal advice concerning the rights to possible sources of water before recommending any for use. The general legal aspects of water law discussed in Chap. 6 apply to waterworks enterprises. The right of eminent domain in the condemnation of private property for public use can be exercised to facilitate the development of a waterworks project. A water system is legally responsible for supplying its customers with water which is safe for drinking. A water free of salts, however, need not be provided.

15-26 Planning a municipal water system In general the following must be done in planning a municipal water system:

1. Estimate the future population of the community, and study local conditions, to determine the quantity of water which must be provided.
2. Locate a reliable source of water of adequate quality.
3. Make provision for the necessary storage of water, and design the works required to deliver the water from its source to the community.

4. Determine the physical, chemical, and biological characteristics of the water.
5. Design any necessary water-treatment facilities.
6. Design the distribution system, including distribution reservoirs, pumping stations, elevated storage, layout and size of mains, and location of fire hydrants.
7. Provide for the establishment of an organization which will maintain and operate the supply, distribution, and treatment facilities.

BIBLIOGRAPHY

Babbitt, H. E.: "Plumbing," 3d ed., McGraw-Hill, New York, 1960.
———, J. J. Doland, and J. L. Cleasby: "Water Supply Engineering," 6th ed., McGraw-Hill, New York, 1962.
Camp, T. R.: "Water and Its Impurities," Reinhold, New York, 1963.
Clark, J. W., W. Viessman, and M. J. Hammer: "Water Supply and Pollution Control," 2d ed., International Textbook, Scranton, Pa., 1971.
Escritt, L. B., and Sidney F. Rich: "The Work of the Sanitary Engineer," McDonald and Evans, London, 1953.
Fair, G. M., J. C. Geyer, and D. A. Okun: "Water and Wastewater Engineering," Vols. I and II, Wiley, New York, 1968.
Hardenbergh, W. A., and Edward R. Rodie: "Water Supply and Waste Disposal," International Textbook, Scranton, Pa., 1961.
Industrial Utility of Public Water Supplies in the United States, *U.S. Geol. Surv. Water Supply Paper* 1300, 1952.
McKee, J. E., and H. W. Wolf: Water Quality Criteria, *Pub.* 3A, The Resources Agency of California, State Water Quality Control Board, Sacramento, 1963.
McKinney, R. E.: "Microbiology for Sanitary Engineers," McGraw-Hill, New York, 1962.
Sawyer, C. N., and P. L. McCarty: "Chemistry for Sanitary Engineers," 2d ed., McGraw-Hill, 1967.
Speller, F. N.: "Corrosion," 3d ed., McGraw-Hill, New York, 1951.
"Standard Methods for Examination of Water and Wastewater," 12th ed., American Public Health Association, New York, 1965.
Steel, Ernest W.: "Water Supply and Sewerage," 4th ed., McGraw-Hill, New York, 1960.

PROBLEMS

15-1 Using census data through 1950, estimate the 1960 and 1970 populations for your community by each of the methods outlined in the text. How do the two estimates check with actual census data? Can you see any reasons for the difference?

15-2 On the basis of the discussion of per capita requirement in the text estimate the current per capita requirement for your community. Compare this with the observed use as reported by the water department. Suggest explanations for the differences.

15-3 On the basis of the estimates of the preceding two problems estimate the maximum hourly and daily requirement for your community. Compare these figures with the actual observed maxima as reported by the water department.

15-4   Tabulated below are census figures for Nashville, Tennessee, for the period 1870 to 1950. Estimate the 1960 population by each of the procedures suggested in the text. How do your results compare with the 1960 census?

1870	25,865	1920	118,342
1880	43,350	1930	153,886
1890	76,160	1940	167,402
1900	80,865	1950	174,307
1910	110,361		

15-5   What pump capacity and what elevated storage capacity would be required for Palo Alto, California (Fig. 15-2), during the maximum summer day if pumping is to continue at a uniform rate for the full 24 hr? What would be required if pumping is to be limited to the hours between 6 A.M. and 8 P.M.? Neglect fire requirements.

15-6   The base of the elevated tank of Fig. 15-4b is 135 ft above ground level and the tank is 25 ft high. The pump is 3 mi from the tank, the average daily use is 1.5 million gallons, power costs 1.6 cents per kilowatthour, pump and motor efficiency is 72 percent, interest rate is 4 percent, and the life of pump and pipe is estimated at 40 yr. The cost of a foot of pipe is found to be $0.28D$ in dollars where D is the pipe diameter in inches. What is the most economical pipe size between the tank and the pump? Assume no difference in ground elevation. $f = 0.025$.

15-7   Prepare a sketch map showing the layout of water mains, valves, and hydrants you would recommend for an area of about six blocks in your community. How does your layout compare with the existing system?

15-8   Using a working stress of 16,000 psi, what plate thickness would you recommend for the lowest circumferential plates in a steel water tank that is 30 ft in diameter and 35 ft high?

15-9   A city is faced with the choice between cast-iron pipe with an estimated life of 100 yr and steel pipe with an estimated life of 40 yr. The installed cost of the cast-iron pipe will be twice that of the steel pipe. If the minimum attractive return is 3 percent, which choice would be better? What rate of return would make the two alternatives equal? What irreducible factors can you see that might influence the decision?

15-10   Make an inspection of a water-treatment plant, and prepare a report showing the flow diagram of the plant. Explain the purpose of each of the treatment processes employed. What alternative processes might have been utilized? Can you see any advantages or disadvantages which these alternatives might have in this particular case?

15-11   Water enters a purification plant with a pH of 6.0. The next day the pH is 8.0. What was the average pH for the 24-hr period, assuming a linear variation of the hydrogen-ion concentration with time?

15-12   In a chemical manufacturing plant 500 lb of $CaCl_2$ are mixed with 300,000 gal of distilled water. What is the hardness (mg/l as $CaCO_3$) of the resulting water?

15-13   A rectangular sedimentation basin is to handle 2 million gallons of water daily. The water is diverted from a river and contains considerable suspended sediment. Specify the dimensions of the basin so that all sediment larger than 0.02 mm in diameter will be removed. Assume specific gravity is 2.65. Compute the basin overflow rate. If the basin depth is 10 ft, compute the detention period.

15-14   A rectangular sedimentation basin is to handle 10.0 million liters of water daily. Specify the dimensions of the basin so that all sediment larger than 0.04 mm in diameter will be removed. Assume specific gravity is 2.65. Compute the basin overflow rate. If the basin is 3.15 m deep, compute the detention period.

15-15   Determine the diameter of a circular sedimentation basin required to handle

1,500,000 gal/day if a detention period of 5 hr is desired. Let the depth of the tank be 10.0 ft at the periphery, and assume a cone-shaped tank bottom with a slope of $1\frac{1}{4}$ in./ft. Equipment for cleaning circular basins is manufactured in standard sizes for tank diameters from 30 ft to 200 ft in 5-ft intervals.

15-16   Determine the diameter of a single circular sedimentation basin required to handle 6.0 million liters per day if a detention period of 4 hr is desired. Let the depth of the tank be 2.80 m at the periphery, and assume a cone-shaped bottom with a slope of 1 on 8. What diameter would be required if two identical tanks were used? Assume same outer depth and bottom slope.

15-17   A rapid sand filter is 12 ft square. The design rate of wash-water rise is to be 30 in./min. How much water is required for a 5-min wash? The wash-water reservoir is 40 ft above the underdrains and is so located that 200 ft of cast-iron pipe with three standard 90° elbows are required to bring water to the filter. If a pressure of 15 psi is required at entrance to the filter, what size pipe is required?

15-18   A surface reservoir is being designed. Estimates indicate that the cost of the floor will be $5 per square foot, the cost of the roof including supports $10 per square foot, and the side walls will cost d^2 dollars per linear foot, where d is the depth of the reservoir. What are the most economical dimensions for this reservoir if its capacity is to be 7 million gallons?

15-19   A water contains 130 mg/l of calcium bicarbonate. To remove all the calcium bicarbonate from 200,000 gal of the water, what amount of lime must be used? Assume the lime contains 90 percent pure CaO.

15-20   A water supply contains 350 mg/l of hardness and is to be softened by the zeolite process to 50 mg/l. What portion of the total water should pass through the softeners?

Chapter 16

Hydroelectric power

Electric power is developed from two types of plants: *hydro* and *thermal*. Hydroelectric generators are driven by water turbines while thermal (or steam) plants derive energy from combustion of fuel. Most thermal plants use steam turbines and fossil (coal, oil, or natural gas) or nuclear fuel. In 1967 world power consumption[1] was 3800 billion kilowatthours, of which about one-third was developed in the United States. While estimates indicate that only about one-quarter of the potential water power of this country has been developed,[2] the most favorable hydroelectric sites in the United States have been used; and future projects will require careful study. Vast power resources still remain to be developed in other parts of the world, particularly in Africa, Asia, and South America.

16-1 Hydroelectric-power development in the United States The first hydroelectric plant in the United States was put into operation at Appleton, Wisconsin, in 1882, a few days after the first thermal plant began operation. This first hydroelectric plant generated 200 kw and transmitted 1 mi at 110 volts. In contrast, the hydroelectric plant at Hoover Dam, generates over 1 million kilowatts and transmits 266 mi at 275,000 volts. Prior to about 1910, the development of thermal plants was more rapid than that of hydroelectric plants because fuel was relatively inexpensive, and the transmission of electricity over great distances (required of most hydroelectric plants) was inefficient. As efficiency of transmission at high voltages was perfected, there was a reversal of this trend, and hydroelectric

---

[1] "United Nations Statistical Yearbook," 20th ed., New York, 1968.

[2] "Hydroelectric Power Resources of the U.S.," Federal Power Commission, Washington, D.C., 1968, indicates that there were 1562 operating plants with total capacity of 46 million kw and 1541 undeveloped sites with a potential capacity of 130 million kw.

developments progressed rapidly. This was further augmented by the fuel shortage during World War I. Recent improvements in the efficiency of thermal plants and the development of the most favorable hydroelectric sites have resulted in a renewed popularity of thermal plants. Most large power-development systems interconnect hydroelectric and steam plants. Formerly the thermal plant served as a standby for use in case of equipment failure, or to supplement hydroelectric power during peak hours. The more recent trend has been toward the use of thermal power to carry a large volume of steady load. As of 1967 about 17 per cent of the electric power in the United States was produced in hydroelectric plants. Approximately 1300 billion kilowatthours of energy were used in the United States in 1967; and demand was doubling every decade.

Many of the older hydroelectric-power developments in this country were made under laws of the individual states. Prior to the Federal Power Act of 1920, private power developments located on federal lands or navigable streams required congressional approval. The Federal Power Commission was created by the act of 1920 to look after improvement of navigation, the development of waterpower, and the use of public lands. The commission was empowered to investigate the cost of waterpower, to determine its fair value, and to issue licenses for private developments when public power development was not warranted. The Public Utilities Act of 1953 gave the Federal Power Commission the responsibility of planning power developments at federal flood-mitigation dams. The commission is charged with the responsibility of seeing that electric-power development proceeds on a sound basis and can order interconnection of power systems and the interchange of power to meet emergency conditions, particularly those which might arise in time of war.

16-2 Comparison of thermal- and hydroelectric-power costs Most modern power systems include thermal plants, and the hydraulic engineer should be acquainted with the relative cost of thermal and hydroelectric power and factors influencing these costs. Before a hydroelectric plant can be economically justified, its cost must be compared with that of an equivalent steam plant. If a plant is to be added to a system, several sites are selected, and on the basis of preliminary designs and cost estimates (including operating costs) the one which develops power at least cost is selected.

The initial cost of a hydroelectric plant is generally higher than that of a comparable thermal plant. The initial cost[1] of the hydroelectric project includes the cost of the dam; diversion works; conduits; land; water rights; railroad, highway, and utilities relocation; and value of improvements

---

[1] In the case of federal multiple-purpose projects some of the total cost may be charged to flood mitigation, navigation, or other functions of the reservoir (Chap. 21).

flooded by the reservoir and the generating plant itself. Often a hydroelectric plant is located at a relatively inaccessible location, which adds to the initial cost because of the expense of hauling materials and equipment and because of long transmission lines to carry the power to market. Thermal plants are usually located near their load centers, eliminating the need for long transmission lines. There are usually several possible sites for a thermal plant, the only requirements being the availability of an adequate supply of cheap fuel and plenty of water for the condensers. As a consequence thermal plants are usually located on a seacoast, or near a lake or river. Consideration of the ecological consequences of discharging large quantities of hot water into a water body is necessary in site selection and outfall design. Heated water has been shown to be useful for irrigation. Cooling towers may also be an alternate solution.

Even though the initial capital investment is greater, taxes on a hydroelectric plant may not be as much as those for a thermal plant because of the lower tax rate in remote locations compared with the higher tax rate in areas where a thermal plant is likely to be located. The cost of insuring a hydroelectric plant may also be less than that for a thermal plant of comparable capacity because a smaller portion of the items comprising the initial cost of a hydroelectric plant are insurable.

The cost of operating a thermal plant is much higher than for a hydroelectric plant, mainly because of fuel costs. A thermal plant is also more difficult to operate and maintain, and the cost of labor, maintenance, and repairs is much higher than for a hydroelectric plant. Air-pollution-control systems will be required for most future plants, and systems for avoiding thermal pollution of streams or lakes will be needed at many thermal plants. Fuel costs for a thermal plant vary with the unit price of fuel and the plant output. The cost of fossil fuels is highly dependent on the distance from the fuel source to the plant, and fuel cost in one locale may be as much as twice that in another area. Because of lack of bulk, the transport costs for nuclear fuel are quite low, and hence nuclear power plants are advantageous in locations where conventional fuels and hydroelectric power are unavailable. As thermal plants get older, their efficiency drops, and they are usually operated at reduced capacity, thus permitting newer, more efficient plants to carry heavier loads. Technical advances in thermal-plant design have reduced average coal requirements[1] from 3.0 lb/kwhr in 1920 to less than 0.9 lb/kwhr in 1960.

Since 1945 nuclear power plants have been developed. In 1967 nuclear power plants accounted for 1 percent of installed capacity in the United

[1]F. L. Weaver, Power Developments in the United States, *Civil Eng.* (*N.Y.*), Vol. 30, no. 9, pp. 70–72, September, 1960.

States. Most of these plants are of low capacity. By the year 2000 nuclear plant capacity is predicted to exceed all other forms of power generation.[1] If properly designed, nuclear power plants can be built so that they are totally safe from the hazards of radioactivity.

Although nationwide average costs for thermal and hydroelectric power differ very little, local conditions may greatly favor one source over the other when a specific installation is considered. In the long run, the best choice in a particular case is the plant which best fits the needs of the system.

16-3   Power systems and load   Some years ago most power systems were served by a single plant. With such an arrangement, one or more reserve units were needed in the plant. Today nearly all power systems are served by several plants, and some systems are interconnected with others. The system need have only enough capacity to supply the expected peak load plus excess capacity at least equal to the largest single generating unit in the system to take care of breakdowns that may occur.

Generally, economy and reliability can be achieved by having a combined power system, i.e., one with both water and steam power. One advantage of having hydroelectric power in a system is that a hydroelectric plant, on standby, can be put into operation in a very short time. The actual time will vary from a few seconds to 3 or 4 min depending on the length of conduit. Thus waterpower is well adapted to provide reserve capacity on short notice, such as might be required if some other unit of the system were to fail suddenly. At least 30 min is required to put most steam units into operation. It is also much more costly to keep steam plants on standby. A system supplied by hydroelectric power alone is at a disadvantage if its carry-over storage is inadequate to meet the demands of a severe drought. Even during a flood a power shortage is possible at low-head plants if high tailwater at the powerhouse greatly reduces the net head.

In planning a power system, an estimate of future power requirements is necessary. Load predictions are usually based on the rate at which use has increased in past years, but variations in this rate as a result of business booms and slumps, technological changes, and other factors make accurate predictions difficult. The problem is quite similar to that of estimating municipal water requirements.

Late afternoon on a dark winter day when manufacturing, commercial, domestic, and street-lighting loads are high has been the traditional time of peak load on power systems. In irrigated areas, however, the operation of pumps brings a maximum about midday at the height of the irrigation season. Increasing use of air conditioning causes a summer maximum load in many metropolitan areas.

The load for the peak day of the year determines the required generating

[1] "The Nation's Water Resources," U.S. Water Resources Council, Washington, D.C., 1968.

capacity, while the power requirements of the peak week or month dictate the amount of power storage required in the form of fuel or water. Typical daily load curves are shown in Fig. 16-1. The *load factor* is defined as the ratio of the average load over a certain period to the peak load during that period. Load factors are expressed as daily, weekly, monthly, or yearly factors. The load factor for a system serving a highly industrialized area may be as high as 80 percent, but in a residential community load factors as low as 30 or 40 percent occur. If the load factor is high, the unit cost of power will be comparatively low because under such conditions the system will be operating near capacity and presumably near the best efficiency. With a low-load factor much of the generating capacity of the system will be lying idle a large part of the time.

During periods of low streamflow, thermal plants in a combined system are operated almost continuously near capacity (Fig. 16-2) to carry the base load, while hydroelectric plants with storage are used intermittently to generate power for peak loads. During periods of high flow, the functions of the steam and hydroelectric plants are reversed, the latter operating continuously near capacity to carry the base load and the former only intermittently to carry peak loads. This mode of operation results in the fullest utilization of the available water power and the minimum consumption of fuel. In any combined system, there is a unique division in capacity between thermal and hydroelectric power which will result in greatest overall

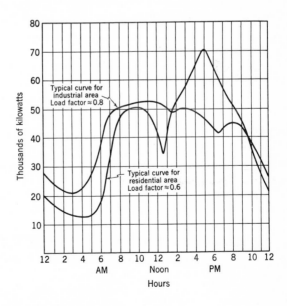

FIG. 16-1 *Typical peak-day load curves for an electric-power system.*

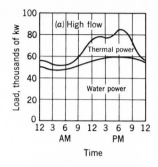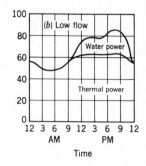

FIG. 16-2   Load distribution in a combined power system.

economy. The optimum point will usually occur when the waterpower capacity is somewhere between one-third and two-thirds of the total capacity.

16-4   Definitions of hydroelectric-power terms   The *gross head* for a hydroelectric plant is the total difference in elevation between the water surface in the stream at the diversion and the water surface in the stream at the point where the water is returned after having been used for power. The *net*, or *effective, head* is the head available for energy production after deducting losses in friction, entrance, unrecovered velocity head in the draft tube, etc. (Chap. 12). The *hydraulic efficiency* of a hydroelectric plant is the ratio of net head to gross head. The *overall efficiency* is equal to the hydraulic efficiency multiplied by the efficiency of the turbines and generators. The overall efficiency of hydroelectric plants operating at optimum conditions will usually be somewhere between 60 and 70 percent. The *capacity* of a hydroelectric plant is the maximum power which can be developed by the generators at normal head with full flow.

The unit of electrical power is the *kilowatt*, which is equivalent to 1.34 hp. The unit of electrical energy is the *kilowatthour*, defined as 1 kw of power delivered for 1 hr. Electrical energy may also be expressed in *kilowatt-days* or *kilowatt-years*.

Firm (or *primary*) *power* is the power which a plant can be expected to deliver 100 percent of the time. For a single hydroelectric plant it corresponds to the power developed when available water, including that derived from storage, is a minimum. *Surplus* (or *secondary*) *power* is all power available in excess of firm power. Some surplus power closely approximates firm power in being available a large percentage of the time and can be sold at rates near those charged for firm power. Much secondary power is sold at very low rates without guarantee as to continuity of service.

16-5   Types of hydroelectric plants   Hydroelectric plants may be classified as *run-of-river, storage,* or *pumped-storage.* A storage-type plant is one with a reservoir of sufficient size to permit carry-over storage from the wet season to the dry season and thus to develop a firm flow substantially more than the minimum natural flow. A run-of-river plant generally has very limited storage capacity and can use water only as it comes. Some run-of-river plants have enough storage (called *pondage*) to permit storing water during off-peak hours for use during peak hours of the same day. Run-of-river plants are suitable only for streams which have a sustained flow during the dry season or where other reservoirs upstream provide the necessary storage.

A pumped-storage plant generates power for peak load, but, at off peak, water is pumped from the tailwater pool to the headwater pool for future use (Fig. 16-3). The pumps are powered with secondary power from some other plant in the system, often a run-of-river plant where water would otherwise be wasted over the spillway. Under certain conditions a pumped-storage plant will be an economic addition[1] to a power system. It serves to increase the load factor of other plants in the system and provides added capacity to meet peak loads. A unique feature of some pumped-storage plants is that very little water is required for their operation. Once the headwater and tailwater pools are filled, only enough water is needed to take care of evaporation and seepage. Pumped-storage plants have been widely used in Europe, especially in Germany. As of 1968 there were eleven such plants in the United States, but about a dozen[2] more were in the planning stage. Pumped-storage plants do not represent the attainment of perpetual motion, for there is a loss of energy in their operation. Their value lies in an "economic efficiency" since they convert low-value off-peak to high-value peak power.

---

[1] Symposium on Pumped Storage, *J. Power Div., ASCE*, Vol. 88, pp. 183–251, July, 1962; Charles Jaeger, Economics of Pumped Storage, *Water Power*, Vol. 9, no. 2, pp. 54–59, February, 1957; Vol. 9, no. 3, pp. 104–112, March, 1957.

[2] "Hydroelectric Power Resources of the United States—Developed and Undeveloped," Federal Power Commission, 1968.

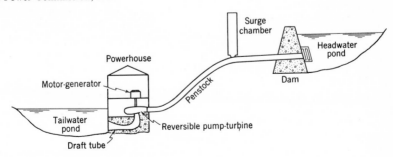

FIG. 16-3   *Section through a typical pumped-storage hydroelectric plant.*

For heads from about 50 to 300 ft, reversible pump-turbines[1] have been devised to operate at a relatively high efficiency as either a pump or a turbine. The same electrical unit serves as generator and motor by reversing poles. Such a machine may reduce the cost of a pumped-storage project by eliminating the extra pumping equipment and pump house. The reversible pump-turbine is a compromise in design between a Francis turbine and a centrifugal pump. Its function is reversed by changing the direction of rotation.

16-6 General arrangement of a hydroelectric project   A hydroelectric development (Fig. 16-4) ordinarily includes a diversion structure, a conduit (*penstock*) to carry water to the turbines, turbines and governing mechanism, generators, control and switching apparatus, housing for the equipment, transformers, and transmission lines to the distribution centers. In addition, trash racks at entrance to the conduit, canal and penstock gates, a forebay, or surge tank, and other appurtenances may be required. A *tailrace*, or waterway, from the powerhouse back to the river must be provided if the powerhouse is situated so that the draft tubes cannot discharge directly into the river. No two power developments are exactly alike,

---

[1] F. E. Jaski, Reversible Pump-Turbines at Niagara Falls, *Mech. Eng.*, Vol. 82, pp. 75–77, 1960; M. Braikevitch, D. Hartland, and R. A. Strub, Development of Reversible Pump-Turbines, *J. Power Div., ASCE*, Vol. 88, pp. 83–105, December, 1962.

FIG. 16-4   The Shepaug hydroelectric plant on the Housatonic River, Connecticut. The dam is 139 ft high and 1412 ft long, tainter gates 35 ft wide and 28 ft high, power produced by a single variable-pitch Kaplan turbine producing 50,000 hp with 4970 cfs flow at a head of 97 ft. (Connecticut Light and Power Co.)

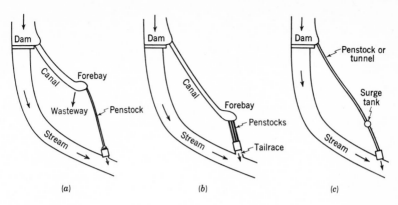

FIG. 16-5   Schematic layout of some divided-fall developments.

and each will have its own unique problems of design and construction. The type of plant best suited to a given site depends on many factors, including head, available flow, and general topography of the area. A *concentrated-fall* hydroelectric development (Figs. 16-4 and 16-7) is one in which the powerhouse is located near the dam and is most common for low-head installations. The powerhouse may be located at one end of the dam, directly downstream from the dam, or even in a buttress-type dam. In a *divided-fall* development (Fig. 16-5) water is carried to the powerhouse at a considerable distance from the dam through a canal, tunnel, or penstock. With favorable topography it is possible to realize a high head even with a low dam. With this arrangement, head variations in the reservoir may be small compared with the total head, and the turbine can operate near optimum head (peak efficiency) at all times.

16-7   The forebay   The forebay serves as a regulating reservoir, temporarily storing water when the load on the plant is reduced and providing water for the initial increments of an increasing load while water in the canal is being accelerated. In many cases the canal itself may be large enough to absorb these flow variations. If the canal is long, its end is sometimes enlarged to provide necessary temporary storage. Occasionally a forebay is formed by building a dam across a natural-flow channel. The forebay is provided with some type of intake structure to direct water to the penstock. Various types of intakes are used, depending upon local conditions.

Intakes should be provided with trash racks to prevent the entry of debris which might damage the wicket gates and turbine runners or choke up the nozzle of impulse turbines. A floating boom is often placed in the storage reservoir to prevent as much ice and floating debris from entering the canal as possible. Where winters are severe, special provision must be made to prevent trouble from ice in the forebay. If the forebay pond is large, with the intake at considerable depth, ice problems are minimized. To prevent ice from clinging to the trash racks, they may be heated electrically. An air-bubbler system which agitates the water in the vicinity of the trash rack and brings warmer water to the surface is also useful.

A forebay must be provided with a spillway, or wasteway, so that excess water can be disposed of safely if the need arises. Siphon spillways are often advantageous for this purpose.

16-8   Penstocks   The structural design of a penstock is the same as for any other pipe. Because of the possibility of sudden load changes, design against water hammer is essential. Long penstocks are usually provided with a surge tank to absorb water-hammer pressures and to provide water to meet sudden load increases. For short penstocks it is generally more economical to forgo the protection of a surge tank and rely on a heavier pipe wall and slow-closing valves.

Penstocks are most commonly made of steel, though reinforced-concrete and wood-stave pipe are also used. If the distance from the forebay to the powerhouse is short, separate penstocks are ordinarily used for each turbine (Fig. 16-5b). Long penstocks are usually branched at the lower end (Fig. 16-5a) to serve several turbines.

Penstocks are usually equipped with head gates at the forebay which can be closed to permit repair of the penstock. An air-inlet valve downstream from the gate is necessary to prevent collapse of the pipe immediately after the head gate is closed. A sufficient water depth should be provided above the penstock entrance in the forebay to avoid formation of vortices which may carry air into the penstock and result in lowered turbine efficiency and undesirable pressure surges.

Sharp bends in a penstock should be avoided because of the head loss and the large forces required to anchor the pipe. In severe climates it is sometimes advisable to bury the penstock to prevent ice formation in the pipe and to reduce the number of expansion joints required. Uncovered penstocks are usually more expensive because of the expansion joints, saddles, anchors, and other appurtenances required, but they have the advantage of being accessible for inspection and repairs. Uncovered penstocks are usually supported on cradles and may be carried across ravines on trestles. It is poor practice to partially bury a pipe as the region near the ground surface will be subject to excessive deterioration. In some cases penstock costs can be reduced by using a tunnel through a hill if a considerable saving in length can be gained.

16-9   The powerhouse   A powerhouse consists of a *substructure* to support the hydraulic and electrical equipment and a *superstructure* to house and protect this equipment. For plants equipped with reaction turbines, the substructure usually consists of a concrete block extending from the foundations to the generator floor with waterways formed within it. The elevation of the turbine with respect to the tailwater level is determined by the necessity of avoiding cavitation (Sec. 12-6). The details of the waterways in the substructure depend on the type of turbine and setting selected. The scroll case and draft tube may be cast integrally with the substructure when it is poured, with steel liners serving as forms.

The superstructure of most powerhouses is a building housing all operating equipment. The generating units (Fig. 16-6) and exciters are usually located on the ground floor. The switchboard and operating room may also be placed on the ground floor but in larger plants is often on a mezzanine overlooking the equipment. Turbines which rotate on a vertical axis are placed just below floor level, while those rotating on a horizontal axis are placed on the ground floor alongside the generators. Vertical-axis units generally require less floor space than those mounted on horizontal axes. The cost of the superstructure can be greatly reduced by providing only individual housing for each generator (Fig. 16-7). This arrangement, the *outdoor powerhouse*, has the disadvantage that the units cannot be disassembled during inclement weather.

Under certain topographic conditions, particularly in narrow canyons where there is no convenient site for a conventional type of powerhouse, it is

FIG. 16-6   *Interior of the generator room of the Cresta Powerhouse.* (*Pacific Gas and Electric Co.*)

FIG. 16-7   *The C. J. Strike power plant on the Snake River in Idaho, a modern 90,000-kw plant with an outdoor station. (Idaho Power Co.)*

advantageous to locate the powerhouse underground. There are many underground powerhouses in Europe;[1] however, there are only a few in North America. One of the world's largest underground plants is the Kemano powerhouse in Canada which has a generating capacity of 1,670,000 kw under a head of 2592 ft.

The generating units of a hydroelectric plant are nearly always placed in a row at right angles to the direction of flow through the powerhouse. This permits the simplest layout of waterways and facilitates the handling of equipment within the plant. Space between individual pieces of equipment should be ample for their removal when the need for repair arises. A working bay should be provided at one end of the powerhouse where equipment may be unloaded and repaired.

---

[1] J. Barry Cooke and Arthur G. Strassburger, Underground Power Plants, *J. Power Div.*, *ASCE*, Vol. 83, *Paper* 1350, August, 1957.

Most elements in a powerhouse have to be repaired occasionally and sometimes replaced. A hydroelectric plant in northern California is designed so that either of two runners, each of different specific speed, may be used on a single unit depending on the available head. Hence, one important feature of a powerhouse is a crane to lift and move equipment. For an outdoor powerhouse a gantry crane is suitable, while in enclosed powerhouses traveling cranes spanning the width of the building and arranged to travel over its entire length are used. The capacity of the crane is determined by the heaviest load it must handle, and the elevation of the crane rail depends on the clearance required for the crane to lift the largest piece of equipment over other machinery.

16-10   The tailrace   The *tailrace* is the channel into which the water is discharged after passing through the turbines. If the powerhouse is close to the stream, the outflow may be discharged directly into the stream, while in other situations there may be a channel of considerable length between the powerhouse and the stream. The tailrace has been a badly neglected feature in many hydroelectric designs. In many low-head plants more of the gross head might have been utilized at very low cost if some attention had been given to tailrace design.

An unlined tailrace may degrade, i.e., the tailwater elevation is lowered by scouring of the channel bottom. In some cases this process has proceeded to the point where the tailwater dropped below the top of the draft-tube outlet and reduced the efficiency of the turbine. To correct this, an overflow dam must be constructed in the tailrace channel to raise the tailwater elevation back to normal. Sometimes the tailrace channel may aggrade, in which case the tailwater elevation rises, with a resulting decrease in the gross head at the plant. This situation may arise where the spillway discharge is close to the tailrace and no adequate training wall is provided. If aggradation is excessive, it may be necessary to remove the material deposited in the tailrace by dredging or other means. In any case a knowledge of tailwater elevation at various flow rates is essential to the hydraulic design of the tailrace. In some instances hydraulic-model studies will give invaluable information.

The outlet of the draft tube should be provided with gates so that the draft tube can be unwatered for repairs. If the span between tailrace piers is not too great, timber stop logs may be used for this purpose. For larger spans, steel stop logs or sectional steel gates are commonly used. A hoisting mechanism must be provided for the gates or stop logs.

The tailraces of underground powerhouses often present unique problems of hydraulic design. In some instances the draft tube discharges into a horizontal tunnel with a free water surface so that no surge tank is required. In other cases there is a tailrace-pressure tunnel with a regular surge tank.

Often the downstream portion of the diversion tunnel can be used as the tailrace tunnel. Such is the case of the Oroville underground powerhouse in California where two 35-ft-diameter tunnels 2200 ft long are provided.

16-11   Electrical equipment   Though the hydraulic engineer is rarely required to select the electrical equipment for a hydroelectric project, he should have some knowledge of the basic requirements and limitations of this type of equipment. The alternating-current synchronous generator is most widely used in the United States. In its simplest form the generator is composed of two elements, a magnetic field consisting of an assembly of electromagnets of alternately opposite polarity and an armature or system of conductors in which an alternating electromotive force is induced when the armature is rotated. When a conductor has passed two adjacent poles of opposite polarity, a complete alternation of the electromotive force occurs, and this is referred to as a *cycle*. The number of cycles per second is termed the *frequency*. In the United States a frequency of 60 cycles per second is the accepted standard, though there are still a few systems operating at other frequencies. A turbine may be governed to operate at constant speed, and generators are designed with the proper number of poles to produce the desired frequency at the selected speed.

Alternating-current generators are rated in *kilovolt-amperes (kva)* for a given temperature rise, usually the rise in temperature which the insulation can safely stand. The apparent output (kva) of a generator differs from the actual output according to the following expression:

$$\text{Actual power (kw)} = \text{apparent power (kva)} \times \text{power factor}   (16\text{-}1)$$

The *power factor* can never be greater than unity. Its value depends on the relationship between the inductance and resistance of the load. A load with very little inductance such as a lighting load will have a power factor close to unity. The usual system load has a power factor ranging from 0.8 to 0.9, but if the load contains a large number of induction motors, the power factor may be as low as 0.5.

Weights and dimensions of generators vary considerably depending on the kilovolt-ampere rating and the selected speed of operation. Alternating-current generators with a rating of 3000 kva vary in weight from about 18 tons for an operating speed of 900 rpm to 55 tons for a speed of 100 rpm. These generators would be about 10 ft high and 8 or 16 ft in diameter, respectively. Special structural problems often arise in providing proper support for generators.

Generators must be adequately ventilated to keep them from overheating. For satisfactory cooling, between 1 and 5 cu ft of cool air per minute are required per kilovolt-ampere of rated capacity. The air for cooling is usually passed through a battery of water jets and dried before using. *Exciters* are direct-current generators usually mounted on the generator shaft. Their function is to energize the field magnets of the main generator. The capacity of an exciter need be only about 1 or 2 percent that of the main generator.

Transformers consist essentially of an iron core about which there are two windings, a primary connected to the power source (generator) and a secondary connected to the receiving circuit (transmission line). The entire assembly is usually submerged in an oil bath which serves as an electric insulator and a cooling medium. The function of a transformer is to step up the voltage for transmission. Transmitting electricity at high voltages reduces the power loss and permits the use of smaller conductors in the transmission line. Transformers are usually placed in switchyards located on high ground above the powerhouse to avoid spray and to remove the hazards of high voltage and fire from the powerhouse.

Nearly all circuits in a hydroelectric plant are provided with circuit breakers so that the current may be switched off or on as the occasion demands. An air circuit breaker is satisfactory at low voltages, while an oil circuit breaker is more commonly used for high voltages. In the latter type of circuit breaker the contacts are submerged in oil, and the arc is extinguished by the cooling effect of the oil. Circuit breakers of small capacity may be manually operated, but large circuit breakers are usually operated electrically or pneumatically by remote control. Most plants are equipped with storage batteries to supply power for emergency lighting and field excitation. Lightning arresters are used to protect electrical equipment from lightning and from abnormally high voltages of all kinds, such as those resulting from switching operations.

16-12 Operation of hydroelectric plants A modern 100,000-kw plant usually requires only one to two operators per shift; however, some plants are operated entirely by remote control. The Mammoth Pool powerhouse of the Southern California Edison Company has a microwave link to Big Creek No. 8 powerhouse for supervisory control, telemetering, visual and audible supervision of operation by means of closed-circuit television, and sound monitoring.

In order to keep a plant in good operating condition, the facilities should be inspected and serviced regularly and repaired when necessary. A complete history of each piece of equipment is helpful in proper maintenance. Operating data such as generator amperes and kilowatts, turbine gate openings, head- and tailwater levels, and spillway gate openings are recorded at least hourly if not more often. These data together with records of the cost of operation and maintenance of the plant are required to determine the relative economy of the plant.

A network of rainfall and river gages and snow surveys to provide data for forecasting streamflow is essential to the sound operation of a hydroelectric plant. The number of gages required depends on the drainage area and the type of plant. For forecasting purposes, reports of rainfall and river flow must be prompt, and hence a good communication system is essential. Successful plant operation is also highly dependent on adequate communication facilities between the individual plants and the central station in order that the load dispatcher is informed at all times as to what is occurring in the system.

The load dispatcher is charged with the responsibility of maintaining continuity of service. He designates the load each plant is to carry. In assigning loads, the overall efficiency of the system is the guiding factor, although requirements for shutting down equipment at various plants for repair must receive consideration. After the loading schedule has been assigned to a plant, it is up to the plant operator to manipulate his equipment for maximum efficiency. The most economic distribution of load among the various units at a plant must be determined by a study of their respective operating characteristics.

Operation of a private hydroelectric-power system is subject to numerous government controls. A state or federal agency will usually specify the minimum flow to be maintained below the dam for water supply, navigation, fish conservation, or sanitary purposes. This flow is often the lowest natural flow that occurred in the stream before any dams were constructed. The Federal Power Commission may also specify other conditions concerning the design or operation of a plant in connection with the issuance of a license. The operating company is legally liable for any increase in flood flows caused by their reservoir operation. As a public service many companies operate their reservoirs for flood reduction insofar as they are able to do so.

One of the difficult problems of hydroelectric operation is scheduling of water use. Spillway discharge may represent loss of power which could have been generated with the spilled water. On the other hand an empty reservoir means substitution of thermal power in peaking service. The nature of the problem depends on the number and kind of plants in the system, hydrologic character of the region, nature of the load curve, and other factors. However, the probability of future flows can be determined; and, for a single reservoir or simple system, these probabilities may be sufficient to indicate the best operating strategy. For more complex systems, a linear-programming solution or a test of operating rules by simulation may be necessary. An intertie with another system may also provide protection by further diversifying system characteristics.

16-13 Planning the hydroelectric-power development There may be several feasible sites for a proposed hydroelectric plant. At each site several different designs may be possible. Hence the selection of a final design for a hydroelectric project often entails comparison of many alternatives. A list of the steps in design follows:

1. Assemble hydrologic data on the streams, and determine the amount of water available and its distribution throughout the year and from year to year.
2. Make preliminary designs for all installations which seem competitive in cost, and determine the most economic design at each site by comparison of costs and estimated power revenues.
3. Determine the requirements to be satisfied (total kwhr/yr and time variations).
4. Select feasible projects as close to the load center as possible.
5. Compare the best designs from the several sites, and select the site or combination of sites which proves best for production of the required amount of power. Often this selection is guided somewhat by estimated future requirements and the possibilities of expansion to meet them.
6. Compare the cost of the hydroelectric-power plant with that of an equivalent thermal plant.
7. If hydroelectric power is competitive with steam, proceed with the detailed design of the hydroelectric installation. This includes design of dam, spillway, sluiceways, canals, penstock, forebay, and tailrace, selection of turbines and generators, and layout of power plant.

BIBLIOGRAPHY

Brown, Guthrie: "Hydroelectric Engineering Practice," Vols. I, II, and III, Blackie, Glasgow, 1958.
Cooke, J. Barry: Haas Hydroelectric Power Project, *Trans. ASCE*, Vol. 124, pp. 989–1026, 1959.
Creager, W. P., and J. D. Justin: "Hydroelectric Handbook," 2d ed., Wiley, New York, 1950.
Doland, James J.: "Hydro Power Engineering," Ronald, New York, 1954.
Electric Power in Relation to the Nation's Water Resources, Senate Select Committee on National Water Resources, *Committee Print* 10, 1960.
Gunby, F. M.: Supply of Water Power in the United States, *Trans. ASCE*, Vol. CT, pp. 461–475, 1953.
Kirchmayer, Leon K.: "Economic Control of Interconnected Systems," Wiley, New York, 1959.
Kyle, John H.: "The Building of TVA," Louisiana State University Press, Baton Rouge, 1958.
Mosony, Emil: "Water Power Development," 2 vols., Publishing House of Hungarian Academy of Sciences, 1957.
Powel, Charles A.: "Principles of Electric Utility Engineering," Wiley, New York, 1955.
Stevens, J. C., and C. V. Davis: Hydroelectric Plants, Sec. 24 in C. V. Davis and K. E. Sorenson (eds.), "Handbook of Applied Hydraulics," 3d ed., McGraw-Hill, New York, 1969.
Uhl, W. F.: Water Power over a Century, *Trans. ASCE*, Vol. CT, pp. 457–460, 1953.

PROBLEMS

16-1 Assume that the two streams whose flow-duration curves appear in Fig. 5-13 have power plants capable of utilizing up to 300 cfs with a head of 500 ft at 78 percent efficiency. Firm power is valued at 0.9 cent per kilowatthour, secondary power available at least half the time can be sold for 0.4 cent per kilowatthour, while dump power (all other power) is worth only 0.1 cent per kilowatthour. What annual revenue could be expected from each plant? How much money could be spent to increase the firm flow by 50 cfs at each plant? Take taxes and insurance at 3 percent, interest at 6 percent, ignore transmission losses, and assume $20,000 per year annual labor and maintenance in connection with the storage facilities.

16-2 Determine from your local power company the data on growth of electric-power use in recent years. Compare this with the increase in water use for the same area. Are the percentage increases comparable? If not, what explanation can you give for the differences?

16-3 A power company is designing a plant which will utilize an average flow of 140 cfs, with a gross head of 800 ft and a turbine and generator efficiency of 81 percent. Two possible routes for the penstock are available. The longer route is 11,000 ft and would be all pipeline. The shorter route is 10,000 ft, but it involves 1000 ft of tunnel. Assuming the pipes are 60 in. in diameter and the tunnel is 6 ft in diameter after lining with concrete, how much could the company afford to pay for the tunnel if steel pipe would cost $24.50 per foot installed? Do you think the longer route using 72-in. pipe at $32.80 per foot would be preferable? Power produced at the plant will yield 0.6 cent per kilowatthour and the interest rate is 5 percent.

16-4 A power company is considering construction of a 100,000-kw power plant. The engineer's estimates are as follows:

Estimates	Thermal	Hydroelectric
First cost	$10,000,000	$24,000,000
Estimated life	25 yr	50 yr
Taxes and insurance	4%	3.5%
Fuel (annual)	$ 1,500,000	$ 0
Labor, maintenance	$ 350,000	$ 140,000
Extra transmission facilities (first cost)	$ 0	$ 400,000

If the minimum attractive return is 6 percent, which plant would be the better selection? What is the annual cost per kilowatt of installed capacity for each plant? What is the cost per kilowatthour with a load factor of 100 percent? What would be the cost per kilowatthour with a load factor of 70 percent?

16-5 A pumped-storage installation uses two 8100-hp centrifugal pumps of 87 percent efficiency to pump water through 500 ft of 72-in. penstock ($n = 0.015$) to an elevation of 375 ft above tailwater level. This water is then discharged through the same penstock to a 33,000-hp turbine whose efficiency is 91 percent. The generator efficiency is 89 percent and that of the pump motors is 85 percent. If all water used in the turbine is pumped, what is the overall efficiency of the arrangement? If one-third of the turbine flow is supplied by natural inflow to the reservoir, what is the overall efficiency of the system? If the pumps utilize dump power with an increment cost of 1 mill per kilowatthour and the generator produces prime power selling at 9 mills per kilowatthour, what is the economic efficiency of the plant when all the water is pumped?

16-6 The 72-in. penstock of Prob. 16-3 will approach the powerhouse on a slope of 70° with the horizontal. Approximately 100 ft from the powerhouse there will be a 40-ft-radius bend which will permit the pipe to enter the powerhouse horizontally. What dimensions would you suggest for a concrete block to resist the forces in this bend? Assume concrete at 150 lb/cu ft.

16-7 A 72-in. penstock carrying 200 cfs of water under a head of 175 ft branches into two 60-in. pipes as it approaches the powerhouse. The branching pipes diverge from one another at an angle of 60°. If the pipes are horizontal at this point, what is the force exerted at the junction?

16-8 A 200-cm penstock carrying 5 cu m/sec of water under a head of 20 m branches into three 135-cm pipes as it approaches the powerhouse. The pipes diverge symmetrically from the penstock; the outer pipes diverge at an angle of 52° from each other. If the pipes and penstock are horizontal at this point, what is the force in kilograms exerted at the junction? What volume of concrete block would you suggest to resist this force? Concrete weighs 2400 kg/cu m.

16-9 The power source for the area represented by the residential load curve of Fig. 16-1 is a run-of-river hydroelectric plant with an effective head of 75 ft and a plant efficiency of 78 percent. What is the minimum average daily flow which could supply the indicated load? What volume of pondage would be required to produce the power required at peak load? What pondage would be required if the daily mean flow were 10,000 cfs? What would be the daily spill?

16-10 The flow from a power plant discharges into a tailrace which is 150 ft wide and has a training wall 200 ft long extending downstream to separate the flow from the powerhouse and the spillway. The plant is equipped with four turbines, each having a capacity of 2000 cfs and equipped with vertical draft tubes (Fig. 12-9a). If the elevation of the bottom of the draft tubes is zero, what should be the elevation of the crest of a broad-crested weir 150 ft downstream from the center line of the draft tubes in order to guarantee a minimum depth of 1.0 ft above the bottom of the draft tube with a single unit operating at full load? What will be the depth over the bottom of the draft tubes with four turbines at full load? Assume the bed of the tailrace to be horizontal and at elevation −15. Use Weir B, Table 10-4.

16-11 What size and type of turbine and number of units would you select for installation in the power plant described in Prob. 16-3? Assume 24-pole generator and 60-cycle electricity.

16-12 What type of turbine and how many units would you select for the installation in the power plant of Prob. 16-9? Specify the number of poles for the generators and their rotative speed. Assume 60-cycle current, and use a wheel diameter of 10 ft.

16-13 How many poles would you recommend for a 60-cycle generator which is to be connected to a turbine operating under a design head of 2000 ft with a flow of 80 cfs?

16-14 What size and type of turbine would you recommend for the plant of Prob. 16-10? Assume the effective head has an average value of 28 ft.

16-15 Determine the cost of power in cents per kilowatthour from a thermal plant where first cost is $125 per kilowatthour of capacity. Assume coal at $6 per ton. Fuel consumption is 1.2 lb of coal per kilowatthour. Plant load factor is 0.65, taxes and insurance 4 percent, interest 7 percent, life of plant 35 yr. Annual maintenance is estimated at 10 percent of first cost.

Chapter 17

River navigation

Waterways have always been important avenues of commerce in world history. They offered comparatively easy routes through unmapped wilderness for the exploration of new lands. As these lands were developed, boats were often the only feasible means of moving cargo too heavy for pack animals or wagon trains. However, rivers in their natural state were not ideal passageways. Rapids and sandbars offered formidable barriers, which were passed only with the utmost labor. Isolated rocks, fallen trees, and other obstructions were a constant hazard. It is reported that 138 steamboats were wrecked on the Ohio River between 1839 and 1942. Since World War II recreational boating has grown rapidly and in some areas constitutes a significant part of the traffic on navigable waters.

17-1   River navigation in the United States   The Mt. Vernon Compact, an agreement between Maryland and Virginia concerning navigation on the Chesapeake Bay and Potomac River, is considered to be a forerunner of the Constitutional Convention of 1787 and of the Commerce clause of the Constitution. In 1807 Congress authorized a study of canals and waterways by Albert Gallatin. By 1850 some 4400 mi of canals had been constructed in the United States at a cost of about 300 million dollars from state and private funds. Many canals were ill-fated from the start because of poor planning, and even the best succumbed to the competition of other methods of transportation. By 1960 only the Erie Canal remained in operation as a reminder of this first great period of waterway development.

A second period of waterway development began about 1910. Almost 95 percent of the 5.8 billion dollars expended by the U.S. Corps of Engineers on waterways improvement between 1824 and 1960 has been spent since 1910. The second period of development came about for much the same reason as the first period, i.e., the need for a cheap means of transporting heavy bulk materials. Steel and iron, oil, sand and gravel, grain, automobiles, and other bulk products can often be shipped more cheaply by water than by other means. The average cost for water transport in the United States in 1960 was 0.4 cent per ton-mile, considerably lower than that for truck or rail.

Representatives of the railroads will argue that the actual cost of water transport is greater than by rail if the federal subsidy in the form of navigation improvements is considered. However, the railroads also received substantial subsidies in their early days. Because of many intangible factors such as recreational benefits and national defense, the actual comparative costs of rail and water transport are difficult to evaluate. There seems little doubt, however, that the waterways are an important component of our present transportation system.[1]

Two big disadvantages of water transport—the slow movement and the large areas not traversed by commercially navigable streams—limit the role of water transport. As of 1967 the inland waterways of the United States (excluding the Great Lakes) carried about 15 percent of the domestic ton-miles[2] of freight compared with about 42 percent carried by rail, 22 percent by motor vehicle, and 20 percent by oil pipeline.[3]

17-2   Requirements of navigable waterways   There are no absolute criteria of navigability and, in the final analysis, economic criteria control. The physical factors which affect the cost of waterborne transport are depth of channel, width and alignment of channel, locking time, current velocity, and the terminal facilities. Commercial inland water transport is, for the most part, accomplished by barge tows consisting of 1 to 10 or more barges pushed by a shallow-draft river tug. The cost for a trip between any two terminals is the sum of the fuel cost and wages, fixed charges, and other operating expenses dependent on the time of transit. The cost per ton-mile is the cost divided by the product of the mileage and tons transported. Low ton-mile costs result from short transit times or large tonnage per tug.

The loading of barges is limited by the depth of the channel at its shallowest point en route. Actually power cost decreases as the depth of water between barge and bottom increases. Because of the increased drag when the barge hugs the channel bottom, fuel costs are about 25 percent lower with 5 ft of water under the barge than with 2 ft.[4] Nine-foot channels are

---

[1] "A Water Policy for the American People," Report of the President's Water Policy Commission, Vol. 1, pp. 197–217, 420–445, 1950.

[2] A ton-mile represents the product of 1 ton transported for 1 mi. It is a widely used unit of measure in transportation.

[3] "Statistical Abstract of the United States," 90th ed., Government Printing Office, Washington D.C., 1969.

[4] Rufus W. Putnam, The Value of Water Transportation, *Trans. ASCE*, Vol. 103, pp. 1279–1333, 1938.

more or less standard in most of the Mississippi River waterway system, but 12-ft minimum depths are projected in the ultimate development.

The number of barges in a tow and the transit time depend on the alignment, width, and current velocity. It is difficult to maneuver a long tow of heavily loaded barges in a channel not much wider than the tow itself. Sharp bends set a definite limit on the length of tow that can pass through them. A fairly wide channel with curves of long radius is necessary for good operation. A minimum width of 300 ft is the goal of the present navigable channel of the Lower Mississippi. A highly irregular alignment also increases the *circuity*, or length in excess of air-line distance, which the barge tow must travel. Present channels have a length about 50 percent greater than the air-line distances.

The speed of most barge tows in still water is around 6 mph or 9 ft/sec. Current velocities on the order of 3 or 4 ft/sec represent a substantial reduction in true speed for tows bound upstream or more specifically in the ton-miles moved per horsepower-hour. The time required for the tow to pass through locks also directly affects transit time. In many cases it is necessary to break up a tow and take it through a lock in portions because the lock is not large enough to accept the entire tow. This increases the time lost in locking.

Since charges for tugs and barges continue while the tow is at a terminal, facilities which permit rapid cargo transfer effectively reduce lost time and decrease cost. Terminal facilities thus are an important factor in the economics of navigation projects.

The actual criteria for depth, width, and alignment of a particular waterway can be determined only by calculating the transportation costs for various degrees of improvement and comparing each increment of annual saving against the annual cost of providing the facilities.[1] Intangible benefits such as the military benefits of a waterway may also affect decisions on improvement in the vicinity of the break-even point. It should be noted that the estimate of savings possible from improved navigation facilities is dependent on an estimate of the traffic which will use the facilities after they are completed.

17-3 Methods of achieving navigability There are three basic methods for improving a river for navigation—*open-channel, lock-and-dam,* and *canalization.* Open-channel methods seek to improve the existing channel to the point where navigation is feasible. Dams create a series of slack-water pools through which the traffic can move, with locks to lift the vessels from one pool to the next. Canalization provides a totally new channel cut

---

[1] C. W. Howe, Mathematical Models of Barge Tow Performance, *J. Waterways and Harbors Div., ASCE,* Vol. 93, pp. 153–166, November, 1967.

by artificial means around an otherwise impassable obstruction or between two navigable waters.

The essential requirements of a stream for open-channel improvement are:

1. Sufficient flow to permit navigation a reasonable portion of the year
2. A channel cross section sufficient (after improvement) for modern barge tows
3. A satisfactory alignment without excessively sharp bends
4. Channel slope sufficiently flat so that velocities are not excessive
5. Bed and bank materials permitting satisfactory treatment by one or more of the open-channel methods

If these requirements are not approximated, open-channel methods should not be considered. Some controls on these factors are possible, however. For example, if annual volume of flow is adequate, reservoirs may sometimes be utilized to augment low natural flows. Some sharp bends may also be eliminated by cutoff channels across the neck of the bend. Such cutoffs, however, result in steeper channel slopes and increased velocities.

Lock-and-dam construction may be indicated where conditions are unfavorable for open-channel methods. A series of locks and dams can be maintained if flow is sufficient to provide for water for lockages, sanitary releases as may be required, and evaporation losses from the pools. This usually requires much less water than open-channel procedures would. The slackwater pools behind the dams submerge rapids and channel bends and because of their relatively large cross section have low flow velocities. Locks and dams are not satisfactory on streams having a heavy sediment load which would fill the pools rapidly. There must also be sites suitable for the construction of the dams, but since navigation dams are not usually high dams, the site requirements are not exacting. The general features of a stream suitable for lock-and-dam construction for navigation improvement are (1) conditions unsatisfactory for open-channel methods, (2) low sediment transport, and (3) suitable dam sites. The use of locks and dams is a second choice to open-channel methods. Cost of locks and dams is usually high compared with open-channel methods, especially if much land is to be submerged or many stream-bank facilities are to be relocated.

Canalization is usually feasible only when a short length of channel opens a large length of navigable waterway. Of the early canals, only the Erie Canal connecting the Hudson River and the Great Lakes remains. The Illinois Waterway connecting the Great Lakes and the Mississippi River system is a modern counterpart of the Erie Canal. The per mile cost of a canal capable of passing modern vessels is normally so high that only short lengths as integral parts of more extensive and less costly waterways are feasible.

17-4   Open-channel methods   Natural channels may be improved for navigation by regulating the flow in storage reservoirs, dredging, contraction works, bank stabilization, straightening, and snag removal. Rarely will any one of these techniques provide all necessary improvement, and most open-channel projects require a combination of methods.

Reservoirs   Storage reservoirs can rarely be justified economically for navigation purposes alone and are usually planned as multiple-purpose projects. Improvement of navigation by reservoirs is possible when flood flows can be stored for release during low flow seasons. The ideal reservoir operation would involve navigation releases so timed as to fill the deficiencies in natural flow without waste (Fig. 17-1). This is possible only if the reservoir is at the head of a relatively short navigable reach. As the distance from the reservoir to the navigable portion of the river is increased, releases must be increased to allow for evaporation and seepage en route to the reach to be served. Moreover, the releases must be made sufficiently far in advance of the need to allow for the travel time to the reach, and in quantity sufficient so that, after reduction by channel storage, the delivered flows are adequate. The water requirement for navigation releases is considerably greater than the difference between actual and required flows. Flows from Ft. Peck Dam require some 3 weeks to reach the Mississippi River.

Dredging   Removal of material from navigable channels is usually accomplished by dredging. Extensive dredging is required to clear bars and badly filled channel sections when a project is initially constructed, and continued maintenance dredging is usually necessary for the removal of bars after the project is finished.

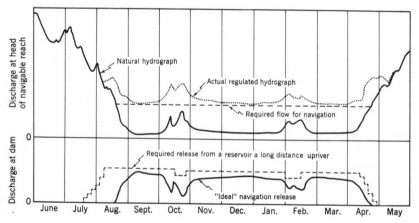

FIG. 17-1   Operation of a hypothetical reservoir for benefit of navigation.

FIG. 17-2   *Yuba No. 20 ladder dredge operating on the Yuba River near Marysville, California. Dredge buckets are 18 cu ft capacity and the dredge can dig to a depth of 124 ft below water level. (Yuba Manufacturing Co.)*

There are three basic types of dredges—dipper, ladder, and suction.[1] A *dipper dredge* is merely a floating power shovel, and such dredges are used only on small jobs. *Ladder dredges* have an endless chain of buckets (Fig. 17-2) which bring the bottom material to the surface. The buckets discharge on a belt conveyor which carries the material to a stacker conveyor at the rear of the dredge for disposal. Because the spoil stackers are limited in length to about 300 ft, ladder dredges cannot be used where the spoil must be discharged at a considerable distance from the dredge. *Suction dredges* pick up the bottom material and water in suction pipes, and the mixture is discharged by pumping through a spoil pipe supported by floats to the desired spoil area (Fig. 17-3). A suction dredge cannot operate in material containing large stones and boulders.

The two common types of suction (or hydraulic) dredges are the dustpan and cutterhead. The *dustpan dredge* has a wide suction head shaped like a dustpan or vacuum-cleaner nozzle. The head is equipped with jets which loosen the bottom material and suction openings through which the material (and water) is drawn into the suction pipe. The dredge works upstream, cutting a path up to 30 ft wide through a bar. A wider channel is achieved by a series of parallel cuts made in the same manner. Propulsion is achieved

---

[1]J. Huston, Dredging Fundamentals, *J. Waterways and Harbors Div., ASCE*, Vol. 93, pp. 45–69, August, 1967.

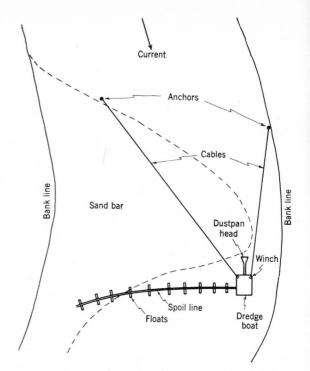

FIG. 17-3 *Schematic layout for a dustpan-dredge operation.*

by winches pulling against anchor cables set upstream of the bar (Fig. 17-3). This type of suction head is suitable for fine-grained, soft material in shallow cuts. It cannot cut through cohesive material or undercut a bank which might collapse on the head and shut off the suction. The specific characteristics of a dredge depend on the pumps and motive power selected. The maximum length of cut for a dustpan dredge is about 3500 ft. Because dustpan dredges move upstream fairly rapidly, shore pipes are not practical and the dredge usually is most efficient with a spoil line under 1000 ft long and a lift of less than 5 ft above the water.

A *cutterhead dredge* (Fig. 17-4) has a roughly hemispherical rotating suction head (Fig. 17-5) equipped with blades which loosen consolidated material so that it can be picked up in the suction pipe which terminates behind the cutterhead. Dredges of this type have two spuds, or heavy vertical posts, at the rear of the dredge hull. The spuds are lowered to the channel bottom to anchor the dredge. Cutterhead dredges operate by swinging about one spud with the head describing an arc up to 250 ft radius. The swing is controlled by winches pulling on cables anchored forward of the dredge (Fig. 17-6). As a swing is completed, the second spud is lowered and the other spud raised and a swing is made in the opposite direction. The dredge thus literally walks forward, cutting a channel up to 300 ft wide as it advances. Large dredges of this type are designed for a total head of 30 ft and a spoil line 2000 ft long or more. They are preferable when the spoil is to be discharged over the riverbank.

If no suitable spoil area is within range of the dredge, spoil may be discharged into barges for disposal elsewhere. *Hopper dredges* are large dredges with hoppers in which the spoil is stored, carried out to sea, and dumped.

Contraction works   When the bed and banks of a stream are of coarse-grained material with little cohesion, a wide shallow channel or, at low water, a number of channels will develop. Under such conditions contraction works may be used to encourage the river to form a narrower and deeper channel. Streams carrying a heavy sediment load are best corrected with permeable dikes, while streams carrying only a little sediment require impermeable groins or jetties. A great many techniques and materials have been employed for contraction works.

A common form of permeable dike is the *pile-clump dike* (Fig. 17-7), which consists of two or more parallel rows of pile clumps. The clumps consist of a tripod of piles cut off at about mid-bank height and with their tops wired together. Clumps are spaced 15 to 20 ft apart in rows which are usually

FIG. 17-4   *Phantom view of a typical heavy-duty* 16-*in. diesel-electric hydraulic pipeline dredge. (Ellicott Machine Corp.)*

FIG. 17-5   *Phantom view of a one-piece spiral basket cutter for a hydraulic dredge.* (*Ellicott Machine Corp.*)

about 5 ft apart. Pile stringers connect the clumps. Piles are usually driven to a penetration of 20 or 30 ft. The piles are driven through a woven willow or lumber mattress 80 to 100 ft wide and extending about 50 ft beyond the terminal clump to prevent scour around the piles. The mattress is ballasted with about 15 lb of stone per square foot, and an additional 500 lb/row-ft is placed along the dike to fill the holes in the mattress made by the pile driving. The bank is graded to a stable slope and covered with rock riprap for a distance of 50 ft both sides of the dike. Dikes of this type are usually placed in series. The upstream dike is turned downstream at a slight angle to the channel to deflect the current. The function of a permeable dike is to slow the current and promote deposition of sediment to fill in the diked area (Fig. 17-8). The concentration of flow in the narrower section also encourages deepening of the channel. Several years must be allowed for the effect of the structures to develop.

Where there is little sediment load to provide fill, a relatively impermeable groin of dumped rock or other material may be used to divert the flow. If used to close a secondary channel, the groin is usually placed at the downstream end, where it serves to trap the suspended and bed loads that are in the channel.

   Bank stabilization   An ultimate requirement for a good channel is that the banks be stable. Meandering of a channel begins with bank caving,

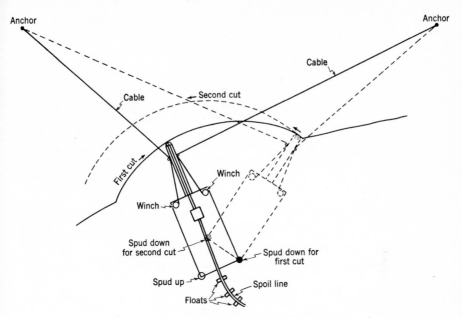

FIG. 17-6   *Schematic layout for a cutterhead-dredge operation.*

FIG. 17-7   *Construction of a pile-clump dike in the Nebraska City area, Missouri River.*
(U.S. Corps of Engineers)

which creates a bend (Fig. 17-9). The cross section in the bend is usually triangular, with the greatest depth at the concave bank. The tangents between the bends develop shallower channels with crossbars formed by the deposition

FIG. 17-8   *Before and after views of a pile-clump dike system at Rock Bluff bend of the Missouri River: (top) during construction (1934), (bottom) in 1942. Note filling along the right bank. Paving to protect new bank line is in place. (U.S. Corps of Engineers)*

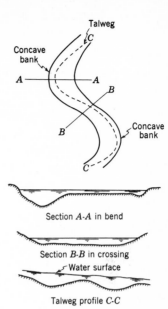

FIG. 17-9   *Topography of a channel bend.*

of material scoured from the upstream bend. It is in these crossings that the controlling depths for navigation occur. Dikes and groins may prove useful as bank-protection structures since, when placed along the concave bank, they promote deposition rather than scour.

A more common method of bank protection is the use of *revetments*. To be effective, the revetment must extend the full length of the concave bank and must extend from the top of the bank to beyond the toe of the underwater slope. If this is not done, scour may undermine the underwater edge of the revetment and cause its failure. The revetment must be flexible, to adapt itself to the surface on which it is placed, relatively impermeable, to avoid washing of fine materials through it, and strong enough to resist the currents encountered. The portion of the revetment below normal low water must be placed underwater, and this is usually accomplished with a mattress. Because the banks of concave bends are usually steep and unstable through the undercutting action of the stream, they must be graded to a stable slope (usually 1 on 3). Early revetments were mattresses woven of willow branches or lumber. The most effective revetment yet designed is the *articulated concrete mattress*. This is formed of units consisting of 20 concrete blocks 4 ft long, 14 in. wide, and 3 in. thick spaced about 1 in. apart on a continuous wire-mesh reinforcing. These units, 25 ft long and 4 ft wide, are joined together by the wires of the projecting reinforcement. The assembly is performed on a launching barge, which moves out from shore as the mattress is assembled and allows it to fall to the bottom (Fig. 17-10).

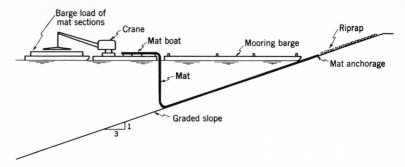

FIG. 17-10   *Placement of articulated concrete mattress.*

The best revetment above the normal low water seems to be riprap on a gravel blanket. Riprap will adjust itself to minor bank sloughing and continue to offer protection. Because of lack of local sources for stone, other types of bank paving are frequently necessary. Articulated mattress on a 4-in. gravel blanket is effective but expensive. Uncompacted asphalt paving has been used along the Lower Mississippi. It consists of bar-run sand and gravel with about 6 percent of asphaltic cement placed to a minimum thickness of 5 in. In this form the paving is as permeable as loose sand and therefore superior to compacted asphalt or monolithic concrete, which might be damaged by uplift pressures.

Straightening   Contraction works and bank stabilization tend to cause some channel straightening or at least to prevent further meandering. Sharp bends are undesirable from the viewpoint of navigation, flood mitigation (Chap. 20), and bank protection. Natural elimination of sharp bends sometimes occurs when a cutoff is formed (Fig. 17-11a) as a result of scour on both sides of a narrow neck of land. Artificial cutoffs have been used in the Lower Mississippi with apparent success provided the banks are stabilized to prevent further meandering. Such cutoffs are formed by dredging a small

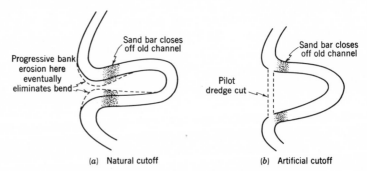

(a) Natural cutoff   (b) Artificial cutoff

FIG. 17-11   *Formation of cutoffs.*

pilot cut through the neck of land. Subsequent river flows scour a full channel along the line of the pilot cut and seal off the old bend by sediment deposits (Fig. 17-11*b*).

Snag removal   Tree trunks, stumps, rocks, and other channel obstructions are hazards to navigation and promote the formation of bars. Their removal is an essential part of the development of a river for navigation. The removal method depends on the circumstances in each case, but tractors and winches on shore, derrick barges, and explosives have proved effective.

17-5   Navigation dams   The general features of structural design of navigation dams do not differ materially from any other type of dam except for the lock facilities. The selection of dam height is determined by the desirable channel characteristics upstream rather than by a required storage capacity. In other words, the dam must be high enough so that adequate depths are provided as far upstream as desired. While a very low dam may provide an adequate but tortuous channel, increased height may provide sufficient depth so that traffic may follow a straighter route across old bends in the river. The cost of the added height must be balanced against the savings from the improved channel. If the minimum flow of the stream is inadequate to provide water for locking, some storage capacity in the pool above navigable depth may be provided. Dredging of the channel may serve as a means of minimizing the height of the dam. Some dredging may be necessary in any case to provide sufficient channel width and access to terminal facilities.

Several dams will be required to open a long waterway to navigation. The planning of the system requires selection of location and height for minimum overall cost. Unless there are reaches of channel which are navigable in their natural state, the dams must be located so that their pools overlap, i.e., each dam backs water up to the dam upstream. Locks are often not large enough to take a complete barge tow, and the tow may have to be split to pass the lock. Even if this is not necessary, lockages increase transit time for navigation and require a staff of operating personnel at the lock. The cost of lockage may become an important factor in determining the number of dams for a stream. Higher dams will reduce the number required but will be individually more expensive. Since the cost of a dam increases with some power of its height and because of the increased area flooded by the higher dams, a large number of low dams may be preferable. Existing developments along the channel may also limit the height of the pools. These and other factors must be weighed to establish the optimum plan of development. Table 17-1 summarizes data on four rivers which have been developed by lock-and-dam methods. It should be noted that the dams of the TVA system are utilized for storage and power production and hence are higher than those of the Ohio and Mississippi Rivers.

A variety of considerations enter into the selection of the precise type of navigation dam for a given site. Both fixed and movable dams are utilized. *Fixed dams* are now usually concrete gravity dams, although earlier structures were frequently rock-filled timber cribs. The dam may consist of a simple overflow weir with a fixed crest, or flashboards or crest gates may be provided to regulate the pool level. The dams of the TVA system are of this type. Fixed dams are usually required if the height is 30 ft or more but are sometimes used for lesser heights.

Movable dams consist of concrete sills slightly above the channel bottom and a damming surface which can be raised above the water or lowered to the channel bed to permit the passage of surplus water. Those movable dams in which the damming surface is raised above the water require massive piers and overhead bridges for hoisting machinery and are *nonnavigable*, i.e., they cannot be passed by barge traffic. The most common types of gates for the nonnavigable dam are roller or Tainter gates (Fig. 9-15) because of the long clear spans between piers which are possible with these types. Both types can provide damming surfaces up to 30 ft high. It is sometimes desirable to be able to depress gates of this type below the water surface so that ice or drift may pass over the gate. In the case of the roller gate this requires only a modification of the sill (Fig. 17-12), but the Tainter gate must have a skin plate extended to form a suitable crest. Vertical-lift gates may also be used if the height is beyond the limits of the other types, but because of the greater lifting load and slower operation, vertical-lift gates are not particularly favored.

Movable *navigable* dams are formed by the use of a gate which folds on the sill when not in use. The earliest of such dams used the Chanoine wicket, a wooden shutter held in position against the water load by a prop (Fig. 17-13). Each wicket is raised by pulling up until the prop seats in the recess provided in the sill and is lowered by pulling forward to disengage the prop. The

TABLE 17-1   Some Lock-and-dam Navigation Systems in the United States

Stream	Length, mi	Number of dams	Lift, ft Aver-age	Lift, ft Maxi-mum	Lift, ft Mini-mum	Average distance between dams, mi
Upper Mississippi River from Alton, Mo., to Minneapolis, Minn.	645	26	13.1	38.2	5.5	24.8
Ohio River from Cairo, Ill., to Pittsburgh, Pa.	981	29	15.1	37.0	7.0	34.1
Tennessee River from its mouth to Knox-ville, Tenn.	648	10	54.0	72.0	9.5	64.8
St. Lawrence Seaway	190	4	36.8	47.9	4.0	12.6

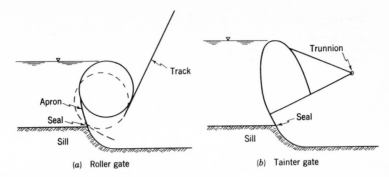

FIG. 17-12   *Submergible gates for navigation dams.*

operation is performed from a maneuver boat operating in the pool above the dam. The Chanoine-Pascaud wicket consists of steel wickets with a rubber seal between adjacent wickets and two stops for the prop so that some wickets may be partially lowered to pass flow over them. A further modification is the Bebout wicket with the horse hinged to fold when the resultant of the water load rises above the fulcrum point. Wickets of this type trip automatically to permit the passage of a flash flood. Bear-trap gates (Sec. 9-15) are also employed for movable dams. Their construction permits more careful regulation of flow than is possible with the one- or two-position wickets.

Movable dams are employed in navigation projects because it is often necessary to minimize the effect of the dam on flood flows. A fixed dam, unless it is very low and completely submerged at flood flows, raises the water levels above those which would occur under natural river conditions. Movable dams cause minimum interference with flood flows. A second advantage of movable navigable dams is that at high flows navigation may proceed unimpeded over the lowered wickets without using the locks. In streams subject to considerable and frequent flow variations the navigable type of dam is impracticable because it cannot be readily adjusted to meet varying flow conditions. The dams of the Upper Mississippi River system

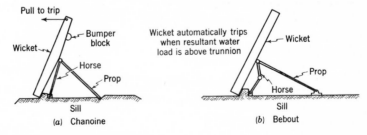

FIG. 17-13   *Wicket dams used for navigation.*

are of the nonnavigable type. Because of ice all dams have at least a portion of their gates submergible. Seventeen of the Ohio River dams have navigable passes varying from 600 to 1200 ft in width, consisting of wicket-type barriers and a section of bear-trap gates for better control of flow and pool levels at low flows. In all cases the overflow sections of the dams are quite long to permit free escape of flood flows. Present plans are to replace the low dams on the Ohio River with high nonnavigable dams by 1980. The reduced velocity in the larger pools offsets the advantage of the navigable passes, especially for upstream traffic.

17-6   Navigation locks   The two major items in the design of navigation locks are the determination of size and the design of the filling and emptying systems. The height of the lock is fixed by the selected upstream and downstream pool levels. The overall height of the lock chamber must equal the maximum expected difference in pool elevations plus the required draft plus freeboard. The highest lock in the United States is the John Day lock on the Columbia River at 113 ft.[1] The elevation of the bottom of the lock chamber must equal the minimum pool level in the downstream pool less the required draft. The size of the lock in plan depends on the traffic expected to pass through it. Most of the locks in the Mississippi, Ohio, and Tennessee River systems are 110 ft wide by 600 ft long, although smaller locks have been installed in some of the upstream dams where less traffic is expected. Modern construction favors a main lock 1200 ft long and an auxiliary lock 600 ft long. Where small boat traffic is heavy a small lock may also be provided to keep small boats and commercial tows separate.

Because the tows must approach the lock at low speed to avoid damage from collision, and since they have little steering control at low speeds, it is important that the lock approaches be protected by guide walls (Fig. 17-14) so that eddies and turbulence in the navigation channel as a result of flow over the dam will be minimized. In many instances the guide walls may be several hundred feet long. In some cases use of spillway gates adjacent to the lock is restricted to emergency needs only.

The structural design of the lock and guide walls depends to some extent on the details of construction. Modern construction practice favors the use of reinforced concrete throughout, but guide walls of steel or timber piling and lock walls of the same material have been used successfully. The greatest drawback in the use of piling for the lock walls is the leakage through such materials. Lock walls of piling act as cantilevers supporting the lateral water pressure (or earth pressure on the shoreward side). Concrete lock walls are usually designed as gravity dams resisting either water or earth pressure with

---

[1] A lock at Ust-Kamengorsk Dam on the Irtish River in the U.S.S.R. has a lift of 42 m or 138 ft.

FIG. 17-14 *Ship passing through the Dwight D. Eisenhower Navigation Lock. (St. Lawrence Seaway Development Corp.)*

the lock empty. Guide walls are normally subjected to equal pressures on both sides, and the only design load of consequence is the force which might be exerted by a colliding barge. However, the landward guide wall may have earth pressure on one side, which would differ in magnitude from the water pressure on the riverward side.

The design of the lock-filling and -emptying system requires a compromise between two differing demands: (1) that the filling time be short (15 min) in order not to delay traffic, and (2) that the disturbances in the lock chamber not cause stresses in mooring hawsers which might cause the tow to break loose and be damaged or damage the lock structure. When the lock is being filled, the vessels are initially in shallow water and water is entering under a relatively high head. This condition is more serious than that encountered in emptying, and therefore the governing criteria are those encountered during the filling operation. If the lock gates were opened instantaneously, an abrupt wave would travel across the lock chamber (Fig. 17-15a). This wave causes an unbalanced force on the vessel in the lock and considerable pull on the mooring hawsers. When the wave reaches the downstream end of the lock, it is reflected back, throwing the vessel in the opposite direction. The entire filling operation would consist of a continuation of this wave motion, alternately in one direction and then the other. Some relief is possible by opening the inlet valve slowly so that a sloping wave (Fig. 17-15b) and smaller unbalanced forces result. The presence of a vessel in the lock causes some reflection of the waves, making a detailed analysis difficult. Hence, design of most lock-filling systems is based on model tests. Ignoring the effect of the vessel on the wave pattern, it can be shown that the unbalanced force on the vessel is proportional to the rate of change of inflow in the lock.

There are a large number of systems for filling lock chambers.[1] American practice favors longitudinal conduits in the lock walls with lateral ports[2] (Fig. 17-16a) to distribute the flow uniformly along the length of the lock and thus avoid the wave action discussed earlier. It is difficult to design a system of this type which distributes the flow uniformly, and the manifold seems to serve more to baffle the flow and thus avoid direct impulse forces

---

[1] M. E. Nelson and H. J. Johnson, Navigation Locks; Filling and Emptying Systems for Locks, *J. Waterways and Harbors Div.*, *ASCE*, Vol. 90, pp. 47–60, February, 1964.

[2] R. A. Elder, J. T. Price, and W. W. Engle, Navigation Locks: TVA's Multiport Lock Filling and Emptying System, *J. Waterways and Harbors Div.*, *ASCE*, Vol. 90, pp. 31–46, February, 1964.

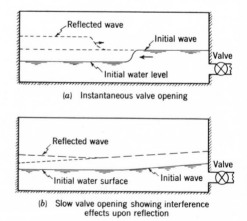

(a)   Instantaneous valve opening

(b)   Slow valve opening showing interference effects upon reflection

FIG. 17-15   *Effect of rate of valve opening on waves in locks.*

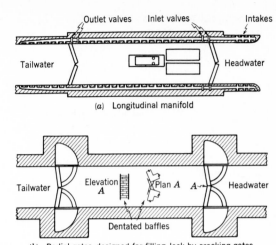

Outlet valves   Inlet valves   Intakes

Tailwater   Headwater

(a) Longitudinal manifold

Tailwater   Elevation A   Plan A   A   Headwater

Dentated baffles

(b) Radial gates designed for filling lock by cracking gates

FIG. 17-16   Lock-filling systems.

which would result from a single entrance. For lifts under 15 ft water may be admitted to the lock chamber by opening the main lock gates.[1] Baffles attached to the inner face of the gate leaves (Fig. 17-16b) help to reduce entrance velocities. The rate of gate opening should be very slow at first and increased as the lock is filled. For high head or large locks a bottom diffuser system is now preferred. Diffuser pipes distributed over the bottom of the lock permit relatively quick filling with a minimum of disturbance.

Preliminary design of the filling mechanism may be accomplished by considering the system as an orifice with coefficient C_d as in Table 17-2 and an

*TABLE 17-2   Equivalent Orifice Coefficients for Lock-filling Systems**

Type of system	Discharge coefficient
Longitudinal conduit and lateral ports	0.85–0.95
Venturi loops	0.75–0.85
Opening main gates†	0.80

*George R. Rich, Hydraulics of Lock Filling Systems, *Military Eng.*, pp. 60–64, January–February, 1933.
†Based on cross-sectional area of water between the gates when level is equalized.

---

[1]F. R. Brown, Navigation Locks: End Filling and Emptying Systems, *J. Waterways and Harbors Div.*, ASCE, Vol. 90, pp. 61–78, February, 1964; G. C. Richardson, Navigation Locks: Navigation Lock Gates and Valves, *J. Waterways and Harbors Div.*, ASCE, Vol. 90, pp. 79–102, February, 1964.

area A equal to the cross-sectional area of the filling conduits at the valves. The filling time t_i for instantaneous valve opening can then be shown to be

$$t_i = \frac{2A_L \sqrt{h}}{C_d A \sqrt{2g}} \tag{17-1}$$

where A_L is the plan area of the lock, and h is the lift. If the valve is opened in such a way that the area increases linearly with time and is fully open before equalization occurs, the filling time becomes equal to half the valve opening time plus t_i. If equalization is obtained before the valve is fully opened, the filling time t' is

$$t' = 2 \sqrt{\frac{A_L t \sqrt{h}}{C_d A \sqrt{2g}}} \tag{17-2}$$

where t is the time required to open the valve. Final determination of the system characteristics is usually accomplished with model tests after a selection of an approximate design on the basis of mathematical analysis.

BIBLIOGRAPHY

Chorpening, C. H.: Waterway Growth in the United States, *Trans. ASCE*, Vol. CT, pp. 976–1041, 1953.

Dann, A. W., and C. B. Janson: River Transportation Development, *Trans. ASCE*, Vol. CT, pp. 1163–1179, 1953.

Elliott, Malcolm, L. E. Feagin, D. B. Freeman, J. F. Friedkin, Charles Senour, and L. G. Straub: Symposium on Open Channel Development of Rivers, *Civil Eng. (N.Y.)*, Vol. 16, pp. 489–495, November, 1946.

Faison, H. R.: Some Economic Aspects of Waterway Projects, *Trans. ASCE*, Vol. 120, pp. 1480–1549, 1955.

Giroux, Carl H.: American Hydraulic Dredges, *Trans. ASCE*, Vol. CT, pp. 1180–1191, 1953.

———: Modern Hydraulic Dredging—a Product of Experience, *Civil Eng. (N.Y.)*, Vol. 22, pp. 156–161, September, 1952.

Haas, Raymond H., and Harvill E. Weller: Bank Stabilization by Revetments and Dikes, *Trans. ASCE*, Vol. 8, pp. 849–870, 1953.

Lane, E. W.: Design of Stable Channels, *Trans. ASCE*, Vol. 120, pp. 1234–1260, 1955.

Leliavsky, Serge: "An Introduction to Fluvial Hydraulics," Constable, London, 1955.

Navigation Dams, chap. 4, Part CXVI in "Engineering Manual, Civil Works Construction," U.S. Corps of Engineers, 1956.

Rich, George R.: Navigation Locks, sec. 32 in C. V. Davis and K. E. Sorenson (eds.), "Handbook of Applied Hydraulics," 3d ed., McGraw-Hill, New York, 1969.

Richardson, George C., and Marvin J. Webster: Hydraulic Design of Columbia River Navigation Locks, *Trans. ASCE*, Vol. 125, pp. 345–364, 1960.

Stratton, J. H., J. H. Douma, and J. P. Davis: Navigation Systems, sec. 31 in C. V. Davis and K. E. Sorenson (eds.), "Handbook of Applied Hydraulics," 3d ed., McGraw-Hill, New York, 1969.

PROBLEMS

17-1 One arrangement of barges in a tow utilizes 15 barges in three rows with outside dimensions of 1120 ft long and 105 ft wide. If the channel is 300 ft wide, what is the minimum radius bend with a 180° included angle which this tow could pass with at least 10-ft clearance at all points?

17-2 What length of barge tow of width 20 m can pass through a channel bend of width 60 m, centerline radius 215 m, and deflection angle 40°, if a clearance of 2 m is desired at all points?

17-3 The Bebout wicket of Fig. 17-13b is 14 ft high (vertically), and its inclination is 30° from the vertical. The fulcrum of the prop is 6 ft along the slope from the bottom end of the wicket. What will be the elevation of the water above the sill when this wicket trips? Neglect the weight of the wicket.

17-4 A 20-m-wide Chanoine wicket (Fig. 17-13a) is 6 m high vertically and its inclination is 20° from the vertical. The fulcrum of the prop is at midheight. What force in kilograms is required to trip the wicket when the water depth is 4 m? 7 m? The wicket gate weighs 13,800 kg.

17-5 The wicket of Prob. 17-3 is 20 ft long and is supported by props at both ends. What size steel channel section will be required for these props?

17-6 What conduit size would be required for a lock whose chamber is 110 by 600 ft in plan and whose lift is 45 ft if the filling time is not to exceed 15 min? Assume a manifold system (Fig. 17-16a) and a uniform rate of valve opening to fully open in 5 min.

17-7 Derive Eq. (17-1) and prove that in lock filling if the valve is opened in such a way that the area increases linearly with time and is fully open before equalization occurs, the filling time equals half the valve opening time plus t_i.

17-8 Derive Eq. (17-2).

Chapter 18

Drainage

"Drainage" is the term applied to systems for dealing with excess water. The three primary drainage tasks are *urban storm drainage, land drainage,* and *highway drainage.* Often considered as minor problems for the hydraulic engineer and frequently designed as if the works were unimportant, these tasks actually involve substantially greater capital investment and probably consume more engineering time each year than all of the flood-mitigation[1] activity undertaken. The primary distinction between drainage and flood mitigation is in the techniques employed to cope with excess water and in the fact that (with the exception of highway culverts and bridges) drainage deals with water before it has reached major stream channels.

Most cities have some form of storm drainage and the total cost of a system for a major metropolitan center far exceeds the cost of even the largest flood-mitigation project. As of 1960 nearly 100 million acres of agricultural land in the United States had been reclaimed by drainage.[2] This is more than twice the amount reclaimed by irrigation. It has been estimated that roughly one-fourth of the cost of highways is spent on culverts and other drainage structures.

URBAN STORM DRAINAGE

The total investment in storm drainage facilities in the United States was estimated at $22 billion in 1966, and annual expenditures between 1967 and 1980 of about $2.5 billion were projected to meet growing needs and to

---

[1] The term "flood mitigation" is used in this text in lieu of the more common term "flood control." For a discussion, see Chap. 20.

[2] Census of Agriculture, U.S. Bureau of the Census, 1960.

allow for replacement and repair of existing works.[1] In cities, storm water is usually collected in the streets and conveyed through inlets to buried conduits which carry it to a point where it can be safely discharged. The collected water is usually discharged into a stream, lake, or ocean, but in some instances it is spread for percolation into the ground. A single outlet may be used, or a number of disposal points may be selected on the basis of the topography of the area. The accumulated water should be discharged as close to its source as possible. Gravity discharge is naturally preferable but not always feasible, and pumping plants may be an important part of a city storm-drainage system.

The design of a drainage project requires a detailed map of the area. Scales of 1 in. representing 100 to 500 ft are usually satisfactory. The contour interval should be small enough to define the divides between the various subdrainages within the system. Ground elevations at street intersections and breaks in grade between intersections should be indicated to the nearest 0.1 ft. Final design requires even more detailed maps of those areas where construction is proposed. All existing underground facilities (gas, water, electricity, telephone, and sanitary sewers) must be accurately located, together with buildings, canals, railroads, or other structures which might interfere with the proposed route. If rock is expected near the surface, rock profiles as determined by borings along the proposed conduit lines are necessary so that the pipe layout can be selected to minimize rock excavation.

18-1   Estimates of flow   The first step in the design of storm-drainage works is the determination of the quantities of water which must be accommodated. In most cases, only an estimate of the peak flow is required, but where storage or pumping of water is proposed, the volume of flow must also be known. Drainage works are usually designed to dispose of the flow from a storm having a specified return period, which should be determined on an insurance basis (Sec. 13-7). It is often difficult to evaluate the damage which results from urban flooding, especially when the "damage" is often merely a nuisance. Hence, the selection of the proper return period is often dependent on the designer's judgment. In residential areas, there may be little harm in filling gutters and flooding intersections several times each year if the flooding lasts only a short time. In a commercial district, such flooding may cause considerable damage and inconvenience, and a greater degree of protection may be warranted. One guide to the selection of the proper return period may be found in reports of trouble in previous storms. Areas of relatively good natural drainage need less protection than low areas which serve as collecting basins for flow from a large tributary area.

---

[1] "Urban Drainage, Practices, Procedures, and Needs," *American Pub. Works Assoc.,* *Spec. Rept.* 31, Chicago, 1966.

It is probable that return periods of 1 or 2 yr in residential districts and 5 to 10 yr in commercial districts are all that can be justified for the average project. In the final analysis, it may be the willingness of the residents to finance the drainage works by taxes or bonds that determines the actual design.

It is almost universally true that drainage projects deal with flows from ungaged areas, so that design flows must be synthesized from rainfall data. For urban drainage the most widely used method has been the rational formula (Sec. 3-11) using a rainfall of the desired frequency (Sec. 5-7). The uncertainties of this method have been discussed (Secs. 3-11 and 5-9). In addition, when applied to urban drainage design, the decrease of peak flows resulting from storage in gutters and sewers is ignored, and peaks from various subareas of the system are assumed to synchronize. In view of the high cost of urban drainage, the rational formula approach cannot be considered an adequate basis for design.

The most satisfactory method for estimating urban runoff is digital computer simulation.[1] In this approach, flows are simulated throughout the system from available rainfall data. For adequate definition of the 10-yr event, at least 30 yr of flow should be simulated. Output is the simulated flow at all key points in the system. From this output annual flow peaks can be selected and subjected to frequency analysis to define the design flow at each point. The effects of small storage reservoirs or pumping plants can be investigated in the simulation, and if more than one rainfall record is available the effect of rainfall variation can be included. Calibration of the simulation model should be made against the nearest gaged stream having soil characteristics similar to those of the urban area under study. Parameter adjustment accounts for the impervious area in the urban area. A preliminary layout of the drainage system is required so that the effect of the improved drainage can be reflected in the simulated flows. The simulation approach avoids the arbitrary assumptions of constant runoff coefficient, uniform rainfall intensity, equal frequency of rainfall and runoff, etc. In addition, snowmelt which may be important in northern cities can be simulated. For small or steep areas where hourly rainfall does not give adequate hydrographic detail, 15- or 30-min rainfall can be used as input.

For areas of a few acres where extensive computer simulation may not be justified, the rational-formula approach using Izzard's estimate of time to equilibrium [Eq. (3-11)] in lieu of time of concentration should give reasonable results. Two computations are suggested—one assuming contribution from only the impervious area with a short time of concentration

---

[1] The simulation approach described here is based on the Hydrocomp Simulation Program developed by Hydrocomp International, Palo Alto, Calif.

and one assuming runoff from the entire area but with a relatively long time of concentration. The one giving the highest flow would be used in design (see Illustrative Example 18-3).

18-2   Gutters   The discharge capacity of gutters depends on their shape, slope, and roughness. Manning's equation may be used for calculating the flow in gutters; however, the roughness coefficient n must be modified somewhat to account for the effect of lateral inflow from the street. Moreover, with the wide, shallow flow and the varying transverse depth common in gutters, the flow pattern is not symmetric[1] and the boundary shear stresses have an irregular distribution. For well-finished gutters, n has a value of about 0.016 when used in the Manning equation. Unpaved gutters or gutters with broken pavement will have much higher values of n. Gutters are generally constructed with a transverse slope of 1 on 20. With such a transverse slope and a 6-in. curb height, the width of flow in a gutter will be 10 ft when there is no freeboard.

The assumption of uniform flow in gutters is not strictly correct since water enters the gutter more or less uniformly along its length. The depth of flow and the velocity head increase downslope in the gutter, and the slope of the energy gradient is therefore flatter than the slope of the gutter. With a gutter slope of 0.01 this error amounts to only about 3 percent, but the error increases rapidly as the gutter slope is flattened, and on very flat slopes the gutter capacity is much less than that computed using the gutter slope in the Manning equation. If the flow in the gutter ponds around the inlet, the depth is controlled by the inlet characteristics rather than by the gutter hydraulics.

18-3   Inlets   Gutter flow is intercepted and directed to buried sewers by drop inlets. There are two main types of inlets, with many commercial patterns available in each type. *Grated inlets* (Fig. 18-1a) are openings in the gutter bottom protected by grates. A *curb-opening inlet* (Fig. 18-1b) is an opening in the face of the curb which operates much like a side-channel spillway. Curb-opening inlets are feasible only where curbs have essentially vertical faces.

The location of street inlets is determined largely by the judgment of the designer. A maximum width of gutter flow of 6 ft has been suggested as a suitable criterion for important highways. Under this rule an inlet is necessary whenever the gutter flow exceeds the gutter capacity within the limiting 6-ft width. In residential sections the ultimate in inlet spacing provides four inlets at each intersection as shown in Fig. 18-2a. With this arrangement flow travels in the gutter only one block before interception. A less

---

[1] Richard J. Wasley, Uniform Flow in Shallow, Triangular Open Channel, *J. Hydraulics Div., ASCE,* pp. 149–170, September, 1961.

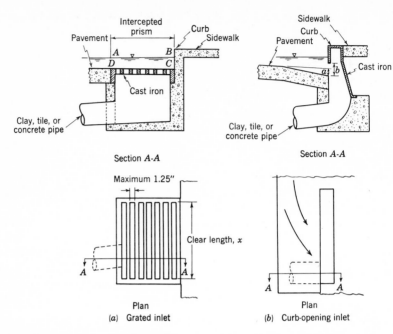

FIG. 18-1   *Some typical storm-drain inlets.*

expensive arrangement provides only two inlets at the uphill corner of the intersection (Fig. 18-2b), and water is allowed to flow around two sides of the block. A much more economical design might place inlets at spacings of several blocks. Not only does this eliminate a considerable number of inlets, but in the higher areas it may considerably reduce the length of sewer lines to be constructed. In residential sections an arrangement such as is shown in Fig. 18-2c may be quite adequate provided inlet spacings are such as to prevent gutter flows so great as to flood sidewalks, lawns, and even houses. In commercial areas with heavy pedestrian and vehicle traffic the arrangement in Fig. 18-2b would be more appropriate.

At one time it was common practice to provide *catch basins* (Fig. 18-3) at inlets to trap debris and sediment and prevent their entry into the drain. Such basins were cleaned at intervals by hand or with a small orange-peel or clamshell bucket mounted on a truck. The expense of cleaning catch basins can be quite large, and it has become common practice to omit them from inlet design and depend on adequate velocities in the drains to prevent trouble from deposition of sediment.

18-4   Grated inlets   Few inlets intercept all the flow that reaches them in the gutter unless they are at low points from which the water has no other route of escape. The most efficient grated inlets have bars parallel to

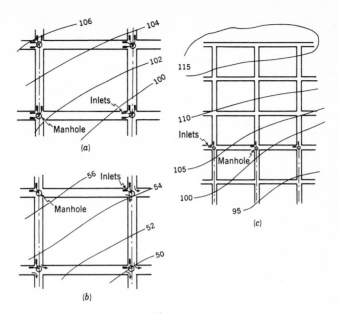

FIG. 18-2   *Some possible arrangements of storm-drain inlets.*

the curb and a sufficient clear length so that water can fall through the opening without hitting a crossbar or the far side of the grate. Experiment has shown that this free length x should be at least

$$x = \frac{V}{2} y^{1/2} \qquad (18\text{-}1)$$

where V is the mean approach velocity in the flow prism intercepted by the grate ($ABCD$, Fig. 18-1a), and y is the drop from the water surface to the underside of the grate. A grate which satisfies these requirements may be expected to intercept all the flow in the gutter prism crossing the grate.

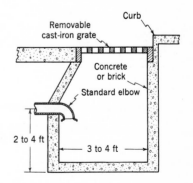

FIG. 18-3   *Inlet and catch basin.*

Unless the grate is considerably longer than the length given by Eq. (18-1) or is depressed below the gutter bottom, any water outside the grate width may be assumed to pass the inlet. On flat slopes, the inflow along the longitudinal side of the grate may be estimated by the method outlined for computing the flow into curb-opening inlets (Sec. 18-5). On steep slopes with high velocities, very little water will cross the longitudinal edge of the grate unless the pavement is warped to encourage this action.

Grates with bars parallel to the curb may be a hazard to bicycle riders unless the space between bars is less than about 1.25 in. However, when the bars are at right angles to the curb, water tends to bounce along the top of the bars, and very little is intercepted. Such grates are more likely to clog with debris. If water ponds over the grate, the orientation of the bars is of little importance. Grates with ponded water, i.e., negligible approach velocity, and heads under about 0.4 ft function as weirs [Eq. (9-1)] with a crest length L equal to the perimeter of the grate over which the water flows and a coefficient C_w of about 3.0. If one side of the grate is adjacent to the curb, it should not be included in the perimeter. If the head on a ponded grate exceeds about 1.4 ft, the grate functions as an orifice with an area equal to the clear opening between bars, a head h equal to the depth from top of grate to surface of pond, and a coefficient of about 0.6. Because of the possibility of trash collecting on a grate where water ponds, it is recommended that at least twice the required grate be provided. Where the depth on a grate is between 0.4 and 1.4 ft, transition conditions exist, and the capacity is intermediate between that of an orifice and a weir.

ILLUSTRATIVE EXAMPLE 18-1   A grated inlet 18 in. wide is to be installed in a gutter on a slope of 0.05. The grate thickness is 2 in. The transverse slope of the roadway is 0.03. How long should the inlet be, and how much water will it intercept when the depth at the curb is 0.3 ft? Use $n = 0.016$.

From Eq. (10-5), the flow in the gutter will be

$$Q = \frac{1.49}{0.016}\left(0.5 \times 0.3 \times \frac{0.3}{0.03}\right)\left(\frac{0.5 \times 0.3 \times 10}{0.3 + \sqrt{0.3^2 + 10^2}}\right)^{2/3}(0.05)^{1/2} = 8.7 \text{ cfs}$$

The flow Q' not crossing the grate is 8.5 ft wide, and the depth of this flow is $d = 0.3 - 1.5/10(0.3) = 0.255$ ft. Thus,

$$Q' = \frac{1.49}{0.016}(0.5 \times 0.255 \times 8.5)\left(\frac{1.09}{8.26}\right)^{2/3}(0.05)^{1/2} = 5.6 \text{ cfs}$$

Hence the flow crossing the grate and intercepted is

$$Q'' = 8.7 - 5.6 = 3.1 \text{ cfs}$$

The mean depth of flow at the grate is

$$\frac{0.3 + 0.3 - (0.03 \times 1.5)}{2} = 0.28 \text{ ft}$$

and the area of flow is therefore

$$0.28 \times 1.5 = 0.42 \text{ sq ft}$$

The velocity of flow is

$$\frac{3.1}{0.42} = 7.4 \text{ ft/sec}$$

and the length of grate should be at least

$$x = \frac{7.4}{2}(0.28 + 0.17)^{1/2} = 2.48 \text{ ft}$$

18-5 Curb-opening inlets   When water enters a curb-opening inlet, it must change direction.[1] If the street has a low crown, the gutter flow will be spread out over a considerable width and a correspondingly greater length of inlet will be required for the change of direction to be accomplished. Curb-opening inlets, therefore, function best with relatively steep transverse slopes. Figure 18-4a indicates the capacity of curb-opening inlets in cubic feet per second per foot of length as determined by experiment. The parameters are the depth of flow d in the approach gutter and the depression a of the gutter bottom at the inlet. A depression of 0.2 ft is common. The inlet length L_a required for full interception is equal to the gutter discharge Q_a divided by the inlet capacity in cubic feet per second per foot. If the length of the inlet is insufficient to intercept the entire flow, the portion Q which it will intercept is indicated by Fig. 18-4b as a function of the ratio of a/d and the ratio L/L_a, where L is the actual length of the inlet.

If the flow is ponded at the inlet and the head is less than the height of the opening in the curb, a curb-opening inlet may be considered as a weir [Eq. (9-1)] with a coefficient of about 3.1. If the depth of the pond exceeds twice the height of the clear opening in the curb, the discharge is that of an orifice, or

$$Q = 5.62 Lbh^{0.5} \tag{18-2}$$

where b is the height of the inlet opening, and h is the depth of water above the mid-elevation of the inlet opening. Equation (18-2) is the usual orifice formula with $C_d = 0.7$.

In many cities the pavement near a curb-opening inlet is warped so that the gutter bottom downstream of the inlet is higher than that upstream in order to force more water into the inlet. The hydraulics of such an arrangement is too complex to permit statement of any general rule for computing inlet capacity.

---

[1] Richard J. Wasley, Hydrodynamics of Flow into Curb-opening Inlets, *Tech. Rept.* 6, Department of Civil Engineering, Stanford University, Stanford, Calif., 1960.

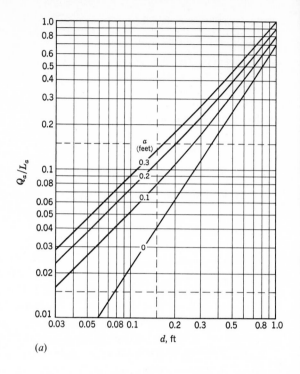

(a)

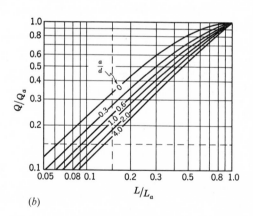

(b)

FIG. 18-4   Curves for computing the capacity of curb-opening inlets: (a) capacity in cubic feet per second per foot of length as a function of depth in approach gutter and depression of the flow line, (b) partial interception ratio. (U.S. Bureau of Public Roads)

ILLUSTRATIVE EXAMPLE 18-2   How much water will be intercepted by a curb-opening inlet 12 ft long in a gutter on a 0.05 slope when the depth at the curb is 0.3 ft and the transverse slope of the street is 0.03? Use $n = 0.016$ and gutter depression $a = 0.2$ ft.

From Illustrative Example 18-1, the gutter flow = 8.7 cfs.

From Fig. 18-4a, $Q_a/L_a = 0.23$; hence the inlet length required for full interception is $8.7/0.23 = 37.8$ ft.

From Fig. 18-4b with $L/L_a = 12/37.8 = 0.32$ and $a/d = 0.2/0.3 = 0.67$,

$$\frac{Q}{Q_a} = 0.47$$

Hence the intercepted flow is $0.47 \times 8.7 = 4.1$ cfs.

Note that in Illustrative Example 18-2 the interception rate for partial interception is 0.32 cfs per foot of inlet length, or about 44 percent greater than for complete interception. When a curb-opening inlet which is not ponded intercepts all the flow, the depth at its downstream end must be nearly zero. Similarly when a grated inlet intercepts all the flow, the depth at its street side must be very small. If a portion of the flow passes the inlet, the depth will be greater than for the condition of complete interception and both types of inlets will intercept more water.

18-6   Manholes   Manholes serve two main purposes in drainage systems. They provide access to the drain for cleaning, and they act as junction boxes for tributary drains. To serve these purposes, a manhole is required wherever a drain changes size, slope, or line; where tributary drains join a main line; and at intervals of not more than 500 ft along a line if the conduit is too small for a man to enter it. This latter spacing is established by the fact that 500 ft of jointed sewer-cleaning rod is about all that can be operated in a conduit.

Manholes are usually constructed of brick or concrete, and occasionally of concrete block or corrugated metal. The general design of brick or concrete manholes is shown in Fig. 18-5. The bottom of the manhole is usually of concrete with a half-round or U-shaped trough for the water. Tributary conduits intersecting above the grade of the main drain may be brought directly into the manhole and flow allowed to drop inside (Fig. 18-5a), or a drop may be constructed outside the manhole (Fig. 18-5b). The latter method is preferred where drops in excess of 2 ft occur in sanitary sewers to avoid splashing which might interfere with work in the manhole. Manholes are usually about 4 ft in diameter at the conduit, decreasing to 20 or 24 in. at the top. If the conduit is less than 4 ft in diameter, the manhole is usually centered over the sewer. For large sewers, the manhole will spring from one wall of the conduit (Fig. 18-5c).

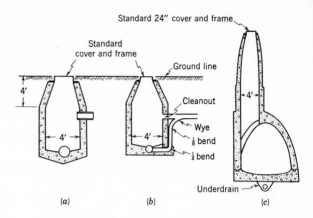

FIG. 18-5   Some typical sewer manholes: (a) inside drop, (b) outside drop, (c) manhole for large sewer.

Manhole covers and cover frames are usually of cast iron, and the combination will weigh 200 to 600 lb. The light frames and covers are used only where subject to negligible traffic loads, while the heavier combinations are employed on major highways. The cover is usually corrugated to improve traction. If a possibility of sewer-gas explosion or of flood waters backing up the sewer exists, some arrangement for bolting the cover in place is desirable.

18-7   Outlet works   In some instances, the terrane does not permit gravity flow of storm water to the point of discharge. In this case, pumping stations[1] will be a necessary part of the system. A small automatic pumping station is shown in Fig. 18-6. Water is collected in the wet well until the level rises and actuates the float switch which starts the motor. When the wet well is emptied, the switch turns the pump off. A system of this kind can be purchased as a complete unit and buried under a street or may be constructed as a major unit at the end of a large outfall.

Storm water is normally discharged into a stream, lake, or tidal estuary. If the level of the water body receiving the sewage varies, it may not always be possible to accomplish this discharge by gravity. In this case the outfall may be equipped with *tide gates* (Fig. 18-7), which operate like check valves to permit gravity flow at low stages in the receiving waters but close to prevent backflow when the elevation of the receiving water is high. Unless the drain itself has sufficient capacity to store the flow until the tide gates reopen, some special provisions must be made. These arrangements may

---

[1] Richard Hazen, Pumps and Pumping Stations, *J. New Engl. Water Works Assoc.*, Vol. 67, pp. 114–129, 1953.

consist of pumps capable of discharging the maximum expected flow against the maximum anticipated head, or they may consist of a storage basin large enough to store the maximum volume of flow expected during a period when gravity discharge is impossible.

A storage basin is often feasible where the discharge is into a tidal estuary in which fluctuations are regular, and the necessary detention time is relatively short. If discharge is into a river which may be subject to high stages for several days, it may be difficult to find adequate storage to meet requirements. On the other hand, a pumping plant capable of discharging the maximum flow may be quite expensive and the standby charges for electric power may be high, even though the plant operates infrequently. The most common alternative is a combination of pumping and storage which permits the use of smaller pumps while the storage absorbs the flow in excess of pump capacity. This arrangement permits the pumps to operate at a more uniform rate and hence more efficiently. The design of discharge works of this type

FIG. 18-6   A small automatic pumping station. (Economy Pumps, Inc.)

involves a study of the joint frequency of storm flows within the drainage system coincident with high water in the water body receiving the storm flow (Sec. 5-9).

18-8   Design of storm drains   The design of storm drains conforms to the principles of flow in open channels. The main problem is the selection of economic pipe sizes and slopes capable of carrying the expected flow. The main rules governing selection of pipe size and slope are:

1. The pipe is assumed to flow full under conditions of steady, uniform flow.
2. To avoid clogging, the minimum pipe diameter should preferably be 10 or 12 in., although 8-in. pipes are used in some cities.
3. The minimum velocity flowing full should be at least 2.5 ft/sec.
4. Pipe sizes should not decrease in the downstream direction even though increased slope may provide adequate capacity in the smaller pipe. Any debris which enters a drain must be carried through the system to the outlet, and the possibility of clogging a smaller pipe with debris which may pass a larger pipe is too great.
5. Pipe slope should conform to the ground slope insofar as possible for minimum excavation. In some cases it may be possible to use a smaller pipe by exceeding the ground slope. If this makes the use of smaller pipe possible for some distance downslope, it may be economical despite increased excavation.
6. Pipe grades are described in terms of the elevation of the invert, or inside, bottom of the pipe. Where pipes of different size join, the tops of the

FIG. 18-7   Tide gates. (Armco)

pipes are placed at the same elevation, and the invert of the larger pipes is correspondingly lower than that of the smaller pipe. This does not apply to tributary drains, which may enter the main sewer through a drop manhole.

The application of these rules and the design criteria of earlier sections is best illustrated by the following example.

ILLUSTRATIVE EXAMPLE 18-3   Design a storm-drain system for the area shown on Fig. 18-8 above the intersection of Maple Avenue and Bay Drive. Use the rainfall curves for Cleveland, Ohio (Fig. 5-7), and a return period of 5 yr. Assume that lots are graded to a 2 percent slope toward the east-west streets. Blocks are 750 by 300 ft. Streets are 40 ft wide. The area is developed as single-family residences. Allow 5-ft cover on crown of buried pipes. Streets have a 5 percent transverse slope and a 6-in. curb. Manning's n for gutters is 0.016 and for drains is 0.013.

From Eq. (3-11) with $L_o = 150$ ft and $k = 0.3$ (Table 3-1)

$$t_c = \frac{41bL_o^{\frac{1}{3}}}{(ki)^{\frac{2}{3}}} = \frac{41 \times 5.32b}{0.45i^{\frac{2}{3}}} = \frac{485b}{i^{\frac{2}{3}}}$$

Since i is a function of t_c, a trial-and-error solution is necessary to determine t_c. Assume

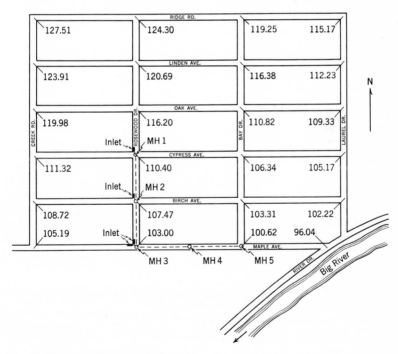

FIG. 18-8   Problem area for storm-drain design.

$t_c = 40$ min; then $i = 1.30/40$ min $= 1.95$ in./hr (Fig. 5-7). From Eq. (3-12) with $c_r = 0.046$

$$b = \frac{(0.0007 \times 1.95) + 0.046}{0.02^{1/3}} = 0.174$$

and

$$t_c = \frac{485 \times 0.174}{1.95^{2/3}} = 54 \text{ min}$$

This is not a good check, and as a second trial assume $t_c = 60$ min. Then $i = 1.60$ in./hr, and from Eq. 3-12

$$b = \frac{(0.0007 \times 1.60) + 0.046}{0.27} = 0.174$$

and

$$t_c = \frac{485 \times 0.174}{1.60^{2/3}} = 61.5 \text{ min}$$

This is a sufficiently close check on the assumed value of 60 min.

For the purpose of computing runoff, the area of a block including one-half the street on all sides is

$$\frac{790 \times 340}{43,560} = 6.16 \text{ acres}$$

Equation (10-5) with transverse slope of 5 percent and $n = 0.016$ becomes

$$Q = 570 S^{1/2} d^{8/3}$$

and the velocity of flow in the gutter is

$$V = \frac{Q}{A} = \frac{Q}{10d^2}$$

The remainder of the computation is shown in Table 18-1. Since flow is to be carried in the gutters until their capacity is exceeded, the first several lines deal with gutter flow. Columns 1, 2, and 3 identify the block under study. Elevations in cols. 4 and 5 are taken from the map, and their differences are entered in col. 6. The distances in col. 7 are measured from the center of the intersections. Slopes in col. 8 are computed by dividing col. 6 by col. 7. The tributary area (col. 9) is computed on the assumption that all runoff is to the east-west streets. Thus for the first and second lines only one-half block contributes. The concentration time (col. 10) is equal to the initial t_c of 60 min for overland flow plus the time of flow in the pipe or gutter (col. 17). The rainfall amount corresponding to the concentration time is read from Fig. 5-7 and divided by the concentration time *in hours* to obtain values of intensity (col. 11). Peak discharge (col. 12) is computed with the rational formula [Eq. (3-10)] with $k = 0.3$. Columns 13 to 15 are the solution for the depth of gutter flow, and the gutter velocity (col. 16) is computed from peak Q and d. Gutter times in minutes (col. 17) are found by dividing the distance (col. 7) by the velocity (col. 16) and dividing again by 60.

TABLE 18-1 Calculations for the Design of a Storm-drain System

Street (1)	From (2)	To (3)	Ground elevation, ft		Differ-ence (6)	Dis-tance, ft (7)	Slope, ft/ft (8)	Tribu-tary area, acres (9)	Concen-tration time, min (10)	Rain inten-sity, in./hr (11)	Peak Q, cfs (12)	$S^{1/2}$ (13)	$d^{8/3}$ (14)	d (15)	Pipe or gutter V, ft/sec (16)	Time, min (17)	Pipe diam, in. (18)	Invert eleva-tion, ft	
			Upper (4)	Lower (5)														Upper (19)	Lower (20)
Ridge	Creek	Rosewood	127.5	124.3	3.2	790	0.004	3.08	60.0	1.60	1.47	0.063	0.041	0.30	1.63	8.1			
Rosewood	Ridge	Linden	124.3	120.7	3.6	340	0.011	3.08	68.1	1.42	1.31	0.105	0.022	0.24	2.27	2.5			
Rosewood	Linden	Oak	120.7	116.2	4.5	340	0.013	9.24	70.6	1.37	3.80	0.114	0.061	0.35	3.09	1.8			
Rosewood	Oak	Cypress	116.2	110.4	5.8	340	0.017	15.40	72.4	1.36	6.29	0.130	0.089	0.40	3.93	1.4			
Rosewood	Cypress	Birch	110.4	107.5	2.9	340	0.009	21.56	73.8	1.33	8.61	0.095	0.166	0.51	Exceeds gutter capacity				
Rosewood	Cypress	Birch	110.4	107.5	2.9	340	0.009	15.40	73.8	1.33	6.15	...	...	...	5.0	1.1	15	104.15	101.25
Rosewood	Birch	Maple	107.5	103.0	4.5	340	0.013	21.56	74.9	1.32	8.54	...	...	...	7.0	0.8	18	101.00	96.50
Maple	Rosewood	Bay	103.0	100.6	2.4	790	0.003	30.80	75.7	1.32	12.18	...	...	...	4.0	3.3	24	96.00	93.60

Between Cypress and Birch on Rosewood, the accumulated flow exceeds the gutter capacity. Hence it is necessary to provide a drop inlet at the intersection of Rosewood and Cypress and to provide sewers to carry the flow beyond this point. In selecting the pipe sizes, cols. 1 through 12 of the table are the same as for gutter flow. Column 18 (pipe diam) is taken from Fig. 10-2, using the slope from col. 8 and the peak flow from col. 12. Pipe velocity (col. 16) is read from the nomogram, assuming the pipe to be flowing full. Time of flow in the pipe is computed as for the gutter. Invert elevations of the pipe are equal to the ground elevations minus $5 + D$, where D is the pipe diameter in feet. Manholes and drop inlets would be provided as shown on the figure.

Assuming a gutter depression of 0.2 ft at the inlets, a side-opening inlet at Cypress and Rosewood would intercept 0.4 cfs/ft for full interception and would have to be $6.15/0.4 = 15.4$ ft long. A shorter inlet might be used and some of the flow permitted in the gutter to the next intersection. In this case a smaller pipe would suffice. A similar calculation might be made for the other inlets.

To make sure that the gutter capacity of the east-west streets is not exceeded, it may be noted that the peak flow per block is $0.3 \times 1.60 \times 3.08 = 1.48$ cfs. From Eq. (10-5) with $d = 0.5$, a slope of $[1.48/(570 \times 0.157)]^2 = 0.00028$ ft/ft is required. This is a difference in elevation of 0.21 ft in 790 ft. By inspection the difference in elevation on all east-west streets between Creek Road and Rosewood Drive is greater than this minimum value. If this had not been the case, an intermediate inlet in mid-block would have been appropriate.

The effect of gutter and conduit storage has not been considered in this computation. It has also been assumed that Creek Road would carry none of the storm flow, although some is sure to use this route. These and other factors are on the conservative side. The computation should be repeated assuming contribution from the impervious area only. For half width of a street ($L = 20$ ft, $S = 0.05$), $t_c = 1$ min. Thus although the area is much less, rainfall intensities are much greater and, at least for the smaller areas, the peaks may be higher. The initial time of concentration should be taken as t_c plus the flow time in the gutter.

LAND DRAINAGE

Land drainage removes excess surface water from an area or lowers the groundwater below the root zone to improve plant growth or reduce the accumulation of soil salts. Land-drainage systems have many features in common with municipal storm-drain systems. Open ditches, which are less objectionable in rural areas than in cities, are widely used for the drainage of surface water, at a considerable saving in cost over that of buried pipe. Under suitable soil conditions ditches may also serve to lower the water table. However, closely spaced open ditches will interfere with farm operations, and the more common method of draining excess soil water is by use of buried drains. Drains usually empty into ditches, although the modern tendency is to use large pipe in lieu of ditches where possible. This frees extra land for cultivation and does away with unsightly and sometimes dangerous open ditches. Since land drainage is normally a problem in very flat or leveed land, a disposal works provided with tide gates and pumping equipment is often necessary for the final removal of the collected water.

18-9 Design flows for land drainage Land drainage is not as demanding in terms of hydrologic design as other types of drainage. The purpose of land drainage is to remove a *volume* of water in a reasonable time. Where subdrainage is installed to remove excess water from irrigated land for salinity control, the volume of leaching water to be applied in each irrigation is known; and the drains should be capable of removing this volume in the interval between irrigations.

Drainage installed for removal of excess water from rainfall is typically designed to remove a specified quantity of water in 24 hrs. Commonly known as the *drainage modulus* or *drainage coefficient*, recommended values are about 1 percent of the mean annual rainfall. The drainage modulus as defined above will usually be in the range of one-quarter to one-half of the 1-yr, 24-hr rainfall and may be taken as an estimate of the infiltration from the 1-yr, 24-hr rain less the water required to bring the soil to field capacity. If the drainage modulus for irrigated lands exceeds the capacity defined by the estimated quantities of leaching water, the higher value should be used. Commonly, however, the minimum tile size (usually 4 in.) will be more than adequate to handle the design flow.

Land protected by levees is often subject to excess water as a result of seepage from the river through or under the levee. The design flow for a drainage system in this case should be based on estimated seepage rates at river stages having a return period of 1 or 2 yr. If the high stages are expected to be concurrent with the rainy season the drainage modulus should be added to the seepage.

18-10 Drainage ditches A ditch-drainage system consists of laterals, submains, and main ditches. Ditches are usually unlined. Small ditches may be constructed with special ditching machines, while larger ditches are often excavated with a dragline. Some very large ditches are constructed with floating dredges. Unless the excavated material (spoil) is needed as a levee to provide additional flow area in the ditch, it should be placed at least 15 ft back from the edge of the ditch so that its weight does not contribute to the instability of the ditch bank. Spoil banks decrease the cultivated area and prevent inflow of water from the land adjacent to the ditch. If possible, this spoil should be spread in a thin layer, but where this cannot be done, openings should be left in the spoil bank wherever a natural drainage channel intersects the ditch and at least every 500 ft along the ditch.

The slope available for drainage ditches is small, and the cross sections should approach the most efficient section as closely as possible. A trapezoidal cross section is most common, with side slopes not steeper than 1 on 1.5. Slopes of 2:1 or 3:1 are required in sand or clay soils with little cohesion. Occasionally, where the drainage water must be pumped, ditches are deliberately made with an inefficient section to create as much storage as possible to minimize peak pumping loads. The slope, alignment, and spacing of ditches are determined mainly by local topography. The minimum practicable slope is about 0.00005 (3 in./mi). The ditches generally follow natural depressions, but where possible are run along property lines. Lateral ditches rarely need to be spaced more closely than $\frac{1}{2}$ mi apart, although on level land a spacing of $\frac{1}{4}$ mi may be necessary. With favorable land slopes a ditch spacing of 1 mi may be feasible.

Ditches are usually between 6 and 12 ft deep. Where tile drainage is to be used, the lateral ditches must be deep enough to intercept the underdrains which are to discharge into it. Similarly submain and main ditches must be deep enough to receive the flow of lesser ditches. If the terrane is flat and the ditches quite long, an excessive depth may be required for the main ditch and it may be advisable to divide the system into two or more portions to shorten ditch lengths. In muck or peat soil, considerable subsidence may occur when the water is drained from the soil, and ditches must be constructed proportionately deeper. In the Florida Everglades, subsidence reduced the depth of lateral ditches from 8 to 5.5 ft within 4 yr of their completion.

Ditch bottoms at junctions should be at the same elevation to avoid drops which may cause scour. This may require some steepening of the last 100 ft or more of laterals before they join a submain. Right-angle junctions encourage local scour of the bank opposite the tributary ditch, and the smaller ditch should be designed to enter the larger at an angle of about 30°. Scour will also occur at sharp changes in ditch alignment, and long-radius curves should be used where a change in line is necessary. With high flows or steep slopes the radius of curvature should be 1200 ft or more, while for small ditches on flat slopes a radius of 300 to 500 ft may be satisfactory. Easement curves are sometimes used on important ditches.

The most important factors controlling the value of Manning's n for drainage ditches are the neatness with which the ditch is constructed and the extent to which vegetation is permitted to grow in the ditch. If brush and weeds are cut and burned annually, a value of $n = 0.040$ is typical for a well-formed channel. If the channel section is very irregular or if several years' growth of weeds and brush develops, n may be 0.100.

18-11 Underdrains Unglazed clay-tile or concrete pipe are the most common materials for underdrains,[1] although wooden box drains and

---

[1] John G. Sutton, Installation of Drain Tile for Subsurface Drainage, *J. Irrigation Drainage Div., ASCE*, pp. 27–47, September, 1960.

perforated steel pipes have also been used. Plain pipes without special joints are used and are placed with the ends of adjacent pipes butted together. Water enters the drain through the space between the abutting sections. The space should not be so large as to permit soil to enter and clog the drain. If the pipe ends are so irregular as to leave $\frac{1}{8}$-in. openings, the joints should be covered with pieces of broken tile, tar paper, or other material to prevent the entry of soil.

Two-inch pipe was widely used in early drainage projects but modern practice favors the use of a 4-in. minimum, and some projects have 6 in. as minimum pipe diameter. The larger size of pipe minimizes the likelihood of clogging with sediment, roots, or other materials. The slope for 4-in. tile should be not less than about 0.2 percent to provide a velocity of about 1 ft/sec when flowing full to avoid sediment deposits in the pipe.

Tile drains are usually spaced 50 to 150 ft apart. Only the most permeable soils permit the wider spacing, while a spacing under 50 ft will usually be too expensive. Where a very close spacing is required, the use of *mole drains* is sometimes feasible. Mole drains are tunnels formed in cohesive soil by pulling a steel ball through it (Fig. 18-9). If the soil is sufficiently cohesive, mole drains may remain effective for many years. Perforated flexible plastic tubing can be installed with a mole to maintain the drain if the added cost can be justified by the longer drain life. Mole drains have been spaced as closely as 5 ft apart in soil of very low permeability.

If the land slope is such that drains can parallel the surface, there is no limit to drain length except that imposed by the topography. On level ground a drain on the minimum slope of 0.002 will drop 1 ft in 500 ft, and this distance becomes about the maximum permissible length. On steeper slopes drains are rarely more than 2000 ft long.

18-12 Flow of groundwater to drains The flow of groundwater to a tile drain or ditch is governed by the same factors controlling the flow to a well. Both ditches and drains create a water table like that shown in Fig.

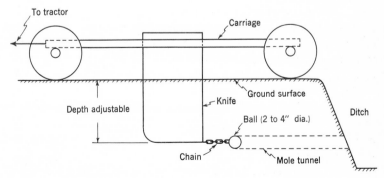

FIG. 18-9 Schematic diagram of a mole-drain machine.

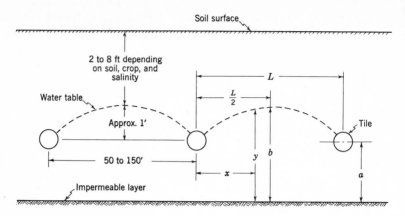

FIG. 18-10   Groundwater in a drained field.

18-10. The cone of depression about a well becomes a trough along the line of the drain. The spacing of drains must be such that the water table at its highest point between drains does not interfere with plant growth. The water table should be below the root zone to permit aeration of the soil. A high water content also reduces soil temperature and retards plant growth. The necessary depth to the water table depends upon the crop, the soil type, the source of the water, and the salinity of the water. In humid regions, a depth of 2 to 3 ft is satisfactory for most crops except orchards, vineyards, and alfalfa, which require a depth of 3 to 5 ft. In general, the water table should be lowered more in heavy clay soils than in light sandy soils. For land under irrigation the depth to the water table should be greater, 3 to 5 ft for ordinary crops and good-quality water, and 5 to 8 ft for deep-rooted crops. If the water contains excessive salts, the minimum depth for any crop should be about 4 ft. It is not desirable to lower the water table far below the recommended minimum depth, for this deprives the plants of capillary moisture needed during the dry season. Where deep ditches are part of the drainage system, check dams should be provided to maintain the water level in the ditch during low flow periods in order to avoid excessive lowering of the water table near the ditch. Drains in peat and muck soils must be installed somewhat deeper than in other soils because of the settlement which takes place as water is removed from the peat.

The variation in water-table elevation in the drained land should be not much over 1 ft. This means that drains should be placed about 1 ft below the desired maximum groundwater level. The spacing of the drains must be such that, with the hydraulic gradient thus established (approximately 1 ft in half the drain spacing), the drains will remove sufficient water from the soil. In homogeneous soil the spacing can be determined by an analysis similar to that for a well in a homogeneous aquifer. In Fig. 18-10 the hydraulic gradient at distance x from the drain is dy/dx, and assuming the flow lines are parallel for a unit length of drain the cross-sectional area of flow is y. Assuming that the flow q toward the drain is inversely proportional to distance from the drain, i.e., when $x = L/2$, $q = 0$ and when $x = 0$, $q = \frac{1}{2}q_D$, where q_D is the design flow per foot of drain, the Darcy equation (Sec. 4-7) for the horizontal flow at x may be written

$$q = \frac{2}{L}\left(\frac{L}{2} - x\right)\frac{q_D}{2} = Ky\frac{dy}{dx} \tag{18-3}$$

where L is the drain spacing and K the coefficient of permeability. Transforming this equation,

$$\frac{q_D}{2KL}(L - 2x)\,dx = y\,dy \tag{18-4}$$

and integrating,

$$\frac{q_D}{2KL}\left(Lx - \frac{2x^2}{2}\right) = \frac{y^2}{2} + C \tag{18-5}$$

When $x = 0$, $y = a$ and hence $C = -a^2/2$. Substituting this value for C in Eq. (18-5) and solving for K,

$$K = \frac{q_D(Lx - x^2)}{L(y^2 - a^2)} \tag{18-6}$$

When $x = L/2$, $y = b$; hence,

$$L = \frac{4K(b^2 - a^2)}{q_D} \tag{18-7}$$

Equation 18-7 is known as the Donnan equation.

The assumption of an inverse linear variation in q with distance from a drain is only an approximation. However, Aronovici and Donnan[1] show good agreement between Eq. (18-7) and field tests in relatively permeable soils. The lower the soil permeability, the greater the curvature of the water table near the drain and the less accurate the assumption. Dagan[2] and Hammad[3] present more sophisticated methods for determining depth and

[1] V. S. Aronovici and W. W. Donnan, Soil Permeability as a Criterion for Drainage Design, *Trans. Am. Geophys. Union*, Vol. 27, pp. 95–101, February, 1946.

[2] G. Dagan, Spacing of Drains by an Approximate Method, *J. Irrigation and Drainage Div.*, *ASCE*, Vol. 90, pp. 41–66, March, 1964.

[3] H. Y. Hammad, Depth and Spacing of Tile Drain Systems, *J. Irrigation and Drainage Div.*, *ASCE*, Vol. 88, pp. 15–36, March, 1962.

spacing of drains which would be more appropriate for soils of low permeability. Hammad also discusses the effect of evaporation from the soil which may reduce the drainage requirement. If no restricting layer is present in the soil Sutton[1] suggests that a be taken equal to the tile depth.

The most important single factor in determining drain spacing is the soil permeability, which may be determined by methods of Chap. 4. Since drainage normally involves only a shallow depth of soil, small-diameter piezometers or open auger holes may serve as test "wells." It is convenient to calculate the permeability from the rate of rise of water in such a hole after a portion of the water is pumped out.[2]

Ditches, because of their larger dimensions, are somewhat more effective than pipe drains in removing soil moisture. However, the increased effectiveness is not sufficient to permit ditch spacing which will not interfere with farm operations. Moreover, ditches considerably reduce the area available for farming. Hence, open ditches for removal of soil water are used only infrequently, and then only in the most permeable soils.

18-13 Layout of a tile-drain system The plan for a system of tile drains is determined largely by the topography of the area. Several possible arrangements are shown in Fig. 18-11. If only isolated portions of the area require drainage, a pattern similar to the natural stream system is most economical. This *natural system* (*a*) is used in rolling topography where drainage is necessary only in small swales and valleys. If the entire area is to be drained, a *gridiron layout* (*b*) is usually more economical. Laterals enter the submain from one side only to minimize the double drainage which occurs near the submain. The gridiron system might be used where the land is practically level or where the land slopes away from the submain on one side. If the submain is laid in a depression, the *herringbone pattern* (*c*) is used. The laterals join from each side alternately. The land along the submain is double-drained but since it is in a depression, it probably requires more drainage than the land on the adjacent slopes. If the bottom of the depression is wide, a *double-main system* (*d*) is often used. This reduces the lengths of the laterals and eliminates the break in slope of the laterals at the edge of the depression. If the main source of excess water is drainage from hill lands, an *intercepting drain* (*e*) along the toe of the slope may be all that is required to protect the bottom land. Roots of trees can easily enter the open joints of underdrains. If possible, only the ends of the laterals (*f*) should be exposed to this hazard. Mains and submains should be kept well away

---

[1] J. G. Sutton, Installation of Drain Tile for Subsurface Drainage, *J. Irrigation and Drainage Div., ASCE*, Vol. 86, pp. 27–50, September, 1960.

[2] C. H. M. Van Bavel and Don Kirkham, Field Measurement of Soil Permeability Using Auger Holes, *Proc. Soil Sci. Soc. Amer.*, Vol. 13, pp. 90–96, 1948.

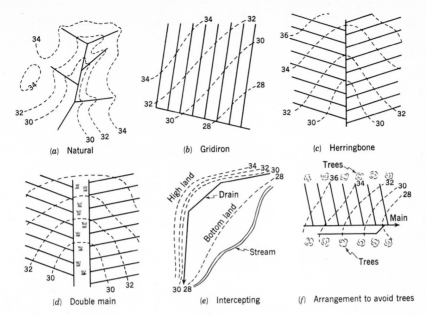

FIG. 18-11   Some arrangements of tile drains.

from trees, or bell-and-spigot joints with joint filler should be used if the drain must run close to trees. Drainage systems often suffer considerable damage from the loads imposed by farm vehicles. Consideration should be given to the loads (Sec. 11-16) in the selection of pipe for drainage.[1] A special heavy-duty grade of clay tile has been developed for severe conditions.

18-14   Design of a land-drainage system   The basic procedure for the design of land-drainage systems is not much different from that for the design of municipal storm drains. A typical farm layout is shown in Fig. 18-12. The steps in design may be summarized as follows:

1. Prepare a detailed contour map of the area. A contour interval of 1 ft is commonly necessary.
2. Select the location of the system outlet. If several outlets are possible, an economy study of the alternatives may be necessary.
3. Determine the drainage modulus for underdrains, and estimate the amount of water the ditches will intercept.
4. Lay out a system of ditches (or pipe mains) of adequate size to carry the expected flows.

---

[1]J. G. Sutton, Installation of Drain Tile for Subsurface Drainage, *J. Irrigation and Drainage Div., ASCE*, Vol. 86, pp. 27–50, September, 1960.

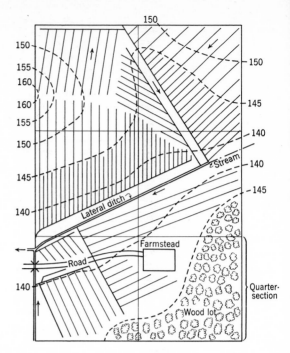

FIG. 18-12   Farm drainage plan.

5. Determine the proper depth for tile drains, and plan the tile-drain layout. Field drains of the customary minimum size (4 in.) will usually have adequate capacity, and it will be necessary to calculate only required sizes for mains and submains.

6. The first trial layout of mains may require revision after the plan for underdrains is completed. The entire system should be planned for minimum cost by use of shortest possible routes for pipes and ditches.

7. Estimate project costs, and proceed with legal steps necessary to undertake the project.

18-15   Drainage by vertical wells   Under some conditions vertical wells may be an effective means of draining land. If a very impermeable subsoil is underlain with permeable sands and gravels, it may be practical to penetrate the impermeable stratum with a well which will permit water to drain from the surface soil into the underlying gravels. A number of small wells may be used (Fig. 18-13a), or a single large well may take the discharge of radial tile lines (Fig. 18-13b). The well may be left open and cased with tile or brick, or it may be filled with gravel or crushed rock.

A second method of lowering the water table is to create a cone of depression by pumping from a well. Because of the pumping costs this method is at a disadvantage when compared with gravity flow of the usual drainage system. Pumped wells are practical only when (1) the surface soil is quite deep and highly permeable so that the effect of the well extends to a large radius, (2) the pumped water can be used, and (3) gravity drainage is impossible or unusually costly. Pumped wells are particularly effective when used to intercept a large groundwater flow approaching an area from an outside source. The design of wells for drainage is the same as for the production of water.

18-16   Legal aspect of drainage   Incorporated cities usually have the necessary legal authority to construct drainage works and to finance them by taxation, bond issue, or special assessment. No comparable authority exists in unincorporated areas, and drainage projects must usually be organized as a drainage district in accordance with state law. Some states permit *mutual drainage districts*, which are formed by the unanimous consent of the landowners involved. All that is usually required is the signing of an agreement describing the work to be done and setting forth the division of cost. The agreement is legally recorded and a board of commissioners formed to direct the work.

Most large drainage districts are *majority-rule districts.* Such districts

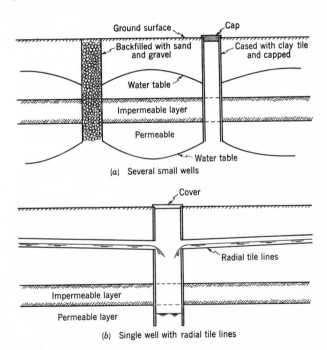

(a)   Several small wells

(b)   Single well with radial tile lines

FIG. 18-13   *Drainage by vertical wells through a thin, impermeable layer.*

are usually initiated by the filing of a petition having the signatures of a specified number of landowners with the clerk of the county court or the board of supervisors. Public notice of the petition is given, and a hearing is held. If the project seems reasonable, commissioners are appointed to supervise the preparation of detailed plans and an assessment roll showing the charges to the various landowners. The commissioners' report is then given a public hearing. In some states the legal authority (court or supervisors) can approve the report and permit work to proceed. In other states a majority vote of the affected landowners is required to approve or reject the plan. The procedure is prescribed by state statutes and varies from state to state. The statutes usually limit the amount of the assessments and specify the maximum repayment period.

Assessments are not taxes in the usual sense but are enforced charges levied by a competent authority to pay for benefits which accrue to the land. A fundamental premise is that the assessments shall not exceed the benefits. The best basis for determining the assessments is usually the increase in value of the land as a result of the drainage works. Some states provide, however, that the land shall be classified into five or more classes according to the degree of benefit received and that the assessment for land in each class shall be the same and in proportion to the benefits received. In states adhering to the Roman civil law, an owner of land having adequate natural drainage cannot be assessed against his will for a drainage district organized under statute law. Under the common-enemy rule, all lands lying within a district are assessable.

HIGHWAY DRAINAGE

Highways occupy long, narrow strips of land and pose two types of drainage problems. Water collecting on the roadway (or on the adjacent land slopes if the road is in cut) must be collected and disposed of without flooding or damaging the highway and adjacent areas. Highways cross many natural drainage channels, and the water carried by these channels must be conveyed across the right of way without obstructing the flow in the channel upstream of the road and causing damage to property outside of the right of way.

18-17 Design flows for highway drainage For the design of longitudinal drainage—roadside ditches and spillways—the design flows are best calculated using Izzard's method (Sec. 3-11) to calculate a time to equilibrium for the transverse flow across the pavement. The peak flow per unit length of pavement (cfs/ft) is given by

$$q = \frac{iL}{43,200} \tag{18-8}$$

where i is rainfall intensity in inches per hour for the duration t_c and desired return period and L is the length of overland flow in feet normal to the contours or

$$L \approx \frac{w\sqrt{r^2 + 1}}{r} \tag{18-9}$$

where w is the width of pavement and r is the ratio of cross slope to longitudinal slope. For short distances and steep slopes the length of pavement multiplied by q can be taken as the design flow for the ditch. For long ditches and flat slopes an allowance for channel storage should be made, possibly by routing with the Muskingum method.

No wholly satisfactory method exists for calculation of design flow for cross-drainage structures—culverts and bridges. For large streams with a gaging station, frequency analysis of the observed peaks will be satisfactory. Only a few of the crossings will be over gaged streams and thus most estimates must be for the flow of ungaged streams. The most widely used procedure is the rational formula, a procedure which is grossly inadequate for the purpose.[1] Regional flood frequency (Sec. 5-5) is a much better approach for basins in the size range from 20 sq mi upward. So few data are available for smaller basins that this method is of doubtful accuracy. Digital simulation methods should provide maximum accuracy for basins of all sizes, but costs would be prohibitive on an individual basin basis. Special adaptations on a regional or state basis could be developed to bring digital simulation within reasonable costs.

The thrust of research for hydrologic design methods for culverts has been to obtain extremely simple methods on the assumption that culverts do not warrant extensive effort in determination of design flows. Actually overdesign may lead to substantial excess costs of construction for culverts, especially for freeways where culverts are necessarily long. Underdesign may result in damages to the culvert and roadway and, by backing up storm flows, may cause damages to property upstream of the right-of-way. Thus substantial savings could result from use of reliable hydrologic methods.

The proper design frequency for culverts should be chosen by the method of Sec. 13-7. The practical difficulty is that of estimating the probable damages resulting from flows in excess of culvert capacity. Except on major highways and in locations with special problems, the design return period should probably be relatively short[2] but no general standards are appropriate

---

[1] L. A. V. Hiemstra and B. M. Reich, Engineering Judgement and Small Area Flood Peaks, *Colo. State Univ. Hydrology Paper* 19, April, 1967.

[2] H. Pritchett, Application of Principles of Engineering Economy to the Selection of Highway Culverts, *Eng.-Econ. Planning Pub.* 13, Department of Civil Engineering, Stanford University, August, 1964.

because of the wide variety of physical and hydrologic situations encountered.

18-18 Longitudinal drainage Unless storm water resulting from rainfall on the highway right of way and tributary slopes is properly controlled, severe erosion of the shoulder and cut slopes may create a costly maintenance problem and dangerous driving conditions. In humid regions erosion can often be controlled by establishing vegetative cover on slopes of cut and fill and along shoulders. In this case the water should be spread out as much as possible since erosion increases with some power of velocity which in turn increases with depth. A good grass cover can withstand velocities up to about 8 ft/sec. Bermuda grass is typical of turf-forming grasses offering good protection. Grass will not withstand immersion for long periods, and slopes of at least 0.005 should be provided so that water will drain off rapidly.

In arid regions where natural rainfall is insufficient to maintain a vegetal cover or in humid regions where the soil is easily eroded, other means of protection are necessary. Slopes of cuts may be protected by an intercepting dike or ditch at the top of the slope (Fig. 18-14). In long cuts the dike should approximately follow a contour and, if possible, should divert the water into a natural drainage way without allowing it to reach the roadway. If it is necessary to direct the water to the roadway ditch before the end of the cut, a paved spillway section down the cut slope will be necessary. Contour furrows at intervals above the top of the cut slope will retard overland flow, and the landowner may benefit from the additional moisture retained on the land.

Roadside ditches are usually constructed in a shallow V shape since this section is easily maintained with graders, is less hazardous to vehicles, and permits the shallow flow necessary to avoid erosion. The ditch should be large enough to carry the design flow with a freeboard of 3 to 6 in. The flow line of the ditch is sometimes placed low enough so that underdrainage from the highway base course may enter the ditch, although where the base course and underlying soil are of low permeability, very little underdrainage can be expected.

Bare earth will withstand velocities between 1 and 4 ft/sec, depending upon the soil type. If the computed velocity for the design flow is too high for the soil, protection may be provided by lining the ditch with asphalt, concrete, dry rubble, or transplanted sod. An alternative to lining the ditch throughout its entire length is drop structures at intervals which permit the remainder of the ditch to be maintained on a grade sufficiently low to prevent excessive velocities. Drop structures may prove to be a traffic hazard. A third alternative is to divert the flow to a natural drainage way by use of a paved chute across the side of the fill before the accumulated runoff in the

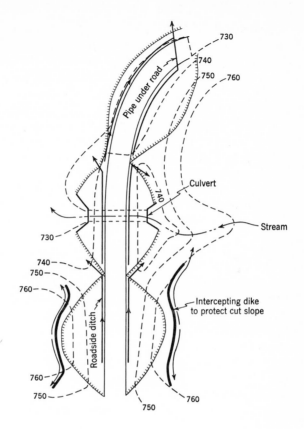

FIG. 18-14 Longitudinal drainage plan for a highway.

ditch is sufficient to cause harmful velocities. Drop inlets discharging the flow into a pipe which carries it to the downstream side of the fill are preferred for intercepting the flow in the uphill ditch unless the intercepted water can be discharged directly into a culvert on the upstream side of the roadway. If interception is necessary in long cuts, a drop inlet discharging into a buried pipe running parallel to the highway is required until a point is reached where the flow can be directed into a natural drainage way.

Depressed highways and underpasses sometimes create a difficult drainage problem. Occasionally the flow which collects at the bottom of an underpass can be carried away by gravity to a storm drain or natural channel. Often, however, a small automatic pumping station set in a sump beneath the roadway is required to lift the water to a point where it can be carried away by gravity flow.

18-19 Cross drainage Since highways cross many natural drainage

channels, provision for carrying this water across the right of way is necessary. It should be noted that a considerable saving is often possible by
locating a highway along ridge lines and eliminating channel crossings.
Since an average of one-fourth of highway-construction costs is for drainage
structures, some saving may be realized even though the ridge route involves
greater length and less satisfactory grades and alignment.

Cross drainage is accomplished with *culverts*, *bridges*, and *dips*. Culverts
and bridges carry the roadway over a stream, and the distinction between
the two types of structures is mainly on the basis of size. Frequently structures with a span in excess of 20 ft are classed as bridges, while structures of
shorter span are called culverts. This is arbitrary, and from the hydraulic
viewpoint it might be more proper to include with culverts all structures
designed to flow with submerged inlet. A dip is a depression of the roadway
so that flow from a cross channel may pass over the road.

18-20   Culverts   The essential features of a culvert are the *barrel*
which passes under the fill; the *headwalls* and *wingwalls* at the entrance
and *endwalls* or other devices at exit to improve flow conditions and prevent
embankment scour; and in some cases *debris protection* to prevent entrance
of debris which might clog the culvert barrel.

Culvert barrels are made of a variety of materials, the main basis of
selection being the cost of the installed culvert. Small culverts may be made
from precast-concrete, vitrified-clay, cast-iron, or corrugated-steel pipe. In
the larger sizes corrugated steel arches, reinforced-concrete arches, or concrete box culverts are usually selected. In some instances rubble masonry
and treated timber have been employed for culverts. Minimum diameter for
culverts should be about 18 to 24 in., although 12-in. pipe might be used for
small flows carrying no debris or trash.

Wherever possible, the culvert barrel should follow the line and grade
of the natural channel (Fig. 18-15a). The length of the conduit can be

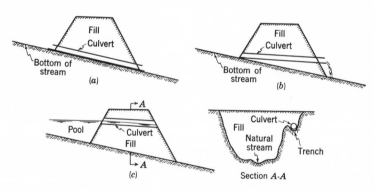

FIG. 18-15   *Various culvert locations.*

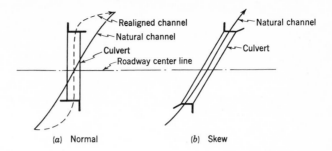

FIG. 18-16   *Typical culvert alignments.*

shortened by raising the outlet above the natural channel bottom (Fig. 18-15*b*), but this requires protection for the downhill fill slope and places the culvert entirely in fill. This location is used mainly for small corrugated-pipe culverts which can be cantilevered out on the downstream end so that no part of the flow strikes the fill, even at low discharges. If the inlet elevation is increased by use of the sidehill location (Fig. 18-15*c*), the culvert may be materially shortened but the embankment must act as a dam and a stagnant pool is formed on the upstream side of the fill. With such an arrangement the culvert is usually placed in a trench or natural bench on the side of the ravine so that it rests on undisturbed earth for most of its length.

It is common to place the culvert axis normal to the highway center line (Fig. 18-16*a*) even though this requires some changes in the natural channel. The alternative is a skew culvert (Fig. 18-16*b*), which will be longer than the normal culvert and will require more complex construction of headwalls and endwalls. However, when the skew of the stream is large, a culvert normal to the alignment of the road creates bends in the channel which may be points of erosion. Moreover, the hydraulic capacity of the culvert will be decreased by the poor entrance and exit conditions. As a general rule, it is best that culvert alignment conform to the natural stream alignment.

18-21   Culvert inlets and outlets   Several entrance structures are illustrated in Fig. 18-17. Entrance structures serve to protect the embankment from erosion and, if properly designed, may improve the hydraulic characteristics of the culvert. The straight endwall (*a*) is used for small culverts on flat slopes when the axis of the stream coincides with culvert axis. If an abrupt change in flow direction is necessary, the L endwall (*b*) is used. The U-shaped endwall (*c*) is the least efficient form from the hydraulic viewpoint and has the sole advantage of economy of construction. Where flows are large, the flared wingwall (*d*) is preferable. The angle of flare is not critical in the case of inlet structures. The flare should, however, be made from the axis of the approaching stream (*e*) rather than from the culvert

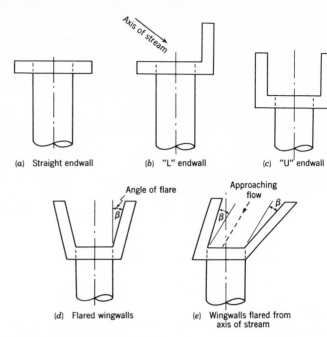

(a) Straight endwall   (b) "L" endwall   (c) "U" endwall

(d) Flared wingwalls   (e) Wingwalls flared from
                              axis of stream

FIG. 18-17   *Culvert endwalls and wingwalls.*

axis. Some hydraulic advantage is gained by warping wingwalls into a smooth transition, but the gain is not usually sufficient to offset the cost of the complex forming required for such warped surfaces.

The purpose of the culvert outlet is to protect the downstream slope of the fill from erosion and prevent undercutting of the culvert barrel. Where the discharge velocity is low or the channel below the outlet is not subject to erosion, a straight endwall or a U-shaped endwall may be quite sufficient. At higher velocities, lateral scour of the embankment or channel banks may result from eddies at the end of the walls, especially when the culvert is much narrower than the outlet channel. With moderate velocities, flaring of outlet wingwalls is helpful, but the flare angle β must be small enough so that the stream from the culvert will adhere to the walls of the transition. Izzard[1] suggests that the flare angle for supercritical flow should be about as indicated by

$$\tan \beta = \frac{1}{2N_F} = \frac{2.85\sqrt{d}}{V} \tag{18-10}$$

[1]C. F. Izzard, Importance of Drainage Design in Highway Construction, *Univ. Calif. Inst. Transportation Traffic Eng. Inf. Circ.* 18, p. 7, March, 1951.

where N_F is the Froude number of the flow $V/\sqrt{gd}$, d is the depth of flow, and V is the mean velocity. For subcritical flow even larger angles might be used.

If the discharge velocity from the culvert is very high or the channel material particularly susceptible to erosion, some provision for dissipating the energy of the outflow may be necessary. This may take the form of a sloping apron to induce a hydraulic jump or a bucket outlet which throws the jet far enough downstream to prevent damage to the road fill (Chap. 9). A stilling pool below the outlet may also be satisfactory, and in many cases such a pool will form naturally by the scouring action of the outflow. If a more elaborate structure seems warranted, a stilling basin such as the SAF basin[1] may be provided.

18-22 Debris barriers The sole function of a debris barrier is to prevent the entrance of material which might clog a culvert. The barrier must be so designed that it does not become blocked with debris and close the culvert entrance. To do this, the spacing of bars in the barrier should be quite wide so that small material will easily pass through them. A bar spacing of one-half to one-third of the least culvert dimension is usually satisfactory. The barrier should not be placed tightly over the culvert entrance where an accumulation of debris might block the culvert. A V-shaped or semicircular rack at the culvert entrance or a straight rack placed at the upstream end of the wingwalls (Fig. 18-18) is effective since if it

---

[1] F. H. Blaisdell, Development and Hydraulic Design, Saint Anthony Falls Stilling Basin, *Trans. ASCE*, Vol. 113, pp. 483–561, 1948.

FIG. 18-18 *Pipe debris rack at end of wingwalls of a reinforced-concrete box culvert entrance. Note warped wingwalls and rounded culvert entrance. (California Culvert Practice)*

should be blocked by debris, it can be overtopped and flow into the culvert can continue. Culvert entrances in a cut are often at roadbed level, and an abrupt break in the channel slope is necessary. This favors debris deposition near the entrance, and a sturdy crib of timber or reinforced-concrete bars with a top to prevent entrance of stones may be necessary (Fig. 18-19).

18-23 Culvert hydraulics The hydraulic design of most culverts consists in selecting a structure which will pass the design flow without an excessive headwater elevation. A secondary hydraulic consideration in some cases is the prevention of scour at the culvert outlet. The maximum head-water elevation must provide a reasonable freeboard against flooding of the highway. It must also be low enough so that there is no damage to property upstream from the highway. Either of these two conditions may control in a specific case. In the majority of culvert designs, the limiting headwater elevation will be well above the top of the culvert entrance, and flow in the culvert will take place under one of the three conditions illustrated in Fig. 18-20. Under conditions (*a*) and (*b*) of the figure the culvert is said to be operating with *outlet control*, and the headwater elevation required to discharge the design flow is determined by the head loss h_L in the culvert.

Submergence of a culvert outlet may result from a reduced channel slope or an inadequate culvert downstream; a winding, brush-grown channel; or

FIG. 18-19   *Extensible concrete debris rack (known as bear trap) over a drop inlet. (California Culvert Practice)*

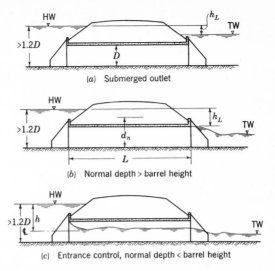

FIG. 18-20   *Flow conditions in culverts with submerged entrance.*

backwater from an intersecting stream. Submergence of the outlet is most likely to occur in flat terrane. If submergence is probable, the tailwater elevation must be computed by backwater calculations from the controlling section downstream. If the inlet and outlet of a culvert are submerged, flow will be in condition (*a*) regardless of culvert slope. If the bottom slope is such that the normal depth for the design flow is greater than the culvert height D and the inlet is submerged, the culvert will flow full[1] (*b*) although the outlet is not submerged. The total head loss[2] h_L is the sum of an entrance loss h_e, friction loss in the barrel h_f, and the velocity head h_v in the barrel.

$$h_L = h_e + h_f + h_v \tag{18-11}$$

Entrance loss is a function of the velocity head in the culvert, while friction loss may be computed with the Manning formula [Eq. (10-8)]. Thus

$$h_L = K_e \frac{V^2}{2g} + \frac{n^2 V^2 L}{2.22 R^{4/3}} + \frac{V^2}{2g} \tag{18-12}$$

This expression can be reduced to

---

[1] If there is a contraction of flow at the culvert entrance, reexpansion will require about six diameters. Hence a very short culvert may not flow full even though $d_n \geqq D$.

[2] The energy grade line at the entrance is above the actual water surface by the velocity head of the approaching water. With water ponded at the entrance this velocity head is usually negligible but should be included in computations if it is of significant magnitude.

$$h_L = \left(K_e + 1 + \frac{29n^2L}{R^{4/3}} \right) \frac{V^2}{2g} \tag{18-13}$$

The entrance coefficient K_e is about 0.5 for a square-edged entrance and about 0.05 if the entrance is well-rounded. If the outlet is submerged, the head loss may be reduced somewhat by flaring the culvert outlet so that the outlet velocity is reduced and some of the velocity head recovered. Tests show that the flare angle should not exceed about 6° for maximum effectiveness.

If normal depth in the culvert is less than the barrel height, with the inlet submerged and the outlet free, the condition illustrated in Fig. 18-20c will normally result. This culvert is said to be flowing under *entrance control*, i.e., the entrance will not admit water fast enough to fill the barrel, and the discharge is determined by the entrance conditions. The inlet functions like an orifice for which

$$Q = C_d A \sqrt{2gh} \tag{18-14}$$

where h is the head on the center of the orifice[1] and C_d is the orifice coefficient of discharge. The head required for a given flow Q is therefore

$$h = \frac{1}{C_d^2} \frac{Q^2}{2gA^2} \tag{18-15}$$

For a sharp-edged entrance without suppression of the contraction $C_d = 0.62$, while for a well-rounded entrance C_d approaches unity. If the culvert is set with its invert at stream-bed level, the contraction is suppressed at the bottom. Flared wingwalls may also effect partial suppression of the side contractions. Because of the wide variety of entrance conditions which may be encountered, it is impracticable to cite appropriate values of C_d, and for a specific design these must be determined from model tests or tests of similar entrances.

Under conditions of entrance control, flaring the culvert entrance will increase its capacity and permit it to operate under a lower head for a given discharge. If the permissible headwater elevation is high enough, some economy may be obtained by using a smaller barrel size and providing a flared entrance so that the barrel flows full. Experiments[2] have shown that the best results are obtained by flaring the entrance of a box culvert to about double the barrel area in a distance equal to four times the offset from the

---

[1] Equation (18-14) applies for an orifice on which the head h is large compared with the orifice height D. When the headwater depth is $1.2D$ (that is, $h = 0.7D$), an error of 2 percent results. In the light of other uncertainties in the design this error can be ignored.

[2] Culvert Hydraulics, *Research Rept.* 15-B, Highway Research Board, Washington, D.C., 1953.

barrel (Fig. 18-21). The entrance of circular culverts should be rounded on a radius of about $0.15D$.

The *drop-inlet culvert* (Fig. 18-22) may be used where there is very little headroom at the fill (a) or where ponding is permissible to reduce peak flow or shorten barrel length (b). The barrel is usually on a flat slope and will flow full at design discharge. This type of culvert is a special case of conditions (a) or (b) of Fig. 18-20, depending on whether the outlet is free or submerged. The required barrel size may be computed from Eq. (18-13), assuming $K_e = 0$ and including a bend loss $K_b V^2/2g$, where K_b varies from a maximum of about 1.5 for a square bend to about 0.45 for a circular bend with radius $r = D$. When the entrance to a drop inlet is submerged, it discharges as an orifice as in the case of the shaft spillway. This is undesirable, and the culvert should be designed for free-entrance conditions. Hence the maximum head loss in the culvert barrel is as defined in Fig. 18-22. The inlet of a drop-inlet culvert functions as a weir as long as it is not submerged, and the rim length must be sufficient to discharge the design flow without an excessive headwater elevation. If necessary, the inlet may be flared (Fig. 18-22d) in order to obtain adequate crest length. If the head loss in the barrel is insufficient to keep the riser filled to the top, the weir coefficient will be approximately 2.7, while with the riser filled negative

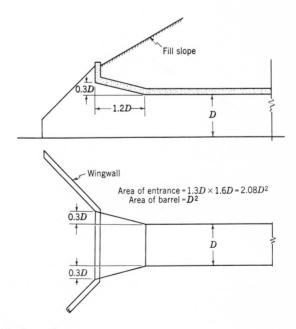

FIG. 18-21   Entrance flare for box culvert.

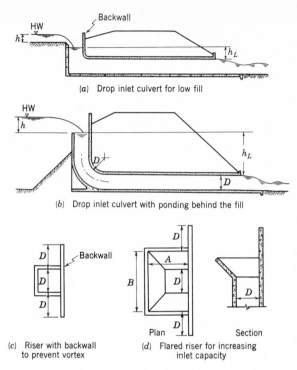

FIG. 18-22   Drop-inlet culverts.

pressures exist under the nappe as long as vortex action is prevented by a backwall (Fig. 18-22c), and the weir coefficient will be about 4.2.

Some box culverts may be designed so that the top of the box forms the roadway. In this case the headwater should not submerge the inlet, and one of the flow conditions of Fig. 18-23 will exist. In cases (a) and (b) critical depth in the barrel controls the headwater elevation, while in case (c) the tailwater elevation is the control. In all cases the headwater elevation may be computed from the Bernoulli principle with an allowance for entrance loss $h_e = K(h_{v_2} - h_{v_1})$, where K will vary from about 0.1 for a smooth transition to 0.5 for an abrupt contraction.

When the culvert is on a steep slope [case (b)], critical depth will occur at about $1.4d_c$ downstream from the entrance. The water surface will impinge on the headwall when the headwater depth is about $1.2D$ if d_c is $0.8D$ or more. Since it would be inefficient to design a culvert with d_c much less than $0.8D$, a headwater depth of $1.2D$ is approximately the boundary between free-entrance conditions (Fig. 18-23) and submerged-entrance conditions (Fig. 18-20).

18-24 Bridge waterways The waterway opening of a bridge must be large enough to pass the design flood without creating excessive backwater. If stream velocities are low or there are only a few piers in the stream, the backwater problem may be of little importance. Massive piers or long-approach embankments encroaching on the waterway of the stream may result in considerable damage from high stages above the bridge and may increase the likelihood of the bridge being overtopped by floodwater. Economic design requires the determination of the minimum clear length of span which will not cause intolerable backwater conditions. The first step in a study is a determination of the permissible backwater heights, based on field investigation of lands and structures along the stream which might be harmed by an increase in stage. The second step is the determination of the stage in the channel downstream from the bridge site at design flow. This may be done by backwater computations from the nearest control section below the bridge. A few measurements of stage and discharge at the bridge site will materially simplify the computations and improve their accuracy.

The drop in water level (Fig. 18-24) through a contracted opening[1] can be computed from

$$\Delta h = K_B \frac{V_2{}^2}{2g} + \frac{SL}{m} - \alpha \frac{V_1{}^2}{2g} \qquad (18\text{-}16)$$

[1] For a detailed discussion see C. E. Kindsvater, R. W. Carter, and H. J. Tracy, Computation of Peak Discharge at Contractions, *U.S. Geol. Surv. Circ. 284*, 1953.

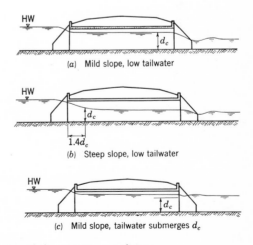

(a) Mild slope, low tailwater

(b) Steep slope, low tailwater

(c) Mild slope, tailwater submerges d_c

FIG. 18-23   Flow conditions in culverts with free entrance conditions.

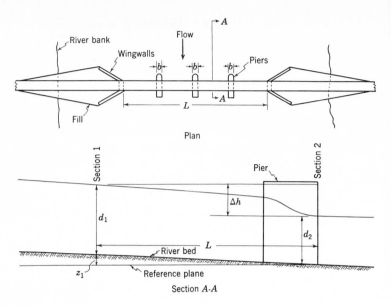

FIG. 18-24   *Backwater caused by bridge.*

where V_2 is the discharge divided by the gross cross section A_2 under the bridge, neglecting piers, S is the normal slope of the unobstructed stream, L is the length of the bridge, disregarding piers, K_B is a head-loss coefficient which is a function of the conveyance ratio m, and $\alpha V_1^2/2g$ is the weighted mean approach velocity head.

The conveyance ratio $m = k_q/k_a$, where k_a is the conveyance of the approach channel and k_q is the conveyance of the approach channel at section 1 for a width equal to the projected area of the bridge opening. The ratio m may be visualized as expressing the fraction of the total flow which is not affected by the contraction. Conveyance is computed from the Manning formula as

$$ k = \frac{1.49}{n} AR^{\frac{2}{3}} \tag{18-17} $$

If the cross section is irregular, it should be subdivided into sections of approximately constant n and d and conveyance for the whole section computed as Σk_i, where k_i is the conveyance of the individual subsections.

Laboratory tests show that the drawdown resulting from the contraction extends upstream approximately one bridge length L. Hence section 1 should be a distance L upstream from the bridge. Assuming the friction loss to increase in proportion to the degree of contraction, the friction loss between sections 1 and 2 is given approximately by SL/m. The head-loss coefficient K_B is a function of m (Table 18-2). It is also a function of the geometry of the bridge, skew, eccentricity of the bridge, and submergence of the superstructure. When m is large, the effect of skew can be accounted for by multiplying K_B by $(1/\cos \beta)^2$, where β is the acute angle between the bridge and a normal section across the stream. If the bridge waterway is at one edge of the flood plain, K_B is increased about 10 percent by the eccentricity. Piers occupying less than one-tenth of the bridge waterway will also increase K_B by about 10 percent at the most.

The factor α accounts for the nonuniformity of flow in the cross section and is given accurately by

$$\alpha = \frac{\displaystyle\int_0^Q V_i^2 \, dQ_i}{V^2 Q} \tag{18-18}$$

where V_i and Q_i are velocity and discharge for a small element of the section. A practical approximation is

$$\alpha \doteq \frac{\Sigma(k_i^3/A_i^2)}{k_1^3/A_1^2} \tag{18-19}$$

where k_i and A_i are the conveyance and area of finite subsections of the total section.

18-25 Dips Dips sometimes offer an economical solution to the cross-drainage problem in arid areas where streamflow is infrequent and of short duration, provided the channel to be crossed is shallow enough to permit construction of the dip without excessive grading of the approaches.

TABLE 18-2   *Values of the Loss Coefficient K_B as a Function of the Conveyance Ratio m*

m	K_B
1.0	1.0
0.9	1.16
0.8	1.36
0.7	1.53
0.6	1.67
0.5	1.79
0.4	1.88
0.3	1.91
0.2	1.92

The upstream edge of the roadbed at the dip should be even with the bottom of the channel to avoid scour which might undermine the road. The downstream edge of the road should be protected with a cutoff wall, paving, or rock riprap for the same purpose. The profile of the dip should be as close to the shape of the stream cross section as possible, to eliminate interference with streamflow. If the stream transports heavy debris, the road surface should be made especially heavy.

Posts should be set to indicate the edges of the road, and suitable warnings should be posted against crossing the dip when flow is occurring. Most vehicles can negotiate water depths of 10 in. to 1 ft, and passage at these depths might be permitted, but there is danger of a vehicle being caught by a swiftly moving floodwave while in transit.

BIBLIOGRAPHY

Babbitt, H. E.: "Sewerage and Sewage Treatment," 8th ed., Wiley, New York, 1960.
"California Culvert Practice," State of California Division of Highways, 2d ed., Sacramento, Calif.
"Design and Construction of Sanitary and Storm Sewers," *ASCE Manual on Engineering Practice* 37, 1969.
Hathaway, G. A.: Design of Drainage Facilities, *Trans. ASCE*, Vol. 110, pp. 697–848, 1945.
Herner, R. C., R. C. Mainfort, and R. L. Pharr: Use of the Rational Formula in Airport Drainage, *Civil Aeron. Adm. Tech. Dev. Rept.* 131, December, 1950.
Luthin, James N.: "Drainage of Agricultural Lands," American Society of Agronomy, Madison, Wis., 1957.
Pillsbury, A. F.: Observations on Tile Drainage Performance, *J. Irrigation and Drainage Div.*, *ASCE*, Vol. 93, pp. 233–241, September, 1967.
Steel, E. W.: "Water Supply and Sewerage," 4th ed., McGraw-Hill, New York, 1960.

PROBLEMS

18-1   An area of 20 acres has an average length of overland flow of 150 ft and a slope of overland flow of 0.05. The equation of the design rainfall is $i = 15/t_R^{0.45}$. The area is in smooth grass sod. Find the peak rate of runoff.

18-2   A triangular drainage ditch parallels an airport runway for a distance of 2000 ft. A 40-ft-wide asphalt-surfaced runway contributes flow to the ditch from one side, and a 200-ft width of grass sod ($k = 0.3$) contributes from the other side. The transverse slope of the runway is 0.005, and that of the sod area 0.010. The design rainfall has been taken as 2.0 in./hr for a duration of 90 min. Ignoring storage effects in the channel, what must be its width at the downstream end if the maximum permissible depth is 1.0 ft, the longitudinal slope is 0.013, and $n = 0.020$?

18-3   What is the capacity of a gutter on a street which is 40 ft wide and has a 6-in. crown, 6-in. vertical curbs, and a longitudinal slope of 0.07 if the water is at the top of the curb? What is the capacity if the maximum width of flow in the gutter is not to exceed 6 ft? Assume the cross slope of the roadway to be uniform. $n = 0.018$.

18-4   A grated inlet 20 in. wide is to be installed in a gutter on a slope of 0.03. The grate thickness is $1\frac{3}{4}$ in., and the transverse slope of the roadway is 0.025. How long should the inlet be and how much flow will it intercept if it accepts all the flow in the gutter? How long should it be and what flow will it intercept when the depth at the curb is 0.4 ft? $n = 0.016$.

18-5   A curb-opening inlet is considered for the installation described in Prob. 18-4. If the depression of the inlet is 0.2 ft, how long must it be to intercept all the flow in the gutter when the depth at the curb is 0.4 ft? How much water would an inlet 6 ft long intercept under these conditions?

18-6   An inlet grate has interior dimensions of 14 by 26 in., with eight longitudinal bars $\frac{3}{4}$ in. wide and two transverse bars 2 in. wide. What is the capacity of this grate if water is ponded to a depth of 0.3 ft? What is the capacity if the pond is 1.8 ft deep at the grate?

18-7   Complete the design of the drainage system for the area shown in Fig. 18-8, using the criteria of Illustrative Example 18-3.

18-8   Design the drainage system for the area shown in Fig. 18-8, using the rainfall equation $i = 35/t_R{}^{0.5}$. All other criteria are as described in Illustrative Example 18-3.

18-9   A farmer plans the installation of tile drains in a field in which the soil permeability is 500 gpd/sq ft. An impermeable stratum exists about 10 ft below the soil surface, and the farmer wants to keep the water level at least 3 ft below the soil surface. What drain spacing would you recommend if the mean annual rainfall is 45 in.?

18-10   What are the legal requirements for the establishment of a drainage district in your state?

18-11   Derive Eq. 18-9. *Hint:* A view normal to the roadway surface is almost identical to a planimetric view.

18-12   A highway drainage ditch has a triangular section with a maximum depth of 8 in. and a top width of 6 ft. The pavement is 60 ft wide with a 6-in. crown. The longitudinal slope of the ditch is 0.004. If the design rainfall is 4 in./hr, at what intervals along the ditch must flow be intercepted so that the freeboard in the ditch will never be less than 2 in.? Is paving or sodding of this ditch necessary? Interception is to be by corrugated pipe ($n = 0.025$) which is to take the flow across under the highway on a slope of 0.05. What size pipe will be required?

18-13   A culvert under a road must carry 150 cfs. The culvert length will be 100 ft, and the slope will be 0.003. If the maximum permissible headwater level is 12 ft above the culvert invert, what size of corrugated-pipe culvert would you select? The outlet will discharge freely. Assume square-edged entrance.

18-14   What is the capacity of a 4- by 4-ft concrete box culvert with rounded entrance if the culvert slope is 0.005, the length 120 ft, and the headwater level 6 ft above the culvert invert? Assume free-outlet conditions.

18-15   What would be the capacity of the culvert of Prob. 18-14 if the water level at the outlet was 1 ft above the top of the box?

18-16   A culvert under a low fill must carry 100 cfs. The maximum possible culvert height is 6 ft, and the slope is 0.002. What must be the width of this culvert if the maximum headwater level cannot be above the top of the culvert entrance?

18-17   The bridge of Fig. 18-24 is 300 ft long, with three piers, each 8 ft wide. The natural slope of the stream is 0.0015 at a design discharge of 60,000 cfs. The channel under the bridge is essentially rectangular. The approach channel can be divided into three subsections. The center section is 400 ft wide and 14 ft deep and has $n = 0.024$, while the two side sections are each 200 ft wide and 3 ft deep, with $n = 0.050$. What is the head loss caused by the bridge?

18-18   An asphalt parking lot ($k = 0.90$ and $c_r = 0.007$) with overall dimensions 400 by 600 ft is graded to drain into a 2-ft-wide rectangular concrete channel which runs longitudinally down the center of the lot. The channel is recessed below grade and has a slope of 0.003. The transverse slope of the asphalt surface is 0.015 in the direction of the 200-ft dimensions. Compute the peak rate of outflow at the outlet from the channel if the design rainfall is as follows:

t_R, min	i, in./hr
5	2.6
10	1.8
20	1.1
30	0.8

18-19 A 120-ft-long corrugated metal pipe ($n = 0.022$) of 30-in. diameter is tested in a laboratory. The headwater is maintained at a level which is 5 ft above the pipe invert at entrance. Assume a square-edged entrance and neglect velocity of approach. Compute values of Q for S values of 0.0, 0.01, 0.03, and 0.08. Whenever $d_n > D$ assume that condition (b) of Fig. 18-20 prevails. As the slope is increased, at what slope does the flow change to condition (c) of Fig. 18-20?

18-20 A concrete box culvert ($n = 0.013$) is 2 ft high, 4 ft wide, 100 ft long, and is on a slope of 0.01. Assume a square-edged entrance and a free jet at exit, Fig. 18-15b. Make the necessary calculations to plot Q versus headwater elevation for conditions of rising head and falling head. Take an elevation range from zero to 8.0 ft above the invert at entrance. Neglect velocity of approach.

18-21 Refer to the data of Prob. 18-20. Determine the culvert slope so that there is no abrupt change in flow rate if the flow changes from entrance control to pipe-full condition when the headwater is 4 ft above the invert at entrance.

Chapter **19**

Sewerage
and wastewater treatment

In areas with a large concentration of population, wastewater which must be disposed of to maintain healthful and attractive living conditions includes *domestic*, or *sanitary*, *wastewater* from toilets, sinks, and other plumbing fixtures and *industrial wastewater* from manufacturing plants. In the modern community, wastewater is removed in underground conduits called *sewers*. Some communities have a *combined system* in which sanitary wastewater, industrial wastewater, and storm runoff are all carried in the same conduit. Other communities have *separate systems*: one system for sanitary wastewater and industrial wastewater which usually delivers these liquids to a wastewater treatment plant and a separate storm-drain system which collects storm runoff and delivers it to a natural watercourse or waterbody for disposal.

Separate systems are more costly than combined systems, but the extra cost is almost always justified. Formerly, many combined systems were built, but now practically all new systems are separate systems. With a combined system a substantial portion of the wastewater must bypass the treatment facilities during storms. Hence untreated wastewater gets into the natural water bodies causing pollution. This is a very serious problem at many of the larger and older cities in the United States that have combined sewers. The cost of installing separate systems in these large cities would be excessive, hence extensive research is being conducted to devise schemes that might economically solve this problem.[1]

---

[1] "Combined Sewer Separation Using Pressure Conduits," American Society of Civil Engineers, New York, October, 1969.

QUANTITY OF WASTEWATER

The details of flow estimation and conduit design for municipal storm-drain systems were discussed in Chapter 18. In the following sections the discussion will deal with the estimation of flows for systems which convey only sanitary wastewater and industrial wastewater. These systems, including conduits and appurtenances, are commonly referred to as *sewerage systems*.

19-1   Amount of domestic and industrial wastewater   The quantity of domestic and industrial wastewater contributed by an area will generally be about one-third less than the water use of the area, i.e., about 60 to 75 percent of the water supplied will reappear as wastewater, the remainder being used in industrial processes, for lawn sprinkling, etc. Hence, if the water use of a community is known, the probable output of sanitary wastewater can be estimated. Design estimates should allow for probable future growth of the area.

Estimates of water supply must include all water from private sources. Industries often get water from their own wells but use public sewers for waste disposal. In this case the combined industrial and domestic wastewater may exceed in volume the water supplied by the public system. Conversely some industries drawing water from the public supply may not discharge their wastes into the public sewers, resulting in a low ratio of wastewater to water consumed. A careful study of local conditions is necessary if an accurate estimate of wastewater flow is to be made.

19-2   Infiltration   There is always some entry of groundwater into the sewers through the pipe joints. The amount of this *infiltration* depends mostly on the groundwater level and the care exercised in making the sewer joints. If the groundwater table is well below the sewer, infiltration will occur only after rain when water is moving down through the soil. If the water table is high, infiltration rates as great as 300 to 1500 gpd per acre of area sewered may occur. Another figure often used is 500 to 2000 gpd per inch of diameter per mile of sewer.

ILLUSTRATIVE EXAMPLE 19-1   A city with a population of 20,000 has an area of 1100 acres. The average water consumption is 160 gpcd, and 70 percent of this water reaches the sewers. If an allowance of 750 gpd/acre is made for infiltration, what is the total quantity of wastewater produced per day? Assume no storm drainage.

Domestic and industrial
wastewater: $20,000 \times 160 \times 0.7 = 2,240,000$ gal
Infiltration: $1100 \times 750 = \underline{825,000}$ gal
                        Total wastewater flow $= \overline{3,065,000}$ gal per day

---

19-3   Variation in the flow of domestic and industrial wastewater
The flow of domestic wastewater and industrial waste varies throughout the day and the year. The daily peak from a small residential area will usually occur in midmorning and be about 225 percent of the average daily flow. Commercial and industrial wastes will be delivered more uniformly throughout the day, with a peak rate of about 150 percent of the average rate. Because of storage and time lag in the sewers, the peak flows expressed as a percent of average flows decrease as the size of the tributary area increases. Peak flow at the outfall of a city sewer system will usually be about 150 percent of the average flow. The minimum flow at the outfall of a city sewerage system rarely falls below 40 percent of the average flow.

Percentage variations in wastewater flow will change with the size of the city, amount of industrial waste, and climate, and the figures quoted above are only approximate. The best source of data in a particular case is actual measurement on the system or on similar systems in the same region.

SEWERAGE

Often in the development of a city, sanitary sewers are built first because of their urgency and because the financing of the additional storm-drain system is impracticable at the time. Moreover, in some areas gutters and natural watercourses can often be relied upon for the drainage of storm waters. At a later date a separate storm-drain system may be constructed.

19-4   Sewer pipes   Clay tile, concrete, and asbestos-cement are the most common materials for sewers. Concrete (and even asbestos-cement) is susceptible to corrosion by sulfuric acid from hydrogen sulfide gas generated in sanitary wastewater or by some industrial chemicals. Sulfide corrosion of concrete pipe is a serious problem in areas where the wastewater is strong, stale, and very warm. Such conditions hasten the bacterial activity that results in the formation of hydrogen sulfide. Where such conditions prevail, it is generally advisable to use clay pipe for diameters smaller than 39 in. and concrete pipe with cast-in-place plastic linings[1] for larger sizes.

The smallest sanitary sewer pipes are house connections, usually of clay or cast-iron 3 or 4 in. in diameter. They are customarily installed from the residence to the property line at the street where they connect to the main sewer system. At downstream regions of a system the sewer-pipe sizes increase to accommodate the flow.

In major cities cast-in-place concrete sewers (Fig. 19-1) 10 ft or more in

---

[1] L. A. Pardee and E. G. Studley, Concrete Sewer Protection, *Water and Sewage Works*, Vol. 104, pp. 145–149, April, 1957.

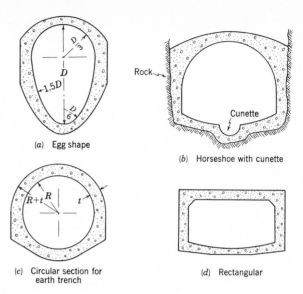

(a) Egg shape

(b) Horseshoe with cunette

(c) Circular section for earth trench

(d) Rectangular

FIG. 19-1   Some cast-in-place sewer sections.

diameter are quite common. Large sewers are known as interceptors, and the final section of sewer leading to the treatment plant or point of disposal is commonly referred to as the outfall.

19-5   Flow in sewers   Sewers rarely flow full and are designed as open channels. Inverted siphons and discharge lines from some wastewater pumping stations flow under pressure. The latter are sometimes called *force mains*. In order to prevent the settlement of wastewater solids, the velocity in a sewer flowing full should be not less than about 2 ft/sec. Such a sewer flowing one-sixth full will have a velocity of 1 ft/sec, which is reasonably adequate. This is especially important in sanitary sewers, for decomposition of settled wastes results in undesirable conditions. Egg-shaped sewers are often used to prevent low velocities at low flows, while large combined sewers sometimes have a small channel (*cunette*) in their invert for the dry-weather flow.

19-6   Sewer appurtenances   Manholes (Sec. 18-6) are the most numerous appurtenance for sanitary sewers. They should be placed wherever there is a change in elevation, size, direction, or slope, at junctions, and at intervals of not more than 500 ft if the sewer is too small for a man to enter. Most cities have standard plans for manholes.

Sewers near dead ends and on flat slopes may not receive sufficient flow to prevent deposition of sediment and may require flushing. Flushing may be accomplished with automatic flush tanks which make periodic releases of water into the sewer. Another method is to fill a manhole with water from a nearby hydrant. Such manholes are provided with stop gates on the outlet sewer which permit filling of the manhole with water or accumulated wastewater. When the outlet gate is opened, water rushes through the sewer. To prevent clogging with grease or oil, many cities require restaurants, service stations, and industries to place *grease traps* (Fig. 19-2) on their waste lines.

If a combined sewer is delivering wastewater to a treatment plant, it may be necessary to bypass storm flow around the plant. Various types of *regulators* have been devised for this purpose. With a *side-flow weir* (Fig. 19-3) the dry-weather flow goes directly to the treatment plant, but when the flow is increased by storm water, the excess flow goes over the weir into a bypass sewer. The weir crest is placed at the level to which the estimated maximum sanitary flow would fill the pipe, while the weir length must be adequate to discharge the expected storm flow. Mechanical devices using float-actuated gates are also used as regulators.

Sewers are usually carried across depressions on a trestle or bridge in steel or cast-iron pipe with special couplings since clay or concrete pipe joints may not remain tight. Sag pipes (Sec. 11-27) are often used for sewers crossing depressions. Because of the variation in wastewater flow it is impossible to design a single sag pipe in which velocities are adequate to prevent deposition of solids at the low point under all flow rates. It is therefore necessary to use two or more pipes in parallel and arranged in such a manner that the number of pipes in operation depends on the flow rate.

When untreated sanitary wastewater is discharged into a body of water, the outlet should be low enough to be covered by water at all times. The outlet should be far enough from shore to prevent accumulation of solid wastes along the shoreline. The distance will vary with the quantity and strength of the wastewater and the size, character, and currents of the water body. The outlet for a combined sewer is often arranged as shown in Fig.

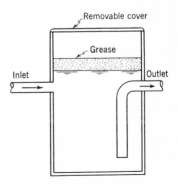

FIG. 19-2   A simple grease trap.

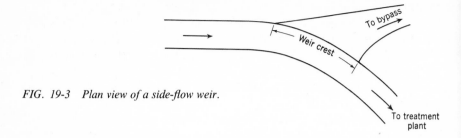

FIG. 19-3   Plan view of a side-flow weir.

19-4. This conveys the sanitary wastewater to a point of suitable disposal with a minimum of expense.

Treated wastewater is often discharged into a river, freshwater lake, or the ocean. Release at the shoreline is rarely satisfactory as the liquid wastes may be poorly dispersed in the receiving waters. Wastewater is warm and generally lighter than the receiving waters, hence discharge at depth will result in moderate dispersion as the wastewater rises up through the water. If the point of discharge is near the center of the river the current will aid in the dispersion process. Discharge of wastewater into a freshwater lake is to be discouraged because of the possibility of eutrophication (Sec. 19-15). In ocean disposal the outfall pipe should be extended a substantial distance from the shore. Dispersion of the liquid wastes can be greatly enhanced through use of a Y-branch near the end of the outfall pipe with multiple ports along each branch near its discharge end.[1]

19-7   Pumping of wastewater   In many communities the topography is such that some pumping of wastewater is necessary. If ground slopes are less than those required to provide adequate velocity in a sewer, the sewer will become progressively deeper and it may become necessary to pump the wastewater to a higher level to avoid excavation. This occurs quite frequently near the point of discharge into the receiving body of water. Pumping will also be required if the wastewater must be conveyed over a hill and is often required at treatment plants to provide sufficient head for plant operation.

---

[1]A. M. Rawn, F. R. Bowerman, and N. H. Brooks, "Diffusers for Disposal of Sewage in Sea Water," *Trans. ASCE*, Vol. 126, part 3, pp. 344–388, 1961.

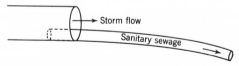

FIG. 19-4   A possible arrangement for the outlet of a combined sewer.

The best type of wastewater pump is a volute pump with two impeller vanes and large passageways to minimize clogging. A wastewater pump with a 6-in. inlet will pass soft 4-in. solids. It is advisable, however, to install a bar screen ahead of the pump, and if the wastewater contains much grit, a grit chamber (Sec. 19-21) will lengthen the life of the impeller. Disassembly of wastewater pumps is usually simplified to make cleaning easy.

A wastewater-pumping station should have at least two pumps and some source of standby power so as to be able to maintain continuous service in event of pump or power failure. Electric motors are the most common power source, and internal-combustion engines are commonly used for standby. If only two pumps are installed at a station, each should have a capacity equal to the maximum anticipated flow. Usually there is a wide variation between the minimum and maximum flows, and it is often advantageous to install three or more pumps to permit more efficient operation. It is desirable to install the pumps in a dry pit with their suction pipes below the lowest wastewater level in the adjacent wet well to eliminate the need for priming. The pump operation may be automatically controlled by a float in the wet well.

Pneumatic ejectors are often used for raising wastewater from building sumps. The ejector consists of an airtight tank into which wastewater flows by gravity and out of which the wastewater is forced automatically whenever sufficient wastewater has accumulated to raise a float and open the compressed air-inlet valve.

19-8 Design of a sanitary sewerage system   The general procedure for the design of storm drains presented in Sec. 18-8 is also applicable to sanitary sewers. A good map of the area to be sewered is essential. Tentative routes for the outfall sewer and its principal tributaries are first selected. If possible, all sewers are planned to slope in the same direction as the ground surface. The layout is governed largely by the topography and the distribution of population and industry. Several trial layouts may be necessary before the most economical arrangement of sewers can be selected.

The depth of a sanitary sewer should be sufficient so that all house sewers connected to it will drain by gravity. A depth of 6 or 8 ft below the ground surface is usually sufficient; in parts of the country where basements are infrequent, a depth of 4 ft may be satisfactory if this provides sufficient protection against superimposed loads. In no case should a sanitary sewer be placed above a water main. After the layout for the sewerage system has been selected and the elevations of controlling points determined, the next step in the design is the calculation of the required capacity and the determination of pipe size and slope for each sewer. The logical method of computing wastewater quantities is to start at the upper end of the uppermost lateral and work downgrade to the outlet. In estimating wastewater flows consideration should be given to possible future growth of the community. It is common practice to design each sewer so that its capacity when flowing full will equal the estimated maximum flow rate at some future date, say 20 yr hence. Such allowance is necessary only when additional area tributary to the sewer can be developed. The calculations for the design of a typical sanitary sewerage system are presented in the following example.

ILLUSTRATIVE EXAMPLE 19-2   Design a sanitary sewer system for the area[1] shown in Fig. 19-5 above the intersection of Oak Avenue and Laurel Drive. The blocks are 750 by 300 ft, and the streets are 40 ft wide. The area is developed as single-family residences with an average population of 25 persons per acre. Assume an average sanitary flow of 80 gpcd and infiltration of 1000 gal/acre/day. Use a minimum sewer diameter of 8 in. and a minimum depth of cover on the crown of the sewer of 6 ft. Assume that a large industrial plant north of Ridge Road discharges a maximum of 1.8 cfs of wastewater (including infiltration) into the sewer on Laurel Drive. Since the area is residential, the maximum rate of sanitary flow may be taken as 225 percent of the average rate. Manholes will be placed at all intersections and at mid-block on the laterals to provide access to the sewer for cleaning.

Table 19-1 shows the design calculations. Columns 1 through 5 summarize the

---

[1] Note that the area shown in Fig. 19-5 is the same as that of Fig. 18-8.

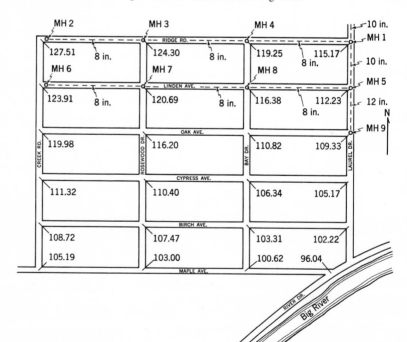

FIG. 19-5   Problem area for sanitary sewer design.

TABLE 19-1 Calculations for the Design of a Sanitary Sewerage System

Street (1)	From (2)	To (3)	Length, ft (4)	Increment of area, acres (5)	Increment of population (6)	Total tributary population (7)	Max wastewater flow, cfs (8)	Infiltration, cfs (9)	Total flow, cfs (10)	Ground elevation, ft Upper manhole (11)	Ground elevation, ft Lower manhole (12)	Drop in elevation, ft (13)	Slope, ft/ft (14)	Pipe diam., in. (15)	Capacity, cfs (16)	Velocity when full, ft/sec (17)	Invert elevation, ft Upper manhole (18)	Invert elevation, ft Lower manhole (19)
Laurel	Northerly	Ridge	...	.....	...	...	.....	.....	1.80	.....	112.2	3.0	0.0088	10	2.0	3.8	.....	105.4
Ridge	Creek	Rosewood	790	3.08	77	77	0.0215	0.0048	0.026	127.5	124.3	3.2	0.0040	8	0.74	2.2	120.8	117.6
Ridge	Rosewood	Bay	790	3.08	77	154	0.0430	0.0096	0.053	124.3	119.2	5.1	0.0065	8	0.90	2.8	117.2	112.5
Ridge	Bay	Laurel	790	3.08	77	231	0.0645	0.0144	0.079	119.2	115.2	4.0	0.0051	8	0.80	2.5	112.5	108.5
Laurel	Ridge	Linden	340	.....	...	231	0.0645	0.0144	1.879	115.2	112.7	3.5	0.0103	10	2.3	4.1	108.4	105.9
Linden	Creek	Rosewood	790	6.16	154	154	0.0430	0.0096	0.053	123.9	120.7	3.2	0.0040	8	0.74	2.2	117.4	114.0
Linden	Rosewood	Bay	790	6.16	154	308	0.0860	0.0192	0.105	120.7	116.4	4.3	0.0054	8	0.82	2.6	114.0	109.7
Linden	Bay	Laurel	790	6.16	154	462	0.1290	0.0288	0.158	116.4	112.2	4.2	0.0053	8	0.82	2.6	109.7	105.5
Laurel	Linden	Oak	340	.....	...	...	0.1935	0.0432	2.037	112.2	109.3	2.9	0.0085	12	3.2	4.2	105.2	102.3

physical data from Fig. 19-5. Note that, unlike storm drains, the sanitary sewers must extend up all streets to permit connections to all houses. The population increment (col. 6) is computed by multiplying the increment of area by 25. Column 7 indicates the total population tributary to the downstream end of the sewer. The maximum rate of sanitary-wastewater flow (col. 8) is computed by multiplying the tributary population by 80×2.25 to obtain the flow in gallons per day and then multiplying by 1.55×10^{-6} to convert to cubic feet per second. The infiltration (col. 9) is found by multiplying the tributary area in acres by 1.55×10^{-3}, and the total flow (col. 10) is the sum of col. 8 and col. 9. The ground elevations (cols. 11 and 12) are taken from the map, and the difference in elevation (col. 13) and the ground slope (col. 14) are computed. The required pipe sizes are estimated using the nomogram of Fig. 10-2. The minimum size (8 in.) is more than adequate for the laterals. The capacity of the selected pipe and the velocity when full are tabulated in cols. 16 and 17 for checking purposes. In all cases the velocity should exceed 2 ft/sec. The invert elevations (cols. 18 and 19) are computed to permit the specified 6 ft of cover on the crown of the sewer. After these elevations are selected, a check should be made to see that sufficient cover is provided to all points along the sewer since a break in ground slope might bring the sewer too close to the surface. The computations of Table 19-1 assume a constant slope in the laterals from one intersection to the next, although a break in the sewer slope at the mid-block manholes could be made to provide adequate cover if necessary.

19-9   Design of a combined sewer system   Theoretically the capacity of a combined sewer should equal the sum of the maximum sanitary-wastewater and storm-water flows. Comparison of Illustrative Examples 19-2 and 18-3 shows that the amount of sanitary wastewater is small compared with the storm flow. In view of the uncertainties in determining the required capacity of storm sewers, a combined system designed to handle storm flows will be quite adequate for the sanitary flow. However, additional capacity must sometimes be provided if there are concentrated discharges of industrial waste into the system.

Thus, in general, the pipe sizes of a combined system will be identical to those for a storm-drain system. However, the pipe layout will be similar to that for a sanitary system since a connection must be provided for each house and business establishment. In some cases, large combined sewers will have an egg-shaped section or a small channel in the bottom of a large pipe for the conveyance of dry-weather flow (sanitary wastewater) at adequate velocities. In addition, if treatment facilities are planned for the sanitary wastewater, provision for diverting excess storm-water flow around the treatment plant will be necessary.

The construction of combined sewers is to be discouraged because untreated wastewater bypassing the treatment plant during storms will pollute the receiving waters. For existing combined sewer systems this problem could be alleviated by constructing detention basins to trap all or most of the storm runoff for later release to the treatment plant. Such a scheme, however, would be quite costly.

19-10 Construction of sewers There are many ways in which the actual construction of a sewer system may be accomplished, depending upon the soil conditions encountered and the construction equipment available for the job. The important thing is that the finished sewer perform the function for which it is intended with a minimum of maintenance cost. To attain this end, three conditions should be met: (1) the pipe should be handled carefully, properly bedded, and backfilled in such a manner that there is a minimum of breakage during and after construction; (2) joints should be made with sufficient care to eliminate excessive infiltration; (3) the line and slope of the sewer should be free of irregularities which might favor the accumulation of solid wastes with resultant clogging of the pipe.

Sewer lines are usually placed in the center of the street so that service connections from either side are of approximately equal length. Where alleyways are provided, it may be advantageous to run the sewer line in the alley to avoid tearing up the street pavement. Such a location may also permit shorter house connections. The exact location of the sewer will depend on the position of other underground utilities, and maps of existing utilities should be obtained prior to design. It is important that accurate maps of the sewer system be prepared as construction progresses so that at a later date repairs, replacement, or expansion of the system or installation of new utilities can be planned intelligently.

19-11 Maintenance of sewers The main problem of sewer maintenance is that of keeping the sewer clear of obstructions. Most stoppages in sewers are caused by tree roots, accumulation of grease, or collapse of the sewer pipe. Roots usually enter through the pipe joints, and once inside the pipe they grow rapidly. Good joint construction is the best preventive measure. Most cities have ordinances requiring the use of grease traps on service connections where wastewater may contain large amounts of grease. Collapse of the pipe is unlikely if adequate cover is provided, and reasonable care is exercised to avoid breakage during and after construction.

Where flushing is inadequate to remove an obstruction, sewers are usually cleaned with special tools attached to cables or jointed rods and pushed or pulled through the sewer from a manhole or other point of entry. The type of tool depends on the cause of the obstruction. Cutting tools are employed to remove roots, scoops or scrapers for grit and sludge, and brushes are effective in removing grease. Power-driven rotary cutters are effective in stubborn cases. The use of a little copper sulfate in a sewer is often effective in killing roots without damaging the tree.

Occasional explosions occur in sewers. The most common sources of explosive gases are inflammable and volatile liquids in the wastewater or leakage of domestic gas from an adjacent main. Practically every city has an ordinance prohibiting the discharge of gasoline, naphtha from dry-cleaning establishments, or other inflammable liquids into the sewers. Gases given off by the decomposition of wastes are rarely the cause of explosions. However, many sewer-maintenance workers have been asphyxiated in gas-filled sewers. In no case should a workman be permitted to enter a sewer until proper tests for the presence of dangerous gases have been made. Whenever a workman enters a sewer, there should be a second man at the surface who can give emergency aid if required.

WASTEWATER TREATMENT[1]

When wastewater is discharged into a stream, floating solids may be washed up on shore near the point of disposal, where they decompose and create unpleasant odors. The large amount of organic matter in the wastewater may seriously deplete the dissolved oxygen in the stream causing fish kills or other undesirable effects. In addition, the waste will contaminate the water with pathogenic bacteria. Thus, even though municipal wastewater is about 99.9 percent water, it requires treatment if nuisance is to be avoided. The required degree of treatment depends on the character and strength of the waste and the disposal facilities. A small community at the seaside might discharge untreated wastewater directly into the ocean without any ill effects, but if this city were located inland on a small stream, a high degree of treatment might be required, particularly during the season of low flow.

In 1925 about 80 percent of the cities in the United States with population greater than 100,000 had no wastewater-treatment facilities. That figure has been greatly reduced, and as of 1970 practically all cities of 100,000 had treatment facilities and almost 90 percent of all municipal wastewaters discharged were treated to some degree. In California a bill passed in 1949 created a State Water Pollution Control Board and prohibited any community from polluting the inland and coastal waters of the state. This has resulted in the construction of many new treatment plants and the improvement and enlargement of many existing ones. Similar statutes exist in many states, while states bordering on many of the large Eastern rivers have formed interstate commissions to regulate waste disposal into those streams. Despite these advances much remains to be done. As of 1962 more than 2000 communities in the United States with a total population of over 14

---

[1] This section on wastewater treatment has been revised with the assistance of Prof. Perry McCarty of Stanford University.

million had no wastewater-treatment facilities.[1] To expedite the construction of waste-treatment facilities, Public Law 660 was enacted in 1956. It provides federal funds to local agencies to help finance the construction of treatment plants. The objective of this program is to upgrade the quality of water along many miles of streams, rivers, and lake fronts.

Urbanization, industrial growth, and the improved standard of living have increased the strength and quantity of municipal wastewaters in recent years to the point where dilution alone can no longer be relied upon to prevent the undesirable effects of pollution. In many areas, more advanced treatment of wastes is essential to prevent undue pollution. Since 1950 a new concept in the field of water-resources management has become prominent, namely that of *water-quality control*. The federal government has established a water-quality network to monitor areas in which water pollution is occurring so that remedial steps can be taken. The economic value[2] of water quality has been recognized, and the need to maintain certain minimum flows in streams and rivers is more clearly apparent than formerly.

19-12 The cycle of decay The organic matter in municipal wastewater is unstable and decomposes readily through chemical and bacterial action. If air or oxygen is supplied in sufficient quantities to the wastewater, the organic matter will undergo aerobic decomposition. Aerobic and facultative bacteria will oxidize the organic matter to the stable and unobjectionable end products of carbon dioxide (CO_2), nitrates (NO_3^-) and sulfates ($SO_4^=$). However, if sufficient oxygen is not available, then anaerobic decomposition (*putrefaction*) will occur. Anaerobic and facultative bacteria will convert the organic matter to simpler organic compounds, carbon dioxide (CO_2), methane (CH_4), hydrogen sulfide (H_2S), and ammonia (NH_3). Some of these materials, if released to the atmosphere, cause obnoxious odors. This is one reason why aerobic rather than anaerobic processes are desirable in streams. Since the oxygen resources of streams are limited, aerobic processes can continue only if the quantity of organic matter discharged is controlled.

The processes just described represent phases in the recirculation of certain chemical elements in nature. Figure 19-6 shows a simplified version of the *nitrogen cycle*. The sulfur and carbon cycles are similar, resulting in the formation of sulfates and carbon dioxide at the end of the oxidation phase of the cycle. The inorganic end products of decomposition are used by plants to form organic compounds which ultimately return again to the

---

[1] A. C. Glass and K. H. Jenkins, Statistical Summary of 1962 Inventory Municipal Waste Facilities in the United States, *U.S. Pub. Health Serv. Pub.* 1165, 1964.

[2] Allen V. Kneese, "Water Pollution—Economic Aspects and Research Needs," Resources for the Future, Washington, D.C., 1962.

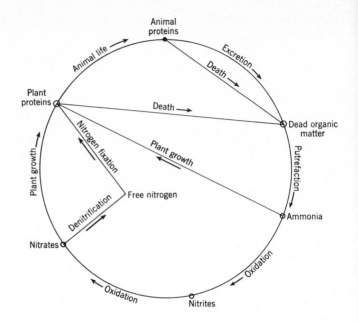

FIG. 19-6   The nitrogen cycle.

decomposition phase of the cycle. An understanding of these cycles permits a determination of the stage of decomposition of wastewater by testing for the products of decay. A well-oxidized waste, for example, will contain nitrates and sulfates but very little ammonia or hydrogen sulfide.

Several types of bacteria found in wastes are dangerous because they produce disease. Most bacteria found in wastewaters, however, are important aids in the process of decomposition. Biological treatment processes rely on an accelerated natural cycle of decay, and the objective of treatment-plant design is generally to provide an environment favorable to the action of the aerobic and anaerobic bacteria that stabilize the organic matter in the wastewater.

19-13 Characteristics of wastewater   Ordinary municipal waste contains between 500 and 1000 mg/l of solid matter. The physical and chemical characteristics of wastewater vary from time to time, and as a consequence frequent tests of the raw waste (*influent*) and of the treated waste leaving the plant (*effluent*) are necessary in order to get satisfactory plant operation.

An accurate determination of the waste characteristics will result if the sample tested is representative. Hence composite samples made up of portions of specimens are collected at regular intervals during a day. The amount utilized from each specimen is proportional to the rate of flow at the time the specimen was collected.

The solids in a wastewater consist of insoluble or suspended-solids and soluble compounds dissolved in the water. The *suspended-solids* content is found by drying and weighing the residue removed by filtering the sample. This residue is then ignited and that which burns is called the *volatile solids*. Volatile solids are presumed to be organic matter, although some organic matter will not burn, and some inorganic solids break down at high temperatures. The organic matter consists of proteins, carbohydrates, and fats. Fats and grease in excessive amount may interfere with the treatment process. The amount of fat or grease in a sample is determined by adding hexane to a sample of solids obtained by evaporation. The solution is then poured off and evaporated, leaving the fats and greases as a residue.

About 40 percent of the solids in an average wastewater are suspended. These solids may float or settle and form objectionable sludge banks if discharged to a river. Some of the suspended solids settle quite rapidly, but those of colloidal size settle slowly or not at all. *Settleable solids* are those which can be removed by sedimentation, and the standard test consists of placing a wastewater sample in a 1-liter conical glass container (Imhoff cone) and noting the volume of solids in cubic centimeters per liter that settles within the detention period of the particular plant. Usually about half of the suspended solids in municipal wastewater are settleable.

The most important test of wastewater is that for *biochemical oxygen demand* (BOD). If adequate oxygen is available, aerobic biological decomposition of waste will continue until oxidation is complete. The amount of oxygen consumed in this process is the BOD. Polluted water will continue to absorb oxygen for many months, and it is impracticable to determine this ultimate oxygen demand. The standard test for BOD consists in diluting a waste sample with water containing a known amount of dissolved oxygen and noting the loss in oxygen after the sample has stood for 5 days at 20°C. The rate at which BOD is satisfied (*deoxygenation*) depends on the temperature and the nature of the organic matter. Oxygen demand resulting from oxidation of organic matter is called *carbonaceous* or *first-stage* demand. Oxygen demand can also result from biological oxidation of ammonia and this is called *nitrogenous* or *second-stage* demand. The latter occurs after about 10 days of standing with raw waste, or sooner with well-treated effluents containing a large population of nitrogen-oxidizing (nitrifying) bacteria. The rate of exertion of carbonaceous demand is usually approximated with the following equation:

$$Y = L(1 - 10^{-K_D t}) \tag{19-1}$$

where Y is the BOD in milligrams per liter exerted after t days, L is the ultimate first-stage BOD, and K_D is a coefficient of deoxygenation at $T°C$. The value of $K_{D_{20}}$ is generally about 0.10 to 0.20 but in highly polluted, shallow streams, may be as great as 0.25. In effluents from biological treatment it is only 0.05 to 0.10. The value of $K_{D_{20}}$ for a given water can be determined experimentally by observing the variation with time of the dissolved oxygen in an incubated sample. If $K_{D_{20}} = 0.10$ the 5-day oxygen demand is 68 percent of the ultimate first-stage demand.

It has been found that K_D varies with temperature as

$$K_D = 1.047^{T-20} K_{D_{20}} \qquad (19\text{-}2)$$

The ultimate first-stage BOD of municipal wastewater varies with temperature as follows:

$$L = L_{20}(0.02T + 0.6) \qquad (19\text{-}3)$$

where L is the value of the ultimate first-stage BOD at $T°C$ and L_{20} is its value at 20°C. These equations make it possible to convert test results from different time periods and temperatures to the standard 5-day 20°C test. Values of the 5-day 20°C BOD of municipal wastes vary from about 100 to 500 mg/l.

It is generally assumed that the average BOD of domestic waste is 0.17 lb/day per capita, excluding industrial wastes. This figure of 0.17 lb/day is used to compute the *population equivalent* of industrial waste. Many cities base their charges for treatment of industrial waste at least partly on the computed population equivalent.

Other tests of wastewaters include tests for pH, free ammonia, organic nitrogen, nitrites, nitrates, phosphates, and chlorides. These tests are used to determine the character of the waste and the state of decomposition. The pH of the waste greatly influences many treatment processes and should be nearly neutral for biological processes. Certain industries discharge wastes containing chemicals which interfere with treatment, and tests for these chemicals may be important. The design of a treatment plant requires a careful study of the characteristics of the raw waste and the waste-assimilating capacity of the receiving stream in order to select the best process.

The biological characteristics of wastes are sometimes of importance. Since there are millions of bacteria per milliliter in raw waste, a total count is rarely made. However, tests for coliforms (Sec. 15-14) in a treatment-plant effluent are sometimes made to indicate the efficiency of the process in reducing bacteria and to determine the suitability for discharge in recreational waters. Some typical municipal wastewater analyses are shown in Table 19-2.

19-14 Self-purification of streams Where waste is discharged into

*TABLE 19-2   Typical Municipal Wastewater Analyses in Milli-
grams per Liter*

	Strength of wastewater		
	Weak	Medium	Strong
Total solids	405	810	1230
Suspended solids:			
Total	130	250	435
Volatile	80	150	270
Settleable solids, ml/liter	3	5	7
BOD	110	225	390
Ether-soluble matter	6	14	20

TABLE 19-3   Solubility of Oxygen in Fresh Water

Temperature		Dissolved oxygen, mg/l	Temperature		Dissolved oxygen, mg/l
°C	°F		°C	°F	
0	32.0	14.6	16	60.8	10.0
2	35.6	13.8	18	64.4	9.5
4	39.2	13.1	20	68.0	9.2
6	42.8	12.5	22	71.6	8.8
8	46.4	11.9	24	75.2	8.5
10	50.0	11.3	26	78.8	8.2
12	53.6	10.8	28	82.4	7.9
14	57.2	10.3	30	86.0	7.6

a body of water, a reserve of dissolved oxygen is necessary if nuisance is to be avoided. If a large supply of diluting water with adequate dissolved oxygen is available, the BOD of the waste can be satisfied without odorous conditions developing. The solubility of oxygen in water (Table 19-3) depends on the temperature. At high temperatures, when the bacterial action is most rapid, the solubility of oxygen is reduced. Hence conditions in a polluted stream will be worse in warm weather, particularly if it coincides with the low-flow season. When waste of concentration C_s flowing at rate Q_s is discharged into a stream with concentration C_R flowing at rate Q_R, the concentration C of the resulting mixture is given by

$$C_S Q_S + C_R Q_R = C(Q_S + Q_R)$$

$$C = \frac{C_S Q_S + C_R Q_R}{Q_S + Q_R} \tag{19-4}$$

This equation is applicable to oxygen content, BOD, suspended sediment, and other characteristic contents of the waste.

ILLUSTRATIVE EXAMPLE 19-3   Wastewater with a dissolved oxygen content of 0.8 mg/l is discharged at a rate of 5 mgd into a stream saturated with oxygen (temp. 55°F) whose flow rate is 40 cfs. What is the dissolved oxygen content of the resulting mixture?
From Eq. (19-4),

$$C = \frac{0.8(5 \times 1.55) + 10.6(40)}{(5 \times 1.55) + 40} = 9.0 \text{ mg/l}$$

If the streamflow had been only 10 cfs and the stream temperature 80°F, the dissolved oxygen of the resulting mixture would have been

$$C = \frac{0.8(5 \times 1.55) + 8.1(10)}{(5 \times 1.55) + 10} = 4.9 \text{ mg/l}$$

Oxygen deficit, D_t, is the difference between the actual oxygen content of the water and the saturation content at the water temperature. When polluted water is exposed to air, oxygen is absorbed to replace the dissolved oxygen that is consumed in satisfying the BOD of the waste. The processes of reoxygenation and deoxygenation go on simultaneously. If deoxygenation is more rapid than reoxygenation, an oxygen deficit results. If the dissolved oxygen content becomes zero, aerobic conditions will no longer be maintained and putrefaction will set in. The amount of dissolved oxygen at any time can be determined if the rates of reoxygenation and deoxygenation are known. These rates have been combined in the Streeter-Phelps equation,

$$D_t = \frac{K_D L}{K_R - K_D} (10^{-K_D t} - 10^{-K_R t}) + D_0 \times 10^{-K_R t} \qquad (19\text{-}5)$$

where D_t is the dissolved oxygen deficit in milligrams per liter at time t, L is the ultimate first-stage BOD at the point of waste discharge in milligrams per liter, D_0 is the initial oxygen deficit in milligrams per liter, K_D is the deoxygenation coefficient, and K_R is the reoxygenation coefficient. When waste is discharged into a stream, the values of D_t, L, and D_0 refer to the characteristics of the mixture. The deoxygenation coefficient K_D can be determined either by laboratory or field test while the reoxygenation coefficient K_R can be determined by field test. It has been found that K_R varies with temperature as

$$K_R = 1.016^{T-20} K_{R_{20}} \qquad (19\text{-}6)$$

where K_R is the reoxygenation constant at $T°C$. Typical values of $K_{R_{20}}$ are presented in Table 19-4.

TABLE 19-4   Values of the Reoxygenation Coefficient $K_{R_{20}}$

Small ponds	0.05–0.10
Sluggish streams	0.10–0.15
Large lakes	0.10–0.15
Large streams	0.15–0.30
Swift streams	0.30–0.50
Rapids	Over 0.50

Curve A of Fig. 19-7 is computed from Eq. (19-5), while curve B is plotted from Eq. (19-1). Curve A is known as the *oxygen-sag* curve, and the differences between curves A and B indicate the effect of reoxygenation. The time t_m at which minimum dissolved oxygen occurs can be found by differentiating Eq. (19-5) and equating to zero,

$$t_m = \frac{1}{K_R - K_D} \log \left(\frac{K_D L - K_R D_0 + K_D D_0}{K_D L} \frac{K_R}{K_D} \right) \tag{19-7}$$

and the critical oxygen deficit is

$$D_c = \frac{K_D L}{K_R} \times 10^{-K_D t_m} \tag{19-8}$$

Knowledge of the rates of deoxygenation and reoxygenation is of practical value in predicting the oxygen content at any point along a polluted stream. It permits an estimate of the degree of waste treatment required, or of the amount of dilution necessary, in order to maintain a certain dissolved oxygen content in the stream. If waste is discharged into a stream with

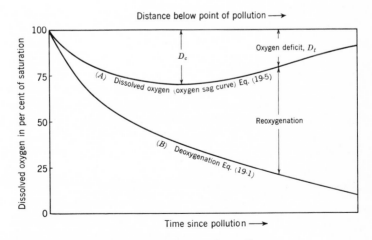

FIG. 19-7   The change in dissolved oxygen in a stream after pollution.

inadequate dissolved oxygen, the water downstream of the outfall will become anaerobic and will be turbid and dark. Sludge will be deposited on the stream bed, and anaerobic decomposition will occur. Over the reach of stream where the dissolved oxygen concentration is zero a zone of putrefaction may occur from which hydrogen sulfide, ammonia, and other odorous gases will arise. Since game fish require a minimum of 4 to 5 mg/l of dissolved oxygen, there will be no fish life in this portion of the stream. Farther downstream the water will become clearer and the dissolved oxygen content will increase until the effects of pollution are negligible.

Aerobic and anaerobic processes transform waste into a stable and unobjectionable form. An ample quantity of diluting water helps to assure the presence of dissolved oxygen. This permits aerobic bacteria to play an active role in the decomposition of the waste. Aerobic conditions will be enhanced if the stream is turbulent and reoxygenation is rapid. In some instances mechanical agitators or air bubblers have been used to put oxygen into streams, rivers, and ponds. These agitators are generally rotating or vibrating paddles; they are capable of putting 2 to 4 lb of oxygen per horsepower-hour into the water.[1] Sedimentation of waste solids on the stream bed permits anaerobic bacterial action.

19-15 Eutrophication This is a problem of current and growing concern. Eutrophication results from the enrichment of a body of water with fertilizing elements which in the presence of sunlight stimulate the growth of algae and other aquatic plants. Some quantities of these plants are helpful as they serve as a source of food for fish. However, large growths of these plants have many undesirable effects. They tend to clog streams and to form floating mats and to decrease water clarity. Large accumulations of algae can decrease property values as they produce unsightly accumulations on shorelines. They form breeding grounds for flies and insects and the decomposition of algae can lead to unpleasant odors. In addition, they have various adverse effects on impounded waters used for water supplies as discussed in Sec. 15-24.

Many elements are essential for the growth of algae, e.g., carbon, cobalt, hydrogen, potassium, magnesium, nitrogen, oxygen, phosphorus, and sulfur. However, in natural waters compounds of nitrogen and phosphorus are frequently in limited supply so that they tend to control the extent of growth that occurs. Compounds of nitrogen and phosphorus are present in the runoff from agricultural lands as well as in waste discharges from many industrial processes. The normal treatment given to municipal wastewaters removes little of the large quantities of nitrogen and phosphorus which they contain.

[1] Roy F. Weston, Advancements in Entrainment Aeration, *Proc. Sixteenth Ind. Waste Conf.*, Purdue University, Lafayette, Ind., May, 1961.

Thus, both treated and untreated domestic wastewaters have high concentration of these fertilizing elements and contribute materially to the rate of eutrophication of waters into which they are discharged. Processes for the removal of nitrogen and phosphorus are currently under development and are being installed in many waste-treatment plants.

19-16 Waste disposal There are two general methods of disposing of wastewaters: (1) *dilution*, or disposal in water, and (2) *irrigation*, or disposal on land. Disposal by dilution is the more common of the two methods. In all cases the larger suspended solids which tend to float or sink and form unsightly sludge banks should be removed from wastewaters prior to dilution. This method of disposal also requires a sufficient quantity of diluting water to maintain adequate dissolved oxygen below the outfall.[1] Just below the outfall the dissolved oxygen content depends on the relative amounts of waste and streamflow and the dissolved oxygen in each. The dissolved oxygen will decrease for some distance downstream because of the BOD of the waste. The minimum permissible dissolved oxygen is commonly taken as 4 mg/l for most fish and 5 mg/l for trout although there is some disagreement as to the proper value. Illustrative Examples 19-3 and 19-4 show typical computations for the determination of possibilities of dilution in a stream.

Disposal by dilution in a lake or estuary or in the ocean depends very much on the currents near the outfall. If these currents are not sufficient to mix the waste with an adequate volume of diluting water, undesirable conditions will develop. Dilution in salt water is less effective than dilution in fresh water because at the same temperature salt water can contain only about 80 percent as much dissolved oxygen as fresh water. On the other hand, the higher specific gravity of the salt water tends to retard sedimentation slightly.

An open, sandy soil is required for disposal by irrigation. A portion of the suspended solids in the raw waste should be removed before disposal on the land to prevent clogging of the soil. Natural drainage is usually depended upon, though drainage may be increased by the construction of tile underdrains. The usual rates for disposal by irrigation range from 5000 to 25,000 gal/acre/day, depending upon the permeability of the soil. The wastewater is usually applied intermittently so that seepage can occur in one area while waste is being applied to another. Organic solids are strained from the waste as it passes through the soil; and later when the bed is idle, aerobic soil bacteria oxidize the organic matter. Wastewater may be used to irrigate cotton, flax, nursery stock, and other crops not for human consumption. Most states have laws limiting the use of wastewater for irrigation of food

---

[1] Water-quality standards under the Water Quality Control Act of 1965 generally require secondary treatment for all wastes discharged to streams or lakes.

crops. Offensive odors, sometimes produced by overdosage or water-logged
soil, can be removed by application of chlorine or permitting the field to lie
idle for some time. The major drawback to disposal by irrigation is the large
land requirement. A city of 100,000 population would need at least 400 acres
of land for satisfactory disposal under the most favorable conditions.

ILLUSTRATIVE EXAMPLE 19-4   Wastewater is to be discharged at a rate of 20 mgd
(31 cfs) into a river with $K_{R_{20}} = 0.25$. During the low-water period, the flow rate in the
river is 350 cfs at a temperature of 24°C. Determine the maximum allowable 5-day BOD
of the waste at 20°C if the dissolved oxygen content below the outfall is to be not less
than 4 mg/l. Assume the waste temperature is 24°C and that the river water above the
outfall is 95 percent saturated with oxygen.

From Table 19-3 the dissolved oxygen content of the river water will be

$0.95 \times 8.5 = 8.1$ mg/l

Assuming that the waste contains no dissolved oxygen, the oxygen content in the stream
just below the outfall will be

$$\frac{(350 \times 8.1) + (31 \times 0)}{350 + 31} = 7.4 \text{ mg/l}$$

The initial oxygen deficit at the point of dilution is

$D_0 = 8.5 - 7.4 = 1.1$ mg/l

From laboratory test $K_{D_{24}} = 0.12$, and from Eq. (19-6) $K_R = 0.27$ at 24°C. As a first
trial, assume the ultimate first-stage BOD just below the outfall to be 20 mg/l. Then
from Eq. (19-7)

$$t_m = \frac{1}{0.27 - 0.12}\log\frac{0.27}{0.12}\frac{(0.12 \times 20) - (0.27 \times 1.1) + (0.12 \times 1.1)}{0.12 \times 20} = 2.13 \text{ days}$$

The critical oxygen deficit [Eq. (19-8)] is

$$D_c = \frac{0.12 \times 20}{0.27} \times 10^{-(0.12 \times 2.13)} = 4.9 \text{ mg/l}$$

The dissolved oxygen in the stream would then be $8.5 - 4.9 = 3.6$ mg/l, which is below
the allowable minimum. A second trial with $L = 18$ mg/l yields $t_m = 2.13$ days and
$D_c = 4.4$ mg/l. This leaves an oxygen residual of $8.5 - 4.4 = 4.1$ mg/l which is adequate.
Thus the ultimate first-stage demand of the diluted waste should not exceed 18 mg/l if
the dissolved oxygen in the river is to be maintained at a minimum of 4 mg/l. Using
Eq. (19-3), the ultimate first-stage demand of the combined flow at 24°C can be converted to first-stage demand at 20°C.

$$L_{20} = \frac{18}{(0.02 \times 24) + 0.6} = 16.7 \text{ mg/l}$$

The value of $K_{D_{20}}$ is found from Eq. (19-2) to be 0.10. Thus the 5-day BOD of the combined flow at 20°C from Eq. (19-1) would be $0.68 \times 16.7 = 11.4$ mg/l. The required

5-day BOD of the waste at 20°C can be found by allowing for the dilution and assuming the BOD of the river water above the outfall to be zero. Thus,

$$(350 \times 0) + 31X = (350 + 31) \times 11.4$$
$$X = 140 \text{ mg/l}$$

Thus, if the raw waste has a BOD greater than 140 mg/l, treatment is required to maintain sufficient dissolved oxygen in the stream.

19-17 Methods of wastewater treatment There are many different ways to treat wastewaters. Treatment processes are often classified as *primary, secondary,* or *tertiary*[1] processes. Primary treatment consists solely in separating a portion of the suspended solids from the waste. This is usually accomplished by screening and sedimentation in settling basins. The separated solids are stabilized by anaerobic decomposition in a digestion tank or incinerated. The residue is used for land fill or soil conditioner. The liquid effluent from primary treatment will ordinarily contain considerable organic material and will have a relatively high BOD.

Secondary treatment involves further treatment (or oxidation) of the effluent from a primary treatment process. This is generally accomplished through biological processes using filters, aeration, oxidation ponds, or other means. The effluent from secondary treatment will usually have little BOD and may even contain several milligrams per liter of dissolved oxygen. Tertiary treatment is for the purpose of removing dissolved and suspended materials remaining after normal biological treatment when required for water reuse or for the control of eutrophication in receiving waters. Various physical and chemical processes are available depending upon the particular requirements. The choice of treatment methods depends on several factors, including the disposal facilities available. Actually, the distinction between primary, secondary, and tertiary treatment is rather arbitrary since many modern treatment methods incorporate physical, chemical, and biological processes in the same operation.

19-18 Screening of wastewater Coarse screens or racks with 2-in. openings or larger are used to remove large floating objects from wastewaters. They should be installed ahead of pumps to prevent clogging. A coarse screen will collect about 1 cu ft of solids per million gallons of waste. This material usually consists of wood, rags, and paper which will not putrefy and may be disposed of by incineration, burial, or dumping.

Medium screens have openings ranging from about $\frac{1}{2}$ to $1\frac{1}{2}$ in. A typical installation of the fixed-bar type is shown in Fig. 19-8. A medium screen will ordinarily collect 5 to 15 cu ft of material per million gallons of

---

[1] Tertiary treatment of wastewaters is often referred to as *advanced treatment*.

*FIG. 19-8   Self-cleaning medium bar screen.
(Link-Belt Co.)*

waste. The screenings usually contain considerable organic material which may putrefy and become offensive and must be disposed of by incineration or burial. Coarse and medium screens should be large enough to maintain a velocity of flow through their openings under 3 ft/sec. This limits the head loss through the screens and reduces the opportunity for screenings to be pushed through the openings.

Fine screens with openings of $\frac{1}{16}$ to $\frac{1}{8}$ in. are often used to pretreat industrial waste or relieve the load on sedimentation basins at municipal plants where heavy industrial wastes are present. They will remove as much as 20 percent of the suspended solids in wastewaters. A fine screen should ordinarily be preceded by a coarse screen or shredder to remove the larger particles.

Screenings usually contain about 80 percent moisture by weight and will not burn without predrying. In most incinerators, the fresh screenings are dried by the heat of the fire before they enter the firebox. Fuel gas, oil, and sludge gas from the treatment plant may be used as fuel for the incinerator.

19-19 Comminutors Comminutors (or shredders) are devices which grind or cut waste solids to about $\frac{1}{4}$ in. size. In one type of comminutor, the wastewater enters a slotted cylinder within which another similar cylinder with sharp-edged slots rotates rapidly. As the solids are reduced in size, they pass through the slots of the cylinders and move on with the liquid to the treatment plant. Comminutors were introduced into the United States in the early 1930's. By 1950 there were more than 2500 in use. Comminutors eliminate the problem of disposal of screenings by reducing the solids to a size which can be processed elsewhere in the plant.

19-20   Grease removal   *Skimming tanks* are used where wastewater contains excessive grease or oil. A basin with detention time of about 10 min is satisfactory. The tank usually includes an aerating device which blows air through the waste at a rate of about 0.1 cu ft of air per gallon of waste. The rising air tends to coagulate the grease and cause it to rise to the surface, where it can be removed.

Another device for grease removal is the Vacuator, which consists of a closed tank into which aerated wastewater flows. The tank is subjected to a vacuum of about 9 in. Hg which causes the air bubbles to expand and move upward through the waste to the surface. The rising bubbles lift grease and the lighter waste solids to the surface, where they are removed through skimming troughs. Heavier solids settle to the tank bottom, where they are collected and carried away for sludge treatment and disposal.

19-21   Sedimentation of wastewater   Sedimentation plays an important part in wastewater treatment. It is used to remove grit from waste and will often reduce the suspended organic material by 50 percent. The basic concepts of sedimentation mentioned in Sec. 15-16 are applicable to wastewater treatment. Types of sedimentation units used include grit chambers, plain sedimentation tanks, chemical sedimentation basins, septic tanks, Imhoff tanks, and other devices. Septic tanks and Imhoff tanks combine sludge digestion with sedimentation (Sec. 19-27).

Grit chambers are intended to remove inorganic particles (specific gravity about 2.65) such as sand, gravel, eggshells, bone, etc., of size 0.2 mm or larger to prevent damage to pumps and to prevent their accumulation in sludge digesters. Typically a grit chamber may have a depth of 3 or 4 ft and a length of about 60 ft with a detention time of 30 to 60 sec. Reference to Fig. 15-11 indicates that a chamber with a depth of 4 ft, a length of 60 ft, and a detention time of 45 sec would remove all particles having a settling velocity greater than 0.088 ft/sec. This corresponds (Fig. 15-9) to all particles larger than about 0.2 mm. If the velocity in the chamber is too low, some of the organic material will settle; if the velocity is too high, the grit will not completely settle. An average municipal wastewater will yield about 2 to 5 cu ft of grit per million gallons. Many grit chambers are provided with automatic grit-removal mechanisms, often of the endless-chain type with scrapers attached. Grit may be used for fill or hauled away and dumped if it does not contain too much organic material.

Flexibility in operation can be achieved by providing several chambers in parallel so that the number in operation can be varied to suit the flow. The preferred design for a grit chamber is one that will permit a constant velocity of flow through the chamber regardless of the rate of flow. This can be achieved by a chamber of rectangular cross section having a proportional weir (Fig. 10-10) at discharge or by a chamber of parabolic cross section with discharge through an overflow weir of rectangular section. Often a trapezoidal section is used to approximate the parabola. Other designs[1] use various types of control flumes.

The function of a plain sedimentation basin in wastewater treatment is to remove the larger suspended material from the waste. The material to be removed is high in organic content (50 to 75 percent) and has a specific gravity of 1.2 or less. The settling velocity of these organic particles is commonly as low as 4 ft/hr, hence a basin detention time of 2 hr will generally give satisfactory removal in a basin 8 ft deep. The flow velocity through a sedimentation basin is very low, often less than 0.005 ft/sec, hence wind blowing across the surface of the basin may disturb the flow and reduce the effectiveness of the basin. Typical overflow rates for wastewater sedimentation basins range from 400 to 2000 gpd/sq ft. The designs of the basins are similar to those used for water purification. Both rectangular and circular tanks are used (Fig. 15-10) with depths in the range of 6 to 10 ft. A sloping bottom facilitates removal of the sludge. In order to get satisfactory performance from a sedimentation basin, the inlet must be designed to cause a uniform velocity distribution in the basin. This may be accomplished by placing baffles just downstream from the inlet. Large amounts of scum usually accumulate on the surface of sedimentation tanks, and skimming troughs, hand-operated scoops, or other devices for scum removal must be provided. A properly designed sedimentation basin will remove 50 to 65 percent of the suspended solids in wastewater. Flexibility of operation is obtained by having several basins in parallel. At low flows the tanks may be used in series, thus increasing the detention time and the efficiency of solids removal. It is good practice to have at least one basin in reserve in case of breakdown or cleaning of one of the other basins.

Chemical precipitation may be used to increase the removal of suspended solids. Such a process often results in over 90 percent removal. Chemical precipitation can also remove phosphates for control of eutrophication. Ferric chloride (Sec. 15-17) is the most commonly used coagulant in wastewater treatment. Lime is often added as an auxiliary chemical to improve the action of the coagulant. Chemical sedimentation is successful only if

[1] L. M. Lee and H. E. Babbitt, Constant Velocity Grit Chamber with Parshall Flume Control, *Sewage Works J.*, Vol. 18, pp. 646–650, July, 1946.

the chemicals and waste are properly mixed. Chemicals are introduced either in dry form or in solution. Mixing is usually accomplished by a 1- or 2-min rapid mechanical agitation. This is then followed by about 20 min of gentle agitation in a flocculating basin before the wastewater is introduced to the sedimentation basin. Chemical sedimentation is particularly advantageous where there is a large seasonal variation in wastewater flow or as an emergency measure to increase the capacity of an overloaded plain sedimentation tank. The main disadvantages of chemical sedimentation are the increased costs and the considerable increase in sludge volume which results.

19-22   Trickling-filter process   The effluent from sedimentation generally contains about 60 to 80 percent of the unstable organic matter originally present in the wastewater. The trickling-filter process is one method of oxidizing this putrescible matter remaining after primary treatment.

A trickling filter (Fig. 19-9) consists of a bed of crushed rock, slag, or gravel whose particles range from about 2 to 4 in. in size. The bed is commonly 6 to 9 ft deep, though shallower beds are sometimes used. Wastewater is applied to the surface of the filter intermittently by sprinklers. The wastewater trickles downward through the bed to underdrains, where it is collected and discharged through an outlet channel. A gelatinous biological film forms on the filter medium, and the fine suspended, colloidal, and dissolved organic solids of the waste collect on this film, where biochemical oxidation of the organic matter is accomplished by aerobic action. A new filter when first put into use will usually be quite ineffective for about 2 weeks until a satisfactory gelatinous film has formed on the particles in the bed. The film eventually becomes quite thick with accumulated organic matter and will

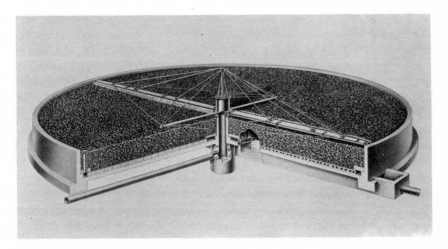

FIG. 19-9   Cutaway view of a trickling filter. (Dorr-Oliver Co.)

slough off (or unload) from time to time and be discharged with effluent. The effluent from trickling filters therefore requires sedimentation to remove the solids which pass the filter.

Raw municipal wastewater should not be applied to a trickling filter. Presedimentation of the applied waste is essential to good performance of the filter, for otherwise the filter will become clogged with the larger suspended solids in the waste. Many trickling-filter installations are designed so that a portion of the effluent can be recirculated back through the filter. Recirculation provides a longer effective contact time in the filter and odors are reduced since the waste is diluted in the recirculation. In addition, the rate of flow through the filters can be kept constant by varying the proportion which is recirculated.

Wastewater may be applied to a trickling filter by means of fixed nozzles or a rotary distributor. Fixed nozzles pointing upward are used for rectangular filter beds, but installation of filters with this shape is not common today. Rotary distributors are used in most current installations of circular filters. They require less head for operation and rotate automatically because of the thrust developed by the directional change in the flow of the waste as it leaves the distributor arm. Intermittent dosing results from the slow rotation of the distributor arm.

The best size for the material in a trickling-filter bed is about 2 in. Smaller sizes give more surface area for the formation of the gelatinous film but have lower permeability, which may result in ponding on the surface of the filter. Ponding is deleterious since successful operation of the filter depends upon a constant circulation of air. The bottom and sides of a filter are usually of reinforced concrete. The bottom is sloped to provide a gradient to carry the effluent to the underdrains which are commonly vitrified clay blocks with slots in the top. Effluent collected in the filter bottom is discharged into collecting channels. The hydraulic loading of settled waste to a standard trickling filter is 25 to 100 gpd/sq ft of filter area; in terms of waste strength, between 5 and 25 lb of BOD per 1000 cu ft of filter medium will be provided. A standard trickling filter with an area of 1 acre will satisfactorily serve a community of 20,000 people.

There are several types of *high-rate trickling filters*, such as the biofilter, accelo-filter, and aero-filter. The flow rate through a high-rate trickling filter may be as high as 250 to 500 gpd per sq ft of filter surface. Since the effluent is always recirculated, the net quantity treated daily is one-half to one-third of the gross flow. High-rate filters may operate singly, although two are often placed in series to form a two-stage system. A high-rate filter is constructed in much the same manner as a standard trickling filter. A biofilter is somewhat shallower, usually being only 3 or 4 ft deep. The effluent from a biofilter is conducted to a secondary settling basin and then recirculated to the primary settling basin or the filter. The rate of recirculation may be as high as 2 to 1, i.e., 2 gal of effluent from the settling tank is recirculated for each gallon of new wastewater introduced to the filter. The flow diagram for a typical two-stage biofilter incorporating some sludge return from the secondary to primary settling tanks is shown in Fig. 19-10. The rates of recirculation are generally adjusted as the wastewater flow changes to maintain approximately constant flow through the filters. An accelo-filter is one in which unsettled filter effluent is returned directly to the filter. The oxidized solids in this effluent intensify the biologic action and improve the effectiveness of the filter. An aero-filter is one in which the wastewater is applied continuously and uniformly over the filter bed by a multi-armed rotating distributor. Some aero-filters are provided with fans which serve to suck air through the filter.

Intermittent sand filters (Sec. 15-18) are occasionally employed in wastewater treatment. The flow rate through a sand filter is much lower than that through a trickling filter. An acre of sand filter is required per 1000 population, and because of the large area requirements and difficulties of winter operation such filters are rarely used except where a very high-quality effluent is required. Sand filters are sometimes used in the treatment of wastewaters from hospitals and other institutions. The sand filter depends upon the straining action of the sand and the action of aerobic bacteria. Wastewater should be presettled before application to the filter. Waste is placed on the filter bed daily to a depth of about 4 in. and allowed to percolate through the bed. Between dosages air enters the interstices between the grains and permits aerobic decomposition of deposited organic solids. Septic conditions may result from clogging of the bed, but this can often be remedied by resting the bed for a week or more. Sand filters are sometimes used for final or tertiary treatment following trickling filters or other secondary treatment processes. Such filters are often called polishing filters. Waste is applied continuously at rates as high as 10 gpd/sq ft, and the

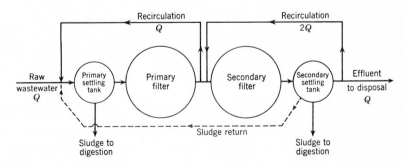

FIG. 19-10   Flow diagram for a typical two-stage biofilter.

straining action of the sand is relied upon to remove most of the remaining suspended solids in the wastewater.

19-23 Activated-sludge process In the activated-sludge process raw or settled wastewater is mixed with 20 or 30 percent of its own volume of activated sludge which contains a large concentration of highly active microorganisms. The mixture enters an aeration tank where the organisms and waste are mixed together with a large quantity of air. Under these conditions the organisms oxidize a portion of the waste organic matter to carbon dioxide and water and synthesize the other portion into new microbial cells. The mixture then enters a settling tank where the flocculant microorganisms settle and are removed from the effluent stream. The settled microorganisms or *activated sludge* is then recycled to the head end of the aeration tank to be mixed again with wastewater. New activated sludge is continuously being produced by this process, and the excess or waste activated sludge must be disposed of along with the sludge collected during primary treatment. The effluent from a properly operated activated-sludge plant is of high quality, usually having a lower BOD than that from a trickling filter. The percentage of removal of suspended solids and BOD may be as high as 90 or 95 percent.

A typical flow diagram for an activated-sludge plant is shown in Fig. 19-11. The wastewater usually undergoes plain sedimentation before aeration. Aeration chambers are commonly 10 to 15 ft deep and about 20 ft wide. The length depends upon the detention period, which generally varies from 4 to 8 hr for municipal wastewaters. There are two basic methods of introducing air to the chamber—air diffusion and mechanical agitation. The air-diffusion method involves introducing air under a pressure of 5 to 10 psi into the chamber through diffusion plates or other devices. In one design the tank bottom is formed into a succession of ridges and furrows,

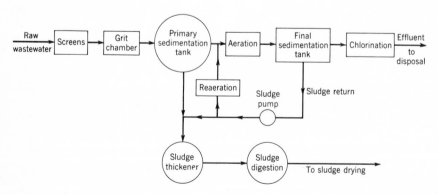

FIG. 19-11   Flow diagram for an activated-sludge plant with a high degree of treatment.

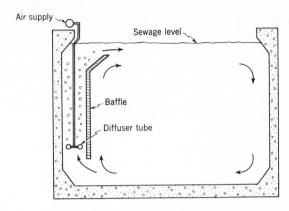

FIG. 19-12   Cross section of one type of spiral-flow aeration chamber.

and air is forced upward through diffuser plates in the bottom of the furrows. Another popular design is the spiral-flow aeration chamber (Fig. 19-12), in which air is introduced near the side of the tank in such a manner that spiral flow results in the tank.

Between $\frac{1}{2}$ and 2 cu ft of air per gallon of municipal wastewater are required to provide intimate mixing of the activated sludge with the incoming wastewater. Only about 5 percent of the oxygen in the air is actually involved in the biochemical action. Approximately 1000 cu ft of air is required per pound of BOD satisfied. Mechanical aeration uses paddles to stir up the waste and introduce air to it.

From the aeration chamber, the effluent passes to a final sedimentation basin with a detention period of about 2 hr and a surface loading of 400 to 2000 gpd per sq ft. The sludge which settles in the final settling basin is referred to as *activated sludge*, and a portion of it is mixed with effluent from the primary settling basin just before it enters the aeration chamber. The remainder of the sludge from the final settling basin is conveyed to sludge digesters or sludge driers. Special provision must be made to remove the activated sludge quickly from the settling tank to prevent it from becoming anaerobic. Industrial wastes with a high carbohydrate content, antiseptic or other properties, may cause bulking of the sludge, i.e., formation of sludge with a high water content. Settled activated sludge usually contains 98 to 99.5 percent water but may contain more moisture under extreme bulking conditions. The *sludge volume index* (volume in milliliters occupied by 1 g of sludge solids after 30 min of settling) should be around 100. When the sludge volume index exceeds 200, it is an indication of bulking. Bulking may sometimes be reduced or eliminated by increased aeration or by adding lime to obtain a pH of 8 or more, or by heavy chlorination to kill off the organisms responsible for bulking. One disadvantage of extreme bulking is that a portion of the suspended solids may not settle and will be discharged in the effluent. Another disadvantage is the large volume of sludge that must be handled. The best remedy for bulking is to determine the cause and exclude it from the wastewater or give it pretreatment at its point of origin.

The concentration of biological mass maintained in the aeration tank normally varies between 1000 and 4000 mg/l. Loadings to aeration tanks vary from over 1 lb of BOD per day per pound of mixed-liquor suspended solids in the high-rate process to less than 0.25 lb/day in the extended-aeration process. Conventional loadings vary between 0.25 and 0.5 lb of BOD per day per pound of mixed-liquor suspended solids. Efficiencies of BOD removal for the highly loaded high-rate process are between 65 and 80 percent, while in the conventional process efficiencies of over 90 percent are usually obtained. The low loadings of the extended aeration process result in high efficiencies of treatment, and more of the waste is oxidized and less synthesized into new biological cells than in the conventional process. For this reason, the extended aeration process requires more air but produces less waste activated sludge for disposal.

There are many modifications of the basic activated-sludge process and the most common are illustrated in Fig. 19-13. The first activated-sludge processes were of the *plug-flow type* in which the mixed influent waste and recycled activated sludge passed as a plug down a long narrow aeration tank. This has a disadvantage in that the largest demand for oxygen by the microorganisms is at the head end of the tank where the mixture first enters. With this design upsets are common and highly skilled operation is required. This adverse condition is relieved somewhat in the *step-aeration plant* as the waste is distributed over the length of the tank, allowing the oxygen demand to be exerted more uniformly. Many plants designed today are of this type. The *complete-mix plan* carries this idea further as the waste is added uniformly and the mixed liquor is also withdrawn uniformly over the length of the tank. Use of complete mix gives much stability to the activated-sludge process and permits its use for many industrial wastes. Inhibitory materials which may enter the tank are diluted over the entire length as is the oxygen demand so that upsets are much less frequent than in a plug-flow process.

The *contact-stabilization process* is designed for colloidal wastes such as domestic wastewater. The raw waste and recycled sludge are mixed and aerated for a period of 15 to 60 min in the contact tank to allow the suspended and dissolved organics to be sorbed to the activated-sludge floc. The sorbed organics and floc are removed in the settling tank and transferred to an aerated stabilization tank where the organics are stabilized over a period of

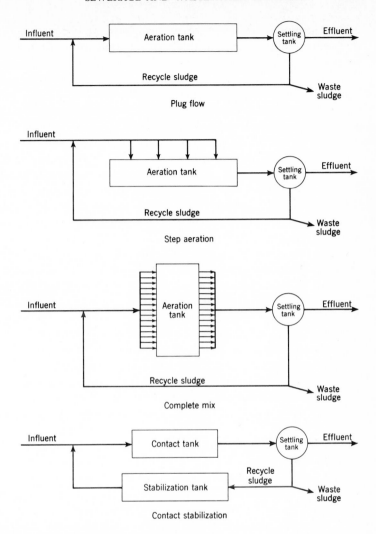

FIG. 19-13   *Modifications of the activated-sludge process.*

4 to 6 hr. The stabilized sludge is then mixed with influent waste again and the process is repeated. The contact-stabilization process requires less tank volume because the stabilization occurs after the sludge is concentrated in the settling tank. Less tank volume is required for a given overall detention time.

The chief advantage of the activated-sludge process is that it offers secondary treatment and an effluent of high quality with a minimum area requirement. The modifications described above have made the process less difficult to operate than formerly with the result that most secondary treatment plants being installed today are of this type. The capital cost is less than for a trickling-filter plant, but the operating cost is greater because of the high power requirements for operation of the air compressors and the sludge-circulation pumps. Generally the power requirements will be between 15 and 30 hp per 1 mgd of wastewater. When a new plant is put in operation, up to 4 weeks is required to form a suitable activated sludge, and during this period almost all the sludge from the final settling basin will be returned through the aeration tank. A plant is sometimes seeded with activated sludge from another plant.

19-24   Stabilization ponds and aerated lagoons   If an adequate area of flatland is available, secondary treatment of settled wastewater can be accomplished by retaining the partially treated wastewater in oxidation or stabilization ponds where organic matter is stabilized through the combined action of algae and other microorganisms. Better efficiency of treatment is obtained if several ponds are placed in series, so that the wastewater flows progressively from one to another until it is finally discharged. The ponds should be at least 3 ft deep to discourage growth of aquatic weeds, but best results are obtained if the depth does not exceed 6 ft as large surface area exposed to sunlight is necessary. The detention time in the ponds is usually 3 to 6 weeks.

In a stabilization pond a symbiotic relationship exists between algae and the other microorganisms such as bacteria and protozoa which oxidize organic matter. Algae, which are microscopic plants, produce oxygen while growing in the presence of sunlight. Oxygen is used by other microorganisms for oxidation of waste organic matter. End products of the process are carbon dioxide, ammonia, and phosphates which are required by the algae to grow and produce oxygen. The result of stabilization pond treatment is the oxidation of the original organic matter and the production of algae which are discharged to the stream. This results in a net reduction in BOD since the algae are more stable than the original matter in wastewater and degrade slowly in the stream. Stabilization ponds should not be constructed upstream of lakes or reservoirs as the algae discharged from the ponds may settle in the reservoirs and cause anaerobic conditions and other water-quality problems.

Properly operated ponds may be as effective as trickling filters in reducing the BOD of wastewater. An oxidation pond an acre in area can stabilize about 50 lb of BOD per day. Assuming an average daily BOD production of 0.17 lb/person, an acre of pond will accommodate a population of about 300. Thus relatively large areas of inexpensive land must be available. The banks of the ponds should be kept clean to avoid mosquito breeding and the ponds should be located sufficiently far from residential areas to avoid complaints from possible odors.

An *aerated lagoon* is a pond system for the treatment of wastewaters, with oxygen introduced by mechanical aerators rather than relying on photosynthetic oxygen production. The ponds are deeper than stabilization ponds and less detention time is required. Aerated lagoons require only 5 to 10 percent as much land as stabilization ponds. Efficiencies of treatment of 65 to 90 percent can be obtained with detention times of 4 to 10 days. Aerated lagoons are frequently used for the treatment of industrial wastes.

19-25 Sludge digestion The liquid *sludge* withdrawn from the bottom of a sedimentation basin contains the solids which have settled in the basin. The quantity of dried solids present in sludge will vary from about 0.1 lb per capita per day with primary treatment only to about 0.2 lb per capita per day with combined primary and secondary treatment. The moisture content of primary sludge will vary from 90 to 95 percent, of trickling-filter sludge from 96 to 98 percent, and of activated sludge from 98 to over 99 percent. Sludge is odoriferous and highly putrescible and must be stabilized before final disposal. A commonly used method of stabilization is *sludge digestion*. Before admission to the digestion tank, sludge from secondary treatment is often passed through a sludge thickener. This is a unit in which the sludge is subjected to prolonged stirring. The mixing action permits the escape of moisture held between the flocs and hence the sludge withdrawn from the thickener has a reduced moisture content and is less bulky. The liquid effluent from the thickener generally discharges into the primary settling basin. In another version, the sludge is thickened by air flotation. Air is mixed with the sludge under pressure and then discharged to a basin. The excess air forms small bubbles on the sludge particles causing them to float and form a concentrated sludge on the top of the basin. The thickened sludge is removed from the top and the underlying liquid is returned to the primary settling basin.

Sludge digestion is a biological process in which anaerobic and facultative bacteria convert 40 to 60 percent of the organic solids to carbon dioxide and methane gases. The organic matter which remains is chemically stable and practically odorless and contains 90 to 95 percent moisture. In older digesters which did not employ mixing, detention times of 30 to 60 days were required for proper stabilization of the sludge; but, with modern high-rate digesters employing mixing, detention times of 15 to 25 days are generally sufficient.

Digestion is accomplished in *sludge-digestion tanks*, or *digesters* (Fig. 19-14), which are ordinarily cylindrical in shape and may be equipped with mixing and sampling devices, temperature recorders, and meters for measuring gas production. Digesters are usually covered to retain heat and odors,

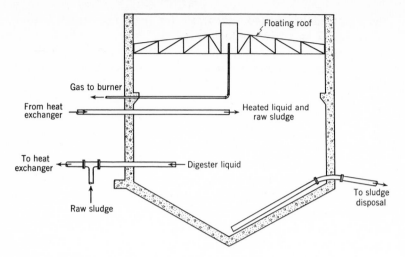

FIG. 19-14   Cross section of a typical sludge-digestion tank.

to maintain anaerobic conditions, and to permit collection of gas. They are commonly built of reinforced concrete with capacities (for ordinary municipal wastewater) between 1 and 3 cu ft per capita and depths of about 20 ft. Digestion proceeds most rapidly at temperatures of 90 to 95°F, and hence digesters require some means of heating. The most common method is to preheat the sludge and continuously circulate digester liquid from the digester to an external heat exchanger. With this system, some mixing of sludge in the digester is accomplished by currents set up in the tank by the circulating liquid. For more positive mixing in high-rate digesters, gas may be recirculated under pressure and discharged near the bottom of the tank. Mechanical stirrers may also be provided within the tank. Mixing brings the bacteria and sludge solids together for more rapid digestion and also maintains digester capacity by breaking up surface scum layers which tend to form.

During digestion volatile organic acids are formed as an intermediate step in the breakdown of organic material. These volatile acids are converted to methane gas by a specialized group of strictly anaerobic and slow-growing bacteria called "methane formers." If the methane formers are not operating properly, an accumulation of volatile acids may occur causing the pH to drop to values as low as 5.0, which will suppress further bacterial action. During digester start-up, care must be taken to prevent a drop in pH below a value of about 6.5. This may be done by adding raw waste slowly and by the addition of lime to the raw sludge. When a tank is first put in operation, it is highly beneficial to *seed* it with digested sludge from another tank. Without seeding, it may take a few months to get a tank operating properly. Raw sludge should be introduced near the top of the tank daily to maintain the digestion process at a uniform rate. The digested sludge is usually withdrawn from the bottom of the tank. With high-rate digesters, this is done daily but may be done less frequently with digesters having a long detention time. The bottom of the tank usually slopes toward an outlet in the center. Rotating sludge rakes are sometimes used to facilitate sludge removal. Partially digested sludge is often withdrawn from a digester and placed in a secondary digester without mixing. Here, the digestion process is completed and the sludge solids settle to the bottom and become concentrated prior to disposal. The liquid which accumulates at the top of the secondary digester is called *supernatant liquor* and is withdrawn at intervals. Since it usually contains about 1500 mg/l of suspended solids, the supernatant liquor is ordinarily returned to the primary settling basin for another journey through the treatment plant.

The gas evolved in a digester contains about 70 percent methane, 30 percent carbon dioxide, and traces of nitrogen, hydrogen sulfide, and other gases. About 10 cu ft of gas will be produced per pound of dried volatile solids added to the digester. This amounts to between 0.5 and 1 cu ft of gas per capita per day. Sludge gas has a fuel value of about 600 Btu/cu ft and may be used for heating digestion tanks and treatment-plant buildings or in gas engines to drive pumps, generators, or air compressors. Sludge gas has also been mixed with natural gas for domestic consumption. Special gas holders are sometimes provided to store gas, or it can be stored in a digestion tank equipped with a floating, gastight cover. A floating cover rises and falls as the volume of sludge, liquid, and gas in the tank changes and thus maintains a constant gas pressure and provides storage to even up production.

19-26 Sludge disposal Wet digested sludge is sometimes discharged at sea from hopper barges or through outfall sewers, but for most other means of disposal drying is a necessary preliminary. The most common method of drying is to spread the sludge to a depth of 12 to 18 in. on a sludge-drying bed, consisting of a 1-ft layer of coarse sand underlain by a layer of graded gravel. The required area for sludge-drying beds is between 0.5 and 2 sq ft per capita. A portion of the moisture drains through the bed, but most of it is evaporated to the atmosphere. It usually takes 1 to 2 months to dry sludge, depending on the weather and condition of the bed. In some instances the required area is reduced by building a glass roof over the beds to give protection against rain and snow. Sludge should never be applied to a bed until the preceding dose has been removed. Hence several drying beds are required.

Digested or raw sludge may also be dried by vacuum filtration. In this

FIG. 19-15   Vacuum filter for conditioned activated sludge. (Dorr-Oliver Co.)

process the sludge is first mixed with a coagulant such as ferric chloride and then conveyed to a vacuum filter (Fig. 19-15) consisting of a hollow drum covered with a replaceable filter cloth. The vacuum within the drum draws liquid through the cloth. The sludge cake which forms on the outside of the drum is removed by a scraper as the drum rotates. High-speed centrifuges are also used for drying of raw or digested sludges and are increasing in popularity because of small area requirements. Vacuum filtration or centrifugation of raw sludge is often used in situations where sludge disposal is by incineration. Vacuum filtration and centrifugation are used where there is insufficient area for sludge-drying beds or as a preliminary to heat drying.

Dried digested sludge has some value as a fertilizer or soil conditioner. Typical sludge may contain about 2 percent nitrogen and 1 percent phosphoric acid. If the sludge is not used for fertilizer, it may be incinerated or used for land fill as a soil conditioner.

ILLUSTRATIVE EXAMPLE 19-5   Raw wastewater entering a treatment plant contains 250 mg/l suspended solids. If 55 percent of these solids is removed in sedimentation, find the volume of raw sludge produced per million gallons of wastewater. Assume that the sludge has a moisture content of 96 percent and the specific gravity of the solids is 1.2. Find the unit weight of the raw sludge. If 45 percent of the raw sludge is changed to liquid and gas in the digester, find the volume of digested sludge per million gallons of wastewater. Assume that the moisture content of the digested sludge is 90 percent.

The amount of dry solids removed per million gallons is

$$\frac{250 \times 0.55}{1,000,000} \times 8.34 \times 1,000,000 = 1150 \text{ lb}$$

The volume of raw sludge per million gallons of wastewater is

$$\frac{1150}{1.2 \times 62.4} + \frac{96 \times 1150}{4 \times 62.4} = 457 \text{ cu ft}$$

The unit weight of the raw sludge is

$$62.4 \times \frac{1.2 \times 0.04 + 1.0 \times 0.96}{1.00} = 62.4 \times 1.008 = 62.9 \text{ lb/cu ft}$$

The weight of the dry solids in the digested sludge is

$$1150 \,(1.00 - 0.45) = 632 \text{ lb per million gallons}$$

The volume of the digested sludge is

$$\frac{632}{1.2 \times 62.4} + \frac{90 \times 632}{10 \times 62.4} = 99.5 \text{ cu ft per million gallons}$$

19-27   Imhoff tanks and septic tanks   An Imhoff tank (Fig. 19-16) provides for both sedimentation and sludge digestion. The raw wastewater enters the upper compartment, where sedimentation occurs. The solids slide down the steeply sloping bottom walls of the upper compartment and enter the lower digestion compartment. The digested sludge settles to the bottom of the lower compartment, from which it can be withdrawn through the sludge removal pipe. One sloping wall of the upper compartment overhangs the other by 6 in. or 1 ft so that the gas produced in the lower chamber

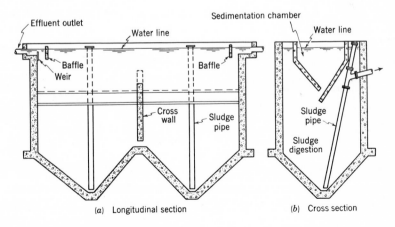

(a) Longitudinal section   (b) Cross section

FIG. 19-16   Simple Imhoff tank.

will not enter the upper compartment. The gas escapes through vents located on either side of the sedimentation compartment.

Before the development of separate sludge digestion, Imhoff tanks were used for many large treatment plants, but most of these have been replaced by more modern and efficient processes. Imhoff tanks are sometimes used for wastewater loads less than 250,000 gpd and will remove about 60 percent of the suspended solids and about 30 percent of the BOD. For small-capacity plants of less than about 1 mgd, various equipment manufacturers provide so-called *packaged units* of this type. The use of Imhoff tanks has greatly decreased and the process is mentioned here mainly for historical reasons.

A *septic tank* (Fig. 19-17) combines the processes of sedimentation and digestion in a single chamber with a detention period of 8 to 12 hr. The effluent from a septic tank is foul, contains 50 to 70 percent of the suspended solids, and has a high BOD. The effluent is run into an underground tile-drain field constructed of plain-end clay pipe which permits the wastewater to percolate through the soil. Septic tanks are suitable only for small communities, institutions, and private residences. Serious problems have resulted from their use in areas where soils have not been sufficiently permeable to carry away the effluent, so engineers should be cautious when specifying this method of disposal.

19-28   Chlorination of wastewater   Chlorination of wastewater is done in the same manner as for water treatment (Sec. 15-19). Chlorination may be used as a final step in the treatment of wastewater when an effluent low in bacterial content is necessary. Such use of chlorine is known as *postchlorination*. The disinfecting properties of chlorine reduce the bacterial count, and the oxidizing characteristics can reduce the BOD somewhat. *Prechlorination* before the wastewater enters the sedimentation tank helps

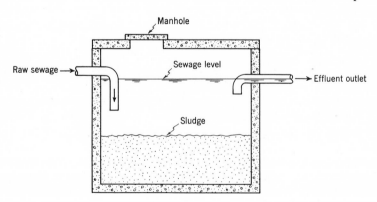

FIG. 19-17   Simple septic tank.

to control odors, may prevent flies in trickling filters, and assists in grease removal. In proper doses, chlorine will destroy the bacteria which break down the sulfur compounds in the wastewater and produce hydrogen sulfide. For this reason, chlorine is sometimes put into the main collecting sewers to prevent the destructive action of hydrogen sulfide on concrete pipe.

The usual chlorine dose varies considerably but is always much greater than in water purification. Prechlorination doses may be as high as 25 mg/l, while postchlorination usually requires at least 3 mg/l. Chlorine is sometimes added both at the beginning and end of the treatment process in what is known as *split chlorination*. Chlorination of wastewater is not always essential and because of the large doses required is relatively expensive. Hence, many treatment plants have no chlorination facilities, and plants with chlorinators often use them only intermittently to relieve overloads during peak periods. Chlorination is advisable when needed to prevent contamination of beaches, shellfish beds, and water supplies.

19-29   Nitrogen and phosphorus removal   Several treatment plants are now being designed for the removal of nitrogen and phosphorus from wastewaters as well as for the removal of organic materials. This is a result of increasing concern over the rapid rate of eutrophication (Sec. 19-15) of many waters receiving normally treated wastewater effluents. Phosphorus is present in municipal wastewaters in organic form, as inorganic ortho-phosphate, or as complex phosphates. The complex phosphates represent about one-half of the phosphates in raw municipal wastewater and result from the use of these materials in synthetic detergents. Complex phosphates are hydrolyzed during biological treatment to the orthophosphate form (PO_4^{-3}). Of the total average phosphorus concentration of about 10 mg/l present in municipal wastewater, about 10 percent is removed as particulate material during primary sedimentation and another 10 to 20 percent is incorporated into bacterial cells during biological treatment. The remainder, or about 70 percent, of the incoming phosphorus is normally discharged with secondary treatment plant effluents.

Many methods have been proposed for the removal of excess phosphorus including most of those listed in Sec. 15-22 for desalination of water supplies.[1] The most promising methods involve chemical precipitation as described in Sec. 19-21. Compounds of phosphorus can be removed by the addition of coagulants such as alum, lime, ferric chloride, or ferrous sulfate though chemical requirements are somewhat high for phosphorus removal as concentrations of these compounds of 100 to 400 mg/l are normally required to remove up to 95 percent of the phosphorus. The chemicals may

---

[1] Eliassen, R. and G. Tchobanoglous, Removal of Nitrogen and Phosphorus, *Proc. Twenty-third Ind. Waste Conf.*, Purdue University, Lafayette, Ind., May, 1968.

be added prior to primary sedimentation, alum and iron salts may be added to the aeration tank during the activated-sludge process, or chemicals may be added in a third stage following biological treatment (Fig. 19-18). When added at the primary treatment stage, a large portion of the organic material is removed as well as phosphorus so that a reduced loading on the biological treatment process results. However, a larger quantity of sludge is produced. When chemicals are added directly to the aeration tank of an activated-sludge plant, chemical treatment and biological treatment take place together and little additional equipment is required. Chemical precipitation, especially using lime, is sometimes practiced in a third stage after biological treatment to accomplish both the removal of phosphorus and the increase in pH of the effluent in preparation for a process of ammonia-nitrogen removal.

As illustrated in Fig. 19-6, nitrogen may be present in wastewaters in various organic forms, as ammonia, nitrites, or nitrates. It is also present as elemental nitrogen but this is of little concern since this form is not readily available to most aquatic plants. Most of the available nitrogen in raw municipal wastewater is in the form of organic or ammonia nitrogen at an average total concentration of around 25 mg/l. About 20 percent of this settles out during primary sedimentation. During biological treatment, a major portion of the organic nitrogen is converted to ammonia nitrogen and a portion of this is incorporated into biological cells which are removed from the waste stream prior to discharge. Thus, another 20 percent of the incoming nitrogen is removed during biological treatment. The remaining 60 percent is normally discharged to the receiving waters. Excess nitrogen can be removed by many of the methods listed in Sec. 15-22 for desalting of water supplies, but the currently most promising methods are *ammonia stripping* and *biological nitrification and denitrification*.

In *ammonia stripping* (Fig. 19-18) the pH of the effluent from secondary treatment is raised to 10 or higher, converting ammonia from the ionized form (NH_4^+) to the gaseous form (NH_3). The effluent is then passed through a packed tower equipped with an air blower such as is commonly used in industry for cooling. By blowing approximately 400 cu ft of air per gallon of wastewater through the tower, the ammonia gas is stripped from the wastewater. The pH must subsequently be reduced to around 8.0 to 9.0 with carbon dioxide to prevent harmful effects on the receiving waters.

In the *biological nitrification and denitrification* process, use is made of the nitrogen cycle (Fig. 19-6). At organic loadings less than 0.25 lb BOD per day per pound of volatile suspended solids, bacteria capable of oxidizing ammonia nitrogen to nitrite and nitrate nitrogen will develop in the activated-sludge process. This is an aerobic process requiring additional oxygen and is called *nitrification*. The next step in the process is the reduction of nitrite

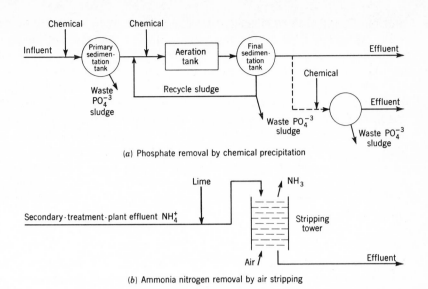

(a) Phosphate removal by chemical precipitation

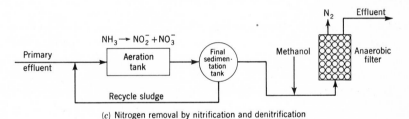

(b) Ammonia nitrogen removal by air stripping

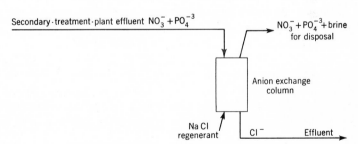

(c) Nitrogen removal by nitrification and denitrification

(d) Nitrate nitrogen and phosphate removal by ion exchange

FIG. 19-18   Processes for removal of nitrogen and phosphorus compounds.

and nitrate nitrogen to elemental nitrogen which escapes from the waste-water as a gas. This is an anaerobic biological process and is termed *denitrification*. It is best carried out in a separate reactor. A reactor similar to a normal activated-sludge system with recycle of denitrifying organisms and with mixing but no aeration may be used.[1] Another type of reactor for denitrification is an anaerobic filter which is similar to a trickling filter except that it is completely submerged so that no air can enter. Wastewater also passes upward rather than downward through the filter. A detention time of about 1 hr is required to remove 90 percent or more of the nitrate and nitrite nitrogen. Denitrification requires the presence of organic matter which the bacteria use for energy while carrying out denitrification. Methanol is a relatively inexpensive organic material which can be used for this purpose. The concentration required can be estimated as follows:[2]

$$C_m = 2.47 \, N_0 + 1.53 \, N_1 + 0.87 \, D_0 \qquad (19\text{-}9)$$

where C_m is the required methanol concentration, N_0 the initial nitrate nitrogen concentration, N_1 the initial nitrite nitrogen concentration, and D_0 the initial dissolved oxygen concentration, all expressed in mg/l. Since denitrification is an anaerobic process, some methanol must be added to biologically remove the dissolved oxygen (0 to 8 mg/l) present in the water to be denitrified.

Another promising method for phosphorus and nitrogen removal is *ion exchange*, similar to the process for water softening described in Sec. 15-21. However, for nitrogen and phosphorus removal, an anion exchange resin (Fig. 19-18) is operated on the chloride cycle so that as the waste passes through the ion-exchange column, both nitrates and phosphates are removed and are replaced with chlorides. After about 200 bed volumes of wastewater are passed through the ion-exchange column, the column must be regenerated. This is done by passing a concentrated sodium chloride solution through the column, removing the nitrates and phosphates and replacing them with chloride. The nitrates and phosphates are now contained in a concentrated brine solution which must be disposed in such a way that the nutrients do not reach a waterway. Other ion-exchange materials are being developed for the selective removal of ammonia, phosphates, and nitrates, and these offer promise for the future.

---

[1] E. F. Barth, M. Mulbarger, B. V. Salotto, and M. B. Ettinger, Removal of Nitrogen by Municipal Wastewater Treatment Plants, *J. Water Poll. Cont. Fed.*, Vol. 38, pp. 1208–1219, 1966.

[2] P. L. McCarty, L. Beck, and P. St. Amant, Biological Denitrification by Addition of Organic Materials, *Proc. Twenty-fourth Ind. Waste Conf.*, Purdue University, Lafayette, Ind., May, 1969.

In stabilization ponds algae incorporate nitrogen and phosphorus into their cell structure. The algae must be removed from the wastewater by chemical precipitation or other means. This has most promise for treatment of agricultural wastewaters where the algae may be used as a supplement for animal feed. Removal of both nitrogen and phosphorus from wastewaters by currently available methods will increase the cost of wastewater treatment by $50 to $200 per million gallons.

19-30   Advanced waste treatment   Advanced waste treatment refers to the use of other than conventional processes to prepare wastewater for direct reuse for industrial, agricultural, and municipal purposes. During a cycle of use for municipal purposes, both inorganic and organic materials are added to water. Normal wastewater treatment removes most of the readily biologically degradable organic material, but leaves from 40 to 100 mg/l of dissolved and biologically resistant or "refractory" organic materials in the effluent. These materials may be end products of normal biological decomposition or man-made products such as synthetic detergents, pesticides, and/or organic industrial wastes. Inorganic materials are also added during a cycle of use so that the concentration of salts such as magnesium, calcium, sodium, sulfate, chloride, and bicarbonate may increase 100 to 300 mg/l. Such salts are also refractory in nature. If wastewater is to be reused, as it must in some water-short areas, these refractory materials may need to be reduced depending on the expected use of the effluent.

Inorganic salts can be removed from treated wastewaters by processes discussed in Sec. 15-22 for desalting of water. Electrodialysis and reverse osmosis are most promising. The major problem with reclaimed wastewater is fouling of the membranes by refractory organic material in the treated waste. The most effective method for removing this refractory organic material is passage through columns containing activated carbon. This material has an enormous surface area per unit weight and is highly effective in removing organic contaminants. Processes for using activated carbon efficiently and for regenerating the carbon when its adsorptive capacity decreases are under development.

Another effective way for reclaiming used wastewater is to treat the water by normal biological methods and then allow the water to percolate through soil to mix with other groundwaters. Approximately 10 mgd of wastewater is being reclaimed for municipal purposes in this way at the Whittier Narrows Treatment Plant operated by the Los Angeles County Sanitation District.[1] Recovery of more water by this method is planned in the future.

19-31   Treatment of industrial wastes   The treatment of industrial

---

[1] J. D. Parkhurst, and W. E. Garrison, Water Reclamation at Whittier Narrows, *J. Water Poll. Cont. Fed.*, Vol. 35, p. 1094, 1963.

wastes may present special problems because toxic chemicals can hinder or destroy the bacterial activity depended upon in most treatment processes. Certain organic wastes impart undesirable tastes and odors. Often it is advantageous to require preliminary treatment of the wastes at the industrial plant before discharge into the public sewer. Such treatment usually includes screening and chemical sedimentation. Occasionally a large or remotely located industry must provide complete treatment of its wastes. In most instances, however, industrial wastes are combined with domestic waste-waters.

Milk-processing wastes often have BOD over 2000 mg/l. Treatment methods for such wastes include chemical precipitation, trickling filters, and activated sludge. Food-processing wastes are generally screened and often are sent through oxidation ponds. Textiles wastes are often highly colored. Their high acidity or alkalinity may be destructive to aquatic life and partial chemical neutralization may be necessary. Paper-mill wastes can be the most troublesome of all industrial wastes, particularly from plants using the sulfite process. Sulfite waste contains lignin which is very resistant to biological oxidation. In some sulfite mills process changes have been introduced to improve the character of the wastes. Radioactive wastes require special handling. Hospitals, research laboratories, water-cooled nuclear reactors, and nuclear-fuel processing plants are the chief contributors of such wastes. Wastes with high radioactivity are generally stored in underground vaults after concentration by evaporation. Low-level radioactive wastes can be removed by a variety of methods, including sand filtration, ion exchange, and biological processes. However, the filter media, ion-exchange beds, or sludges become radioactive and must be safely disposed of.

19-32 Résumé of wastewater-treatment processes   The design of a treatment plant is influenced by the effluent quality required for disposal. A well-designed plant provides for efficient expansion if the load increases or an improved effluent becomes necessary. An outline of the basic treatment methods follows:

1. Preliminary preparation ahead of the principal treatment processes
   a. Racks, screens, and comminutors
   b. Grit chambers
   c. Grease removal
2. Primary treatment processes for removal of settleable solids
   a. Septic tanks for institutions or private homes
   b. Plain or chemical sedimentation with separate sludge digestion for large cities

3. Secondary treatment processes for partial or complete oxidation
   a. Trickling filters
   b. Intermittent sand filters
   c. Activated-sludge process
   d. Oxidation ponds
4. Nitrogen and phosphorus removal for control of eutrophication
   a. Phosphorus removal by chemical precipitation or ion exchange
   b. Nitrogen removal by gas stripping, denitrification, or ion exchange
5. Supplementary treatment to improve effluent
   a. Polishing filters
   b. Chlorination
   c. Activated carbon.
   d. Electrodialysis or reverse osmosis.

BIBLIOGRAPHY

Babbitt, H. E., and E. R. Baumann: "Sewerage and Sewage Treatment," 8th ed., Wiley, New York, 1958.

Clark, J. W., W. Viessman, and M. J. Hammer: "Water Supply and Pollution Control," 2d ed., International Textbook, Scranton, Pa., 1971.

Design and Construction of Sanitary and Storm Sewers, *ASCE Design Manual* 37, 1959.

Eckenfelder, W. W., and D. J. O'Connor: "Biological Waste Treatment," Pergamon Press, New York, 1961.

Ehlers, V. M., and E. W. Steel: "Municipal and Rural Sanitation," 6th ed., McGraw-Hill, 1964.

Fair, G. M., J. C. Geyer, and D. A. Okun: "Water and Wastewater Engineering," Vols. I and II, Wiley, New York, 1968.

Hardenbergh, W. A., and E. R. Rodie: "Water Supply and Waste Disposal," International Textbook, Scranton, Pa., 1961.

Imhoff, K., and G. M. Fair: "Sewage Treatment," 2d ed., Wiley, New York, 1956.

Isaac, P. C. G. (ed.): "Waste Treatment-Proceedings of Second Symposium on the Treatment of Waste Waters," Pergamon Press, New York, 1960.

McGauhey, P. H.: "Engineering Management of Water Quality," McGraw-Hill, New York, 1968.

McKinney, R. E.: "Microbiology for Sanitary Engineers," McGraw-Hill, New York, 1962.

Nemerow, N. L.: "Theories and Practice of Industrial Waste Treatment," Addison-Wesley, Palo Alto, 1963.

Phelps, E. B.: "Stream Sanitation Engineering," Wiley, New York, 1961.

Rich, L. G.: "Unit Operations of Sanitary Engineering," Wiley, New York, 1961.

Sawyer, C. N., and P. L. McCarty, "Chemistry for Sanitary Engineers," 2d ed., McGraw-Hill, New York, 1967.

Sewage Treatment Plant Design, *ASCE Design Manual* 36, 1959.

Steel, E. W.: "Water Supply and Sewerage," 4th ed., McGraw-Hill, New York, 1960.

Units of Expression for Waste and Waste Treatment, *Sewage Ind. Wastes*, Vol. 30, pp. 709–716, May, 1958.

PROBLEMS

19-1 Compare the per capita sewage flow for your town with the per capita water use. Explain the differences between the two quantities. Compute the ratio of peak wastewater flow to average flow.

19-2 An 8-ft square reinforced-concrete combined sewer ($n = 0.015$) is on a slope of 0.0003. The maximum sanitary flow is estimated at 5 cfs. What opening should be provided in a leaping weir to permit a maximum of 5 cfs to drop into the interceptor and the balance to discharge into the bypass sewer if the downstream lip is 1.5 ft below the upstream lip of the weir?

19-3 A 24-in. cast-iron pipe has an outside diameter of 25.8 in. and a wall thickness of 0.63 in. and weighs 165.3 lb/ft. Can an 18-ft length of this pipe safely support its own weight plus the weight of wastewater (full) as a simply supported beam if its allowable working stress in tension is 12,000 psi?

19-4 A sag pipe must be constructed to carry the flow of a 72-in. circular reinforced-concrete sewer across a valley. The sewer is on a slope of 0.0008. The estimated maximum dry-weather flow is 7 cfs, and storm flows of 35 cfs are quite frequent. Design a three-pipe system to carry the full flow of the sewer with a minimum head loss. The bottom of the valley is 75 ft below the invert elevation of the sewer and the overall length of the sag pipes is estimated at 935 ft.

19-5 Design the sanitary sewer system for the area from Oak Avenue south on Fig. 19-5. The line on Laurel Avenue should be designed to take the flow into manhole 9 as computed in Illustrative Example 19-2. Other design criteria are as in Illustrative Example 19-2.

19-6 Investigate your state laws concerning stream pollution and prepare a summary of those provisions which should be considered in the design of a municipal sewerage system. Is your state a party to any interstate pollution-control compacts?

19-7 One hundred milliliters of water, saturated with oxygen at 68°F, is mixed with 200 ml of water saturated with oxygen at 50°F. Find the temperature of the resulting mixture and its percentage saturation with oxygen.

19-8 In a suspended-solids determination, 150 ml of wastewater is filtered and the collected solids are found to weigh 0.0358 g. Find the concentration of suspended solids in this sewage in mg/l.

19-9 In a BOD determination 5 ml of wastewater are mixed with 245 ml of diluting water containing 8.4 mg/l of dissolved oxygen. After incubation the dissolved oxygen content of the mixture is 6.2 mg/l. Calculate the BOD of the wastewater. Assume initial oxygen content of the wastewater is zero.

19-10 Find the 5-day, 20°C BOD of a wastewater if its 3-day, 10°C BOD is 100 mg/l. Assume $K_{D_{20}} = 0.10$.

19-11 A sanitary wastewater has a 5-day 20°C BOD of 400 mg/l. What is its 20-day, 25°C BOD?

19-12 An industrial plant discharges 250,000 gal of wastewater daily with a BOD of 315 mg/l. What is the population equivalent of this wastewater?

19-13 A city discharges 1.8 mgd of wastewater with a BOD of 295 mg/l into a stream. If the BOD of the stream is zero, what is the BOD of the diluted wastewater after discharge if the streamflow is 5 cfs? What stream discharge is required to reduce the BOD of the diluted wastewater to 40 mg/l? To what value must the BOD of the wastewater be reduced by treatment so that the BOD of the diluted wastewater is 40 mg/l if the streamflow is 2 cfs?

19-14 A city discharges 3.5 mgd of 15°C wastewater with a 5-day, 20°C BOD of 145 mg/l into a stream whose discharge is 37 cfs and whose water temperature is 18°C. If $K_R = 0.2$, what is the critical oxygen deficit and the time at which it occurs? Assume the stream is saturated with oxygen before the wastewater is added.

19-15 A stream with $K_R = 0.25$, temperature of 25°C, and minimum flow of 300 cfs receives

8 mgd of wastewater from a city. The river water is initially saturated with oxygen. What is the maximum permissible BOD of the wastewater if the dissolved oxygen content of the stream is never to go below 4.0 mg/l?

19-16 A raw wastewater has a suspended solids content of 250 mg/l. A sedimentation tank is expected to remove 60 percent of the solids. What volume of sludge will result from a million gallons of wastewater if the water content of the sludge is 95 percent? What sludge volume will result if its water content is 93 percent?

19-17 A wastewater-treatment plant for a city of 100,000 includes a medium screen, grit chamber, sedimentation basin, trickling filter, sludge digester, and sludge-drying beds. Assuming average conditions, specify the required sizes of the units for this plant.

19-18 The average flow at a wastewater treatment plant is 2.9 mgd. The raw wastewater contains an average of 300 mg/l of suspended solids and has a BOD of 350 mg/l. The plant serves a population of approximately 27,000. Assume total solids content is 1100 mg/l.

(*a*) What percentage of the raw wastewater (by weight) is water?
(*b*) What weight of suspended solids enters the plant per 24 hr?
(*c*) What is the volume of wastewater per capita per day?
(*d*) How many pounds of suspended solids are there entering the plant per capita per day?
(*e*) What is the oxygen requirement per person per day?
(*f*) There is a 24-in.-diameter pressure pipe leading to the plant from a pumping station. If the peak flow is 6.5 mgd and the minimum 1.6 mgd what are the maximum and minimum velocities in the pipe?

19-19 Seventy percent of the suspended solids in the wastewater of Prob. 19-18 are removed in the treatment process as raw sludge having a water content of 96 percent. Determine the volume of raw sludge formed each day. Laboratory tests indicate that the raw sludge solids are 30 percent fixed and 70 percent volatile on a weight basis. In the digestion process 60 percent of the total volatile solids are transformed into liquid or gas. Assume specific gravity of suspended solids is 1.20.

(*a*) If the water content of the digested sludge is 88 percent, what volume of digested sludge is formed each day?
(*b*) The tank has a diameter of 60 ft and average liquid depth of 16 ft. What is the average detention time?
(*c*) Gas weighing 0.06 lb/cu ft is evolved at the rate of 12,000 cu ft/day at standard temperature and atmosphere. How many pounds of gas are formed each day and how many pounds of sludge are liquefied each day?
(*d*) If the gas has a heating value of 600 Btu/cu ft how much horsepower could one develop using a gas-burning internal-combustion engine which has an overall efficiency of 35 percent?

Chapter 20

Flood-damage mitigation

Flood-damage mitigation was distinguished from drainage (Chap. 18) as embracing methods for combatting the effects of excess water in streams. More commonly called flood control in the United States, the terminology *flood-damage mitigation* has been adopted in this text from the Australian practice to emphasize that absolute control over floods is rarely feasible either physically or economically. What we seek to do is to reduce flood damage to a minimum consistent with the cost involved. For conciseness the term *flood mitigation* is used as a shorthand for the longer term *flood-damage mitigation*.

This broad definition encompasses many possible mitigation measures. A flood is the result of runoff from rainfall and/or melting snow in quantities too great to be confined in the low-water channels of streams. Man can do little to prevent a major flood, but he may be able to minimize damage to crops and property within the flood plain of the river. The commonly accepted measures for reducing flood damage are:

1. Reduction of peak flow by *reservoirs*
2. *Confinement* of the flow within a predetermined channel by *levees, flood walls*, or a *closed conduit*
3. Reduction of peak stage by increased velocities resulting from *channel improvement*
4. Diversion of flood waters through *bypasses* or *floodways*
5. Temporary *evacuation* of the flood plain
6. *Floodproofing* of specific properties
7. Reduction of flood runoff by *land management*
8. Flood insurance

Most flood-mitigation projects involve combinations of these measures.

20-1   Flood-mitigation jurisdiction   Man has always tried to avoid flood damage by one method or another, but the increase in population and property values in flood-threatened lands has brought the problem into sharper focus in recent years. Most federal flood-mitigation activity in the United States dates from about 1936. Federal authority for flood mitigation stems from the Commerce clause of the Constitution (Chap. 6). Responsibility for planning federal flood-mitigation projects rests with the U.S. Corps of Engineers and the U.S. Department of Agriculture.[1] Flood-mitigation features of projects built by other federal agencies are planned and operated under the direction of the U.S. Corps of Engineers. Recognizing a measure of local responsibility, Congress has provided that local interests (states and municipalities) must provide the right-of-way for local protection works (levees, flood walls, and reservoirs protecting a single locality) and must assume the maintenance and operation after the works are completed. Works protecting a large area (reservoirs and bypasses) are retained under federal control.

Nothing prevents local governments or private interests from undertaking their own protection works, but if these works are on a navigable stream, approval by the U.S. Corps of Engineers is necessary. Many state governments exercise supervisory control over local and private projects in order to assure that the projects fit properly into an overall plan for the stream. Without such caution, it is possible that protection works for one site may aggravate flood conditions elsewhere on the stream. Any person or organization altering the course of a natural stream is responsible for any undesirable effects, but it is preferable that these problems be anticipated and avoided.

20-2   The public view of flood mitigation   It is not unusual for a newspaper to describe a new flood-mitigation project as one which will "prevent floods for all time." This thought is a sedative which can lull the public to sleep behind inadequate protection, and their awakening may come too late to minimize the damage. Instances of preventable loss of life and property during a flood are plentiful. These losses often occur when people and property are not evacuated because it is assumed that adequate protection is available. Levees can fail, reservoirs can be full when a flood occurs, trash jammed against a bridge may create unexpected stages, or floods greater than the design flood can occur. In many such cases there might well be less loss of life or property if less faith were placed in the flood-mitigation works. This situation places a heavy burden on an engineer to explain clearly to the public the degree of protection intended. He must also provide evacua-

---

[1] The U.S. Department of Agriculture has authority to carry out projects for reduction of flood flow by watershed-management methods and by "upstream" reservoirs less than 5000 acre-ft capacity.

tion warnings at the earliest possible time if for any reason flooding of the protected area is imminent.

Alvord and Burdick[1] state:

> When works are concerned intended to protect a large population, a halfway measure is nothing less than a death trap. Anything less than to adequately protect a place of human habitation is worse than no protection at all, for it creates a false sense of security and multiplies the consequences of failure.

The hazard of halfway measures is always present. In the spring of 1913 work had just commenced on a project to protect Dayton, Ohio, from a flood of 90,000 cfs. Before work was well under way, a flood peak of about 250,000 cfs occurred, causing great damage and the loss of about a hundred lives. It is important therefore that the designer make a careful study of the situation before proposing a flood-mitigation plan.

The physical facts of most flooding situations are against the planner who attempts to provide absolute flood control, and talk of "adequate protection" can imply no more than a calculated risk. The planner must, therefore, make a very thorough study of each situation to derive the best alternative plan.

20-3   Flood-plain management   The flood plain of a river is formed by sediment deposition or removal accompanying intermittent overflows of the stream above its low-water channel. Most streams occupy some part of the flood plain every 2 or 3 years. Flood plains occupy about 5 percent of the land in the United States, and because they are nearly level they are attractive sites for railroads, highways, and cities. Historically the use of the streams for transport and water power encouraged occupance of the flood plain. Occupance of a flood plain subject to periodic flooding is not necessarily uneconomic. The advantages of the flood-plain location may outweigh the cost of periodic flooding. If, however, the extra cost of occupying a nearby flood-free location is less than the flood costs, flood-plain occupation is economically inefficient.

In the 30 years between 1936 and 1966 the United States government spent over $7 billion for flood mitigation. Nevertheless, during this same period flood damages increased steadily to a figure of about $1 billion[2] annually. Large annual expenditures for flood protection will be required to prevent the potential damage from increasing still further unless a systematic approach to flood-plain management is adopted.

Flood-plain management should, therefore, undertake to minimize the

[1] J. W. Alvord and C. B. Burdick, "Relief from Floods," p. 54, McGraw-Hill, New York, 1918.

[2] A Unified National Program for Managing Flood Losses, 89th Cong., 2d Sess., House Doc. 465, Government Printing Office, Washington, D.C., 1966.

costs of flood-plain occupance which consist of (1) the initial cost of development, (2) the cost of flood protection, (3) the residual flood damage, and (4) costs of relief and rehabilitation. Because federal authority over flood plains extends only to federal installations, the primary responsibility for flood-plain management falls on state and local jurisdictions. A federally subsidized program of flood protection encourages poor flood-plain management.

The tools of flood-plain management are many. *Flood-hazard surveys* indicating the extent of flooding at various probability levels provide the basic information for defining flood risk. *Flood-plain zoning* is a legal device through which local jurisdictions can restrict occupancy of the flood plain to uses which will suffer little or no damage during floods. Development of flood-plain lands can be prohibited, leaving them in their natural state or in agricultural uses. *Floodproofing* (Sec. 20-18) can be used by individual property owners to protect their property. Federal or state *mortgage insurance* could be refused to prospective developers in flood-hazard areas. Already developed areas subject to frequent flooding tend to be slums, and *urban renewal* might be used to clear the area and convert it to uses not threatened by flooding. Purchase of flood-plain land for parks and recreational use is an option open to local interests. A program of *flood insurance* in which premiums accurately reflect the risks would be an effective way of informing prospective builders or buyers of the true cost of floods. However, if premiums were too low to properly cover the flood hazard, insurance would become a form of subsidy encouraging unwise flood-plain use.

Specific recommendations for flood-plain management should be developed from a careful study of each flood plain. The best solution will vary from stream to stream depending on the geometry of the flood-plain area, the hydrologic characteristics of the stream, and the nature of the prospective uses of the flood plain. Such a study should precede or be a part of a study of possible flood-protection measures. Management studies should not concern themselves solely with residential uses but must consider industrial and commercial use problems, transportation, water-supply and wastewater facilities, solid-waste disposal, and all other activities associated with urban and agricultural use of land.

20-4   The design flood   The U.S. Corps of Engineers uses a *standard project flood* as the basis of its studies. This flood is defined as the "discharges that may be expected from the most severe combination of meteorologic and hydrologic conditions that are considered reasonably characteristic of the geographical region involved, excluding extremely rare combinations." The standard project flood is usually about 50 percent of the probable maximum flood for the area. However, because of the extreme rarity of the probable maximum flood, the standard project flood will not have been exceeded by more than a few percent of the floods within the general region. The relation between the standard project flood and the project design flood is best expressed by a quotation:[1]

In the design of a flood control project it would of course be desirable to provide protection against the maximum probable flood, if this were feasible within acceptable limits of cost. However it is seldom practicable to provide absolute flood protection by means of local protection projects or reservoirs; usually the costs are too high, and in many cases the acquisition of adequate rights-of-way for the purpose would involve unreasonable destruction or modification of properties along the floodway. As a rule, some risk must be accepted in the selection of a design flood discharge. A decision as to how much risk should be accepted in each case is of utmost importance and should be based on careful consideration of flood characteristics and potentialities in the basin, the class of area to be protected, and economic limitations.

The "design flood" for a particular project may be either greater or less than the standard project flood, depending to an important extent upon economic factors and other practical considerations governing the selection of the design capacity in a specific case. However, the selection should not be governed by estimates of the average annual benefits of a tangible nature alone, nor should construction difficulties that may prove troublesome but not insurmountable be allowed to dictate the design flood selection, particularly where protection of high class urban or agricultural areas is involved. Intangible benefits resulting from provision of a high degree of security against floods of a disastrous magnitude, including the protection of human life, must be considered in addition to tangible benefits that may be estimated in monetary terms.

The standard project flood is intended as a practicable expression of the degree of protection that should be sought as a general rule in the design of flood control works for communities where protection of human life and unusually high value property is involved. Inasmuch as standard project flood estimates are to be based on generalized studies of meteorologic and hydrologic conditions in a region, the standard project flood estimate provides a basis for comparing the degree of protection provided by a flood control project in different localities, thus promoting a more consistent policy with respect to selection of design floods giving a comparable degree of protection for similar classes of properties.

The standard project flood is usually determined by transposing the largest rainstorm observed in the region surrounding the project and converting the storm to flow by use of a rainfall-runoff relation and unit hydrograph. Thus the probability of the resulting flow is unknown and the standard project flood does not really serve the purpose of providing a comparable basis of design for projects in different regions.

---

[1] Section 1-08, *Civil Eng. Bull.* 52-8, rev. ed., Department of the Army, Chief of Engineers, March, 1965.

FLOOD-MITIGATION RESERVOIRS

20-5   Purpose of flood-mitigation reservoirs   The function of a flood-mitigation reservoir is to store a portion of the flood flow in such a way as to minimize the flood peak at the point to be protected. In an ideal case, the reservoir is situated immediately upstream from the protected area and is operated to "cut off" the flood peak. This is accomplished by discharging all reservoir inflow until the outflow reaches the safe capacity of the channel downstream (point *S*, Fig. 20-1). All flow above this rate is stored until inflow drops below the safe channel capacity, and the stored water is released to recover storage capacity for the next flood. Since the reservoir is situated immediately upstream from the point to be protected, the hydrograph at that point is the same as that released at the dam, and the peak has been reduced by an amount *AB* (Fig. 20-1).

If there is some distance between the reservoir and the protected area

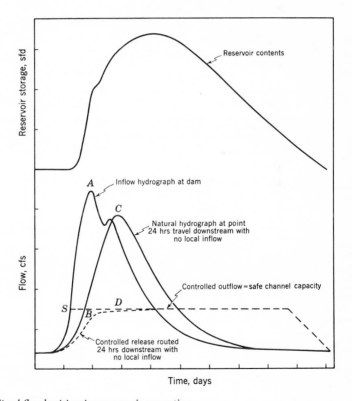

FIG. 20-1   *Idealized flood-mitigation reservoir operation.*

but no local inflow between these points, the reservoir operation will be quite similar. However, the natural hydrograph would be reduced by storage in the reach downstream from the reservoir so that a reduction AB at the dam is decreased to CD (Fig. 20-1) at the control point. If there is a substantial local inflow between the dam and the control point, the reservoir must be operated to produce a minimum peak at the protected area rather than a minimum peak at the dam. If, as is the usual case, the local inflow crests sooner than the inflow from upstream, the operation usually requires low releases early in the flood, with relatively higher releases timed to arrive after the peak of the local inflow (Fig. 20-2).

20-6   Location of reservoirs   The previous section indicates that the most effective flood mitigation is obtained from an adequate reservoir located immediately upstream from the point (or reach) to be protected. Often such a reservoir would be located in a broad flood plain where a very long dam would be necessary and a large area of valuable bottom land would be flooded. Sites farther upstream require smaller dams and less valuable land but are less effective in reducing flood peaks. The loss in effectiveness results from the influence of channel storage (Fig. 20-1) and from the lack of control over the local inflow between the reservoir and the protected city (Fig. 20-2). If the local area is sufficiently large, it may be capable of producing a flood over which the reservoir would have little or no control. It should be noted also that a single reservoir cannot give equal protection to a number of cities located at differing distances downstream. A significant, although largely qualitative, criterion for evaluating a flood-

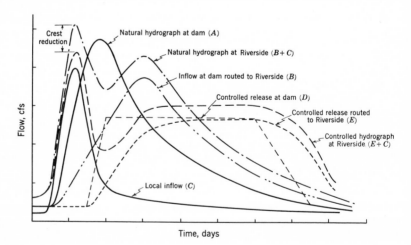

FIG. 20-2   *Hypothetical reservoir operation for the protection of the town of Riverside located 1 day's travel downstream.*

mitigation reservoir or a system of reservoirs is the *percent of the total drainage area controlled by the reservoirs.* Table 20-1 presents pertinent data on several flood-mitigation systems constructed or proposed in the United States. In general, at least one-third of the total drainage area should be under reservoir control for effective flood reduction.

Economic analysis and other factors often favor the upstream site despite its lesser effectiveness. Often several small reservoirs are indicated in preference to a single large reservoir. No general rules can be set forth because each problem is unique, and the several alternatives must be evaluated by economic analysis. The use of several small reservoirs offers the possibility of developing initially only those units of the system which yield the highest economic return and constructing the additional units as the development of the area increases the potential benefits.

20-7 Size of reservoir The potential reduction in peak flow by reservoir operation increases as reservoir capacity increases, since a greater portion of the flood water can be stored. For this reason a second criterion for evaluation of a flood-mitigation reservoir is its *storage capacity, usually expressed in inches of runoff from its tributary drainage area.* If this value is compared with the possible storm rainfall over the area, one obtains a rough idea of the potential effectiveness of the reservoir. The capacities provided in some flood-mitigation projects are indicated in Table 20-1.

It must not be presumed from the foregoing that the basic rule of design

TABLE 20-1   Data on Some Typical Flood-mitigation Projects

Drainage basin	Total area, sq mi	Number of reservoirs*	Controlled area		Capacity, in.	Cost† per acre-ft
			Sq mi	Per-cent		
Merrimack R. (N.H. and Mass.)‡	5,000	4	1,622	33	4.2	$ 89
Yazoo R. (Miss.)	13,076	4	4,425	34	16.3	13
Ohio R. above Pittsburgh, Pa.§	19,045	12	6,438	33	7.7	99
Arkansas R. above Pueblo, Colo.	4,800	1	4,800	100	1.0	20
Miami R. above Dayton, Ohio	2,513	4	2,433	97	5.5	26
Tennessee R. above Chattanooga, Tenn.	21,400	26	21,400	100¶	8.7	35
Muskingum R. (Ohio)	8,038	14	4,267	53	5.8	30
Willamette R. above Salem, Oreg.	7,280	11	3,303	45	10.1	121
Los Angeles R. (Calif.)	584	10	476	82	6.3	207

*Includes proposed reservoirs as well as those already constructed.
†Cost figures are very approximate and are not adjusted to present price levels.
‡There is additional natural storage in lakes in this basin.
§Major multipurpose projects not solely for flood mitigation.
¶Substantially controlled though more dams will be built.

is "the bigger, the better," for economic factors control the decision. The maximum capacity required is the difference in volume between the safe release from the reservoir and the design-flood inflow. As the reservoir size is increased, the law of diminishing returns may come into play. Because the hydrograph is wider at the low flows, more water must be stored to reduce the peak a given amount as the total peak reduction is increased. Assuming the hydrograph to be a triangle, the storage capacity required to achieve a given peak reduction varies as the square of the reduction. Moreover, the benefits achieved by a unit of peak reduction are usually less as the reduction is increased since more marginal area is being protected. In some instances the unit cost of storage in the reservoir will increase as the size is increased, although usually this cost decreases slightly. These factors are illustrated in Fig. 20-3. Curve A shows the variation in cost of a unit of storage as capacity is varied. Curve B shows the amount of flood-peak reduction which can be achieved for various capacities. These curves are used to plot curve C, which shows the total cost of a given peak reduction. When this cost is compared with the present worth of the estimated benefits of the reduction (curve D), it is evident that the project with a peak reduction of 60,000 cfs (storage capacity of 280,000 acre-ft) is the optimum as it gives the greatest net benefit. Investment beyond that point results in incremental benefits that are smaller than the incremental costs, hence from an economy viewpoint it is uneconomic to build beyond the 280,000 acre-ft storage capacity.

20-8   Operation problems   The idealized reservoir operation illustrated in Fig. 20-1 was determined solely by the limiting downstream channel capacity. If the flood volume had approached or exceeded the storage capacity of the reservoir, the operation would have necessarily been different. This could have been foreseen only with an accurate forecast of reservoir inflows. Similarly, an operation involving local inflow (Fig. 20-2) cannot be effectively planned without forecasts of the local inflow. Thus, streamflow forecasts become necessary in planning reservoir operations for flood mitigation. These forecasts are usually made on the basis of reports received by telephone, telegraph, or radio from a network of rainfall and river gages in the basin. In some instances automatic remote gages have been used. These reports permit the use of rainfall-runoff relations, unit hydrographs, and flood routing. Such forecasts may be quite accurate (± 10 percent) under favorable conditions, but if rain occurs after a forecast is made, it may be greatly in error.

A flood-mitigation reservoir has its maximum potential for flood reduction when it is empty. After a flood has occurred, a portion of the flood-mitigation storage is occupied by the collected floodwaters and is not available for use until this water can be released. A second storm may occur before the drawdown is complete. Consequently it is often necessary to reserve a portion

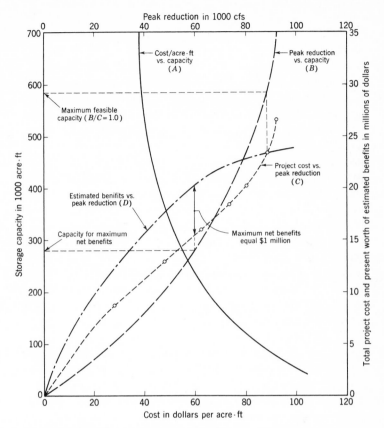

Peak reduction in 1000 cfs

Cost/acre·ft
vs. capacity
(A)

Peak reduction
vs. capacity
(B)

Maximum feasible
capacity (B/C = 1.0)

Project cost vs.
peak reduction
(C)

Estimated benifits vs.
peak reduction (D)

Capacity for maximum
net benefits

Maximum net benefits
equal $1 million

Storage capacity in 1000 acre·ft

Total project cost and present worth of estimated benefits in millions of dollars

Cost in dollars per acre·ft

FIG. 20-3   Cost analysis for a flood-mitigation reservoir.

of the storage capacity as protection against a second flood, i.e., the full
capacity of the reservoir cannot be assumed to be available for the control
of any single flood. If a second flood should occur while the reservoir is full,
the effect of the reservoir might be to make this flood worse.[1] These two
effects—uncertainty as to future inflows during the flood and the need to
reserve storage against a possible second flood—mean that a flood-mitigation
reservoir cannot be fully effective. Rutter[2] concluded that the TVA reservoirs

---

[1] The celerity of a flood wave through a reservoir is approximately equal to $\sqrt{gd}$, where d
is the depth of the water in the reservoir. This celerity is usually greater than the celerity of
the same wave in the natural channel without the reservoir. Hence, a full reservoir accelerates
the flood peak, and this may result in a closer synchronization with peak inflows to the stream
below the dam.

[2] E. J. Rutter, Flood-control Operation of Tennessee Valley Authority Reservoirs, *Trans.
ASCE*, Vol. 116, pp. 671–707, 1951.

were only about 50 percent effective during three floods on the Tennessee River at Chattanooga. The potential flood reduction to be expected from a flood-mitigation reservoir should be taken as substantially less than that calculated on the basis of ideal operation, except for small floods which require only a small portion of the reservoir capacity for adequate control.

A third operational problem develops when flows in excess of natural flows are released from a reservoir and synchronize at some point downstream with flood flows from a tributary. The resulting flows below this tributary may be greater than the natural flood flows would have been. This situation has occurred many times and is one of the hazards of flood-mitigation operation, especially on the larger rivers. It can be minimized only by weather forecasts several days or even weeks in advance. Releases from reservoirs in the upper Ohio or Missouri Rivers require 2 to 4 weeks to reach the lower Mississippi.

20-9   Types of reservoirs   There are two basic types of flood-mitigation reservoirs—*storage reservoirs* and *retarding basins*—differing only in the type of outlet works provided. The discharge from a storage reservoir is regulated by gates and valves operated on the basis of the judgment of the project engineer. Storage reservoirs for flood mitigation differ from conservation reservoirs only in the need for a large sluiceway capacity to permit rapid drawdown in advance of or after a flood.

A retarding basin is provided with fixed, ungated outlets which automatically regulate the outflow in accordance with the volume of water in storage. The outlet usually consists of a large spillway or one or more ungated sluiceways. The Pinay retarding basin in France (Fig. 20-4) consists of two wing dams partially closing the river but with a gap between them for discharge. The type of outlet selected depends on the storage characteristics of the reservoir and the nature of the flood problem. Generally the ungated sluiceway functioning as an orifice is preferable because its discharge equation ($Q = C_d A \sqrt{2gh}$) results in relatively greater throttling of flow when the reservoir is nearly full than would a spillway operating as a weir. A simple spillway is normally undesirable because storage below the crest of the spillway cannot be used. However, a spillway for emergency discharge of a flood exceeding the design magnitude of the outlets is necessary in any case.

The discharge capacity of a retarding basin with full reservoir should equal the maximum flow which the channel downstream can pass without causing serious flood damage. The reservoir capacity must equal the flow volume of the design flood less the volume of water released during the flood (Fig. 20-5). As a flood occurs, the reservoir fills and the discharge increases until the flood has passed and the inflow has become equal to the outflow. After this time,

FIG. 20-4   *The Pinay Dam on the Loire River in southern France (Courtesy Arthur E. Morgan)*

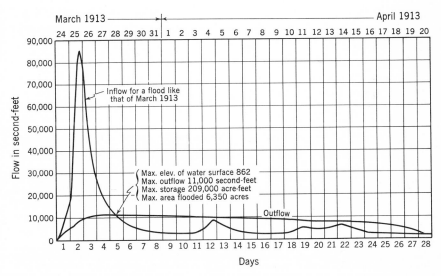

FIG. 20-5   *Inflow and outflow hydrographs for the Stillwater River at Englewood Retarding Basin, Ohio, for a flood like that of March, 1913. (Courtesy of Arthur E. Morgan)*

water is automatically withdrawn from the reservoir until the stored water is completely discharged.

The outstanding examples of retarding basins in the United States are the four reservoirs of the Miami Conservancy District in Ohio. Retarding basins were selected for this project because the small streams rise so rapidly that it would be difficult to operate storage reservoirs effectively. Moreover, the retarding basin assures the drawdown of the reservoir after a flood and prevents use of the reservoir for conservation purposes at the expense of flood control. Much of the land below the maximum water level in the reservoirs will be inundated only very occasionally and can be successfully used for agriculture, although no permanent habitation can be permitted on this land. Land near the highest proposed reservoir elevation will be flooded so infrequently that it can be farmed with almost no risk. At lower elevations the risk increases, until near the bottom of the reservoir the only practicable use may be for pasture. Thus usually only a small amount of land is permanently removed from use by the construction of retarding basins.

The planning of a system of retarding basins must assure that the basins will not make a flood worse by synchronizing the increased flow during drawdown with flood peaks from tributaries. When the entire drainage area is small, such an event is most unlikely. However, separate tributaries within a large basin may be subjected to wholly independent storms, and the probability of such synchronization is greater. Hence retarding basins are preferable for relatively small streams, and storage reservoirs are preferable for large streams.

LEVEES AND FLOOD WALLS

One of the oldest and most widely used methods of protecting land from floodwater is to erect a barrier preventing overflow. Some 2000 miles of levees exist in the lower Mississippi River Valley alone. Over 725 million cubic yards of earth were placed in this levee system between 1928 and 1940.

20-10 Structural design of levees and flood walls   Levees and flood walls are essentially longitudinal dams erected roughly parallel to a river rather than across its channel. A levee is an earth dike, while a flood wall is usually of masonry construction. In general, levees and flood walls must satisfy the same criteria of design as regular dams.

Levees are most frequently used for flood mitigation because they can be built at relatively low cost of materials available at the site. Levees are usually built of material excavated from borrow pits paralleling the levee line. If dry construction is used, material should be placed in layers and compacted. The least pervious material is placed along the riverside of the levee. Usually there is no suitable material for a core, and most levees are homo-

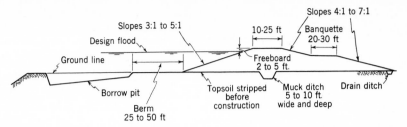

FIG. 20-6   Typical levee cross section.

geneous embankments. If a dredge is used for construction, the hydraulic fill method of construction is employed, and some gradation of the fill results if there is any gradation in the excavated material. Draglines or scrapers are commonly used for dry construction.

Levee cross sections must be adjusted to fit the site and the available materials. Details of a typical levee are shown in Fig. 20-6. Material is excavated from a borrow ditch parallel to the levee, and an adequate berm is left between the toe of the levee and the ditch to avoid collapse of the ditch bank. Top width of levees is usually determined by the use to which they are to be put, with a minimum width of about 10 ft to permit movement of maintenance equipment. Bank slopes are usually very flat because of the relatively poor construction materials. The slopes should be protected against erosion by sodding, planting of shrubs and trees, or use of riprap. For aesthetic reasons levees may be constructed with slopes even flatter than required for stability. This makes the levee less obvious and, if adjacent to a park area, makes it easier for people to cross the levee for access to the river bank.

Even though a levee does not fail during a flood, prolonged high water may raise the line of saturation to the point where seepage through the levee may cause extensive shallow flooding of the protected land. A drain ditch or tile line along the back toe of the levee is advisable, and the back slope should be flat enough to contain the seepage line if possible. Some saving in fill is possible by constructing the back slope with a berm (*banquette*) which increases base width without requiring heavy fill above the line of saturation. Topsoil should be stripped before the levee fill is placed. A cutoff trench (*muck trench*) extending 5 to 8 ft below the bottom of the fill and backfilled with the best available material is often used. If seepage threatens to be a serious problem, a sheet-pile cutoff may be used. In some cases a narrow trench is excavated in the completed levee with a ditcher and backfilled with a puddled-clay core.[1] The trench is filled with a thin slurry such

---

[1] Hans Kramer, Deep Cutoff Trench of Puddled Clay for Earth Dam and Levee Protection, *Eng. News-Record*, Vol. 136, pp. 76–80, June 27, 1946.

as drilling mud during excavation to prevent the walls from collapsing. A core must penetrate to a fairly impermeable horizon if any appreciable seepage reduction is to occur.

Because of the flat side slopes of levees, a levee of any considerable height requires a very large base width. Real estate costs for levees may be reasonable in rural areas, but in cities it is often difficult to obtain enough land for earth dikes. In this case masonry flood walls (Fig. 20-7) may be a preferable solution. Flood walls are designed to withstand the hydrostatic pressure (including uplift) exerted by the water when at design flood level. If the wall is backed by an earth fill, it must also serve as a retaining wall against the earth pressures when stages are low.

Flood walls and levees may cross the lines of railroads or highways. In some cases the roadbed can be elevated above the flood wall, but in many instances this represents an unwarranted cost in approach fill, higher bridges, and excessive slopes. Another solution is to leave a gap in the flood wall for the road or railroad. In this latter case a closure structure is necessary so that the gap may be blocked during high water. Stop logs or needles are commonly used for narrow openings, and large timber or steel gates are employed for wide openings. The design in each case is made to suit local conditions, the primary requirement being that the gate can be closed rapidly enough to prevent flooding. When closures of this type are used, the road or railroad becomes temporarily inoperative, but if the closures are

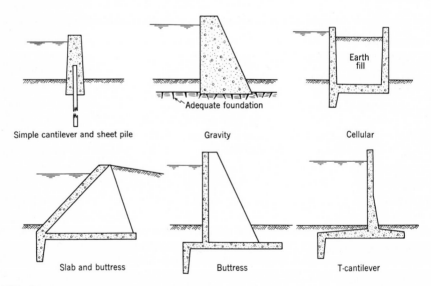

FIG. 20-7   Some typical floodwall sections.

infrequent and of short duration, this inconvenience may be small as compared with the cost of an elevated crossing.

20-11   Location of levees   A sufficient channel must be provided to transmit the design flow with a reasonable freeboard against wave action. The channel width between levees and the height of the levees are closely related. If the flood plain of the river is flat, an increase in channel width will permit lower levees. The cost of a levee system consists of the cost of the land for levees and channel plus the cost of levee construction. Hence, it is necessary to determine by trial the channel-width and levee-height combination which offers the minimum cost. In natural situations the conditions are rarely as simple as those just suggested. Most wide alluvial valleys have natural levees or high ground along the edge of the channel as a result of deposition of sediment when the stream overflows. It is often cheaper to place the levees along this high ground. In any case, full advantage should be taken of ridges, which permit lower levees and often provide better foundation conditions.

A city or agricultural district may be protected by a *ring levee* which completely encircles the area (Fig. 20-8a). The alternative to a ring levee is to carry the levee line back until it can be terminated in high ground (Fig. 20-8b). Without such a tie-in, the levee ends might be flanked and the levee would be useless. Minor tributaries are not leveed but are treated as problems in interior drainage (Sec. 20-12).

If a river channel is reasonably straight and land values are about equal on both sides of the stream, the levees will usually be spaced equidistant from each side of the river. Usually, however, the river is not straight, and the levee lines skirt the outside of the bends so that the leveed channel is less tortuous than the natural low-water channel.

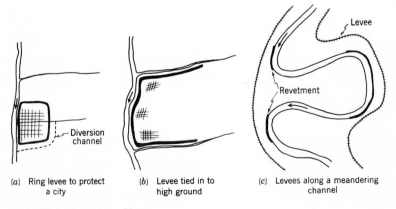

(a)   Ring levee to protect   (b)   Levee tied in to   (c)   Levees along a meandering
a city   high ground   channel

FIG. 20-8   Location of levees.

In no case should the levee be so close to a bend that bank caving will undermine the levee. At points where bank erosion can be expected, the levee should be firmly revetted and perhaps protected by permeable dikes as assurance against failure (Sec. 17-4). The importance of bank-protection work associated with levee construction cannot be overemphasized. Between 1935 and 1945 it was necessary to construct 135 mi of setback levees along the lower Mississippi River[1] to replace levees lost by bank caving. The Yazoo Levee District with a total levee length of 178 mi lost 305 mi of levee between 1880 and 1943. Less than 27 mi of levee now stand on the original line of 1867. It is usually true that levee construction for flood mitigation goes hand in hand with bank protection and channel improvements, since a combination of these two techniques usually provides the most protection for the least cost.

20-12 Interior drainage A levee line must inevitably cross tributary channels, and the designer has two alternatives: (1) to carry the levees upstream along the tributaries to tie in to high ground or (2) to block the channel and create an interior drainage problem. The selection between these alternatives is primarily an economic one, but obviously many small streams cannot be treated economically by alternative 1, and hence the problem of interior drainage (the local drainage problem behind a levee) is present in almost all levee designs. Most streams (and sewers) can discharge by gravity during low flows if tide gates or other closures are provided to prevent backflow of river water during floods. Fairly large streams can be treated with a battery of tide gates, but in some cases it is necessary to provide one or more large gates to be closed at the discretion of local authorities.

Four general solutions to the problem of interior drainage have been used. The water may be collected at some low point and pumped over the levee during floods when gravity flow through outlet gates is impossible (Fig. 20-9a). The water may also be collected in an open channel on the land side of the levee and diverted downstream to some point where gravity discharge is always possible (Fig. 20-9b, c, d). Tributary streams are sometimes enclosed in a pressure conduit whose upstream end is at such a level as to permit gravity flow into the main stream at all times (Fig. 20-9e). A final possibility is to collect the water in a storage basin until gravity discharge to the stream is possible. The best solution for a given problem depends on the local topography and the stream characteristics. Storage of water is impractical if periods of high water in the main stream are of long duration. Use of a pressure conduit is feasible only if high land is not so far from the main stream as to make the conduit costs excessive. The most widely used

[1]Charles Senour, New Project for Stabilizing and Deepening the Lower Mississippi River, *Trans. ASCE*, Vol. 112, pp. 277–297, 1947.

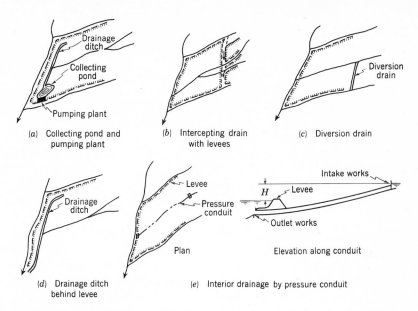

FIG. 20-9   *Some plans for the interior drainage of leveed areas.*

solution is that of a pumping plant with a limited amount of storage to reduce variations in pumping load.

20-13   Levee maintenance and flood fighting   Foundation conditions and building materials for levees are rarely fully satisfactory, and thus, even with the best construction techniques, there is a hazard of failure. Caving of the stream bank may result in undercutting of the riverward toe of the levee. Gopher holes or channels left by decaying roots may permit the start of destructive erosion, leading to an eventual levee failure, or *crevasse*. Seepage through the foundation material at high stages of the river can cause a sand boil (Fig. 20-10), and the removal of foundation material by piping through the boil may create a channel which can collapse under the weight of the levee. There are many possible causes of levee failure, and no levee can be assumed to be safe during a flood. Nor can failure be prevented by a finger in the dike in the tradition of the little Dutch boy.

Levees should undergo regular annual inspection by competent engineers who will look for evidence of dangerous bank caving, weak spots created by animals or vegetation, foundation settlement, bank sloughing, erosion around the outlets of sewers or other pipes passing through the levee, and other possible sources of danger. Any alarming condition should be corrected promptly. During floods a continuous patrol of the levee should be maintained. Patrolmen should have arrangements for immediate communication

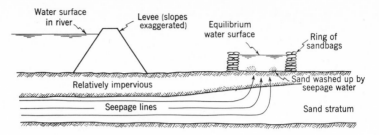

FIG. 20-10   Sand boil with a protective ring of sandbags.

with flood-fighting forces and equipment for immediate repair of minor danger spots.

Flood fighting is the term applied to the work which is necessary during a flood to maintain the effectiveness of a levee. Fast work and considerable ingenuity are sometimes necessary to prevent failures, and the necessary materials and equipment should be kept immediately available at all times. A sand boil is, in effect, an artesian spring in the aquifer under the levee, with a velocity sufficient to move the foundation material. Boils are checked by a ring of sandbags (or other material) to create a pool causing a back pressure sufficient to reduce the net head to the point where the flow velocity is too low to cause washing of the soil (Fig. 20-10). Bank caving may proceed unnoticed under the floodwater but, if detected, may sometimes be controlled by dumping rock, sandbags, fascines of timber, or other material into the caving area. If seepage causes a slide on the land side of the levee, this slope may be strengthened with brush or timber weighted with sandbags. As the river rises, low spots in the levee will become apparent and, if overtopping is threatened, these areas must be raised. A levee can be raised a foot or two with earth taken from the land side of the embankment or by sacks filled with soil (Fig. 20-11). If further raising is necessary, a timber wall supported by earth or sandbags or a mud box filled with earth is usually necessary. Plastic sheeting weighted with sandbags is an effective temporary protection against erosion and seepage on emergency fills. If the possibility of defending the existing levee seems poor, a *setback levee* may be constructed to contain the floodwaters which would enter through a break in the main levee. The location of the setback levee should be selected primarily on the basis of ease and speed of construction, although as much land and property as possible should be protected without endangering the entire leveed area.

Flood walls are usually less susceptible to failure than are levees, but high water may overtop a wall or sand boils may occur. Consequently, flood walls should also be patrolled, and material and equipment should be available to raise their height or to control sand boils.

20-14   Effect of levees on river stages Levees usually restrict the channel width by preventing flow on the flood plain, and this results in increased stages in the leveed reach. Channel improvements, which usually accompany levee construction, will increase velocities and may offset some or all of this increase. If stages in the leveed reach are increased, stages will also be higher upstream from the leveed reach. Downstream from the leveed area, peak flows will be increased because of lessened channel storage as a result of the increase in flow velocity. The net result of levee construction depends very much on the physical characteristics of the situation. Usually, however, levee construction and associated flood-mitigation works result in a general increase in flood stages along a river unless reservoirs or extensive channel improvements are provided.

The increase in stage following levee construction has sometimes led to unfortunate consequences. Often an area protected by levees has suddenly found itself endangered and perhaps flooded because of new levees constructed in the vicinity. The best program of flood control for a river will be obtained if a master plan is developed at an early date.

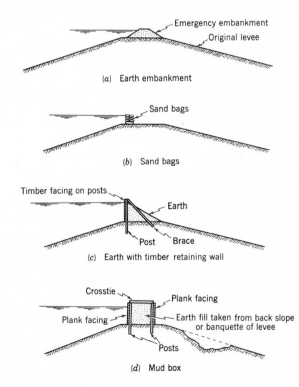

FIG. 20-11   *Methods for raising levee height in emergencies.*

FLOODWAYS

20-15   Floodways serve two functions in flood mitigation. First, they create large, shallow reservoirs which store a portion of the floodwater and hence decrease the flow in the main channel below the diversion. Second, they provide an additional outlet for water from upstream, increasing velocity and decreasing stage for some distance above the point of diversion.

Opportunities for the construction of floodways are limited by the topography of the valley and the availability of low-value land which can be used for the floodway. The floodway is ordinarily used only during major floods, and the land in the floodway may be used for agriculture, although usually no fixed improvements of any great value are permitted in the floodway area. The bypass system of the Mississippi River is shown in Fig. 20-12. Floodways should be used only when absolutely necessary. Much sediment is deposited in river channels during low flows, and unless periodic floods are permitted to scour these deposits, the channel tends to aggrade, or build up. In addition, the purchase of flowage rights for the floodway will be less costly if flooding is infrequent.

Admission of water to a floodway is achieved in a number of ways. In many cases flow occurs over a low spot in the natural bank or a gap in the levee line. In some cases a *fuseplug levee* is provided. This is a low section of levee which, when once overtopped, will wash out rapidly and develop full discharge capacity into the floodway. In other locations a concrete sill, or weir, is provided so that overflow occurs at a definite river stage. This is advantageous when overflow occurs quite frequently since it obviates the need for replacement of a levee section each time flow into the bypass occurs. Admission of water to the bypass may also be accomplished by dynamiting a section of levee when the flood situation warrants. Even more complete control is obtained with a spillway section with openings closed by stop logs or needles. With this latter arrangement diversion may be limited to the amount required to reduce flows to the capacity of the leveed channel. Controlled spillways such as the Bonnet Carre Spillway above New Orleans and the Sacramento Weir near Sacramento, California, are relatively costly and are ordinarily used only for the protection of major cities.

At the point of diversion to a floodway, the capacity of the two channels (main and diversion) is greater than that of the original channel, and a lower stage suffices to discharge a given flow. The water-surface slope above the diversion point is increased, while that below the diversion point is decreased. If the channel slope is relatively steep, the change in water-surface slope caused by the diversion may be negligible. If the channel slope is quite flat, the change in water-surface slope may be relatively large and the

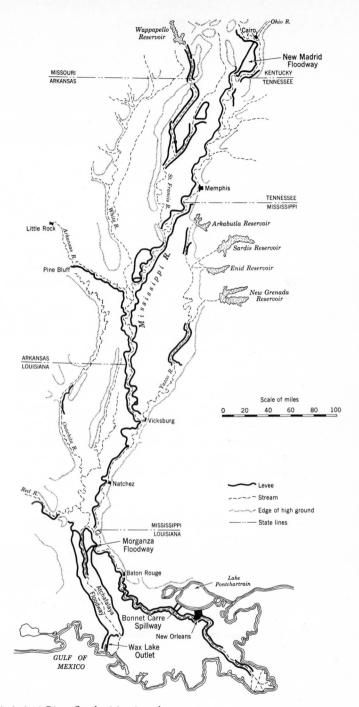

FIG. 20-12   *Mississippi River flood-mitigation plan.*

capacity of the channel above the weir considerably increased, with a corresponding decrease in stage for a given discharge. The flattening of the slope downstream decreases the channel capacity at a given stage, but, with a substantial amount of water diverted, stages may still be lowered for some distance downstream. The stage lowering will not, however, be as large as would be expected on the basis of the stage-discharge relation existing before the diversion.

The hydraulic design of the diversion works for a floodway is considerably more complicated than a simple determination of the amount of flow to be diverted. If a specific point such as a city is to be protected, the diversion works should be above the city if the channel is relatively steep and below the city if the channel is flat. The actual design requires the determination, largely by trial, of a weir length and crest elevation which will achieve the desired effect. This is done by assuming several weir designs and computing the water-surface profiles above and below the weir at design discharge. The proper design must satisfy the condition that the stage computed at the weir for the given total inflow upstream must equal the stage required to discharge this same flow over the weir and down the stream channel. This is a statement of the principle of continuity, i.e., for a given stage at the weir, the sum of the weir discharge and the flow in the main channel below the weir must equal the total inflow from upstream.

The effect of a bypass on the stage downstream from the point where the bypass rejoins the main stream is only that produced by the increased channel storage in the floodway. If the hydrographs of flow via the main channel and floodway are adjusted for the effects of storage and added together, the sum will represent the flow at a point below the junction of the two channels.

CHANNEL IMPROVEMENT

20-16   Marked reduction in stage at a specific point on a stream can often be achieved by merely improving the hydraulic capacity of the channel. Removal of brush and snags, dredging of bars, straightening of bends, and other devices (Chap. 17) can be effective, though care must be taken not to make the channel susceptible to bank erosion. Carried to its ultimate, the channel may be completely lined and greatly straightened as has been done for the Los Angeles River. These methods achieve their purpose by decreasing Manning's n for the reach, increasing hydraulic radius by increasing depth, and increasing the channel slope by shortening the channel length. The effect of such improvements on flood heights can be computed by usual hydraulic procedures.

As indicated in the discussion of levees (Sec. 20-14), measures for improving channel capacity are essentially local protection measures which may increase flood magnitudes at downstream points. Like levees, channel improvements should be considered as items in an overall plan for the stream and so planned and executed that their benefits at one point are not offset by increased damages elsewhere.

A lined channel with vertical sidewalls is dangerous because a person falling into the channel when it is dry can be seriously injured and when it is flowing will have no means of escape. If vertical walls are the only possible solution because of space limitations, they must be fenced when they pass through or near residential areas. Such a channel is not especially attractive. However, sloping sidewalls of smooth concrete are only slightly safer. Stepped sidewalls offer a chance for a person in the water to escape and reduce the risk of someone falling in.

In some areas lining of channels has a further disadvantage that it may eliminate natural recharge to the groundwater, thus providing flood mitigation at the cost of reducing water supply. Elimination of the bottom slab, use of some form of precast grill for the bottom, or use of quarry rock for bottom protection may be appropriate alternatives depending on the hydraulic conditions.

EVACUATION AND FLOODPROOFING

20-17   Emergency evacuation   Under certain circumstances, one of the most effective means of flood-damage reduction is the emergency evacuation of the threatened area. With reliable flood forecasts this technique is adapted to sparsely settled areas where property values do not justify other controls and loss of life can be prevented by prompt evacuation. River bottom lands can be used for grazing, and the livestock can be removed when a flood threatens. Contractors working in the river channel may prefer removal of their equipment in the event of a flood to the cost of a cofferdam capable of protecting the equipment. Warehouses can sometimes be emptied of their stock at costs well below that which would be required to protect them from flooding.

A good flood-forecasting service is relatively inexpensive and can often provide adequate warnings sufficiently far in advance to permit orderly and complete evacuation. The success of such a plan depends heavily on the hydrologic characteristics of the stream in question. Generally speaking, the smaller the drainage area of the stream, the more difficult it is to provide warnings in time to permit removal or protection of property.

20-18   Floodproofing individual units   In instances where only isolated units of high value are threatened by flooding, they may sometimes be individually floodproofed. An industrial plant comprising buildings, storage yards, roads, etc., may be protected by a ring levee or flood wall. Individual buildings sufficiently strong to resist the dynamic forces of the floodwater are sometimes protected by building the lower stories (below the expected high-water mark) without windows and providing some means of watertight closure for the doors. Thus, even though the building may be surrounded by water, the property within it is protected from damage, and many normal functions can be carried on. Several telephone exchanges in the Ohio River Valley have been so protected. Important public utilities might be flood-proofed as added insurance against loss of vital services even after failure of other flood-mitigation works.

LAND MANAGEMENT AND FLOOD MITIGATION

20-19 Vegetation and floods A lengthy argument over the influence of vegetal cover on floods has been in progress for many years. Noted authorities have asserted that deforestation is the basic cause of present flood problems, while others of equal distinction feel that vegetal cover has virtually no effect on floods. It is well established that vegetal cover removes moisture from the soil by transpiration and that it also promotes loose organic soil, which is favorable for the infiltration of rainfall. A heavy vegetal cover also means a high interception loss during storms. One could therefore expect less flood runoff from a well-vegetated area than from an area bare of vegetation. The vegetal cover creates a sort of retarding basin which stores a portion of the runoff which might otherwise contribute to floods.

However, while this storage may be an important quantity in minor storms, it may be quite negligible during a major flood particularly if the flood-producing storm is preceded by other rains which fill the storage space. As a result, one might expect that a good vegetal cover on a basin will result in a reduction in the frequency and severity of minor floods but will have relatively little effect on major floods. This is borne out by the evidence of major floods on many streams long before modern civilization began deforestation and land development.

20-20 Water conservation and flood mitigation Since about 1930 there has been a rapid increase in the use of water-conservation measures in agriculture. Contour plowing and terracing are used to retard surface runoff and promote infiltration of water into the soil. Farm ponds retain the flow of small creeks for irrigation and stock water. In addition cover crops are used in fields to avoid bare, fallowed ground during the nongrowing season. There is no argument about the value of these measures for reduction of soil erosion and preservation of soil moisture, but there is debate as to their value from the viewpoint of flood mitigation.

Water- and soil-conservation measures create reservoirs in the soil (or farm ponds), and if the reserve capacity of these reservoirs is large, i.e., ponds empty and soil moisture very low, many inches of rainfall may be stored. Since the purpose of the work is to conserve water, the normal situation will find the reservoirs at least partly full. Nevertheless, the storage may be sufficient to create a substantial reduction in surface runoff and thus to reduce the lesser floods on the small streams of the vicinity. It has been noted that reservoirs in the extreme headwaters of a basin are relatively inefficient protective devices for an area far downstream. Moreover, the farm ponds control only a small percentage of the drainage area, and the capacity of the soil as a storage reservoir is not large unless it is initially very dry. Hence, water- and soil-conservation methods are useful in reducing flood flows in small streams but are not very beneficial in protection of areas along major streams or in the control of unusually large floods. The value of this work for other purposes warrants its continuation, any flood reduction constituting an incidental benefit. It should be noted that land-drainage operations tend to increase floods by accelerating runoff of soil water and by eliminating natural water storage in ponds and swamps.

ECONOMICS OF FLOOD MITIGATION

20-21   Combined projects   The term *combined project* includes those projects in which several flood-mitigation methods are utilized jointly. Projects in which flood mitigation is combined with other functions such as navigation and power production are known as multiple-purpose projects. It is rare to find a stream in which flood mitigation is concentrated in a single form of protective measure. Economic analysis usually suggests that a combination of procedures is most desirable. Thus reservoirs are often combined with levees and channel works at key points along the stream. The cost balance for such a situation is illustrated in Fig. 20-13. Several possible reservoir sites are available, and the more reservoirs constructed, the smaller the regulated peak flow. The cost of reservoirs to reduce the design flood of 240,000 cfs to various lesser values is indicated by curve *A*. Reservoirs would not provide complete protection, and levees and other channel improvements would be necessary. Curve *B* indicates the costs of these works for protection against various flows. The sum of these curves indicates the total cost of protection against a flood of 240,000 cfs. The minimum point in the total cost curve represents the least expensive combination, i.e., reservoirs to reduce the peak to about 100,000 cfs and channel improvement to provide protection against this flow.

The construction of a diagram such as Fig. 20-13 requires the design of channel works for protection against several different peak flows to define curve *B* and the determination of the effect of several combinations of reservoirs on peak flow to establish curve *A*. A similar sort of analysis would be possible for other combinations of protective measures.

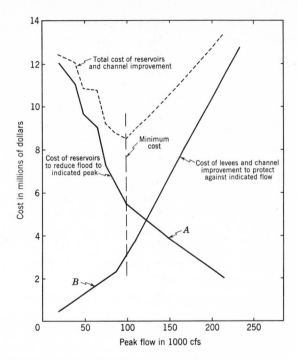

FIG. 20-13   *Example of cost analysis for a combination of flood-mitigation measures.*

20-22   Estimating the benefits of flood mitigation   *Tangible benefits* from flood mitigation are of two kinds: (1) those arising from prevention of flood damage and (2) land enhancement from more intensive use of protected land. The primary benefit from the prevention of flood damage is the difference in expected damage throughout the life of the project with and without flood mitigation. *Primary benefits* may include:

1. Cost of replacing or repairing damaged property
2. Cost of evacuation, relief, and rehabilitation of victims, and emergency flood-protection measures
3. Losses as a result of disruption of business
4. Loss of crops, or cost of replanting crops

The probable flood damages in an area must be estimated from a careful survey of the area of flooding. This is often done after a major flood when the damage is clearly evident. The data collected by the survey are classified by stage at which the damage occurs, so that a curve or table showing the damage from various river stages can be prepared (col. 2, Table 20-2). In general, market prices expected to be applicable during the anticipated life of the project are used in these estimates. In some instances there will be a seasonal variation in potential flood damage which must be considered. For example, damage to farm crops will be a maximum for floods during the height of the growing season. Allowance must also be made for any changes in damage potential as a result of the floodproofing of an individual unit or the removal of a large industry from the flood plain.

Land which is protected from floods may often be utilized for more productive purposes than when it is subject to flood hazards. Idle bottom land may be converted to use by agriculture or industry. The preferred method for determining the benefit gained by improved-land use is to estimate the difference in net income from the property with and without flood protection. An alternative method is to estimate the increase in market value of the land and convert this to an average annual benefit by use of a rate of return appropriate to private enterprise in the type of activity under consideration. There must be a real need for improved land use to justify any claim for benefits from this source. If an alternative flood-safe location is available for the same use, the benefits which can be credited to flood mitigation are only those which can be expected over and above those obtainable at the flood-safe location.

Secondary benefits from flood mitigation may arise in activities which

TABLE 20-2   Cost and Benefit Analysis for a Flood-mitigation Project
(All costs in millions of dollars)

Peak stage, ft (1)	Total damage below indicated stage (2)	Increment of loss (3)	Return period, yr (4)	Annual benefit for protection of increment (5)	Total benefits with protection below given stage (6)	Project cost (7)	Ratio of benefit to cost (8)
20	0				0	0.4	0
		4	10	0.4			
22	4				0.4	0.6	0.67
		6	15	0.4			
24	10				0.8	0.8	1.00
		10	22	0.45			
26	20				1.25	1.0	1.25
		12	30	0.4			
28	32				1.65	1.3	1.27
		13	70	0.19			
30	45				1.84	1.6	1.15
		15	150	0.10			
32	60				1.94	1.8	1.08
		20	300	0.07			
34	80				2.01	2.0	1.00

stem from use or processing of products and services directly affected by floods. For example, if a steel mill is closed by a flood and steel reserves are short, factories far removed from the flood area might have to close until steel production is again underway. Secondary benefits are difficult to assess and are, in part at least, offset by secondary costs not usually included in the cost estimates. Thus secondary benefits are not normally included in estimates of flood-mitigation benefits. *Intangible benefits* of flood control include prevention of loss of life and reduction in disease resulting from flood conditions. It is difficult to place a monetary value on these intangibles, although it has been suggested[1] that the death benefits payable under workmen's compensation laws be taken as the economic value of human life.

Average annual benefits may be computed by multiplying those benefits expected by prevention of flooding at a given stage by the probability of occurrence of that stage in any year $(1/t_p)$. The computation is illustrated in Table 20-2, where the increment of prevented loss in 2-ft intervals of stage is divided by the expected return period of a flood at the mid-height of the interval. The summation of these increments indicates the total benefits with protection against floods of various sizes. Note that even though the total damage for large floods is great, the annual increment of benefits is small because of the long return period for such floods.

The final step in the economic analysis of a project is a comparison of the benefits and costs. This is accomplished by estimating the cost of various degrees of protection. In the example of Table 20-2, the costs are those for a levee affording protection below the given stage. Protection against the rare floods is uneconomical because of the large investment and infrequent flood occurrence. In the example, levees designed to withstand a 28-ft stage would be most satisfactory.

The preceding analysis ignores possible intangible benefits and costs which may warrant the selection of another design stage. Actually the determination of the design criteria for many flood-mitigation works cannot be based on a direct economic analysis. Nevertheless, any project should show economic feasibility before it is constructed. Only those portions of a project which are in themselves beneficial should be constructed. Thus, levees and channel works for the protection of an agricultural district might not be justified by the expected benefits. However, if a proposed reservoir will have some benefit for the agricultural area, these savings should be included in considering the justification of the reservoir. Reservoirs or channel improvements change the flood-frequency curve at the protected point and increase

---

[1] "Report to the Inter-Agency Committee on Water Resources. Proposed Practices for Economic Analysis of River Basin Projects," prepared by Subcommittee on Evaluation Standards, Government Printing Office, Washington, D.C., May, 1958.

the expected return period for floods of all magnitudes. Benefits from such works are computed by comparing the damage expected before and after the works are constructed.

20-23 Summary In general, the steps in the design of a flood-mitigation project are as follows:

1. Determine the project design flood and the flood characteristics of the area.
2. Define the areas to be protected, and, on the basis of a field survey, determine the flood damages which can be expected at various stages.
3. Determine the possible methods of flood protection. If reservoirs or floodways are considered feasible, select possible sites and determine the physical characteristics of these sites.
4. Design the necessary facilities for each method of mitigation in sufficient detail to permit cost estimates and an analysis of their effect on flood flows.
5. Select the facilities or combination of facilities that offer the desired protection at minimum cost.
6. Compare the costs and benefits of the project to determine whether the project is economically justified.
7. Prepare a detailed report setting forth the possibilities explored, the protection recommended, and the degree of protection which will be provided.

BIBLIOGRAPHY

Barrows, H. K.: "Floods—Their Hydrology and Control," McGraw-Hill, New York, 1948.

Bernard, Merrill: Flood Forecasts That Reduce Losses, *Eng. News-Record*, Vol. 141, pp. 64–66, Nov. 25, 1946.

Brown, C. B.: Flood Prevention through Watershed Planning, *Agr. Eng.*, Vol. 34, pp. 159–162, March, 1953.

Elam, W. E.: "Speeding Floods to the Sea," Hobson Book Press, New York, 1946.

Feringa, P. A., and C. W. Schweizer: One Hundred Years of Improvement on Lower Mississippi River, *Trans. ASCE*, Vol. CT, pp. 1100–1124, 1953.

Flood Walls, Part CXXV, "Engineering Manual, Civil Works Construction," U.S. Corps of Engineers, January, 1948.

Hoyt, W. G., and W. B. Langbein: "Floods," Princeton, Princeton, N.J., 1955.

Leopold, L. B., and Thomas Maddock: "The Flood Control Controversy," Ronald, New York, 1954.

Morgan, Arthur: "The Miami Conservancy District," McGraw-Hill, New York, 1953.

President's Water Resources Council: Policies, Standards, and Procedures in the Formulation, Evaluation, and Review of Plans for Use and Development of Water and Related Land Resources, 87th Cong., 2d Sess., Sen. Doc. 97, Government Printing Office, Washington, D.C., 1962.

Reservoir Regulation, Part CXXXVI, "Engineering Manual, Civil Works Construction," U.S. Corps of Engineers, August, 1951.

Schneider, George R.: The History and Future of Flood Control, *Trans. ASCE*, Vol. CT, pp. 1042–1099, 1953.

"Bank and Shore Protection in California Highway Practice," State of California Department of Public Works Division of Highways, Sacramento, November, 1960.

Task Force on Federal Flood Control Policy, A Unified National Program for Managing Flood Losses, 89th Cong. 2d Sess., House Doc. 465, Government Printing Office, Washington, D.C., 1966.

PROBLEMS

20-1   Assume the hydrograph of Prob. 3-2 is the inflow hydrograph to a flood-control reservoir with a storage capacity of 15,000 sfd. (*a*) What is the maximum possible flood-crest reduction immediately downstream from the dam? (*b*) What would be the crest reduction at a point 12 hr downstream without reservoir, ignoring local inflow? Assume Muskingum $K = 12$ hr and $x = 0.1$.

20-2   Make a study of the flood-mitigation facilities for some stream in your area, and prepare a report indicating the methods of flood mitigation employed, the magnitude of the design flood, etc. Describe the structures used. For reservoirs indicate storage capacity and percent of area controlled. Are additional flood-mitigation works possible and desirable?

20-3   A retarding basin has an elevation-storage curve which closely fits the equation $V = 5d^{2.5}$, where V is in second-foot-days and the maximum value of d is 100 ft. Discharge from the basin is through two 84-in.-diameter reinforced-concrete sluiceways which are 350 ft long and discharge freely to the atmosphere. The horizontal sluiceways have bellmouth entrances with their center lines at elevation 10 ft. Assume that the inflow to this reservoir is as given in Prob. 3-2, and compute the peak outflow.

20-4   Solve Prob. 20-3, using the inflow hydrograph of Prob. 3-27 as the reservoir inflow.

20-5   Assume that a sandbag holds 1 cu ft of earth and is about 14 in. wide when filled. How many sandbags would be required to raise the level of a mile of levee by 2 ft if two rows of sandbags were used?

20-6   If the mud box of Fig. 20-11 is made of two-by-eights on edge with 6-ft posts of four-by-four timbers every 6 ft and two-by-two crossties, how many board feet of lumber and cubic yards of earth are required for a mile of levee if the mud box is 2 ft high and 4 ft wide?

20-7   A river of rectangular cross section is 400 ft wide and carries a flow of 50,000 cfs at a depth of 35 ft. If a side-channel diversion discharges 15,000 cfs from this river, compute the lowering of river stage downstream of the diversion at a point where the flow is uniform.

20-8   For the river of Prob. 20-7 find the depth 1 mi upstream from the weir if the depth of flow at the weir is 30 ft. Assume n is 0.025.

Chapter 21

Planning for
water-resources development

Planning can be defined as *the orderly consideration of a project from the original statement of purpose through the evaluation of alternatives to the final decision on a course of action.* It includes all the work associated with the design of a project except the detailed engineering of the structures. It is the basis for the decision to proceed with (or to abandon) a proposed project and is clearly a most important aspect of the total engineering for the project. Because each water-development project is unique in its physical and economic setting, it is impossible to describe a simple process which will inevitably lead to the best decision. There is no substitute for "engineering judgment" in the selection of the method of approach to project planning even though each individual step toward the final decision should be supported by quantitative analysis rather than estimates or judgment.

One often hears the phrase "river-basin planning." It is important to note that the planning phase is no less important in the case of the smallest project. The planning for an entire river basin involves a much more complex planning effort than the single project, but the difficulties in arriving at the correct decision may be just as great for the smaller project.

This chapter presents some guidelines for the planning effort and attempts to point out some of the pitfalls which have been encountered in past efforts. The definition of planning includes the evaluation of alternatives by the principles of engineering economy but since this topic is discussed in Chap. 13 it will not be included in this chapter.

The term "planning" carries another connotation which is different than the meaning described in the previous paragraphs. This is the concept of the city or regional *master plan* which attempts to define the most desirable future growth pattern for an area. If the master plan is in reality the most desirable pattern of development, then all future efforts should be directed toward this pattern. Unfortunately, the concept of "most desirable" is a subjective one, and it is difficult to assure that any master plan meets this high standard when first developed. Subsequent changes in technology, economic development, and public attitude often make a master plan obsolete in a relatively short time.

The concept of a master plan for water development has been employed in some areas.[1] To the extent that such an effort is a carefully worked out, long-range development plan to serve as a base for the timing of project construction, it may be worthwhile, but it is always subject to change as a result of changing technological, economic, and social environment. A master water plan which is no more than a catalog of all physically feasible projects may prove to be a wasted effort and can, by fixing in the mind of local interests the concept of certain projects, prove to be a deterrent to the most efficient long-run development of an area.

21-1 Statement of objective   An essential preliminary to project planning is a clear statement of objectives.[2] A public utility might have as its objectives the comprehensive development of electric power utilizing hydropower as it best fits into the total system and maximizing total return to the company. A flood-mitigation district might express its objectives as the reduction of flood damage within its district boundaries to the extent economically justified in terms of benefits to district residents. A major government unit might have the broad goal of maximizing the total economic and social benefits of water development. Government has, however, other functions—highways, education, welfare, defense, etc.—and unless financing is unlimited some decision must be made relative to the expenditures which can be permitted (Chap. 13). This does not mean that a plan should not be prepared but, rather, that the whole plan may not be implemented at once. Planning does not necessitate implementation by a single agency—private groups and various government agencies can and should carry the development forward in accord with the plan.

Table 21-1 outlines the steps required to prepare a plan for water-resources development. It is presumed that all possible water uses—irrigation, municipal and industrial supply, hydroelectric power, navigation, recreation,

---

[1] The California Water Plan, *Bull.* 3, State of California Department of Water Resources, Sacramento, 1957.

[2] The U.S. Water Resources Council has proposed that federal water projects be evaluated against four objectives: (1) national efficiency, (2) regional economic development, (3) environmental preservation, and (4) well-being of the people. If adopted, this plan will require four separate accounts and the choice between alternatives will be made by consideration of all accounts.

enhancement of fisheries and wildlife, and the control functions, i.e., flood and pollution mitigation—are pertinent to the problem. One or more of these ends may be obviously inappropriate for any particular project and can be dropped from consideration at the beginning of the study.

21-2 Data needed for planning Reliable decisions on water development must be based on facts if emotion, supposition, or political ambition are to be prevented from governing the decisions. Step 2 of Table 21-1 outlines the required data. Note that this goes far beyond the hydrologic and site data required for engineering design. A substantial body of economic, social, and resource data are required and the services of specialists in many fields are needed to collect and interpret these data. These data items have been discussed in more detail in Chaps. 14 through 20.

Virtually all data items are current, and the required facts can be gathered by a survey at the start of planning. Hydrologic data, however, are historical and the longer the period of record, the greater the value of the data. This emphasizes the importance of a basic data program in hydrology even before the idea of a specific project is conceived. When funds for planning are authorized, intensive hydrologic data-collection programs should get underway immediately.

21-3 Projections for planning The historical and current data collected for project planning have only one purpose—to serve as a base for projection of future conditions. All projects must be planned to meet future needs. The projections should be more than a simple extrapolation of past growth curves, for social, economic, and technological developments may cause significant changes in trends. The most important item under each subheading of step 3 of Table 21-1 is the item on technological change. The planner must make every effort to anticipate such change.

Among possible technological advances are the possibilities of conserving or augmenting natural water supplies (Sec. 21-9). Developments in nuclear-power production (or other means of producing electricity), improved agricultural techniques which increase crop production on existing farmlands, new systems of transportation, changing patterns of living, and new modes of recreation may all play a role in the economy of a project. In most cases the exact impact of any future technological breakthrough will not be obvious, but it may be possible to plan with sufficient flexibility so that expansion of the project may be governed by advances achieved when the expansion is undertaken. Stage construction of all or part of the project features is an example of planned flexibility.

The final projections required for water planning are the "demand curves." Traditionally the engineer has used a single curve of expected growth in water use as a function of time for each project. Such curves are unrealistic and dangerous because of the many factors that may disturb the trends.

TABLE 21-1   The Steps in Planning

1. Statement of objective
2. Collection of data
   A. General
      a. Water resources
         (1) Hydrology—precipitation, streamflow, evapotranspiration, water quality, sediment
         (2) Geology—groundwater, soil survey, erosion
         (3) Cartographic—maps
      b. Other basic resources
         (1) Geologic—minerals
         (2) Ecologic—vegetation, fish and wildlife
         (3) Demographic—people and institutions
         (4) Economic—industry, transportation, markets, tourism, recreation, land, taxes
      c. Constraints
         (1) Law—water rights, pollution control, land zoning, land ownership, administrative patterns, Indian lands, interstate compacts, treaties
         (2) Public opinion
         (3) Existing projects
   B. Special data
      a. Agriculture
         (1) Land classification
         (2) Crop water requirements—quantity and quality
         (3) Climatic limitations
      b. Municipal uses
         (1) Industrial water needs—quantity and quality
         (2) Population water needs
      c. Hydro power
         (1) Projected needs
         (2) Alternate sources
      d. Flood mitigation
         (1) Extent of past flooding and damages
         (2) Local storm drainage requirements
      e. Navigation
         (1) Present water traffic patterns
         (2) Alternatives—roads, railroads, airways
      f. Recreation
         (1) Natural attractions
         (2) Present recreation patterns—types, place, time
      g. Pollution control
         (1) Existing waste discharges—location, time, character of waste
         (2) Water pollution regulations or quality standards
3. Projections
   A. General
      (1) Population—place and time
      (2) Land use—place and time
      (3) Economic—markets, tourism, etc.
   B. Specific
      (1) Agriculture
         (*a*) Markets
         (*b*) Crops
         (*c*) Technological development
         (*d*) Water demand

TABLE 21-1   The Steps in Planning (continued)

    (2) Municipal
       (*a*) Domestic water demand
       (*b*) Industrial water demand
       (*c*) Technological changes
    (3) Power
       (*a*) Market and demand
       (*b*) Growth of alternate sources
       (*c*) Technological improvements
    (4) Flood control
       (*a*) Zoning possibilities
       (*b*) Projected flood damage—place and frequency
       (*c*) Possibilities of flood warning
    (5) Navigation
       (*a*) Growth of demand
       (*b*) Growth of competing facilities
       (*c*) Transportation patterns—point to point
       (*d*) Technological advance—navigation and competing facilities
    (6) Recreation
       (*a*) Growth of demand
       (*b*) Changes in recreation preferences
    (7) Pollution
       (*a*) Anticipated quantities and characteristics of waste
       (*b*) Technological advances
4. Project formulation
  A. Defining boundary conditions
  B. Listing all possible land-use plans and their water requirements
  C. Listing all possible projects* which could meet the needs as projected under 3 and 4*B*
  D. Preparing preliminary designs and cost estimates for projects listed under 4*C*
  E. Estimating benefits for all projects
  F. Rejecting all projects which do not appear to be individually economic†
  G. Forming of all combinations of projects which are not mutually exclusive‡ and totaling the costs and benefits of all combinations
  H. Selecting the two or three combinations which offer prospect of maximum return for detailed consideration and reanalyzing costs and benefits—selecting the "final alternative plans"
  I. Preparing reports on final alternatives showing costs, benefits, staging, financing, and intangible factors for review by appropriate authority
5. Project authorization
  A. Reviewing results of step 4 by appropriate authority
  B. Selecting one of the plans for authorization

Project is used here as any solution which will help to meet estimated needs including reservoirs, levees, diversion works, water-treatment plants, forecasting services, flood-plain management, etc. In many cases there will be mutually exclusive alternatives, i.e., dams of different heights at a given site. Each such alternative is considered to be a "project."

†In some cases the benefits of a project are dependent on other projects, i.e., a single reservoir and power plant may not be economic but in combination with other storage is economic. Such interrelations should be considered here.

‡Consider that some projects may be mutually exclusive in place but not in time; e.g., where two heights of dam are considered, the low dam may be built now and subsequently raised.

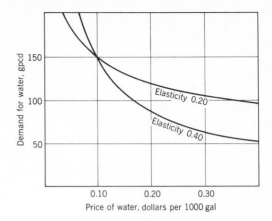

FIG. 21-1   Demand-curve relation between water use and price.

The economist defines *demand* as the relation between water use and price (Fig. 21-1). This recognizes that price has an effect on the amount of water used, and that given the opportunity, water users will adjust their use in relation to cost. The relative change in use with a change in price is known as the *elasticity*.[1] Thus if a doubling of price would decrease use by 20 percent, the elasticity is said to be 0.2. If the demand is inelastic, use and price are not related. If demand is elastic, use is dependent on price. A considerable amount of elasticity does exist for water sales and, hence, pricing policy considerations are important in planning.

Many terms are used quite loosely in connection with water studies. Typical of these terms are "water needs" and "water requirements." Generally, under present connotation, these terms should be translated as the "amount of water which could be used," rather than the "amount that will be used." Often the "water requirement" is estimated on the basis of assumed "ultimate requirements" when all land in the service area is fully developed and requires water in maximum quantities. Such estimates are based on projections of population, industry, and agriculture using unit water rates per person or per acre which are often also projections from present use rates. The role of cost or of technology in modifying the actual future water use is rarely considered. Estimates may be quite unrealistic and somewhat equivalent to saying that every family "needs" a yacht. Large quantities of water could be used to irrigate the Sahara desert, but this does not prove that a project should be built to provide this water.

Unrealistic values for "water needs" lead to overdesign and excessive costs and, if the supply is limited, to the diversion of water from other uses

---

[1] *Elasticity* as used here is actually negative, but the minus sign is often omitted.

which are also economic but of lower priority. When excessive allowances are accepted as values of the real need, conservation of water is discouraged. Often bargain prices are adopted to encourage water use up to the capacity of the overdesigned works. Rational planning should attempt to estimate true economic demand rather than a fictitious need.

It will be evident that accurate long-range projections are difficult, if not impossible. The planning process must allow for this by retaining flexibility so that the plans can be adjusted to meet changing conditions. Several alternate projections[1] should be considered in developing a plan which can be adapted as conditions change. Differing flexibility of alternate plans may be an intangible factor of importance in the final decision.

21-4   Project formulation   Once the basic data and the projections of future conditions are assembled, actual formulation of the project can commence. This is a phase of planning where imagination and skill are required. The important first consideration is the compilation of a list of alternatives. Such a list should be comprehensive. Often a preliminary project plan is in hand as the first phase of the planning operation and the operation is reduced to attempting to justify this plan. This approach is wrong. The planning process should be an evaluation of all possible alternatives with respect to project features and water use. It is not even sufficient to start with the assumption that a specified quantity of water is required at point X and to evaluate the merits of two or more plans for supplying this quantity. Planning should consider alternative and competing uses[2] for water as well as the various possibilities for control and delivery of the water. The planning process may involve considerable feedback. As project formulation proceeds, it may be evident that new data or projections are required or that some revision of background data is needed.

The first step in project formulation is the definition of the boundary conditions which restrict the project. For example:

1. One or more of the possible aspects of water development can be eliminated on the basis of physical limitations, i.e., no navigation on torrential mountain streams.
2. Certain problems may be fixed in location, i.e., flood mitigation for an existing city.
3. The available water may be limited or subject only to minor changes.
4. Maximum land areas usable for various purposes may be definable. This does not exclude possibility of alternative uses for a given parcel of land.

---

[1] For an example of use of the method of alternative futures in projecting domestic water use, see Peter Whitford, "Forecasting Demand for Urban Water Supply," *Rept. EEP*-36, Department of Civil Engineering, Stanford University, September 1970.

[2] Nathaniel Wollman, "The Value of Water in Alternative Uses," The University of New Mexico Press, Albuquerque, 1962.

5. A policy decision may reserve certain lands for specific purposes, i.e., parks and recreation areas.
6. Possible sites for water storage (both surface and underground) can be defined and their limiting capacities evaluated.
7. Certain existing foci of water use exist and must continue to be supplied.

These restraints may simplify further steps by eliminating some alternatives from consideration. They should not, however, be imposed unless careful consideration shows that no unnecessary restraint bars an economic project from consideration.

The second step in project formulation is to define alternative land-use plans for the study area insofar as they influence water requirements. In this connection it is worth noting that agricultural water use and urban water use are usually about the same when expressed on a per acre basis. Unusually heavy industrial water use or a requirement of considerable water for leaching salts in farm land may upset this ratio. However, all possibilities for urban expansion into agricultural land need not be considered at this point as far as water use is concerned. It is possible, however, that quite different power requirements would exist and this could affect planning. Hence, two or three alternate land-use plans might be devised to bracket the range of possibilities. Reservations of land for wildlife sanctuaries, parks, recreation, timber reserves, wilderness areas, etc., should be considered and defined. The result of this step will be a series of land-use plans designated I, II, III, IV, etc.

The next step in project formulation is the cataloging of possible project units (Fig. 21-2). Each possible unit site should be located and designated by number, i.e., site 1, 2, 3, etc. Each alternative at a given site may be designated by a letter suffix A, B, C, etc. Thus at a damsite designated as site 14, three alternative dams of different heights might be $14A$, $14B$, and $14C$. In addition to storage sites, conveyance works should be listed and designated. There may be many possible alternatives for some units but for practical purposes three or four which bracket the range of possibilities at a site should be sufficient for preliminary planning. Alternatives which will be unusually costly and appear to offer no special advantages should be ignored. Channel-improvement works (levees, straightening, dredging, revetment) should be listed, again with alternatives when they appear feasible. The land-use plans can serve as a guide to units which may have some utility. Units which serve no useful purpose should not be included. In preparing the catalog of units, careful study should be given to defining *all* possible alternatives. This may call for some "brainstorming" to visualize the less obvious plans.

When the unit catalog is complete, preliminary cost estimates should be made for each unit. Obviously, many units will prove undesirable and will

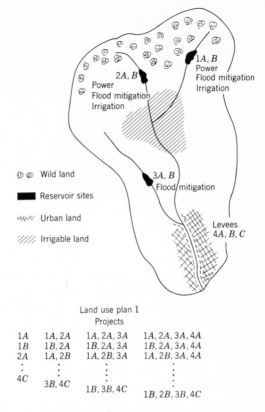

Land use plan I
Projects

1A	1A, 2A	1A, 2A, 3A	1A, 2A, 3A, 4A
1B	1B, 2A	1B, 2A, 3A	1B, 2A, 3A, 4A
2A	1A, 2B	1A, 2B, 3A	1A, 2B, 3A, 4A
⋮	⋮	⋮	⋮
4C	3B, 4C	1B, 3B, 4C	1B, 2B, 3B, 4C

FIG. 21-2   *Project formulation for a hypothetical river basin.*

ultimately be discarded. Consequently, cost estimates should not be made in such detail as to be excessive in time or cost. However, all estimates should be based on similar unit costs so that they will be comparable.

21-5   Project analysis   The procedure outlined in the previous section leads to the listing of a large number of individual alternatives and their associated costs. From this list the planner must select the unit or combination of units which is economically most efficient and environmentally satisfactory. If this most efficient unit (or units) meets the established standards, i.e., benefit-cost ratio above a specified minimum value, the unit or units may be recommended for construction.

An occasional unit will be completely independent both physically and economically of all other units and may be tested alone. To be physically independent, there can be no other units either upstream or downstream of the unit in question which would affect the inflows to the unit or be affected by the outflows from it. In addition, no other unit can contribute to the same end purpose. Thus, two reservoirs on different streams but both providing flood mitigation for a single downstream point are physically interdependent. To be economically independent, a unit must have no economic interconnection with any other unit. Thus, two power plants on different streams might have no physical interrelation as far as streamflow is concerned, but if they serve the same power system, they are economically interdependent in that their power production must be integrated into the system. The comments above apply also to the case of a group of mutually exclusive units at a given site provided all of the mutually exclusive units would be independent if constructed.

For an independent, single-purpose unit the testing is relatively simple. One need only establish a sufficient number of points on a curve of benefits vs. costs (Fig. 13-1) to determine the level of project development which produces the largest net benefits.

For an independent, multipurpose project, the testing requires the definition of a net-benefit surface (Fig. 21-3) if two purposes are being considered. An $n + 1$ dimensional space is required if there are n purposes. Usually, however, the two major purposes will govern, and the optimum can be closely approximated with a three-dimensional surface by holding the other purposes at a constant level. The problem is to test enough combinations of the two primary purposes to determine location of the maximum. In the case of the independent project there will usually be some intuitive guides which define the general location of the maximum, and after two or three trials it is usually possible to estimate the direction of shift to move closer to the maximum.

In the case of the multiunit system the problem is also one of defining the maximum net benefits within a multidimensional response surface. Because of the relatively large number of combinations which are possible if many units are involved, it is difficult to make an intuitive estimate of the location of the maximum if the physical interrelationships are at all complex. Hufschmidt[1] discusses this problem in some detail. In essence a detailed operation study (Chap. 7) must be performed for each alternative system. In addition to estimating water availability as discussed in Chap. 7, the study must determine power production, irrigation use, etc., so that the benefits in dollars may be computed. With many units and various possible combinations of water uses at each unit, a rather large number of studies is indicated. A high-speed digital computer can remove some of the drudgery from these studies, but unless many hours of computer time are to be utilized, some system of sampling is essential. From the work reported by Huf-

---

[1] M. M. Hufschmidt, Analysis by Simulation: Examination of Response Surface, chap. 10 in Arthur Maass and others, "Design of Water Resource Systems," Harvard, Cambridge, Mass., 1962.

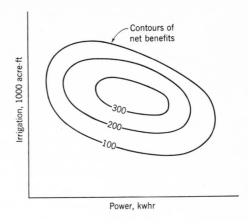

FIG. 21-3   Typical net-benefit surface for multipurpose project.

schmidt, it appears that a random sampling procedure may be the most efficient means of defining the general nature of the response surface. Once the region of the peak is defined, a second random sample or a uniform grid sample in the region may be used. The random sampling procedure seems most effective when the number of variables is large.

If the use of a digital computer is not possible, the many trials must be made manually. The natural tendency is to try to minimize the work by studies of the various units as independent single-purpose projects in an attempt to narrow down the possible range of variation which must be considered in the system. The results of the Harvard study indicate, however, that such an approach may be quite misleading and that much better results may be obtained with fewer trials by treating the systems *in toto* from the start. Because of the advantages offered by the use of computers, only the simplest projects should be considered for manual analysis.

After the optimum level of development for the total system is defined, computations should be performed deleting each separable unit from the optimum combination to verify that the increment of benefits from each unit warrants its cost.

21-6   Environmental considerations in planning   The decade of the 1960's will be remembered as a period of increasing concern about man's environment. The inevitable arithmetic of uncontrolled population growth combined with an increased per capita production of waste products threatened severe pollution of air, water, and land with attendant damage to flora and fauna and possible danger to man himself. Population growth and technological capability combined to drastically alter the natural landscape with cities, highways, dams, and other engineering works. These changes often led to changes in the ecology of an area bringing about local extinction of some species of flora and fauna, sometimes with replacement by less desirable species. The rapidly growing population also raised for many the spectre of famine first suggested by Malthus almost two centuries earlier (1798).

The growing environmental concern poses a dilemma for engineers faced at one extreme by demands that all construction cease and at the other extreme by pressure to get on with building as rapidly as possible. A new set of social values—moral, philosophic, and aesthetic—suddenly joined technical standards and economic evaluation as decision factors in the planning process. Although many planners were taken by surprise, these changes were long overdue. The basic problem of population control will be met, either by man or nature, but the water-resources planner of the future must give more thought to the environmental problems.

To play his new role, the planner of the future must be more innovative, more broad-gaged, and more critical of evaluations of "need." He cannot look to quantitative measures of beauty or ecology to develop his plans. He cannot rely solely on a showing of economic benefits. Innovation may require such steps as devising ways to lower water requirements, encouraging nonstructural solutions in flood mitigation, and finding better ways to treat wastes and reclaim wastewater. A broad-gage viewpoint must recognize the interrelationships among water pollution, air pollution, and solid-waste disposal; the role of water supply in population dispersion; the consequences of water project construction on local ecologic relationships; and the effect of projects on water pollution. Most important of all, however, is the critical evaluation of the real need for a project.

It is a reasonable assumption that public works necessary to maintain needed services will continue to be constructed. Water projects needed to maintain public health and safety and the accepted amenities will be included in these public works. Power projects which add insignificant amounts to current and projected output can hardly be considered essential. Irrigation projects which could be replaced by increased production in humid regions, flood-mitigation projects which are substitutes for good land management, storage to modify pollution by dilution when better treatment could serve more effectively, water supply projects to encourage growth of major metropolitan areas, and recreation projects which compete with nearby projects or natural areas are also nonessential. The planner will have to be more alert to alternatives than ever before.

Where projects seem to be essential, the planner will find it necessary to consider carefully the ecologic impact[1] on the stream and adjacent areas and try to develop a plan which will have a minimum of detrimental effects.

---

[1] "A Procedure for Evaluating Environmental Impact," *U.S. Geological Survey Circular* 645, 1971.

Finally, in the architectural design of structures special thought must be given to appearance. Special treatment of surfaces to avoid large expanses of smooth concrete, coloring to blend with the surroundings, planting of grass, shrubs, or trees to enhance visual feeling and other similar measures should be considered.

A partial list of some environmental consequences of water-resource projects might include:

1. Degradation of downstream channel or coastal beaches by loss of sediment trapped in a reservoir
2. Loss of unique geological, historical, archaeological, or scenic sites by flooding under a reservoir
3. Flooding of spawning beds for migratory fish preventing their reproduction or destruction of spawning gravel by channel dredging or lining
4. Change in stream water temperature as a result of a reservoir leading to changes in aquatic species in the stream
5. Release of reservoir bottom water which may be high in dissolved salts or low in oxygen resulting in a change in aquatic species
6. Drainage of swamps, potholes, etc., decreasing the opportunity for survival of aquatic or amphibious animals
7. Change in water quality as a result of drainage from an irrigation project which may encourage growth of algae in the receiving water or lead to a change in aquatic species as salinity of the receiving body increases
8. Creation of a barrier to normal migration routes of land animals by a reservoir
9. Altering aquatic species by increased turbidity from man-induced erosion or from dredging operations
10. Damage to higher species by reason of toxic materials (pesticides, toxic metals, etc.) discharged to a stream and concentrated in the food chain
11. Damage to fish by passage through pumps or turbines
12. Damage to stream-bank vegetation by alteration of flow patterns in a stream

Many more items could be added to this list and there are, no doubt, other subtle effects which have not yet been identified. A clear distinction also should be made between damage which is strictly temporary (construction operations, tree clearing, sanitary landfill, etc.) and effects which are long-term and irreversible.

21-7 Systems analysis in water planning Since it is the role of the planner to select the best of all possible alternatives, the various methods of optimization collectively called *systems analysis* are obvious tools for his use. Optimization problems are encountered at three levels in water planning. The first level deals with individual features of the project. For instance, given the cost-diameter and head loss-diameter functions for a penstock, it is possible to use the methods of engineering economy to accurately determine the optimum (least-cost) solution. Many similar problems in suboptimization are encountered in water-project planning.

The second level involves the single project. To some extent suboptimization of project units leads to optimization of the whole project. Usually, however, there are issues of project scale which must be separately resolved. How large a reservoir? What design flood for a levee system? What mix of purposes for a multipurpose project? Questions of this type can usually be answered by appropriate mathematical analysis with two limitations. The aspects of project scope which are not quantitatively measurable—environmental and ecologic consequences, value of human life, public preference, and others—cannot be included in the solution. Secondly, decisions on project scope rest heavily on the projections and other estimates involved in planning, and a mathematical solution can be no better than the data employed. Despite these constraints, mathematical solutions are valuable in defining one possible optimum point which provides a point of departure for judging adjustments required to satisfy the nonquantifiable factors.

The third level of planning involves systems of projects—multiple reservoirs, canals, levees. Evidence to date indicates that optimization is not practical because of the very large number of constraints involved. Solutions based on simplifying assumptions may indicate a first approximation to the best configuration of the system, but the only method demonstrated successfully involves simulation. A simulation (operation study) can be programmed for computer solution and many alternative combinations tested to determine which alternative offers the maximum net benefits. The same constraints which existed at the second level apply also at the third level. In addition, the number of possible alternatives may be so large that a systematic test of all alternatives is impossible and a sampling on a multidimensional response surface is required (Sec. 21-5). Despite these constraints, systematic simulation should lead to a solution closer to the optimum point than any other method.

Space limitations do not permit a discussion of specific methods of systems analysis as applied to water-project planning. A selected bibliography which offers examples of many specific applications will be found at the end of this chapter.

21-8 Some common pitfalls in project planning   A study of project reports by various agencies indicates that there are many pitfalls which may substantially negate an otherwise excellent effort to make a sound project evaluation. Some of the most common pitfalls are discussed in the following paragraphs. In general, the only protection against these difficulties is to be constantly alert against incorporating them in the analysis.

The survey report   Because of the large number of possible projects it is fairly common practice to prepare a survey or preliminary report on a project as the first phase of its analysis. If this report is unfavorable, the project can be dropped without excessive expenditure of funds. If it is favorable, the decision to proceed on more thorough studies is made. In theory, the project can be rejected even after a favorable survey report, but for public projects political considerations may work to avoid any such rejection. Consequently the survey report becomes a very important step in the economic analysis of projects, and it is important that the survey report not be based on approximations and short-cut procedures.

In particular the study of hydrology and those factors related to expected water use should be carried to the level of refinement appropriate for final design. Anything less may incur the hazard of a serious error. Accurate maps are also essential in a survey report. While this may increase the cost of preliminary studies, it should reduce the cost of the final design for projects which reach that phase.

Agency standards   Many agencies have developed standards of design as the result of tradition or administrative rule. These may not, however, always be appropriate to the particular problems of a given project. For example, a spillway which will flow only once in 50 yr might profitably be built of soil cement or other similar construction since repair or replacement on the average of once every 50 yr might be less costly than more permanent construction. Many agencies would not, however, consider anything less than massive or strongly reinforced concrete for such construction.

Early construction   Assuming that the economy is functioning fully, a dollar investment postponed for 10 yr at 5 percent interest is equivalent to $1.63. Construction of a project substantially before it is needed constitutes a waste of funds which might profitably be employed gainfully in the interim period. Even increased construction costs over the 10-yr period need not necessarily represent a loss resulting from the delay, since increased costs will be at least partially offset by a relative increase in benefits.

Stage construction or investment may be appropriate. If a project is known to be needed 10 yr hence it may be quite desirable to buy right-of-way at once to prevent development that would greatly increase cost. There are numerous examples of dams which have been planned for stage construction as growing water requirements increase the required reservoir size. In some cases where a water supply is required immediately, it may be prudent to install conduits designed to meet anticipated future demands and perhaps, even, a dam large enough for the long-range future. In each case, however, this need should be demonstrated by analysis and not assumed *a priori*.

The practice of building flood-mitigation works on the basis of anticipated future growth of the area to be protected is especially erroneous. Flood mitigation should not be provided until it is justified by existing values.

A priori decisions   Decisions regarding certain features of a project are sometimes made before economic analysis and are never checked. Thus in one flood-mitigation project the decision was made to limit maximum flows to 3000 cfs in the protected reach to prevent bank erosion although flood damage did not begin until flows of over 4000 cfs occurred. No study was made of the incremental cost of the additional 1000 cfs reduction in peak flows to determine whether the cost was justified by the relatively small losses incurred by bank erosion. *A priori* decisions of this type can lead to incorporation of uneconomic increments in many projects. No such decision should stand until its incremental benefit-cost ratio justifies it.

Failure to consider all alternatives   Proably the most common pitfall is the failure to consider all alternatives. In particular, nonengineering alternatives such as flood-plain zoning may be overlooked by the engineering designer. In addition, simple engineering alternatives are often overlooked especially when they depart from the traditional solutions. When it became necessary to reduce the seepage through a small earth dam, consulting soils engineers recommended construction of a clay cutoff wall and blanket on the upstream face. While successful, this precluded the possibility of relief wells on the downstream side which, in this instance, would have been a useful source of filtered water as well as a seepage-control measure. In the same vein, one rarely hears water conservation suggested as a solution to a water shortage when it is feasible to construct works to provide an additional supply.

Use of market prices   Market prices of water do not always reflect its real value. Hence, market prices are not necessarily a satisfactory basis for computing benefits. Wherever possible, the real benefit should be estimated and utilized in the economic analysis.

Use of next best alternative   In lieu of any other estimate of the benefits of a project, it is not uncommon to use the cost of the next best alternative as a measure of benefits. By judicious selection of the next best alternative, the benefits can be made very large indeed. Thus in developing the analysis of the California aqueduct system, the cost of desalted seawater was taken as the value of water delivered in southern California. This is no better measure of value than market price and perhaps worse, for one must ask whether the consumers would be willing to pay the high cost of desalted water if this was their only alternative. The only alternative acceptable as a basis for estimating benefits is an alternative which would be built if the project under study were dropped. This approach is simply a search for the least-cost solution.

21-9   Augmentation of water supplies   Section 21-3 emphasizes the need to consider technological development in water-resources planning. One phase of such development is of special interest because it relates directly to the important variable—water. World supplies of fresh water are not well defined but are undoubtedly adequate for world needs for many decades in the future. These supplies are, however, poorly distributed, and some regions are already short of water, and other areas will be in short supply in a relatively short time. One obvious alternative is to limit the development of such areas to that which can be accommodated within the available water supplies. A second alternative is to import water from some nearby region of surplus. For many reasons these alternatives may be undesirable, and efforts to find other ways of augmenting water supplies are already under way.

Augmentation may include methods for providing an entirely new supply of fresh water but should also be considered to include techniques for increasing the utility of available supplies by conservation or regulation which permits greater use. Techniques for providing completely new supplies include weather modification which may include many possible plans for altering the atmospheric circulation to provide a more favorable distribution of precipitation or to decrease evapotranspiration. Such plans will have many political implications which require that they be approached with the utmost caution. For the present the main effort has been devoted to localized precipitation increases by cloud seeding. Recent studies[1] suggest that under favorable conditions increases in precipitation averaging about 10 percent can be expected. Lumb[2] has demonstrated that the increase in annual runoff ΔR resulting from a small increase in annual precipitation ΔP is given approximately by

$$\Delta R = \Delta P\left(0.29 + 1.2\frac{R_{avg}}{P_{avg}}\right) \tag{21-1}$$

where R_{avg} and P_{avg} are the mean annual runoff and precipitation respectively. Equation (21-1) is based on the analysis of seven rural basins. It should be used only for preliminary estimates, where $0.1 < R_{avg}/P_{avg} < 0.5$. The study was performed by using digital simulation to compare runoff from observed precipitation. The portion of the added runoff which can be put to use depends on the facilities available in the basin. Generally, relatively large reservoir storage is required to make the added flow useful.

Desalting of seawater (Sec. 15-22) is a feasible but costly method of augmenting fresh-water supplies. Cost depends on plant size and the possibility of combining power production with water production. With very

[1] Weather and Climate Modification Problems and Prospects, *Pub.* 1350, National Academy of Sciences, Washington, D.C., 1966; Weather Modification Law, Controls, Operations, Report to the Special Commission on Weather Modification, National Science Foundation, Washington, D.C., 1966.

[2] Alan M. Lumb, Hydrologic Effects of Rainfall Augmentation, *Tech. Rept.* 116, Department of Civil Engineering, Stanford University, November, 1969.

large nuclear power plants as a source of power and heat, costs are within acceptable ranges for urban water supply but far beyond reasonable costs for irrigation. Few coastal locations are in need of plants in the billion gallon per day range where the scale economies would become effective. Since the production is at sea level, pumping costs to inland locations at higher elevations greatly increase water costs. While distillation techniques are most feasible for seawater, brackish water from estuaries or brackish groundwater can be desalted by reverse osmosis or electrodialysis. For these processes cost is a function of salinity and can be reasonable for water under 5000 mg/l concentration. However, disposal of the residual brine which may equal half of the total volume of input and be twice as concentrated may create serious problems. The brines cannot be permitted to return to the source or to pollute other water sources.

Conservation of available supplies can also be achieved by a number of methods. Very little consideration is now given to methods of meeting present needs with less water. Sewage disposal by water transport and dilution degrades vast amounts of pure water, and techniques which will use less water in the home and industry could result in substantial savings of water. In many areas agricultural use of water could be markedly reduced by reducing conveyance losses and wasteful irrigation practices. The use of surface agents to reduce evaporation from reservoirs[1] and even to reduce evapotranspiration by crops[2] is under study. Present costs are high but will probably be reduced by further work. Large parts of our freshwater supply are utilized for evapotranspiration by native vegetation. Substantial amounts of water can be saved by elimination of phreatophytes in arid regions.[3] There is also evidence that selective thinning of forests or elimination of brush and chaparral will increase stream flow. The effect is most marked on the low flows so that regulation for greater utility is also a product of such forest management. Studies of the effect of forest management on flow in areas with seasonal snow cover are under way.[4]

Since substantial amounts of water are made unfit for use by the addition of relatively small amounts of pollutants, reclamation and reuse of the wastewater is another method by which existing water supplies can be augmented.

---

[1] Joseph B. Franzini, Evaporation Suppression Research Progress Report, *Water and Sewage Works*, Vol. 108: pt. I, pp. 167–172, May, 1961, pt. II, pp. 221–225, June, 1961.

[2] W. J. Roberts, Reduction of Transpiration, *J. Geophys. Res.*, Vol. 66, pp. 3309–3312, October, 1961.

[3] Water Resources Activities in the United States—Evapotranspiration Reduction, Senate Select Committee on National Water Resources, *Committee Print* 21, February, 1960.

[4] Henry W. Anderson, Prospects for Affecting the Quantity and Timing of Water Yield through Snow-pack Management in California, *28th Ann. Proc., Western Snow Conf.*, pp. 44–50, 1960.

Reclamation can be accomplished by direct reuse for industrial processes with a minimum of treatment or more elaborate reclamation plants could be constructed to make the water biologically and chemically suitable for higher uses. Cost again is a controlling factor, but continuing research should result in lower costs while the value of water will probably increase. At some point in time the two curves should cross.

Since the groundwater is generally free from evapotranspiration losses, increased use of groundwater storage in conjunction with moderate surface storage will reduce the large evaporation losses occurring from reservoirs in arid regions. Urban development with its impervious areas actually increases runoff but also accelerates it so that it is more difficult to regulate and store. When urban development is accompanied by extensive works for lining natural channels, considerable amounts of natural groundwater recharge may be prevented. This is only one of many examples of the need for careful planning of the entire complex of human effort for optimum use of resources.

Sealing the soil surface with chemicals or asphalt can greatly increase water yield in arid regions. Since nearly all the rain which falls on the protected area is recovered, the yield is easily estimated as the mean annual precipitation over the area sealed. If the runoff water is conveyed to a covered tank, evaporation loss is prevented. Stock water, irrigation water for kitchen gardens, water for domestic use on the farm, and even water supplies for small villages may be developed in this fashion. Cost is dependent on the cost of surface sealing but will generally be higher than securing water from natural sources.

21-10   Multiple-purpose projects   The previous sections of this chapter have discussed water-development projects serving several purposes. Many hydraulic projects can serve more than one of the basic purposes— water supply, irrigation, hydroelectric power, navigation, flood mitigation, recreation, sanitation, and wildlife conservation. Multiple use of project facilities may increase benefits without a proportional increase in costs and thus enhance the economic justification for the project. A project which is designed for single purpose but which produces incidental benefits for other purposes should not, however, be considered a multiple-purpose project. *Only those projects which are designed and operated to serve two or more purposes should be described as multiple-purpose.*[1] The first real multiple-purpose project in the United States was the Sacandaga reservoir in the Hudson River basin, New York (completed in 1931), which combines flood mitigation with storage of water for power and industrial use downstream.

Since 1931 many major multiple-purpose projects have been built in the

---

[1] Symposium on Multiple-purpose Reservoirs, *Trans. ASCE*, Vol. 115, pp. 789–908, 1950.

United States. Because private interests rarely have need for multiple functions, most multipurpose projects have been constructed by government agencies. Some private projects have incorporated multiple-use features through cooperative arrangements with government units. It is imperative that maximum use be made of our water resources and the concept of multiple use is essential to such maximum use. Indeed, with the best project sites already utilized, many projects cannot be justified economically without multiple-purpose goals. The preceding chapters discussed the special problems of most of the possible water uses from the single-purpose viewpoint. Multiple-use planning, however, involves substantially more than the simple combination of single-purpose elements. Therefore, this and the following sections discuss the special problems of multiple-purpose planning. None of the possible uses of a reservoir is entirely compatible with any other use, although under many circumstances it is possible to bring some uses into good agreement.

The basic factor in multi-purpose design is *compromise*. A working plan must be devised which permits reasonably efficient operation for each purpose although maximum efficiency is not necessarily attained for any single purpose. The physical elements of a multiple-purpose project (dam, spillway, sluiceways, gates, power plant, etc.) differ in no way from those for a single-purpose project. The unique feature in multiple-purpose design is the selection of physical works and an operation plan which is an effective compromise among the various uses.

Two possible extremes in allocation of reservoir storage are (1) to assume that no storage is jointly used and (2) to assume that all storage is jointly used. In the first instance storage requirements for all functions are pyramided to create a large total storage requirement which can be economically attained only when unit cost of storage is constant or decreases as total storage increases. The other extreme results in maximum economy since the required storage is no greater than that necessary for any one of the several purposes. Such a situation is relatively rare, and the usual multipurpose project is designed to fall somewhere between these two extremes.

21-11 Functional requirements in multiple-purpose projects The success which can be obtained in achieving joint use of storage space in a multipurpose project depends upon the extent to which the various purposes are compatible. It is helpful, therefore, to review the requirements of the various uses and to consider the ways in which these uses may be coordinated.

Irrigation Water requirements for irrigation are usually seasonal, with a maximum during the summer months and little or no demand during the winter (Fig. 21-4). Water requirements do not vary greatly from year to year, although low-rainfall years usually create a greater irrigation demand than high-rainfall years. Unless the project area is increased, the average

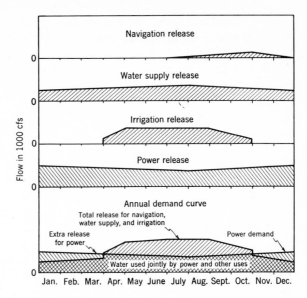

Navigation release

Water supply release

Irrigation release

Power release

Annual demand curve
Total release for navigation,
water supply, and irrigation
Extra release   Power demand
for power
Water used jointly by power and other uses

Flow in 1000 cfs

Jan. Feb. Mar. Apr. May June July Aug. Sept. Oct. Nov. Dec.

FIG. 21-4 Synthesis of annual demand curve from the estimated requirements of the various uses in a multiple-purpose project.

annual demand will remain nearly constant. Since irrigation storage is insurance against drought, it is desirable to maintain as much reserve storage as possible consistent with current demands.

Water supply   Requirements for domestic water are more nearly constant throughout the year than are irrigation requirements, but a seasonal maximum is usually experienced in the summer. The demand normally increases slowly from year to year, and provision must be made for this increase. Maintenance of an adequate reserve to avoid water shortage during droughts is necessary. Sanitary precautions may preclude use of the reservoirs for recreational purposes.

Hydroelectric power   Power demand usually has a marked seasonal variation depending on the type of area served. Most power plants are parts of an interconnected system, and considerable flexibility is possible in co-ordinating power needs with other uses of water. Power production is not a consumptive use of water and is therefore more compatible with other uses. Water released from a reservoir for irrigation use downstream may be passed through a turbine to produce power as well. In the extreme, power production may be limited to those times when releases are necessary for other uses or to discharge excess water. However, this may result in a low load factor for the plant, with a loss in overall efficiency. Most in-tegrated power systems do not have need for many low-load-factor plants.

Navigation   Reservoirs designed to provide water to sustain downstream flows for navigation represent a very marked seasonal water requirement, with peak releases required during the latter portion of the dry season. Reservoirs for slackwater improvement of streams for navigation must be limited in height because of the need for locks and hence cannot have large storage allocations for other purposes.

Flood mitigation   The basic requirement of flood mitigation is sufficient *empty* storage space to permit withholding of floodwater during the flood season.

Recreation   It is rarely practicable to design a large reservoir for recreational purposes only, and any recreational benefits are usually incidental to the other functions of the project. The ideal recreational reservoir is one which remains nearly full during the recreation season to permit boating, fishing, swimming, and other water sports. A reservoir subject to large drawdowns is usually unsightly and creates problems of maintaining docks, boat moorings, beaches, and other waterfront facilities in usable conditions.

Fish and wildlife   The problems of fish and wildlife in large reservoirs are mainly ones of protection. Reservoir construction results in a major change in habitat for existing wildlife and may result in a decrease in one species and an increase in others. Protection of existing fisheries may require the provision of fish ladders or other arrangements (Sec. 9-26) permitting migratory fish to travel upstream. Large and rapid fluctuations in water level are harmful to fish, particularly at critical periods such as the spawning period. Complete stoppage of flow below the dam is also destructive of fish and wildlife. Dams which flood spawning grounds of migratory fish will also injure the fish runs. Unless sufficient natural spawning grounds are left, installation of a hatchery may be necessary to maintain the fish runs.

Pollution control   A reservoir can be used for *low-flow augmentation*, the release of water during the low-flow season to provide dilution water so that the stream can better assimilate the wastewater that is discharged into it. Releases from a reservoir, however, may contribute to pollution because of water-quality changes that often occur in the reservoir. Dissolved salts may be increased by evaporation and leaching from the soils or rocks of the bed and banks. Also, algae growth or decay of vegetation within the reservoir may depress the dissolved oxygen content, especially in the lower levels of the reservoir.

Mosquito control   If desired, a reservoir can be operated to control mosquito growth through rapid fluctuations of the water level which strand larvae on the shore.

21-12 Compatibility of multiple-purpose uses   Irrigation, navigation, and water supply all require a volume of water which cannot be jointly used, and hence a project combining these functions must provide a clear allocation of storage space to each (Fig. 21-4). Since power development is not a consumptive use, any water released for the other purposes may be used for power. If the power plant can be operated as a base-load plant, its water requirements may fit in well with the relatively uniform releases for other purposes. If it is proposed to use the plant for peaking, an afterbay or downstream reregulating dam can be provided to smooth out the fluctuations of the power release. The afterbay storage capacity need be sufficient only to regulate flows for a few days at a time, in the same manner that a distribution reservoir smooths out the load on a pumping plant. A low-head power plant at the reregulating dam may produce a small amount of base-load power. Since the seasonal variations in power demand may not coincide with the requirements of other uses, it is usually necessary to allocate a certain amount of storage for power use (Fig. 21-4) or to provide other plants in the system which can pick up the load during the winter months.

A dam creating a pool for slackwater navigation can often be used for power production. The dam will usually be low and the controllable storage fairly small so that the plant may operate as a run-of-river plant. The modern tendency has been toward higher navigation dams because of the need for fewer locks and the improvement in power head. The area of such high pools may be sufficient to provide some storage for power and/or flood mitigation by the addition of a few feet to the height of the dam.

Flood mitigation with its requirement for empty storage space is the least compatible of all uses. Storage for flood mitigation can be obtained in two ways, either (1) by permanent allocation of space exclusively for flood mitigation or (2) by the seasonal allocation of space for flood storage. In the first case, the space is usually allocated at the top of the pool, and water is allowed to encroach on this space only during floods. Often the allocated space is above spillway level and is made available when required by closing spillway crest gates. Any additional space available when a flood occurs is used but is not dependable. In evaluating the benefits of storage for flood mitigation, only that storage in excess of the natural-channel storage in the reach occupied by a reservoir should be considered.

Seasonal allocations to flood mitigation can be made where a definite flood season occurs. In the Western states many basins experience floods only during the spring thaw from April to July. It is possible to predict the runoff volume during this period with considerable accuracy (Sec. 3-8), and necessary storage may be allocated to provide adequate flood protection with the assurance that the space will be filled for other uses by the time the flood hazard is past. Where long-range forecasts are impossible, a plan of seasonal allocation may be developed on the basis of a calculated risk. A study of historic floods will indicate the storage space which would have been needed

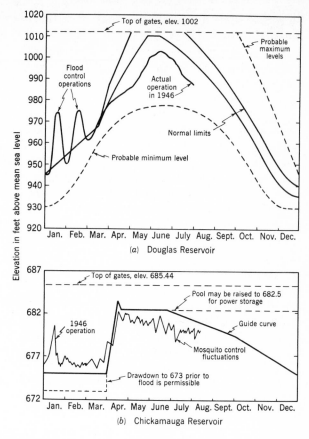

FIG. 21-5   *Rule curves for a typical storage reservoir (a) and a mainstream navigation reservoir (b). (After Bowden)*

on any given date to permit satisfactory regulation of flood flows. From this information a *rule curve* (Fig. 21-5) can be constructed to show the amount of space to be left vacant on any date and the dates during which the reservoir may be safely filled if adequate water is available. Often some storage above spillway level is reserved for flood mitigation at all times since serious floods have occurred in every month of the year in much of the country. A rule curve allocates a certain amount of storage to flood mitigation with the proviso that during the periods when floods are unlikely this space may be used for other purposes *provided there is sufficient inflow to fill it*. There is no guarantee that this space will be filled every year, and hence the water which might be held there must be considered as secondary water.

Recreational benefits of a reservoir are usually taken as opportunity permits. In some cases it may be possible to plan system operations so that the reservoir need not be drawn down excessively during the height of the recreation season. Picnic grounds, access roads, boating facilities, etc., add to the usefulness of a lake for recreation only if the operation is such as to create pleasant conditions. Where heavy drawdown is unavoidable, an arm of the reservoir may be closed off by a small dam to create a subimpoundment which can be maintained at a constant level for swimming. Usually such subimpoundments are too small for boating or fishing.

What has been said regarding recreation also applies for fish and wildlife production. The operation plan should provide for a minimum drawdown and the maintenance of adequate flow in the channel downstream. Subimpoundments may provide spawning grounds for fish in reservoirs subject to rapid drawdown. Use of chemicals to control algae or weeds should be regulated to avoid poisoning of wildlife. The use of subimpoundments for either recreation or wildlife is a permanent allocation of storage for these purposes. If a minimum flow must be maintained downstream, this too may require a separate allocation of storage unless releases for other purposes will be adequate. If the permissible drawdown is limited to some minimum level for fish and wildlife protection, the storage below this level is not useful except to provide sediment storage and power head. Thus some of this space might become a charge against the fish and wildlife aspects of the project.

21-13   Cost allocation for multiple-purpose projects   Because a multiple-purpose project serves several different groups of beneficiaries, it is often necessary to allocate the cost among the several uses to fix the prices of water or power or to determine the contribution required of flood-mitigation beneficiaries.

There is no really satisfactory method of cost allocation in the sense that such a method would be equally applicable to all projects and would yield allocations which are unquestionably correct. Any method of allocation must first set aside the *separable costs* which are clearly chargeable to a single project function, such as the cost of the powerhouse, navigation locks, or fish ladders. The separable costs for a single function are usually estimated by computing the cost of the project with that function omitted. The separable cost of the function is the total project cost less the estimated cost with that function omitted. The real problem in cost allocation is the division of the *joint costs* (total cost less the sum of the separable costs) among the project functions. The two methods which seem most applicable to a multiple-purpose water project are the *remaining-benefits method* and the *alternative justifiable-expenditure method.*[1] The application of these methods is illustrated by Table 21-2. The first line of the table presents the separable costs,

---

[1] "Proposed Practices for Economic Analysis of River Basin Projects," Government Printing Office, Washington, D.C., 1958; T. B. Parker, Allocation of the Tennessee Valley Projects, *Trans. ASCE*, Vol. 108, pp. 174–216, 1943.

TABLE 21-2   *Cost Allocation for a Multiple-purpose Project Costing $1,765,000*
(All items in thousands of dollars)

Line	Item	Flood miti-gation	Power	Irri-gation	Navi-gation	Total
1	Separable costs	380	600	150	50	1180
2	Estimated benefits	500	1500	350	100	2450
3	Alternate single-purpose cost	400	1000	600	80	2080

Remaining-benefits method

Line	Item	Flood miti-gation	Power	Irri-gation	Navi-gation	Total
4	Benefits limited by alternate cost	400	1000	350	80	1830
5	Remaining benefits	20	400	200	30	650
6	Allocated joint costs	18	360	180	27	585
7	Total allocation:					
	Dollars	398	960	330	77	1765
	Percent	22.5	54.4	18.7	4.4	100

Alternative justifiable-expenditure method

Line	Item	Flood miti-gation	Power	Irri-gation	Navi-gation	Total
8	Alternate cost less separable cost	20	400	450	30	900
9	Allocated joint costs	13	260	292	20	´585
10	Total allocation:					
	Dollars	393	860	442	70	1765
	Percent	22.2	48.7	25.0	4.0	100

which total $1,180,000. Since the total cost to be allocated is $1,765,000, the joint cost is $585,000.

In the remaining-benefits method, the joint costs are assumed to be distributed in accordance with the differences between the separable costs (line 1) and the estimated benefits of each function (line 2). In no case, however, are the benefits assumed to be greater than the cost of an alternate single-purpose project which would provide equivalent benefits. Thus the remaining benefits (line 5) are the differences between the lesser value of lines 2 or 3 and the separable costs in line 1. The joint costs are distributed in proportion to these remaining benefits and added to the corresponding separable costs to obtain the total allocation (line 7).

Under the alternative justifiable-expenditure method the joint costs are assumed to be distributed in accordance with the differences between the separable costs and the estimated cost of a single-purpose project which would provide equivalent services and would itself be economically justifiable. These costs (line 8) are the differences between lines 1 and 3. The distributed joint costs (line 9) are added to the separable costs to obtain the total allocated costs (line 10). It should be noted that a large percentage difference in the allocation of joint costs is only a moderately small percentage difference in the allocated total cost because the separable costs are normally a large part of the total. The greatest difficulty with both methods is that of estimating the benefits or alternative costs. From a practical viewpoint both methods give results which are within the probable limits of accuracy obtainable in any allocation.

In order to establish a pricing policy it is sometimes necessary to allocate costs between several classes of users. When a long pipeline is constructed, it may be proper to charge users near the head of the line less for water than is charged to users at the far end of the line. In this case the allocation is often based on the proportional use of facilities.

No single method of cost allocation can properly be described as the "best" method. For establishing prices or allocating charges to various beneficiaries, any method which is agreeable to all concerned can be considered acceptable. Because of the arbitrary nature of cost allocation, it must be viewed as an accounting device totally unrelated to the economic evaluation of a project. Costs allocated to a specific project purpose should *not* be compared with expected benefits to justify the inclusion of the particular purpose in the project plans. If the remaining-benefits method is used, it will automatically make each purpose appear to have a benefit-cost ratio ratio in excess of one (Table 21-2).

BIBLIOGRAPHY

Bowman, James S.: Multipurpose River Developments, *Trans. ASCE*, Vol. CT, pp. 1125–1131, 1953.

Dixon, J. W.: Planning an Irrigation Project Today, *Trans. ASCE*, Vol. CT, pp. 357–387, 1953.

Eckstein, Otto: "Water Resource Development: The Economics of Project Evaluation," Harvard, Cambridge, Mass., 1958.

Hirshleifer, Jack, James C. De Haven, and J. W. Milliman: "Water Supply: Economics, Technology, and Policy," The University of Chicago Press, Chicago, 1960.

James, L. D., and R. R. Lee: "Economics of Water Resources Planning," McGraw-Hill, New York, 1971.

Kneese, Allen V.: "Water Resources Development and Use," Federal Reserve Bank of Kansas City, 1959.

Krutilla, John, and Otto Eckstein: "Multiple Purpose River Development: Studies in Applied Economic Analysis," Johns Hopkins, Baltimore, 1958.

Schad, Theodore: Perspective on National Water Resources Planning, *J. Hydraulics Div. ASCE*, Vol. 88, pp. 17–42, July, 1962.

Symposium on Multiple-purpose Reservoirs, *Trans. ASCE*, Vol. 115, pp. 790–908, 1950.

SPECIAL BIBLIOGRAPHY: Systems Analysis in Water Resource Planning

Bather, J. A.: Optimal Regulation Policies for Finite Dams, *J. Soc. Indus. Applied Math.,* Vol. 10, pp. 395–443, 1962.

Beard, L. R.: Optimization Techniques for Hydrologic Engineering, *Water Resources Research,* Vol. 3, pp. 809–815, 1967.

Buras, N.: Conjunctive Operation of Dams and Aquifers, *J. Hydraulics Div.,* ASCE, Vol. 89, pp. 111–131, 1963.

———: A Three Dimensional Optimization Problem in Water-Resources Engineering, *Operations Res. Quart.,* Vol. 16, p. 479, 1965.

Castle, E. N.: Programming Structures in Watershed Development, chap. 12 in G. S. Tolley and F. E. Riggs (eds.), "Economics of Watershed Planning," Iowa State Univ. Press, Ames, 1961.

Chow, V. T.: Systems Design by Operations Research, sec. 26–11 in V. T. Chow (ed.), "Handbook of Applied Hydrology," McGraw-Hill, New York, 1964.

Dracup, John A.: "The Optimum Use of Groundwater and Surface-Water Systems; A Parametric Linear Programming Approach," *Cont.* 107, University of California Water Resources Center, Berkeley, 1966.

Flinn, J. C.: Development and Analysis of Input-Output Relations for Irrigation Water, *Australian J. Agr. Econ.,* Vol. 11, pp. 1–19, 1967.

Gysi, M., and D. P. Loucks: A Selected Annotated Bibliography on the Analysis of Water Resources Systems, Pub. No. 25, Cornell University Water Resources and Marine Sciences Center, Ithaca, New York, August 1969.

Hall, W. A.: Aqueduct Capacity under an Optimum Benefit Policy, *J. Irrigation and Drainage Div.,* ASCE, Vol. 87, pp. 1–11, 1961.

———: Optimum Design of a Multiple-Purpose Reservoir, *J. Hydraulics Div.,* ASCE, Vol. 90, pp. 141–149, 1964.

——— and N. Buras: Optimum Irrigation Practice under Conditions of Deficient Water Supply, *Trans. Amer. Soc. Agri. Engrs.,* Vol. 4, pp. 131–134, 1961.

——— and ———: The Dynamic Programming Approach to Water Resources Development, *J. Geophys. Res.,* Vol. 66, pp. 517–520, 1961.

——— and John A. Dracup: "Water Resource Systems Engineering," McGraw-Hill, New York, 1970.

——— and T. G. Roefs: Hydropower Project Output Optimization, *J. Power Div.,* ASCE, Vol. 92, pp. 67–79, 1966.

——— and R. W. Shephard: Optimum Operation for Planning of a Complex Water Resources System, *U.C.L.A. Engineering Rept.* 67–54, University of California Water Resources Center, 1967.

Hufschmidt, M. M., and Myron Fiering: "Simulation Techniques for the Design of Water-Resource Systems," Harvard, Cambridge, Mass., 1966.

Loucks, D. P.: Computer Models for Reservoir Regulation, *J. Sanitary Div.,* ASCE, Vol. 94, pp. 657–669, 1968.

———: A Comment on Optimization Methods for Branching Multi-Stage Water Resources Systems, *Water Resources Research,* Vol. 4, pp. 447–450, 1968.

Maass, Arthur, M. M. Hufschmidt, Robert Dorfman, H. A. Thomas, S. A. Marglin, and G. M. Fair: "Design of Water-Resource Systems," Harvard, Cambridge, Mass., 1962.

Manne, Alan S.: Product-Mix Alternatives: Flood Control, Electric Power, and Irrigation, *Cowles Foundation Paper* 175, Yale University, New Haven, Conn., 1962.

Mannos, Murray: An Application of Linear Programming to Efficiency in Operation of a System of Dams, *Econometrica,* Vol. 23, pp. 335–336, 1955.

McKean, Roland: "Efficiency in Government through Systems Analysis, with Emphasis in Water Resources Development," Wiley, New York, 1958.

Meier, W. L., and C. S. Beightler: An Optimization Method for Branching Multistage Water Resources Systems, *Water Resources Research*, Vol. 3, pp. 645–652, 1967.

"Proceedings of the National Symposium on the Analysis of Water-resource Systems," American Water Resources Association, Urbana, Illinois, 1968.

Schweig, Z., and J. A. Cole: Optimum Control of Linked Reservoirs, *Water Resources Research*, Vol. 4, pp. 479–498, 1968.

Thomas, H. A., and R. P. Burden: Operations Research in Water Quality Management, *Final Rept. for Contract* Ph86–62–140, Harvard University, Cambridge, Mass., 1963.

PROBLEMS

21-1 For a proposed water development the following units have been identified: $1A$, $1B$, $2A$, $2B$, $2C$, $3A$, and $3B$. None is independent. How many possible combinations are there?

21-2 Cost and benefit data for the units of a small development are tabulated below. The costs are mutually exclusive at each site. Assuming a project life of 50 yr and an interest rate of 4 percent, rank the projects or combinations in order of preference. Repeat for a rate of 7 percent.

Project unit	First cost	Annual operation cost	Annual benefits
$1A$	$1,000,000	$50,000	$100,000
$1B$	800,000	45,000	85,000
$2A$	1,550,000	85,000	140,000
$2B$	600,000	22,000	62,000
$1A, 2A$			220,000
$1A, 2B$			150,000
$1B, 2A$			235,000
$1B, 2B$			147,000

21-3 Using the data given below (all items given in thousands of dollars) calculate the allocations to each project purpose by the remaining-benefits method and by the alternative justifiable-expenditure method. Assume total cost of project is $2 million.

Item	Flood mitigation	Power	Irrigation
Separable costs	250	670	550
Estimated benefits	350	1200	970
Alternate single-purpose cost	400	870	850

21-4 A small reservoir and 20 mi of canal are proposed to provide irrigation water to five separate irrigation units.

Unit	Distance from reservoir, mi	Annual water use, acre-ft	Incremental annual cost of canal
A	5	1000	$9500
B	9	2000	6800
C	13	2500	6400
D	18	1500	4700
E	20	1000	1400

In the table the incremental annual cost of canal is the annual charges on the increment of canal between the reservoir and unit *A* or between any unit and the diversion point for the previous unit. Annual charges for the reservoir are estimated at $46,000. Calculate the cost of water delivered to each unit at canalside by the proportional use of facilities method.

21-5 Using the data given below calculate the allocations to each project purpose by the remaining-benefits method and by the alternate justifiable-expenditure method. The estimated total cost of the project is $12,300,000.

Item	Water Supply	Flood Mitigation
Separable costs	$4,000,000	$ 6,500,000
Estimated benefits	5,100,000	10,200,000
Alternate single-purpose cost	5,800,000	8,600,000

21-6 In a particular region the mean annual precipitation is 26 in. and the mean annual runoff is 4.5 in. According to Eq. (21-1) how much additional runoff will occur if a cloud-seeding program is in effect during a year when the annual precipitation would have been 20 in.? Assume the cloud seeding increased the annual precipitation by 10 percent.

Physical constants, conversion tables, and equivalents

Conversion Table for Volume

Unit	Equivalents							
	Cu in.	Gal.	Imperial gal	Cu ft	Cu yd	Cu m	Acre-foot	Second-foot-day
Cubic inch	1	0.00433	0.00361	5.79×10^{-4}	2.14×10^{-5}	1.64×10^{-5}	1.	$\ldots \times 10^{-5}$
U.S. gallon	231	1	0.833	0.134	0.00495	0.00379	3.0	$\ldots \times 10^{-4}$
Imperial gallon	277	1.20	1	0.161	0.00595	0.00455	$3.68 \times$	$\ldots \times 10^{-4}$
Cubic foot	1,728	7.48	6.23	1	0.0370	0.0283	2.30×10	
Cubic yard	46,656	202	168	27	1	0.765	6.20×10^{-4}	
Cubic meter	61,000	264	220	35.3	1.31	1	8.11×10^{-4}	4.09×10^{-4}
Acre-foot	7.53×10^7	3.26×10^5	2.71×10^5	43,560	1610		1	0.504
Second-foot-day	1.49×10^8	6.46×10^5	5.38×10^5	86,400			1.98	1

Conversion Table for Discharge

Unit	Gpd	Cu ft/day	Gpm	Acre-ft/day	Cfs	Cu m/sec
U.S. gallon per day	1	0.134	6.94×10^{-4}	3.07×10^{-6}	1.55×10^{-6}	4.38×10^{-8}
Cubic foot per day	7.48	1	5.19×10^{-3}	2.30×10^{-5}	1.16×10^{-5}	3.28×10^{-7}
U.S. gallon per minute	1,440	193	1	4.42×10^{-3}	2.23×10^{-3}	6.31×10^{-5}
Imperial gallon per minute	1,728	231	1.20	5.31×10^{-3}	2.67×10^{-3}	7.57×10^{-5}
Acre-foot per day	3.26×10^5	43,560	226	1	0.504	0.0143
Cubic foot per second	6.46×10^5	86,400	449	1.98	1	0.0283
Cubic meter per second	2.28×10^7	3.05×10^6	15,800	70.0	35.3	1

Properties of Water

Temp, °F	Specific gravity	Unit weight, lb/cu ft	Heat of vaporiza- tion, Btu/lb	Viscosity		Vapor pressure	
				Absolute, lb-sec/sq ft	Kinematic, sq ft/sec	Millibars	Psi
32	0.99987	62.416	1073	0.374×10^{-4}	1.93×10^{-5}	6.11	0.09
40	0.99999	62.423	1066	0.323	1.67	8.36	0.12
50	0.99975	62.408	1059	0.273	1.41	12.19	0.18
60	0.99907	62.366	1054	0.235	1.21	17.51	0.26
70	0.99802	62.300	1049	0.209	1.06	24.79	0.36
80	0.99669	62.217	1044	0.180	0.929	34.61	0.51
90	0.99510	62.118	1039	0.160	0.828	47.68	0.70
100	0.99318	61.998	1033	0.143	0.741	64.88	0.95
120	0.98870	61.719	1021	0.117	0.610	.....	1.69
140	0.98338	61.386	1010	0.0979	0.513	.....	2.89
160	0.97729	61.006	999	0.0835	0.440	.....	4.74
180	0.97056	60.586	988	0.0726	0.385	.....	7.51
200	0.96333	60.135	977	0.0637	0.341	.....	11.52
212	0.95865	59.843	970	0.0593	0.319	.....	14.70

Variation of Pressure, Temperature, and Boiling Point with Elevation (U.S. standard atmosphere*)

Elevation ft msl	Pressure			Air temp, °F	Boiling point, °F
	In. Hg	Millibars	Ft of water		
−1,000	31.02	1050.5	35.12	62.6	213.8
0	29.92	1013.2	33.87	59.0	212.0
1,000	28.86	977.3	32.67	55.4	210.2
2,000	27.82	942.1	31.50	51.8	208.4
3,000	26.81	907.9	30.35	48.4	206.5
4,000	25.84	875.0	29.25	44.8	204.7
5,000	24.89	842.9	28.18	41.2	202.9
6,000	23.98	812.1	27.15	37.6	201.1
7,000	23.09	781.9	26.14	34.0	199.2
8,000	22.22	752.5	25.16	30.6	197.4
9,000	21.38	724.0	24.20	27.0	195.6
10,000	20.58	696.9	23.30	23.4	193.7
11,000	19.79	670.2	22.40	19.8	191.9
12,000	19.03	644.4	21.54	16.2	190.1

*The data of this table are based on average conditions and must be adjusted to the particular meteorological conditions prevailing at any time.

Miscellaneous Physical Constants

---

1 liter = 0.001 cu m
1 hp = 0.746 kw = 550 ft-lb/sec
1 millibar = 0.0143 psi   1 bar = 14.3 psi
1 sec-ft-day/sq mi = 0.03719 in.
1 in. of runoff per sq mi = 26.9 sec-ft-days = 53.3 acre-ft
1 cfs = 0.9917 acre-in./hr
e = 2.71828
$\log_{10} e$ = 0.43429
$\log_e 10$ = 2.30259
1000 Btu/lb = 555 cal/g
1 ft/day = 7.48 gpd/sq ft
1 m/day = 22.9 gpd/sq ft

Values of the Reduced Variate b Corresponding to Various Values of Return Period and Probability of Exceedance [*Eq.* (5-5)]

Reduced variate b	Return period t_p	Probability of Exceedance P
0.000	1.58	0.632
0.367	2.00	0.500
0.579	2.33	0.429
1.500	5.00	0.200
2.250	10.0	0.100
2.970	20.0	0.050
3.902	50.0	0.020
4.600	100	0.010
5.296	200	0.005
6.000	403	0.0025

1 bar = 14.5 psi; average atmosphere pressure at sea bar

Table of K Values for Use with the Log Pearson Type III Distribution

Skew coefficient (g)	Recurrence interval in years					
	2	10	25	50	100	200
	Percent chance					
	50	10	4	2	1	0.5
3.0	−0.396	1.180	2.278	3.152	4.051	4.970
2.5	−0.360	1.250	2.262	3.048	3.845	4.652
2.0	−0.307	1.302	2.219	2.912	3.605	4.298
1.8	−0.282	1.318	2.193	2.848	3.499	4.147
1.6	−0.254	1.329	2.163	2.780	3.388	3.990
1.4	−0.225	1.337	2.128	2.706	3.271	3.828
1.2	−0.195	1.340	2.087	2.626	3.149	3.661
1.0	−0.164	1.340	2.043	2.542	3.022	3.489
0.9	−0.148	1.339	2.018	2.498	2.957	3.401
0.8	−0.132	1.336	1.993	2.453	2.891	3.312
0.7	−0.116	1.333	1.967	2.407	2.824	3.223
0.6	−0.099	1.328	1.939	2.359	2.755	3.132
0.5	−0.083	1.323	1.910	2.311	2.686	3.041
0.4	−0.066	1.317	1.880	2.261	2.615	2.949
0.3	−0.050	1.309	1.849	2.211	2.544	2.856
0.2	−0.033	1.301	1.818	2.159	2.472	2.763
0.1	−0.017	1.292	1.785	2.107	2.400	2.670
0	0	1.282	1.751	2.054	2.326	2.576
−0.1	0.017	1.270	1.716	2.000	2.252	2.482
−0.2	0.033	1.258	1.680	1.945	2.178	2.388
−0.3	0.050	1.245	1.643	1.890	2.104	2.294
−0.4	0.066	1.231	1.606	1.834	2.029	2.201
−0.5	0.083	1.216	1.567	1.777	1.955	2.108
−0.6	0.099	1.200	1.528	1.720	1.880	2.016
−0.7	0.116	1.183	1.488	1.663	1.806	1.926
−0.8	0.132	1.166	1.448	1.606	1.733	1.837
−0.9	0.148	1.147	1.407	1.549	1.660	1.749
−1.0	0.164	1.128	1.366	1.492	1.588	1.664
−1.2	0.195	1.086	1.282	1.379	1.449	1.501
−1.4	0.225	1.041	1.198	1.270	1.318	1.351
−1.6	0.254	0.994	1.116	1.166	1.197	1.216
−1.8	0.282	0.945	1.035	1.069	1.087	1.097
−2.0	0.307	0.895	0.959	0.980	0.990	0.995
−2.5	0.360	0.771	0.793	0.798	0.799	0.800
−3.0	0.396	0.660	0.666	0.666	0.667	0.667

Approximate Particle Size and Permeability of Various Soils

Material	Particle size,* mm	Approximate permeability, gpd/sq ft
Clay	0.0001–0.005	10^{-5} to 10^{-2}
Silt	0.005–0.05	10^{-2} to 10
Very fine sand	0.05–0.10	10 to 50
Fine sand	0.10–0.25	50 to 250
Medium sand	0.25–0.50	250 to 1,000
Coarse sand	0.50–2.00	1,000 to 15,000

*1 mm = 0.03937 in.

Soil Types
(U.S. Bureau of Soils)

Material	Clay, percent	Silt, percent	Sand, percent	Approximate permeability, gpd/sq ft
Clay	30–100	0–50	0–50	10^{-4}
Silty clay	30–50	50–70	0–20	10^{-3}
Sandy clay	30–50	0–20	50–70	10^{-3}
Silty clay loam	20–30	50–80	0–30	10^{-2}
Clay loam	20–30	20–50	20–50	10^{-2}
Sandy clay loam	20–30	0–30	50–80	10^{-2}
Silt loam	0–20	50–100	0–50	10^{-1}
Loam	0–20	30–50	30–50	10^{-1}
Sandy loam	0–20	0–50	50–80	1
Sand	0–20	0–20	80–100	Over 10

Combining Weights of Common Elements and Radicals*

Calcium	20	Carbonate (CO_3)	30
Chlorine	35	Bicarbonate (HCO_3)	61
Boron	3.6	Sulfate (SO_4)	48
Hydrogen	1	Sulfite (SO_3)	40
Magnesium	12	Nitrate (NO_3)	62
Oxygen	8	Nitrite (NO_2)	46
Potassium	39		
Sodium	23		

*The combining weight is equal to the atomic weight divided by the valence.

Dimensions of Standard Screw Pipe

Nominal size, in.	Diameter, in. External	Diameter, in. Internal	Wall thickness, in.	Weight per foot,* lb
$1/8$	0.405	0.269	0.068	0.244
$1/4$	0.540	0.364	0.088	0.424
$3/8$	0.675	0.493	0.091	0.567
$1/2$	0.840	0.622	0.109	0.850
$3/4$	1.050	0.824	0.113	1.130
1	1.315	1.049	0.133	1.678
$1\frac{1}{4}$	1.660	1.380	0.140	2.272
$1\frac{1}{2}$	1.900	1.610	0.145	2.717
2	2.375	2.067	0.154	3.652
$2\frac{1}{2}$	2.875	2.469	0.203	5.793
3	3.500	3.068	0.216	7.575
$3\frac{1}{2}$	4.000	3.548	0.226	9.109
4	4.500	4.026	0.237	10.790
5	5.563	5.047	0.258	14.617
6	6.625	6.065	0.280	18.974
8	8.625	7.981	0.322	28.554
10	10.750	10.020	0.365	40.483
12	12.750	12.000	0.375	45.557

*Plain ends.

Manufacturers' Standard Gage for Sheet Steel

Gage number	Thickness, in.	Weight, psf
4	0.2242	9.3750
5	0.2092	8.7500
6	0.1943	8.1250
7	0.1793	7.5000
8	0.1644	6.8750
9	0.1495	6.2500
10	0.1345	5.6250
11	0.1196	5.0000
12	0.1046	4.3750
13	0.0897	3.7500
14	0.0747	3.1250
15	0.0673	2.8125
16	0.0598	2.5000

Manufacturers' Standard for Plate Steel

Thickness Fraction	Thickness Decimal	Weight, psf
$1/2$	0.500	20.40
$15/32$	0.46875	19.13
$7/16$	0.4375	17.85
$13/32$	0.40625	16.58
$3/8$	0.375	15.30
$11/32$	0.34375	14.03
$5/16$	0.3125	12.75
$9/32$	0.28125	11.48
$17/64$	0.265625	10.84
$1/4$	0.2500	10.20
$15/64$	0.2391	9.76

Name index

Ackerman, E. A., 8
Addison, H., 22n, 372
Allen, C. M., 312n
Altouney, E. G., 429n
Alvord, J. W., 604n
Ambursen, N., 210
American Concrete Pipe Assoc., 336
American Public Health Assoc., 440n, 441n, 464
American Public Works Assoc., 509n
American Society of Civil Engineers, 46n, 115, 185, 192n, 193n, 267, 272n, 550, 553n, 559, 651n, 659
American Society of Mechanical Engineers, 310n
American Water Resources Assoc., 661
Anderson, E. A., 56n
Anderson, E. R., 29n, 30n
Anderson, H. W., 650n
Anderson, L. J., 29n
Aronovici, V. S., 529n
Ashbel, D., 11n

Babbitt, H. E., 464, 550, 578n, 599
Ball, J. W., 256n
Bandini, A., 401n
Banys, R., 258n, 289n
Barnett, M. P., 168n
Barrows, H. K., 631
Barth, E. F., 596n
Bather, J. A., 660
Baumann, E. R., 599

Bays, C. A., 109n
Beard, L. R., 126n, 660
Beck, L., 596n
Beightler, C. S., 661
Benton, G. S., 10n
Berg, P. H., 411n
Bernard, M. M., 134n, 137n, 631
Bishop, A. A., 411n
Biswas, A. K., 8
Blackburn, R. T., 10n
Blaisdell, F. H., 541n
Bodhaine, G. L., 129n
Boissonault, F. L., 245n
Bondurant, D. C., 174n
Bonin, C. C., 315n
Bouet, G. A., 357n
Bowen, I. S., 30n
Bowerman, F. R., 558n
Bowman, J. S., 659
Bradley, J. N., 239n, 264n, 267
Braikevitch, M., 474n
Brater, E. F., 73, 267, 273n, 294, 279n
Brooks, N. H., 448n, 558n
Brown, C. B., 185, 294, 631
Brown, F. R., 505n
Brown, G., 483
Brown, J. G., 372
Brune, G. M., 175n
Bullinger, C. E., 390
Buras, N., 660
Burden, R. P., 661
Burdick, C. B., 604n

Subject index